BIOCHEMICAL SOCIETY SYMPOSIA

No. 38

# NITROGEN METABOLISM IN PLANTS

*BIOCHEMICAL SOCIETY SYMPOSIUM No. 38*
*IUB–IUBS Joint Symposium held in the*
*University of Leeds, July 1972*

# Nitrogen Metabolism in Plants

ORGANIZED AND EDITED
BY

T. W. GOODWIN
AND
R. M. S. SMELLIE

1973
LONDON: THE BIOCHEMICAL SOCIETY

The Biochemical Society,
7 Warwick Court,
London WC1R 5DP, U.K.



ISBN: 0 9501972 3 8



*Printed in Great Britain by William Clowes & Sons Limited,
London, Colchester and Beccles*

# List of Contributors

J. E. Allende (*Departamento de Biología, Facultad de Ciencias, and Departamento de Bioquímica y Química, Facultad de Medicina, Universidad de Chile, Casilla* 6671, *Santiago*-4, *Chile*)

G. E. Blair (*Division of Biological Sciences, University of Warwick, Coventry CV*4 7*AL*, *U.K.*)

G. Blobel (*The Rockefeller University, New York, N.Y.* 10021, *U.S.A.*)

L. Bogorad (*Biological Laboratories, Harvard University, Cambridge, Mass.* 02138, *U.S.A.*)

G. Burkard (*Laboratoire de Chimie Biologique, Université Louis Pasteur, Rue Descartes, Esplanade*, 67000 *Strasbourg, France*)

D. Chen (*Department of Biophysics, Weizmann Institute of Science, Rehovot, Israel*)

M. D. Chisholm (*National Research Council of Canada, Prairie Regional Laboratory, Saskatoon, Sask., Canada*)

N.-H. Chua (*The Rockefeller University, New York, N.Y.* 10021, *U.S.A.*)

E. E. Conn (*Department of Biochemistry and Biophysics, University of California, Davis, Calif.* 95616, *U.S.A.*)

L. S. Dure, III (*Department of Biochemistry, University of Georgia, Athens, Ga.* 30601, *U.S.A.*)

R. J. Ellis (*Division of Biological Sciences, University of Warwick, Coventry CV*4 7*AL*, *U.K.*)

M. Gatica (*Departamento de Biología, Facultad de Ciencias, and Departamento de Bioquímica y Química, Facultad de Medicina, Universidad de Chile, Casilla* 6671, *Santiago*-4, *Chile*)

P. Guillemaut (*Laboratoire de Chimie Biologique, Université Louis Pasteur, Rue Descartes, Esplanade*, 67000 *Strasbourg, France*)

M. A. Harmey (*Department of Botany, University College, Dublin, Irish Republic*)

M. R. Hartley (*Division of Biological Sciences, University of Warwick, Coventry CV*4 7*AL*, *U.K.*)

R. Krauspe [*Institute of Plant Biochemistry, Research Centre of Molecular Biology and Medicine, German Academy of Sciences*, 401 *Halle* (*Saale*), *German Democratic Republic*]

A. B. Legocki (*Institute of Biochemistry, College of Agriculture, Poznan, Poland*)

C. J. Leaver (*Department of Botany, University of Edinburgh, Edinburgh EH*9 3*JH*, *U.K.*)

S. Litvak (*Departamento de Biología, Facultad de Ciencias, and Departamento de Bioquímica y Química, Facultad de Medicina, Universidad de Chile, Casilla* 6671, *Santiago*-4, *Chile*)

A. Marcus (*Institute for Cancer Research, Fox Chase, Philadelphia, Pa.* 19111, *U.S.A.*)

M. Matamala (*Departamento de Biología, Facultad de Ciencias, and Departamento de Bioquímica y Química, Facultad de Medicina, Universidad de Chile, Casilla* 6671, *Santiago*-4, *Chile*)

L. J. Mets (*National Institute of Arthritis, Metabolic and Digestive Diseases, National Institutes of Health, Bethesda, Md.* 20014, *U.S.A.*)

O. Monasterio (*Departamento de Biología, Facultad de Ciencias, and Departamento de Bioquímica y Química, Facultad de Medicina, Universidad de Chile, Casilla* 6671, *Santiago*-4, *Chile*)

K. P. Mullinix (*Laboratory of Chemical Biology, National Institutes of Health, Bethesda, Md.* 20014, *U.S.A.*)

J. M. Ojeda (*Departamento de Biología, Facultad de Ciencias, and Departamento de Bioquímica y Química, Facultad de Medicina, Universidad de Chile, Casilla* 6671, *Santiago*-4, *Chile*)

B. Parthier [*Institute of Plant Biochemistry, Research Centre of Molecular Biology and Medicine, German Academy of Sciences,* 401 *Halle* (*Saale*), *German Democratic Republic*]

D. L. Rayle (*Department of Botany, California State University at San Diego, Calif.* 92115, *U.S.A.*)

S. N. Seal (*Institute for Cancer Research, Fox Chase, Philadelphia, Pa.* 19111, *U.S.A.*)

P. Siekevitz (*The Rockefeller University, New York, N.Y.* 10021, *U.S.A.*)

F. Skoog (*Institute of Plant Development, Birge Hall, University of Wisconsin, Madison, Wis.* 53706, *U.S.A.*)

H. J. Smith (*Biological Laboratories, Harvard University, Cambridge, Mass.* 02138, *U.S.A.*)

A. Steinmetz (*Laboratoire de Chimie Biologique, Université Louis Pasteur, Rue Descartes, Esplanade,* 67000 *Strasbourg, France*)

G. C. Strain (*French Scientific Mission, Washington, D.C.* 20006, *U.S.A.*)

A. Tarragó (*Departamento de Biología, Facultad de Ciencias, and Departamento de Bioquímica y Química, Facultad de Medicina, Universidad de Chile, Casilla* 6671, *Santiago*-4, *Chile*)

E. W. Underhill (*National Research Council of Canada, Prairie Regional Laboratory, Saskatoon, Sask., Canada*)

D. P. Weeks (*Institute for Cancer Research, Fox Chase, Philadelphia, Pa.* 19111, *U.S.A.*)

J. H. Weil (*Laboratoire de Chimie Biologique, Université Louis Pasteur, Rue Descartes, Esplanade,* 67000 *Strasbourg, France*)

L. R. Wetter (*National Research Council of Canada, Prairie Regional Laboratory, Saskatoon, Sask., Canada*)

F. Wightman (*Department of Biology, Carleton University, Ottawa, Ontario K*1*S* 5*B*6, *Canada*)

M. H. Zenk (*Lehrstuhl für Pflanzenphysiologie, Ruhr University,* 463 *Bochum, German Federal Republic*)

# Preface

It has always been the policy of the International Unions of Biochemistry and of Biological Sciences to hold joint Symposia whenever it seemed useful to bring together biochemists and those interested in the broader aspects of biology. Scientists who work on the biochemistry of plants often feel that their important work tends to be neglected at biochemical congresses, and it seemed appropriate, therefore, that I.U.B. should organize a symposium on some aspects of plant biochemistry. This also provided an opportunity to bring into the discussions those biologists whose interests are represented by the I.U.B.S.

Professor T. W. Goodwin, who is the Symposium Organiser for the Commission of Biochemistry of I.U.B.S., served as Chairman of the Organizing Committee to decide on the scientific contents of the Symposium. The Committee consisted of J. E. Allende (Santiago), P. N. Campbell (Leeds), M. Florkin (Liège), L. Fowden (London), T. W. Goodwin (Liverpool), A. M. Marcus (Philadelphia), M. Zenk (Bochum) and J. B. Harborne (Reading). The Phytochemical Society and The Biochemical Society also agreed to sponsor the Symposium and lend their support. It was further agreed that the Proceedings should be published by The Biochemical Society as one of their series of Symposia.

T. W. GOODWIN

*Department of Biochemistry,*
*University of Liverpool,*
*P.O. Box* 147,
*Liverpool L*69 3*BX*,
*U.K.*

R. M. S. SMELLIE

*Institute of Biochemistry,*
*University of Glasgow,*
*Glasgow G*12 8*QQ*,
*U.K.*

# Contents

*Biochem. Soc. Symp.* (1973) **38**, 1–15
*Printed in Great Britain*

# Early Synthesis of Nucleic Acids in Germinating Wheat Embryos

By DAVID CHEN

*Department of Biophysics, Weizmann Institute of Science, Rehovot, Israel*

## Synopsis

The synthesis of $G_1$ proteins is suggested as mediating the initiation of DNA replication in germinating wheat embryos. Sequential activation of the genome was studied by analysing tRNA, rRNA and mRNA transcription. The synthesis and nature of newly synthesized ribonucleoprotein particles is described.

## Introduction

The germinating wheat embryo is a partially differentiated organism whose life processes were arrested during the ripening processes of the seed (Mayer & Polyakoff-Mayber, 1963). Nuclear control over cell function is arrested in the quiescent stage together with most of the metabolic process, and water imbibition does not trigger an immediate overall activation of biochemical functions (Marcus & Feeley, 1964; Waters & Dure, 1966; Chen *et al.*, 1968). If the quiescent stage really reflects a suppressed stage of the genome, then the early stage of germination of the wheat embryo provides an excellent opportunity for the study of the control of initiation of replication and transcription processes at the molecular level.

In the present paper I describe a few experiments designed to study the early events associated with early synthesis of nucleic acids in germinating wheat embryos and discuss their interrelationships.

## Experimental

Wheat embryos (*Triticum vulgare* var. Florence) were prepared according to the procedure of Johnstone & Stern (1957). Viable embryos germinated in darkness at 23°C on sterile plates containing agar (1%) and sucrose (1%, w/v) for various periods in the presence of 2ml of germination medium [10mM-Tris–HCl buffer, pH7.6, containing 20mM-KCl and 1% (w/v) sucrose]/g of dry embryos. Plate 1 depicts embryos germinated on agar plates and Plate 2 shows a cross-section of a germinated wheat embryo. For labelling purposes the embryos were washed thoroughly with germination medium, and the medium used was germination medium supplemented with the radioactive precursor in the concentrations indicated in the appropriate figure legends for the time specified for each experiment.

RNA was prepared as described by Chen *et al.* (1971), and polyacrylamide-gel electrophoresis was carried out as described by Loening (1967). Ribonucleoprotein particles were prepared by homogenizing 1 g of embryo in 20 ml of PSM buffer [10mM-potassium phosphate buffer, pH 7.6, containing 10mM-$MgCl_2$, 10$\mu$M-2-mercaptoethanol and 1% (w/v) sucrose]. The homogenate was centrifuged for 20min at 20000$g$ in the Sorvall RC2 centrifuge and the supernatant spun for 2h at 160000$g$ in the Spinco L-50 preparative ultracentrifuge.

The precipitate was resuspended in PM buffer (10mM-potassium phosphate buffer, pH 7.6, containing 10mM-$MgCl_2$ and 10$\mu$M-2-mercaptoethanol) and re-centrifuged for 2h at 160000$g$, and then dispersed in 3ml of PM buffer and clarified by centrifuging for 10min at 160000$g$ in the Sorvall RC2 centrifuge. All steps were performed at between 0° and 4°C. The supernatant contains the ribonucleoprotein fraction.

Sedimentation equilibrium in CsCl and sedimentation velocity in 5–20% (w/v) sucrose in TM buffer (10mM-Tris–HCl buffer, pH 7.6, containing 20mM-KCl and 10mM-$MgCl_2$) were performed as described by Chen *et al.* (1971).

## Initiation of DNA Replication

The initiation of DNA replication in germinating wheat embryos is delayed (Chen & Osborne, 1970), and entrance of the embryos into S phase does not begin before 15h of germination (Mory *et al.*, 1972).

The pattern of DNA replication is shown in Fig. 1. Wheat embryos were 'pulse'-labelled with [$^{14}$C]thymidine (2$\mu$Ci/ml; 58mCi/mmol) for 1h at different

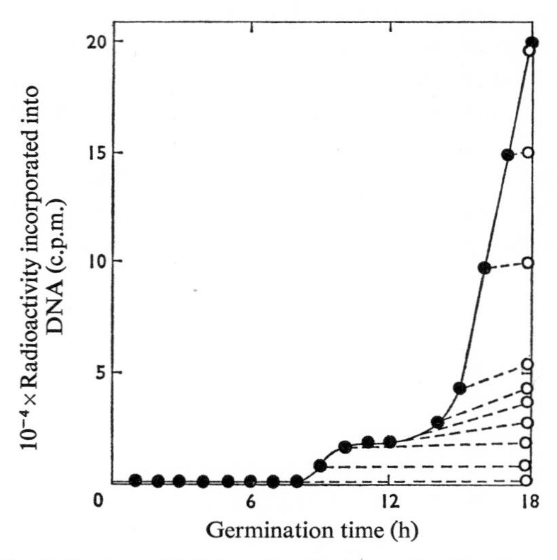


Fig. 1. *Pattern of DNA replication in germinating wheat embryos*

●—●, Batches (500mg) of embryos were germinated for the specified times and transferred for 1h into germination medium containing 10$\mu$Ci of [$^{14}$C]thymidine/ml. The embryos were washed with germination medium and homogenized in 8ml of cold 5% (w/v) trichloroacetic acid, and samples of the homogenates were filtered on Whatman GFC glass-fibre filter paper, washed with ethanol–ether (3:1, v/v) and their radioactivites counted. ●--○, Batches (500mg) of embryos were germinated as described above and then transferred into a medium containing blasticidine S or actidione (15$\mu$g/ml). The embryos were allowed to continue germination up to 17h and were then 'pulsed' for 1h with [$^{14}$C]thymidine, processed and their radioactivities counted as described above.

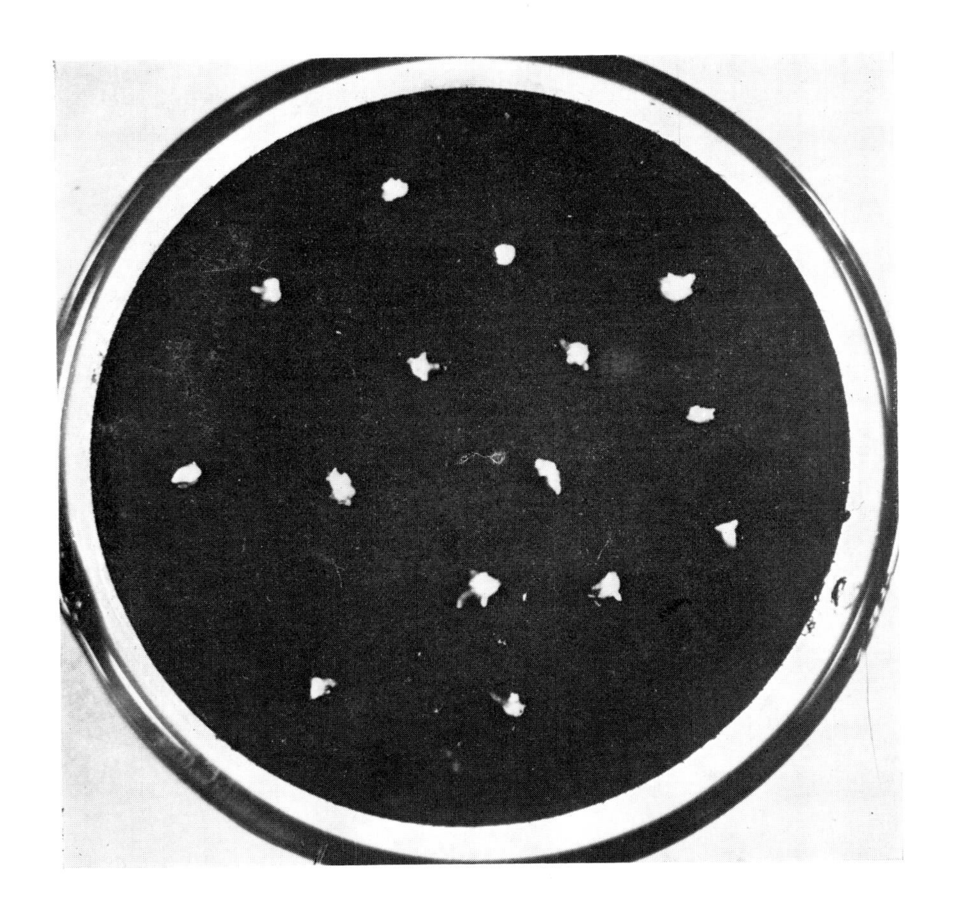

EXPLANATION OF PLATE I

*Wheat embryos germinated on a* 1% *agar plate in the presence of* 1% (*w*/*v*) *sucrose for* 18 *h*

D. CHEN

(*Facing p.* 2

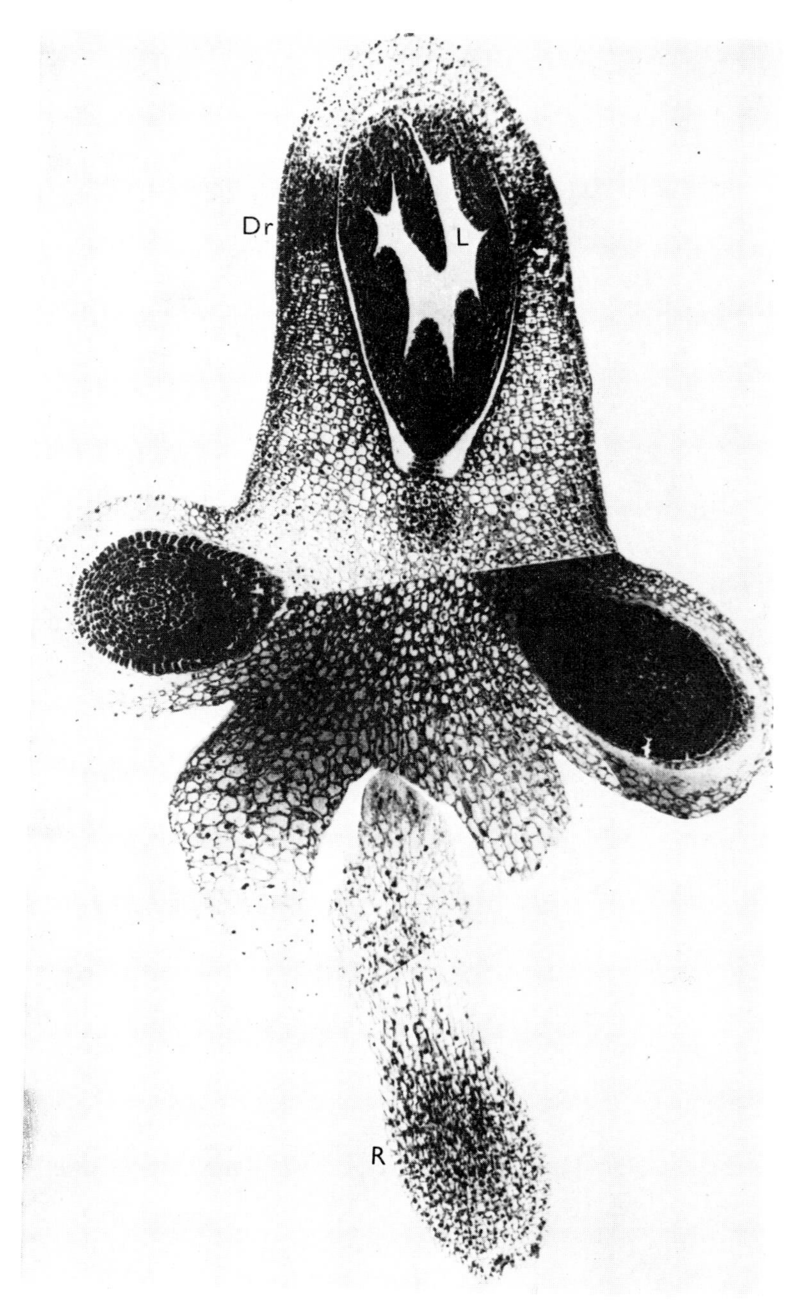


EXPLANATION OF PLATE 2

*Germinated wheat embryo: a cross-section*

The embryo was germinated for 30h and fixed in ethanol, and serial sections were photographed in the Carl Zeiss light-microscope. Key: Dr, dividing region; L, leaf; R, root.

D. CHEN

periods of germination, and the radioactivity incorporated into DNA was measured. The blocking of protein synthesis with the drug blasticidine S at different times of germination followed by a 1 h 'pulse' of [$^{14}$C]thymidine incorporation at 18 h of germination discloses three growth phases (Fig. 1).

Phase I occurs between 0 and 9 h of germination. Blocking of protein synthesis at any time of this phase results in a complete inhibition of DNA replication at 18 h.

Phase II occurs between 9 and 15 h of germination. During this period a semi-quantitative relationship is established between the time-span allowed for protein synthesis and the rate of DNA replication at 18 h.

Phase III occurs at 18 h of germination, when DNA replication is not dependent on concomitant protein synthesis, neither for initiation nor for sustaining the replication process.

Evidence for the dependence of DNA replication on protein synthesis has been reported for *Escherichia coli* (Kogoma & Lark, 1970), *Vicia faba* (Jakob & Bovey, 1969) and *Chlorella pyrenoidosa* (Wanka & Moors, 1970).

When the germinating wheat embryos were analysed for mitotic figures, practically none were found during the first 15 h. It is thus possible to conclude that no cell transitions of the type $G_2$(4C) → $G_1$(2C) take place before the entrance into the S phase, and that the events preceding the initiation of replication are taking place in cells that are at the $G_1$(2C) configuration. This is in accordance with the finding that *Triticum durum* embryos undergo depletion of cells at the $G_2$(4C) configuration before dehydration (Avanzi & Deri, 1969; D'Amato, 1972). The above findings justify the designation of the protein(s), synthesized during phases I and II of germination of the wheat embryo, that are relevant to the initiation of DNA replication as $G_1$ proteins.

### Partial Isolation of $G_1$ Proteins

The fact that DNA replication in 18 h-germinated embryos is not dependent on concomitant protein synthesis implies that such an embryo contains the full complement of $G_1$ proteins required for the completion of the cell cycle. A crude preparation of $G_1$ proteins was prepared from 18 h-germinated embryos in the following manner. A 1 g batch of embryos was homogenized in 20 ml of TM buffer, and the homogenate was centrifuged at 18000 ***g*** for 30 min in the Sorvall centrifuge. The supernatant was dialysed against GB buffer [10 mM-Tris–HCl buffer, pH 7.6, containing 10 mM-KCl, 1 mM-$MgCl_2$, 1 mM-2-mercapto-ethanol and 10% (w/v) glycerol]. The extract was then centrifuged in the Spinco ultracentrifuge at 160000 ***g*** for 60 min. The supernatant was subjected to $(NH_4)_2SO_4$ fractionation and the fraction precipitated at between 30% and 50% saturation was collected, dissolved in 10 ml of GB buffer and kept frozen at −20°C.

This crude fraction has the following characteristics. (*a*) The proteins are soluble in a low-salt buffer and stable at 0°C for several weeks. (*b*) This fraction contains at least six molecular components, as detected by polyacrylamide-gel electrophoresis (Raymond, 1964), isoelectric focusing and DNA–cellulose chromatography (Litman, 1968). (*c*) The fraction is devoid of deoxyribonuclease

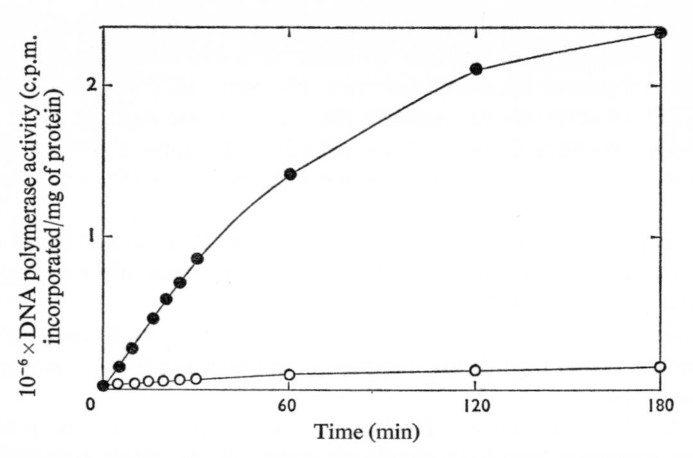


Fig. 2. *DNA-dependent DNA polymerase activity in extracts prepared from dry and* 18*h-germinated wheat embryos*

Extracts were prepared from ungerminated (○) and 18h-germinated (●) embryos as described in the text. The reaction mixture (0.25ml) contained 50mM of each of the four deoxyribonucleoside triphosphates (dNTP), 1$\mu$Ci of [$^3$H]dTTP (17$\mu$Ci/mmol), 5mM-$MgCl_2$, 12mM-KCl, 20mM-Tris–HCl buffer, pH7.6, 5$\mu$g of native wheat DNA and 40$\mu$g of $G_1$ proteins (crude extract).

activity. (*d*) The fraction catalyses the formation of polydeoxyribonucleotides, the reaction being dependent on the presence of all four deoxyribonucleoside triphosphates and the presence of template DNA. Double-stranded DNA is as active as single-stranded DNA. The standard reaction mixture (0.25ml) contained 50mM of each of the four deoxyribonucleoside triphosphates (dNTP), 1$\mu$Ci of [$^3$H]dTTP (17.4Ci/mmol), 5mM-$MgCl_2$, 12mM-KCl, 20mM-Tris–HCl buffer, pH7.6, 5mg of native wheat DNA and 40$\mu$g of $G_1$ proteins (crude extract). (*e*) Further fractionation of the crude fraction by the above-mentioned methods resulted in an 80–100% loss of activity. Full activity was regained when the fractions isolated by $(NH_4)_2SO_4$ fractionation were reconstituted back. (*f*) No such fraction could be isolated from ungerminated embryos (Fig. 2). The detailed characterization of the polymerization product and kinetics of the system have been published elsewhere (Mory *et al.*, 1973).

Regulation of the initiation of early DNA replication can be positively controlled through direct de-repression of repressed template synthesis or activation of the functional catalytic proteins. Another possibility is the negative control through the lack of precursors or catalytic proteins. It seems highly likely that the initiation of replication in germinating wheat embryos is mediated via the synthesis of $G_1$ protein(s). Only further characterization of the $G_1$ protein(s) would enable further elucidation of interrelationships between the various functional proteins of the initiation of DNA replication.

## tRNA Transcription

Proof of the formation of a transcription product of tRNA genes requires the following: (*a*) labelling of a low-molecular-weight RNA component; (*b*) the

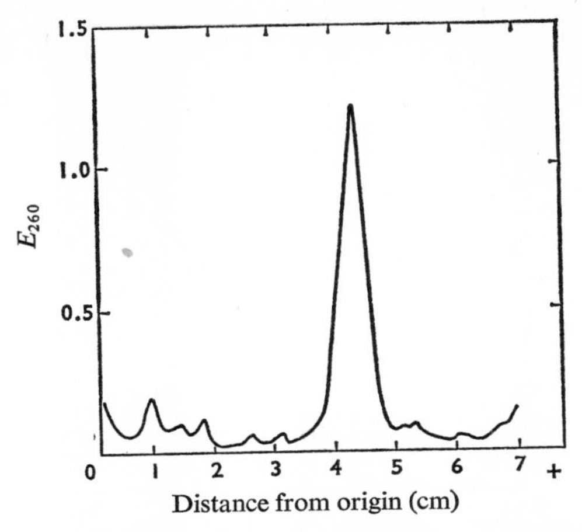


Fig. 3. *Polyacrylamide-gel electrophoresis of tRNA from germinating wheat embryos*

Electrophoresis was performed in 0.6cm × 9cm tubes at 5mA/tube for 1h. Gels were made up in buffer containing 30mM-$NaH_2PO_4$, 36mM-Tris–HCl buffer, pH7.8, 1mM-EDTA and 0.2% sodium dodecyl sulphate. After the electrophoresis the gels were scanned on a Gilford spectrophotometer.

capacity of the newly transcribed low-molecular-weight RNA to be charged with amino acids.

For investigation of this problem the following procedure was utilized. Wheat embryos were germinated for different times and 'pulse'-labelled with [$^3$H]-uridine (100$\mu$Ci/ml; 5Ci/mmol) for 1h before being harvested. Total RNA was prepared and this RNA preparation was further dispersed in 3M-KCl for 24h at 4°C. The soluble fraction was dialysed against water, freeze-dried and kept frozen dry. Electrophoresis of the RNA on polyacrylamide gel and scanning on a Gilford spectrophotometer gave the result presented in Fig. 3.

The tRNA was then charged with $^{14}$C-labelled amino acid mixture (50$\mu$g/ml; 52mCi/mg-atom) in a mixture (1ml) containing 30mM-Tris–HCl buffer, pH7.6, 1mg of tRNA, 2mM-ATP, 2mM-$MgCl_2$, 0.7mM-2-mercaptoethanol and 50$\mu$g of protein of S-100 extract (see Chen *et al.*, 1968). The reaction mixture was incubated for 15min at 27°C and then cooled, and the RNA was re-extracted with a mixture of 4ml of 0.5M-$NaClO_4$, 0.5ml of chloroform and 3ml of 80% (w/v) phenol. The RNA was precipitated with 2vol. of cold ethanol and re-dissolved in water. A sample was then monitored for absorbance at 260nm and for labelling with $^{14}$C and $^3$H. This preparation now contained the following: (1) pre-existing charged tRNA ($^{14}$C-labelled in the amino acid residue); (2) pre-existing non-charged tRNA (not labelled); (3) newly synthesized charged tRNA ($^{14}$C-labelled in the amino acid residue and $^3$H-labelled in the RNA); (4) newly synthesized non-charged tRNA ($^3$H-labelled in the RNA). The mixture was then allowed to react with $\beta$-benzyl-*N*-carboxy-L-aspartate anhydride, in 1,4-dioxan–water mixture (Katchalski *et al.*, 1966). The water-insoluble poly-peptidyl-tRNA was trapped on a Whatman GFC glass-fibre filter paper and its radioactivity was counted. Whereas about 60% of the charged amino acids were recovered on the filter, only 18% of the $^3$H radioactivity could be recovered.

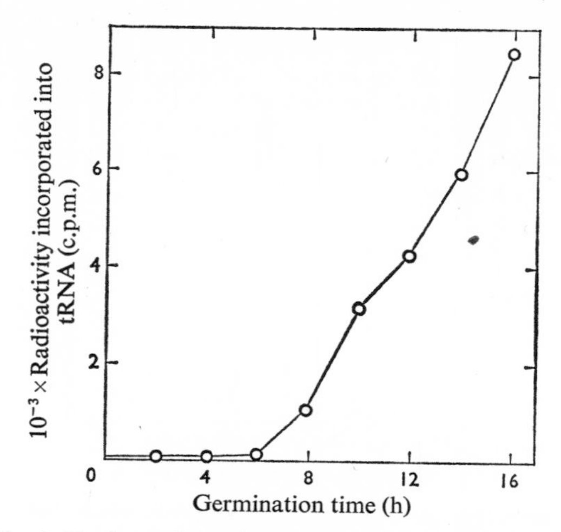


Fig. 4. *Newly synthesized tRNA in germinating wheat embryos*

Batches (1 g) of embryos were germinated for different times and 'pulse'-labelled for 1 h with 100 $\mu$Ci of [$^3$H]uridine/ml. Low-molecular-weight RNA was extracted by the phenol method and charged with $^{14}$C-labelled amino acid mixtures, and the newly synthesized charged tRNA was trapped on a Whatman GFC glass-fibre filter paper in the form of polypeptidyl-tRNA.

However, this is sufficient to prove that the third species is present on the filter, i.e. a newly synthesized low-molecular-weight chargeable RNA. The results presented in Fig. 4 suggest that tRNA genes are activated after 8 h of germination.

## tRNA Transcription

A newly synthesized precursor ribosomal RNA can be detected in the nuclei of wheat embryos already after 2h of germination (Chen *et al.*, 1971). The final products of processing of the precursor rRNA can already be detected at 3h of germination. The processing scheme is basically similar to other schemes suggested (Loening, 1967; Grierson & Loening, 1972):

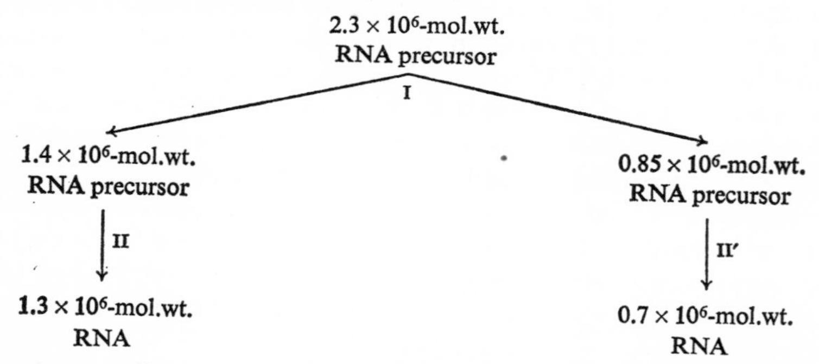


Since no intermediate can be found in the cytoplasm, the processing site for steps I and II must be within the nucleus. Since processing steps II and II′ are independent they must be well co-ordinated since the $1.3 \times 10^6$-mol.wt. and $0.7 \times 10^6$-mol.wt. rRNA species appear in a 1:1 ratio within the nucleus.

There are four chromosomes carrying 8000 copies of ribosomal genes in the

Table 1. *DNA-dependent RNA polymerase activity in extracts of germinated wheat embryos*

The reaction mixture (0.25ml) contained 10 μg of wheat-embryo DNA, 5mM-Tris–HCl buffer, pH 7.8, 0.5 mM-$MgCl_2$, 0.1 mM-$MnSO_4$, 1 mM of each of the four ribonucleoside triphosphates (NTP), 1 μCi of [$^3$H]UTP and 50 μg of protein in wheat S-100 extract. After 15 min at 30°C the reaction was stopped, and the incorporation of [$^3$H]UTP was assayed as described in the text.

| Germination time (h) | [$^3$H]UTP incorporated (c.p.m.) |
|---|---|
| 1 | 8272 |
| 2 | 8305 |
| 3 | 6705 |
| 4 | 6210 |
| 5 | 5687 |
| 6 | 5816 |

hexaploid wheat. Usually only one chromosome displays an activated nucleolar organizer, but as yet there is no way of assessing how many of the copies of the activated chromosome are transcribed at any given time or during germination.

## Dependence of rRNA Transcription on Protein Synthesis

The low rate of early ribosomal transcription suggested the presence of a limiting factor. Such a factor could be the enzyme DNA-dependent RNA

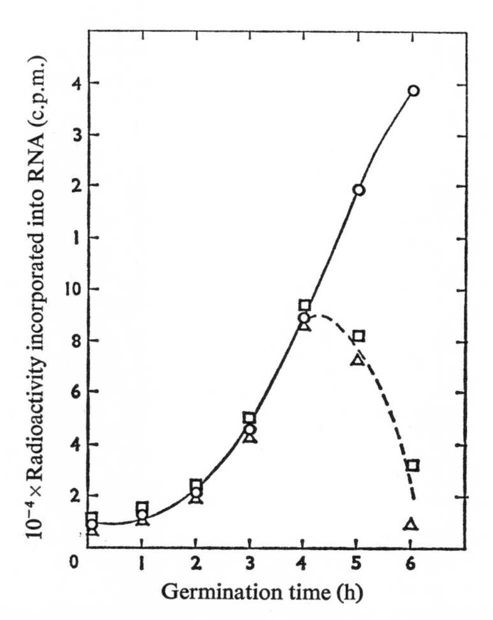


Fig. 5. *Effects of blasticidine S and actidione on the incorporation of uridine into RNA of germinating wheat embryos*

Batches (500 mg) of embryos were germinated for the different times in the presence of actidione (□) or blasticidine S (△) (15 μg/ml) (o, control). At 1 h before harvest [$^3$H]uridine (100 μCi/ml; 5 Ci/mmol) was added to the medium. The embryos were then washed with 50 ml of germination medium and homogenized in 8 ml of cold 5% (w/v) trichloroacetic acid, and 3 ml portions of the homogenate were filtered on Whatman GFC glass-fibre filter paper. The filters were washed with ethanol–ether (2:1, v/v) dried and their radioactivities counted.

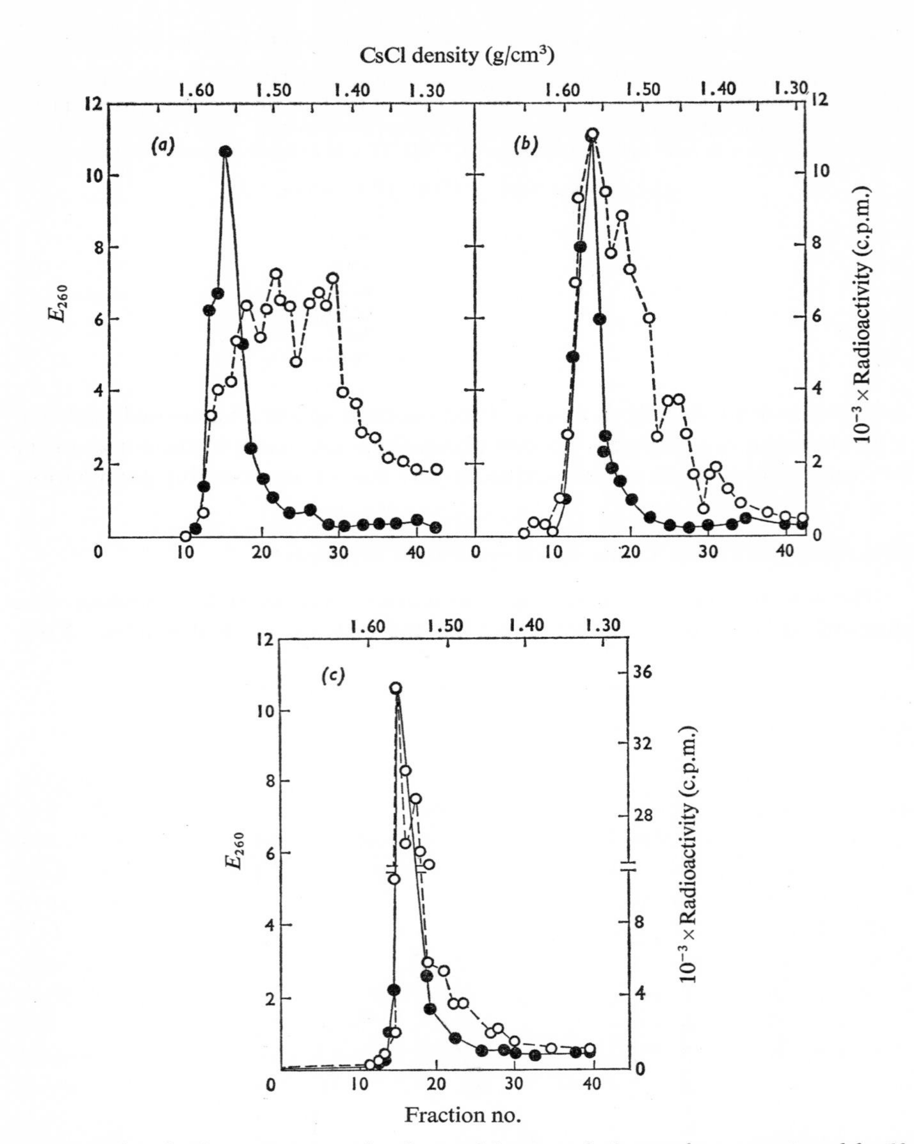


Fig. 6. '*Pulse–chase' experiment on the ribosomal fraction of wheat embryos germinated for* 6*h*

Dry embryos (1g/experiment) were germinated for 5h, washed and labelled with 100μg of [³H]uridine (30Ci/mmol) for 1h. After being labelled, the embryos were washed and either collected and homogenized immediately (*a*) or germinated in the presence of unlabelled uridine at a concentration 20 times that of the labelled uridine for 2h (*b*) and 6h (*c*) before collection and homogenization. After homogenization, the ribosomal fraction was prepared as described in the text and the resultant ribosomal suspension was fixed in 6% (w/v) formaldehyde for 4 days at 4°C. The fixed ribosomes were applied to CsCl density gradients and centrifuged to equilibrium. Fractions were collected and analysed spectrophotometrically at 260nm (●) and precipitated on Whatman GF/C glass-fibre filter papers with 5% (w/v) trichloroacetic acid, washed, dried and their radioactivities counted (○). Buoyant density was determined by measuring the refractive index of a few samples across the gradient. (Note the break in the radioactivity scale for the 6h 'chase' experiment.) The figures are reproduced from Chen *et al.* (1971).

polymerase. The activity of this enzyme was measured in post-ribosomal supernatants prepared from wheat embryos after 0, 1, 2, 3, 4, 5 and 6h of germination. The assay system (0.25ml) contained 10$\mu$g of wheat-embryo DNA, 5mM-Tris–HCl buffer, pH7.8, 0.5mM-$MgCl_2$, 0.1mM-$MnSO_4$, 1mM of each of the four ribonucleoside triphosphates (NTP), 1$\mu$Ci of [$^3$H]UTP and about 50$\mu$g of protein from the post-ribosomal extract in TM buffer. Incubation was for 15min at 30°C and was stopped by the addition of 1ml of 10mM-EDTA, pH7, 1% (w/v) sodium dodecyl sulphate and 50$\mu$g of wheat RNA in 5% (w/v) trichloroacetic acid. After being cooled the mixture was filtered on a Whatman GFC glass-fibre filter paper, which was washed with ethanol–ether (2:1, v/v), dried and its radioactivity counted. The results (summarized in Table 1) suggest that, in contrast with DNA polymerase, the dry embryo contains a relatively high specific activity of DNA-dependent RNA polymerase as compared with the germinated embryo.

In experiments where protein synthesis was completely blocked by the presence of actidione or blasticidine S (15$\mu$g/ml) in the germination medium, RNA transcription *in vivo* was not affected at all during the first 4h of germination. It was only then that the rate of transcription was slowed down, and it was blocked completely at 6h of germination (Fig. 5). Since the polymerizing activity does not seem to be limiting, the control of early transcription must be looked for at a different level.

## Ribonucleoprotein Particles

The first transcription products appear in the cytoplasm of the wheat embryos after 6h of germination (Chen *et al.*, 1971). The newly transcribed RNA is detected in the cytoplasm in the form of ribonucleoprotein particles with a buoyant density, $\rho$, of 1.42–1.52g/ml. These particles require a further 6h to become converted into particles that co-sediment with pre-existing ribosomes in an equilibrium gradient (Fig. 6).

Similar particles have been reported in other organisms (Perry & Kelly, 1968; Spirin, 1969), and it has been suggested that the ribonucleoprotein particles are in fact mRNA packages.

In order to study the nature of ribonucleoprotein particles in the germinating embryos, my colleagues and I have studied particularly the first newly synthesized ribonucleoprotein particles appearing in the cytoplasm between 5 and 6h of germination.

When the ribonucleoprotein fraction is prepared from embryos 'pulse'-labelled with [$^3$H]uridine (30Ci/mmol; 100$\mu$g/ml) and fractionated by centrifugation at 120000$g$ on a linear 5–10% (w/v) sucrose density gradient for 2.5h (Spinco rotor SW40, 12cm tubes), the resulting distribution of the particles is as presented in Fig. 7. The labelled ribonucleoprotein particles occur in a distinct sedimentation band, which sediments more slowly than the main ribosomal band. Pretreatment of this fraction with Pronase (5$\mu$g/ml) at 27°C or washing with 1M-$NH_4CL$ (Fig. 8) results in an increase in the density of the newly synthesized ribonucleoprotein particles. Both experiments suggest the association of an excess of protein with the newly synthesized ribonucleoprotein particles.

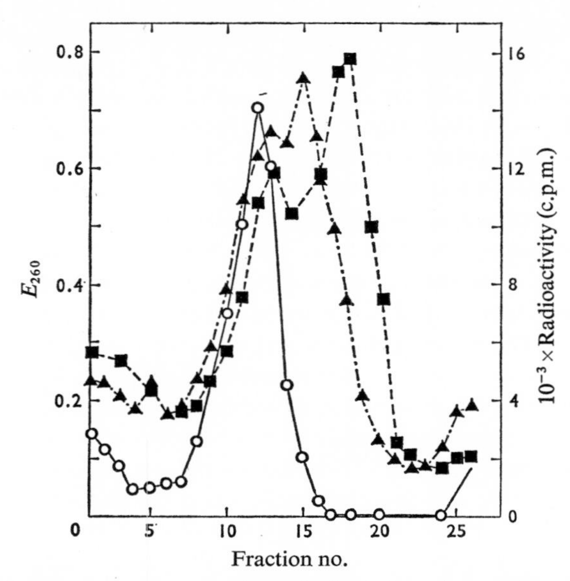


Fig. 7. *Effect of Pronase treatment on newly synthesized ribonucleoprotein particles appearing in the cytoplasm of wheat embryos between 5 and 6h of germination*

Ribonucleoprotein particles were prepared from embryos 'pulse'-labelled between 5 and 6h of germination and divided into two fractions. One fraction was incubated for 10min at 27°C with Pronase (5$\mu$g/ml) and both fractions were centrifuged separately in 5–20% (w/v) sucrose gradients as described in the text. ○, $E_{260}$; ■, radioactivity in control preparation; ▲, radioactivity in Pronase-treated preparation.

When the ribonucleoprotein fraction, prepared as in the above experiment, was fractionated on an $Mg^{2+}$-free sucrose density gradient for 3.5h (Fig. 9) the newly synthesized ribonucleoprotein particles occur in a band behind that of the small ribosomal subunit (34S).

The three gradient fractions designated L (large ribosomal subunit), S (small ribosomal subunit) and N (newly synthesized ribonucleoprotein particles) were pooled, dialysed against 0.1M-potassium phosphate buffer, pH7.5, fixed with 6% (w/v) formaldehyde for 3 days at 4°C and centrifuged to equilibrium in a CsCl density gradient. The results (Fig. 10) indicate clearly that the pre-existing components, as measured by the absorbance at 260nm, are similar in fractions S and N, whereas both newly synthesized ribonucleoprotein particles and radioactively labelled small ribosomal subunits show a difference in buoyant density. The densities recorded in a typical experiment are: large subunit, 1.560g/ml; small subunit, 1.545g/ml; ribonucleoprotein particle, 1.535g/ml. The fact that the newly synthesized ribonucleoprotein particles can be separated on a sucrose density gradient enabled us to prepare enough material for analysis of the RNA component of the ribonucleoprotein particle. A 12g batch of embryos was 'pulse'-labelled between 5 and 6h of germination with [$^3$H]uridine (30Ci/mol; 100$\mu$g/ml), and the ribonucleoprotein fraction was prepared and fractionated as indicated in Fig. 9. Fractions L, S and N were pooled and washed with TM buffer on Aminco Xm-100 ultrafiltration membranes, and RNA was extracted by the phenol method.

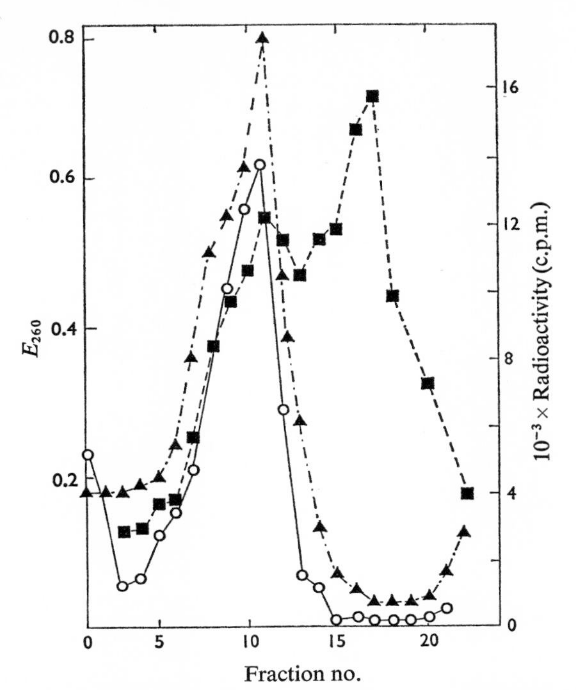


Fig. 8. *Effect of* 1M-$NH_4Cl$ *on newly synthesized ribonucleoprotein particles appearing in the cytoplasm of wheat embryos between* 5 *and* 6*h of germination*

Ribonucleoprotein particles were prepared as described in Fig. 7 and divided into two fractions. One fraction was dispersed in 1M-$NH_4Cl$ in TM buffer for 20min at 0°C, and both ribonucleoprotein fractions, recovered by re-centrifuging in the Spinco ultracentrifuge, were separated on 5–20% (w/v) sucrose gradients as indicated in Fig. 7. ○, $E_{260}$; ■, radioactivity in control preparation; ▲, radioactivity in $NH_4Cl$-treated preparation.

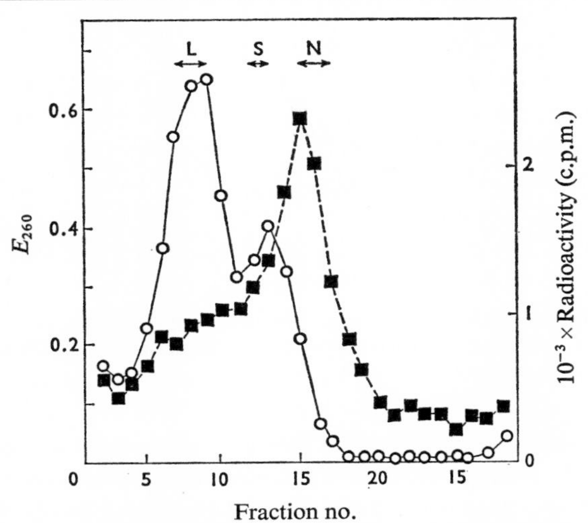


Fig. 9. *Distribution pattern of the ribonucleoprotein fraction from* 5–6*h-germinated wheat embryos in an* $Mg^{2+}$*-free sucrose density gradient*

Ribonucleoprotein particles were prepared as described in Fig. 1, and they were centrifuged at 120000*g* for 3.5h on a 5–20% (w/v) sucrose gradient ($Mg^{2+}$-free). ○, $E_{260}$; ■, radioactivity.

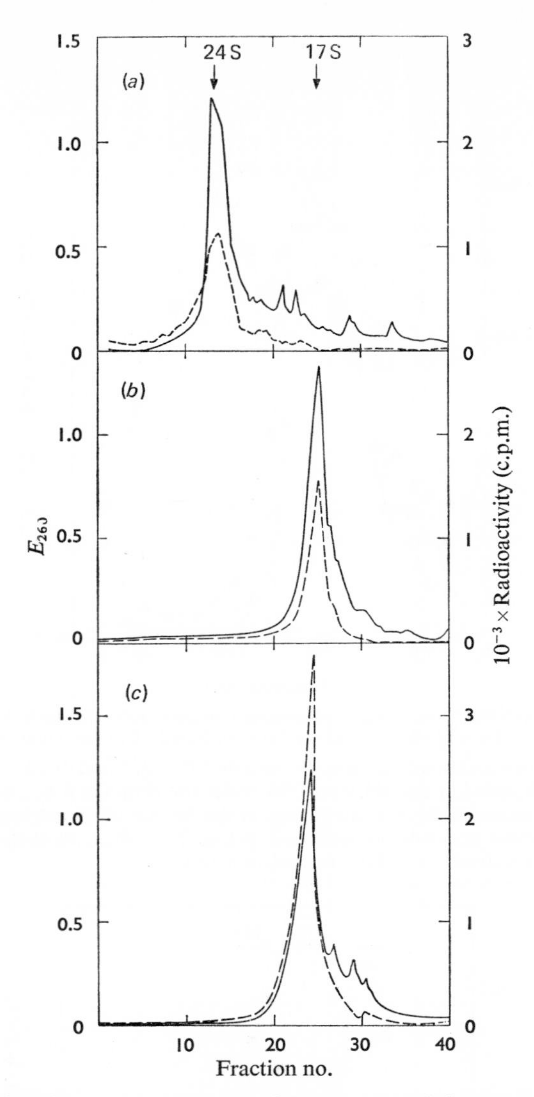


Fig. 10. *Polyacrylamide-gel electrophoresis of RNA prepared from sucrose-density-gradient fractions L, S and N* (*see Fig.* 9) *of the ribonucleoprotein fraction from* 5–6*h germinated wheat embryos*

Experimental details are given in the text. (*a*) Fraction L (large ribosomal subunit); (*b*) fraction S (small ribosomal subunit); (*c*) fraction N (newly synthesized ribonucleoprotein particles). ——, $E_{260}$; ----, radioactivity.

Samples containing about 100 μg of RNA were subjected to electrophoresis on a 2.3% polyacrylamide gel, scanned at 260 nm, frozen over solid $CO_2$, sliced into 1 mm sections, hydrolysed overnight at 37°C in Nuclear–Chicago NCS solubilizer and their radioactivities counted in toluene scintillator. The results are given in Fig. 10, and show that fraction L contains newly synthesized RNA that migrates in the electrophoretic field to a position corresponding to that of rRNA of molecular weight $1.3 \times 10^6$. The newly synthesized RNA in fractions S and N has a molecular weight of $0.7 \times 10^6$, i.e. exactly that of the RNA component of the small ribosomal subunit. No other molecular species could be detected in the cytoplasm. The hybridization pattern of newly synthesized RNA obtained from fraction N with wheat DNA was identical with that expected from rRNA sequences (0.3% saturation). 'Melting' analysis indicated a G + C content of about 53% (Edelman *et al.*, 1970).

On the basis of the above data we may conclude that the newly synthesized ribonucleoprotein particles are mostly precursors for the small ribosomal subunits. The precursor consists of a complete $0.7 \times 10^6$-mol.wt. rRNA component, and differs structurally on the basis of density, which probably reflects structural differences such as excess of protein.

**Initiation of mRNA Transcription**

The required evidence for an RNA to be mRNA is its capacity to code for a well-defined protein. Any other method of characterization would be limited to an indication that the RNA is not rRNA or tRNA.

However, a legitimate and relevant question can be: is there any newly synthesized RNA associated with the ribosomal complex engaged in early protein synthesis?

It has already been demonstrated that before 5h of germination no newly synthesized RNA can be detected within the ribosomal population, including the small ribosomal subunits. Such newly synthesized RNA does appear between 5 and 6h of germination, and this was shown to be strictly ribosomal RNA packed in the small ribosomal subunit. Even if, somehow, newly synthesized ribonucleoprotein particles have an undetectable proportion of mRNA packages, they do not participate in supporting protein synthesis at that time, since newly synthesized ribonucleoprotein particles do not enhance peptide-bond formation. An additional 6h is required for these to be converted into particles that sediment in the region typical of ribosomes.

It therefore seems highly unlikely that the transport of new RNA to the cytoplasm is taking place during the 0–6h germination period, either coupled to ribosomal transport (McConkey & Hopkins, 1965) or as an independent package (Spirin, 1969).

It is therefore possible to conclude that newly transcribed message does not have a role at the early stage of germination. The question whether information is transcribed at that stage has to wait for improved techniques for detection of mRNA copies.

**Summary**

Some of the early events during germination of wheat embryos associated with nucleic acid synthesis are summarized schematically in Fig. 11. The quiescent stage of an ungerminated wheat embryo seems to be characterized by a suppressed genome. The activation of the gene seems to be a well-regulated process. Activation of the genome is not dependent directly on water availability. The

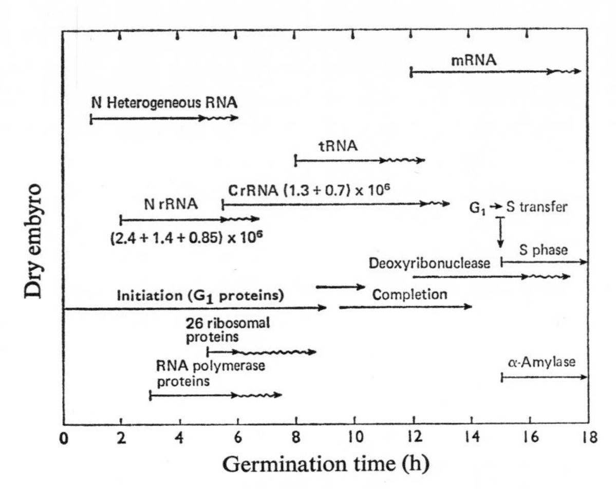


Fig. 11. *Schematic representation of molecular events during early stages of the germination of wheat embryos*

Key: N, in the nucleus; C, in the cytoplasm; ⇝, continued synthesis.

first genes are activated between 2 and 3h of germination and they are the ribosomal genes. Genes for rRNA are turned on after about 6h of germination and genes for tRNA after about 8h of germination.

RNA complementary to DNA and that contains sequences other than those of rRNA or tRNA can be detected after 12h of germination. However, newly synthesized mRNA, if transcribed at all at the early stage of germination (0–6h), does not seem to participate in coding for early protein synthesis. The early protein synthesis is coded for by conserved mRNA (Chen *et al.*, 1968), which includes at least the information for $G_1$ proteins and the 26 ribosomal proteins. The initiation of replication is controlled via the translation of $G_1$ protein, whose exact mode of action has yet to be studied.

The initiation of early transcription is not dependent on concomitant protein synthesis. The first transcriptional product that appears in the cytoplasm after 5h of germination is the RNA fragment of the small ribosomal unit packed in a ribonucleoprotein particle.

## References

Avanzi, S. & Deri, P. L. (1969) *Caryologia* **22**, 187–194
Chen, D. & Osborne, D. J. (1970) *Nature* (*London*) **225**, 336–340
Chen, D., Sarid, S. & Katchalski, E. (1968) *Proc. Nat. Acad. Sci. U.S.* **60**, 902–909
Chen, D., Schultz, G. & Katchalski, E. (1971) *Nature* (*London*) *New Biol.* **231**, 69–72
D'Amato, F. (1972) *The Dynamics of Meristen Cell Population*, pp. 149–154, Plenum Publishing Corp., New York
Edelman, M., Verma, I. M. & Littauer, U. Z. (1970) *J. Mol. Biol.* **49**, 67–83
Grierson, D. & Loening, U. E. (1972) *Nature* (*London*) **235**, 80–82
Jakob, K. J. & Bovey, F. (1969) *Exp. Cell Res.* **54**, 118–126
Johnstone, F. B. & Stern, H. (1957) *Nature* (*London*) **179**, 160–161
Katchalski, E., Yankofski, S., Novogrodsky, A., Galenter, Y. & Littauer, U. Z. (1966) *Biochim. Biophys. Acta* **123**, 641–643
Kogoma, T. & Lark, K. G. (1970) *J. Mol. Biol.* **52**, 143–164
Litman, R. M. (1968) *J. Biol. Chem.* **243**, 6222–6233

Loening, U. E. (1967) *Biochem. J.* **102**, 251–257
Marcus, A. & Feeley, J. (1964) *Proc. Nat. Acad. Sci. U.S.* **51**, 1075–1079
Mayer, A. M. & Polyakoff-Mayber, A. (1963) *The Germination of Seeds*, p. 236, Pergamon Press, Oxford
McConkey, E. H. & Hopkins, J. W. (1965) *J. Mol. Biol.* **14**, 257–270
Mory, Y. Y., Chen, D. & Sarid, S. (1972) *Plant Physiol.* **49**, 20–23
Mory, Y. Y., Sarid, S. & Chen, D. (1973) *Plant Physiol.* in the press
Perry, R. P. & Kelly, D. E. (1968) *J. Mol. Biol.* **35**, 37–59
Raymond, S. (1964) *Ann. N.Y. Acad. Sci.* **121**, 350–365
Spirin, A. S. (1969) *Eur. J. Biochem.* **10**, 20–35
Wanka, F. & Moors, J. (1970) *Biochem. Biophys. Res. Commun.* **41**, 85–89
Waters, L. C. & Dure, L. S. (1966) *J. Mol. Biol.* **19**, 1–27

## Discussion

**J. W. Davies**: You suggested that early protein synthesis is required for subsequent nucleic acid synthesis, and that the polymerase may be one of these protein synthesis products. Polymerases, however, may be 'multicomponent' enzymes, or may require a 'factor' for activity. How do you know that polymerase subunits are not already present before germination, and it is the activating factor that needs to be synthesized?

Have you ever tried taking individual proteins (e.g. from polyacrylamide-gel separation) from a late-stage extract, and adding these to extracts from an early stage or ungerminated seeds, to see whether polymerase can be activated?

This may also apply to other processes that you say are switched on at certain times, but in fact could have been synthesized earlier.

**D. Chen**: Our experiments have not reached the stage where the subunits or molecular components of the enzymic complex were purified. It is therefore impossible to ascribe the regulating role to any simple molecular species that has to be synthesized *de novo*. It might be an activating factor, but it might as well be any other component of the replicating proteins. This is why we prefer to talk about the $G_1$ proteins rather than of DNA-dependent DNA polymerase. Only after the $G_1$ proteins are properly characterized will the experiments suggested by Dr. Davies be feasible.

*Biochem. Soc. Symp.* (1973) **38**, 17–41
*Printed in Great Britain*

# Possibilities for Intracellular Integration: The Ribonucleic Acid Polymerases of Chloroplasts and Nuclei, and Genes Specifying Chloroplast Ribosomal Proteins

By LAWRENCE BOGORAD, LAURENS J. METS,* KATHLEEN P. MULLINIX,† HARRIET JANE SMITH and GUSTAVE C. STRAIN‡

*Biological Laboratories, Harvard University, Cambridge, Mass.* 02138, *U.S.A.*

## Synopsis

Two approaches to studying mechanisms of intracellular integration in eukaryotic cells are described. First, we have investigated the DNA-dependent RNA polymerases in chloroplasts and in the nucleo-cytoplasmic system of maize to try to get an understanding of the basis for differential control of RNA synthesis. The chloroplast enzyme and one of the nuclear polymerases have been brought to sufficient purity for analysis of their polypeptide subunit compositions. Another approach has been to examine the genetics of chloroplast ribosomal proteins in *Chlamydomonas reinhardi*. By working with erythromycin-resistant mutants, genes which alter proteins of the 52S subunit of the chloroplast ribosome have been traced to two different linkage groups in the nuclear genome. Another gene which affects another protein has been traced to the uniparentally transmitted genome.

Finally, possible mechanisms for selection and evolution within a single cell are discussed. The endosymbiont and cluster-clone hypotheses of the origin of organelles are examined and discussed. Possible mechanisms for the transfer of genetic control of a protein from one component of the cell to another are presented.

## Introduction

Chloroplasts contain DNA, RNA species and ribosomes. They grow in a medium selected by the plasma membrane and elaborated by the cytoplasmic elements according to instructions in the nuclear genome. The medium is the sum of the expressed nuclear genes and it may differ in each cell of an organism. Which constituents of the medium are essential for chloroplast growth and maturation? How complex is the minimal medium for plastid growth? What does the plastid do for itself?

This range of problems can be recast into a set of more limited questions. Parts of some of these questions have been answered already. Additional solutions will require further experiments.

* Present address: National Institute of Arthritis, Metabolic and Digestive Diseases, Room 107, Building 6, National Institutes of Health, Bethesda, Md. 20014, U.S.A.

† Present address: Laboratory of Chemical Biology, Building 109-N-312, National Institutes of Health, Bethesda, Md. 20014, U.S.A.

‡ Present address: French Scientific Mission, 2011 I Street, N.W. Washington, D.C. 20006, U.S.A.

(*a*) Where are the genes for the RNA species and proteins of the organelle? Are all the genes in a single genome? Do some of the same genes occur in two genomes of the cell? In more?

(*b*) How are these genes transcribed and how is their transcription regulated? Is transcription controlled in the same way in all the genomes and for all the genes in the cell? Is there a single RNA polymerase in the cell? Many?

(*c*) Where are mRNA species for specific organellar proteins translated? Can 'nuclear mRNA species' be translated by the plastid? And the converse?

(*d*) Is the translational machinery of the organelle different from that of the cytoplasm? How? What is the functional significance of any differences in terms of the regulation of organelle development and activity?

Least is known about question (*c*). Most is known about question (*d*): chloroplast ribosomes are smaller than cytoplasmic ribosomes and contain smaller RNA species; the ribosomal proteins are different; there are differences between cytoplasmic and plastid ribosomes in their susceptibility to certain antibiotics; there are at least some tRNA species which are limited to the chloroplast. Various aspects of these differences have been reviewed within the past few years by Smillie & Scott (1969), Kirk (1970), Woodcock & Bogorad (1971) and Boulter *et al.* (1972). Many of these items will be discussed and amplified by various other participants in this symposium. Some of these differences have been utilized in experiments to be described here.

This paper will deal largely with approaches to questions (*a*) and (*b*) in the list above. First, studies of the RNA polymerases in the nuclei and chloroplasts of *Zea mays* will be described. Second, analyses of the sites of genes which specify certain chloroplast ribosomal proteins will be presented.

## The RNA Polymerases of Maize

Kirk (1964) showed that purified broad-bean chloroplasts could form polyribonucleotides. The synthesis required the presence of all four nucleotide triphosphates and was abolished by DNAase (deoxyribonuclease) or actinomycin and thus had all the characteristics of enzymic DNA-dependent RNA synthesis. Further, RNA-synthesizing activities of plastids and nuclear fractions differed somewhat in their responses to $Mg^{2+}$ and $Mn^{2+}$ ions. These data suggested that there might be different enzymes present in the plastid and nuclear fractions although the effect observed could probably be explained by other unidentified properties of the different fractions which might easily affect this complex system of templates, substrates, ions, etc. At about the same time Mans & Novelli (1964) described a soluble RNA polymerase in homogenates of maize leaves. Then in the summer of 1965, at another meeting organized by Professor Goodwin, RNA polymerase activity tightly associated with *Z. mays* chloroplasts was described (Bogorad, 1967*a*,*b*) and was shown to increase very soon after leaves of etiolated plants were illuminated. The soluble RNA polymerase of maize, by contrast, did not change in activity during greening of etiolated maize leaves. Along the same line, increases in plastid RNA polymerase activity during greening of etiolated pea apices has been reported recently by Ellis & Hartley (1971). The increase in activity of maize plastid RNA polymerase was consistent with the observed increase in incorporation of [$^{32}$P]phosphate into chloroplast rRNA during greening of etiolated maize (Bogorad, 1967*a*). However, again the differences in response of the plastid-bound and soluble RNA polymerases during greening suggested, but did not prove, that more than one enzyme was present.

The changes in the activity of chloroplast RNA polymerase noted during development constituted one important impetus toward our more thorough examination of all the RNA polymerases of maize. Another impetus to RNA polymerase studies has come from the work of Burgess (1969) which showed that the *Escherichia coli* enzyme, after chromatography on phosphocellulose, is composed of three different polypeptide chains, $\beta$, $\beta'$ and $\alpha$ of molecular weights 155000, 165000 and 39000 respectively. Burgess (1971) has also reviewed the growing literature on some other proven (and some imagined?) specific transcription factors, smaller proteins which may control the transcription of particular kinds of DNA by the RNA polymerase complex. These findings have fed the growing idea that RNA polymerases may play a role in the selective expression of genes.

We set out to see if there were several RNA polymerases in a maize leaf cell and how they might differ, in order to understand plastid development and to explore the possibility that the nuclear–cytoplasmic system might regulate chloroplast growth via these enzymes. The prospect of subunit structures, 'finger prints' which could help us tell the enzymes apart physically or which might tell us whether there were common parts (an exciting possible mechanism for integration of plastid development by the expression of the nuclear genome), as well as the prospect of being able to compare initiation signals and other properties once the enzymes were free of one another and were completely dependent on added DNA stimulated us in our search for information about these enzymes.

*Nuclear RNA polymerases of maize*

The soluble RNA polymerase activity first described by Mans & Novelli (1964) in crude extracts of homogenates of frozen leaves was subsequently partially enriched by $(NH_4)_2SO_4$ fractionation and by chromatography on DEAE-cellulose by Stout & Mans (1967). A single peak of activity was eluted from a column of DEAE-cellulose. We re-investigated somewhat similar, but not identical, soluble enzyme preparations (Strain *et al.*, 1971). Chromatography on DEAE-cellulose revealed a more complex pattern when elution was with a gradient of $(NH_4)_2SO_4$ and when activity was assessed with both double- and single-stranded DNA templates. Some RNA polymerase activity eluted with about 0.08 M-$(NH_4)_2SO_4$; another peak of activity appeared in the eluent centred at about 0.2 M-$(NH_4)_2SO_4$ (Fig. 1*a*). RNA polymerase activities which elute in these two regions have been identified in extracts of calf thymus nuclei (Roeder & Rutter, 1969, 1970).

Enzyme I, the polymerase which elutes with about 0.08 M-$(NH_4)_2SO_4$, is much more active with native than with denatured (i.e. melted, single-stranded) maize nuclear DNA. [Enzyme I of calf thymus nuclei is thought to be localized in nucleoli by Roeder & Rutter (1970)]. Maize enzyme II, the RNA-synthesizing

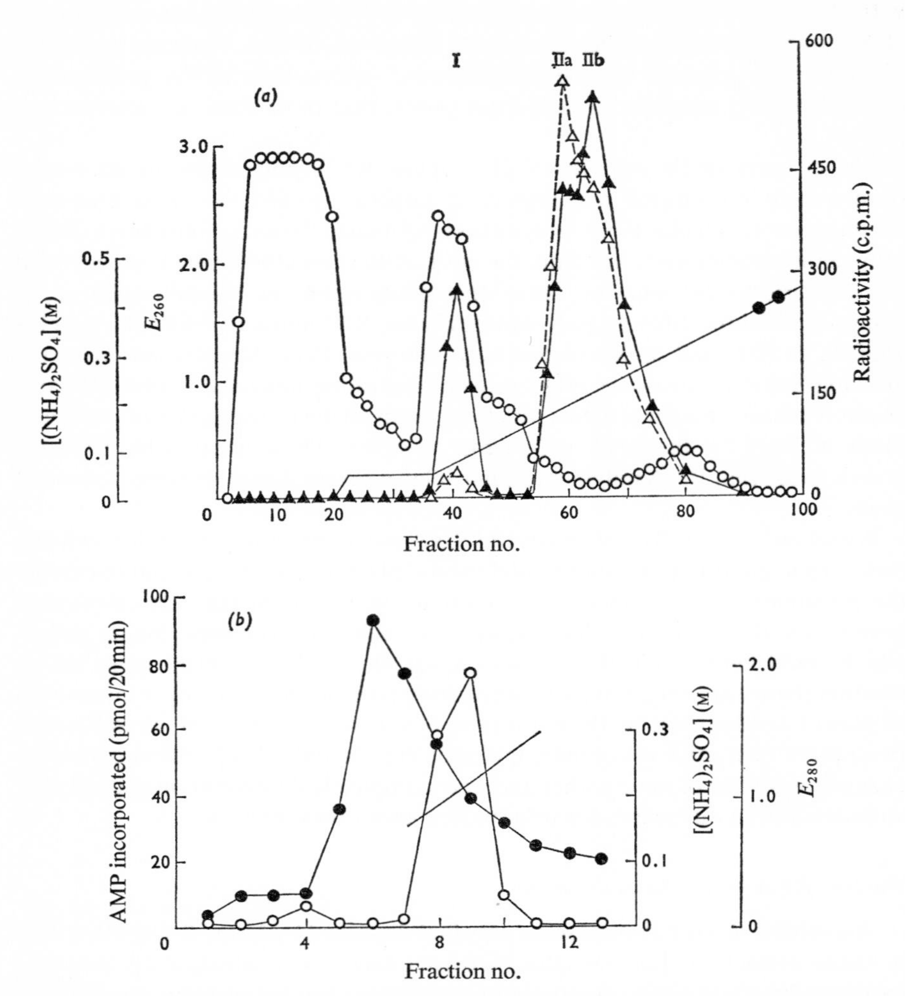


Fig. 1. *DEAE-cellulose chromatography of (a) soluble nuclear RNA polymerase and (b) solubilized maize chloroplast RNA polymerase*

(*a*) A 450mg sample of protein from the 30–50% $(NH_4)_2SO_4$ fraction of an extract of maize leaves was applied to a column (2cm × 20cm) of DEAE-cellulose equilibrated with Tris–glycerol–2-mercaptoethanol buffer. The column was washed with the buffer, followed by a wash with the buffer and 0.05M-$(NH_4)_2SO_4$. The enzyme activities were eluted with a linear gradient of 0.05–0.5M-$(NH_4)_2SO_4$ (●) in 0.05M-Tris (pH8.0)–20% (v/v) glycerol–0.01M-2-mercaptoethanol. Fractions (5ml) were collected and 20μl portions were assayed for RNA polymerase activity with native (▲) and denatured (△) maize nuclear DNA (Strain *et al.*, 1971). (*b*) The enzyme was extracted from sucrose-density-gradient-purified chloroplasts by incubation at 37°C for 15min in a solution of 0.05M-Tris–HCl buffer, pH8.0; 40mM-2-mercaptoethanol; 4mM-EDTA. The supernatant obtained by centrifugation at 12000*g* for 10min was applied to a column of DEAE-cellulose. The profile of activity (with maize nuclear enzyme as the template) (○) and $E_{280}$ (●) during elution with a gradient of 0–0.3M-$(NH_4)_2SO_4$ (——) in 0.05M-Tris–HCl (pH8.0)–40mM-2-mercaptoethanol–20% (v/v) glycerol buffer are shown (Bottomley *et al.*, 1971*a*). The peak of activity is eluted with 0.21M-$(NH_4)_2SO_4$, which is similar to the behaviour of maize nuclear RNA polymerase IIb. The two enzymes differ markedly in other respects.

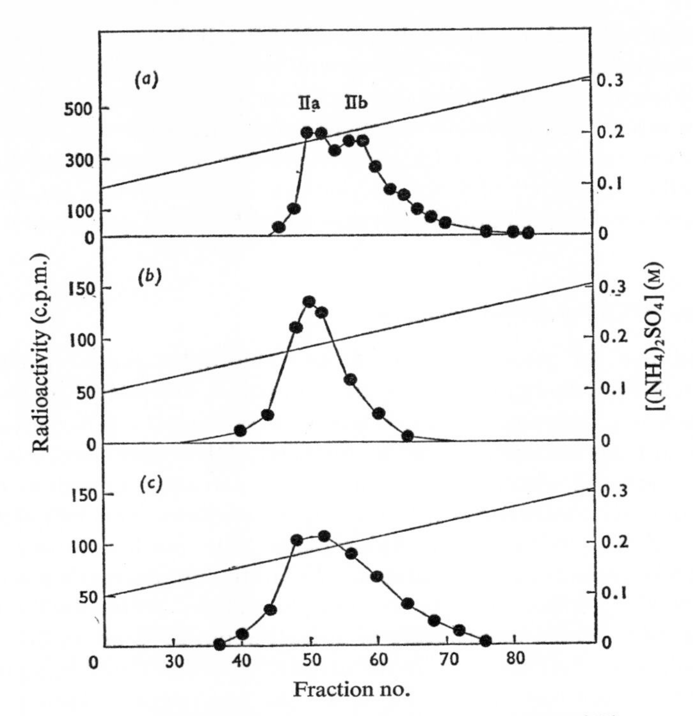


Fig. 2. *Rechromatography of enzyme activities IIa and IIb*

(*a*) A sample of the 30–50%-$(NH_4)_2SO_4$ fraction was run on a column (0.9cm × 10cm) of DEAE-cellulose as described in Fig. 1(*a*). (*b* and *c*) Activities IIa and IIb respectively were rechromatographed separately on columns (0.9cm × 10cm) of DEAE-cellulose under identical conditions. The activities (●) of the fractions were assayed with denatured maize DNA (Strain *et al.*, 1971). ——, [$(NH_4)_2SO_4$].

activity which elutes at higher concentrations of $(NH_4)_2SO_4$ includes two distinguishable components (Fig. 1*a*). Component IIa, an enzyme activity which is eluted from the DEAE-cellulose with about 0.18 M-$(NH_4)_2SO_4$, has a very strong preference for denatured nuclear DNA whereas the activity which is eluted later, i.e. with 0.22 M-$(NH_4)_2SO_4$ (component IIb) is more active when native rather than denatured maize nuclear DNA is provided as a template. The enzyme which prefers native DNA (component IIb) can be converted into enzyme IIa by rechromatography on DEAE-cellulose (Fig. 2), by chromatography on Sepharose 6B, by glycerol-density-gradient centrifugation, and by salt treatments. It is not yet clear whether a cofactor required for transcription of native DNA is lost from enzyme IIb under these various conditions, but this possibility is being investigated further.

Another conspicuous difference between enzymes of types I and II is that only the latter is inhibited by α-amanitin.

We (K. P. Mullinix, G. C. Strain & L. Bogorad, unpublished work) have now developed a new procedure for purifying enzyme IIa about 3000-fold from extracts of maize leaves clarified by high-speed centrifugation. The yield is about 5%. This procedure includes, in series: $(NH_4)_2SO_4$ fractionation, heat treatment, gel filtration, and chromatography on DEAE-cellulose and DNA-cellulose. Electrophoresis of the highly purified enzyme on polyacrylamide gels containing sodium dodecyl sulphate reveals the presence of two major polypeptides in equimolar ratio. Their molecular weights are approx. 200000 and 160000. There are additional smaller polypeptide components (see 'Note added in proof').

Some other properties of RNA polymerases of maize nuclei are included in Table 1 and are reviewed after the discussion of the RNA polymerase of maize plastids.

*Chloroplast RNA polymerase of maize*

A good deal of information about the chloroplast-bound partly DNA-dependent RNA polymerase was obtained in studies of isolated highly purified chloroplasts (e.g. Bogorad, 1967*b*; Bogorad & Woodcock, 1970) but it seemed necessary to bring the enzyme into a completely DNA-dependent soluble state in order to proceed with purification and characterization as well as to permit more meaningful comparisons with the nuclear enzymes. This was achieved by Bottomley *et al.* (1971*a*). Maize chloroplasts were purified through sucrose density gradients and then incubated at 37°C in the presence of 4mM-EDTA in 0.05M-Tris–HCl buffer, pH 8.0, containing 0.04M-2-mercaptoethanol. The absence of $Mg^{2+}$ and a temperature of at least about 25°C are both critical for the solubilization. Solubilization in the absence of EDTA could be achieved but the temperature optimum was then at 49°C. The 'crude soluble' enzyme preparation had some slight activity in the absence of added DNA, but was stimulated two- to five-fold by the addition of either denatured calf thymus DNA or denatured maize plastid DNA. After further purification the enzyme was more active with the latter template.

The following purification procedures were useful. Chromatography on phosphocellulose and elution in 0.6M-KCl provided an enzyme which was completely dependent on added DNA. Chromatography on columns of DEAE-

Table 1. *Properties of chloroplast and nuclear RNA polymerases of maize*

The data are taken largely from Bottomley *et al.* (1971*a*) and Strain *et al.* (1971). SSP, single-stranded plastid DNA; SSN, single-stranded nuclear DNA; DSN, double-stranded nuclear DNA.

| Property | Chloroplast | Nuclear | | |
|---|---|---|---|---|
| | | I | II | |
| | | | (a) | (b) |
| DEAE-celluose chromatography elution with: | | | | |
| $(NH_4)_2SO_4$ (M) | 0.21 | 0.08 | 0.18 | 0.22 |
| KCl (M) | 0.26 | | | |
| Phosphocellulose chromatography, elution with KCl (M) | 0.6 | | Lost | |
| Optimum [$Mg^{2+}$] (mM) | 15–40 | 25 | 25 | |
| Optimum [$Mn^{2+}$] (mM) | 8 | 8 | 8 | |
| $\frac{\text{Activity with 25 mM-}Mg^{2+}}{\text{Activity with 8 mM-}Mn^{2+}}$ | 5 | 2.5 | 2.5 | |
| % inhibition with α-amanitin (10 μg/ml) | 0 | | 90 | |
| Preferred template | SSP | DSN | SSN | DSN |
| Temperature optimum (°C) | 48 | 44 | 44 | |

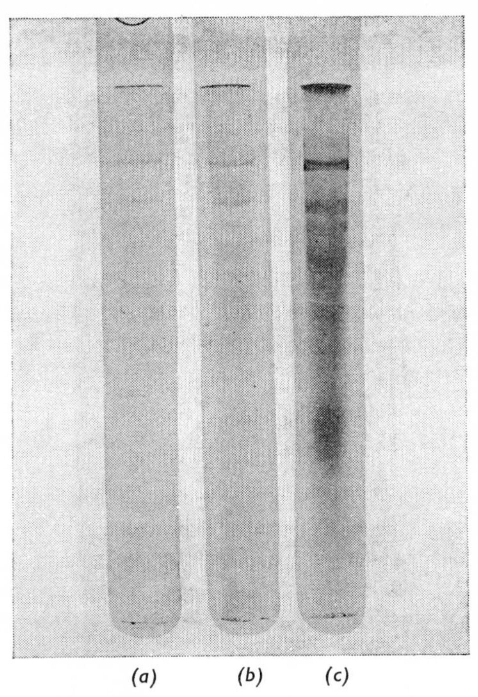


Fig. 3. *Comparison of the large subunits of maize nuclear RNA polymerase IIa and maize chloroplast RNA polymerase after electrophoresis on sodium dodecyl sulphate–5% polyacrylamide gel*

(*a*) Maize nuclear enzyme IIa; (*b*) maize chloroplast polymerase; (*c*) both enzymes.

cellulose yielded a single peak of activity which eluted with a salt gradient in 0.26M-KCl or 0.21M-$(NH_4)_2SO_4$ (Fig. 1*b*) and the product was completely dependent on added DNA. Similarly, an almost completely DNA-dependent enzyme was obtained after glycerol-density-gradient centrifugation in the presence of 0.5M-KCl. Chromatography on DNA-cellulose has proved useful as a secondary purification step.

The chloroplast enzyme has now been purified further by chromatography on DNA-cellulose and has been examined on sodium dodecyl sulphate–poly-acrylamide-gel electrophoresis (H. J. Smith & L. Bogorad, unpublished work). Two large polypeptide chains persist through all the purification steps we have employed and constitute an increasingly large fraction of the protein as the specific activity of the preparations rises. The largest of these subunits is about the same size as the 200000-dalton polypeptide of the maize nuclear enzyme IIa (Fig. 3). The next larger polypeptide of the chloroplast enzyme is about 130000 daltons, considerably smaller than the 'corresponding' polypeptide of maize nuclear enzyme IIa. There may also be some smaller polypeptides in the chloroplast polymerase. The large subunits are, in a way, more like the nuclear enzyme than bacterial enzymes in absolute size as well as in size relative to one another. However, maize nuclear enzyme IIa, the chloroplast polymerase and *E. coli* RNA polymerase all have two polypeptide chains larger than 130000 plus one or more smaller polypeptides.

Bogorad & Woodcock (1970) have shown that rifamycin SV inhibits chlorophyll production during illumination of etiolated maize by about 50% whereas rifampin [rifamycin, the 3-(4-methylpiperazinyl)iminomethyl derivative of rifamycin SV; see Wehrli & Staehelin, 1971] had virtually no effect. When etiolated maize leaves are exposed to light in the presence of [$^{32}$P]phosphate the label is incorporated very rapidly into chloroplast rRNA. At concentrations of up to about 50$\mu$M there is virtually no effect of rifampin or rifamycin SV on this incorporation. However, at 100$\mu$M virtually all incorporation is abolished. Bottomley *et al.* (1971*b*) failed to observe such inhibition but the conditions of the administration of rifampin were not identical with those used by Bogorad & Woodcock (1970). The reasons for the differences are not known.

Purified maize chloroplasts are generally partly dependent on added DNA for activity (e.g. Bogorad, 1967*a*) and up to about 30% of the activity which is dependent on the exogenous DNA was found to be inhibited by rifamycins at concentrations as high as 50$\mu$M (Bogorad & Woodcock, 1970). In contrast with this, Bottomley *et al.* (1971*b*) did not observe inhibition with preparations of maize and other plastids. The assays in the experiments of Bogorad & Woodcock (1970) always included $NH_4^+$ whereas neither these ions nor $K^+$ were included in the assay employed by Bottomley *et al.* (1971*b*). Further, the RNA polymerase activity of the plastids prepared by Bottomley *et al.* (1971*b*) was not stimulated by the addition of DNA.

Table 2. *Inhibition of solubilized maize plastid RNA polymerase by rifamycin SV*

Each 0.25ml of assay mixture contained: 10$\mu$mol of Tricine buffer, pH 8.5; 10$\mu$mol of mercaptoethanol; 5$\mu$mol of $MgCl_2$; 30$\mu$g of denatured calf thymus DNA; 0.6$\mu$mol each of UTP, CTP and GTP; 0.03$\mu$mol of ATP including 0.3$\mu$Ci of [8-$^{14}$C]ATP (specific radioactivity 53.2Ci/mol, Schwarz/Mann, Orangeburg, N.Y., U.S.A.); 10–25$\mu$l of enzyme solution at the stage of purification noted. '+$MgCl_2$' assays contained 8$\mu$mol of $MgCl_2$. Assays were performed by incubating in the absence of the nucleoside triphosphates for 5min at 37°C. After addition of nucleoside triphosphates samples were incubated for 20min at 47°C. Rifamycin SV (Calbiochem) concentrations were: crude soluble series 0.5mM (350$\mu$g/ml), glycerol-density-gradient series 0.29mM (200$\mu$g/ml). Each value represents the mean of three assays. Five additional triplicate assays were performed on the $NH_4Cl$ + $MgCl_2$ series and a mean inhibition of 7% in the absence of the salts and 33% in their presence was found.

| | Radioactivity (c.p.m.) | Salt inhibition (%) | Rifamycin inhibition (%) |
|---|---|---|---|
| Crude soluble enzyme | | | |
| Control | 2228 | — | — |
| +Rifamycin SV | 2187 | — | 4 |
| +0.125M-$NH_4Cl$ | 1490 | 34 | — |
| +0.125M-$NH_4Cl$ + rifamycin SV | 989 | — | 33 |
| +$MgCl_2$ | 1632 | 28 | — |
| +$MgCl_2$ + rifamycin SV | 1559 | — | 3 |
| +0.125M-$NH_4Cl$ + $MgCl_2$ | 873 | 61 | — |
| +0.125M-$NH_4Cl$ + $MgCl_2$ + rifamycin SV | 582 | — | 33 |
| +0.125M-KCl + $MgCl_2$ | 1257 | 46 | — |
| +0.125M-KCl + $MgCl_2$ + rifamycin SV | 951 | — | 24 |
| Glycerol-density-gradient-purified enzyme | | | |
| Control | 2290 | — | — |
| +Rifamycin SV | 2392 | — | 0 |
| +0.125M-$NH_4Cl$ | 272 | 82 | — |
| +0.125M-$NH_4Cl$ + rifamycin SV | 179 | — | 35 |

The data in Table 2 (obtained by H. J. Smith, W. Bottomley & L. Bogorad, unpublished work) show that the inhibition of either crude solubilized or glycerol-density-gradient-purified maize chloroplast RNA polymerase by rifamycin SV is dependent on the presence of $NH_4^+$ ($K^+$ is also effective) which at 0.125 M has an inhibitory effect alone. Neither $Mg^{2+}$ nor $Cl^-$ change the sensitivity of the enzyme to rifamycin SV. These results demonstrate how easily sensitivity to rifamycin can be altered and point to the importance of carefully defining isolation and assay conditions when studying the effects of rifamycins and probably other inhibitors on enzymes from diverse sources. They also accentuate the obvious again, it is very difficult to be sure of the effects of inhibitory agents on enzymes *in vivo* from experiments *in vitro*. It is pertinent to mention that Georgopoulos (1971) has shown that the RNA polymerases of $gro^+$ and of $gro^-$ strains of *E. coli* respond differently to rifamycin *in vitro* depending on the presence or absence of $NH_4^+$.

In the case of maize the data of Bogorad & Woodcock (1970) show that rifamycin SV and rifampin do some things *in vivo*, both arrest light-induced plastid rRNA production but, if anything, stimulate cytoplasmic rRNA production and rifamycin SV blocks chlorophyll production. The experiments *in vitro* (Table 2, and Bogorad & Woodcock, 1970) show that plastid RNA polymerase activity can be partly inhibited by these rifamycins at concentrations comparable with those required for inhibition of plastid rRNA synthesis *in vivo*. But we do not know how conditions near the RNA polymerases *in vivo* compare with any of our conditions *in vitro* nor what other enzymes these antibiotics may be able to inhibit *in vivo*.

*Comparison of the properties of maize RNA polymerases*

Properties of the various maize RNA polymerases we have examined are enumerated in Table 1. Some of these attributes have already been described. This, in summary, is a rudimentary succinct key to this group of three enzymes. Our preparations of nuclear enzyme I do not use denatured DNA templates effectively, but are very active with native DNA templates, are unaffected by α-amanitin and are eluted from DEAE-cellulose with much lower concentrations of salt than either the plastid enzyme or the type II enzymes. The plastid enzyme elutes from DEAE-cellulose at about the same concentration of salt as the type II enzymes, but the two behave differently on phosphocellulose and only the type II enzymes are inhibited by α-amanitin. The nuclear enzymes are most active with maize nuclear DNA species as templates whereas the chloroplast enzyme prefers denatured plastid DNA. Both enzyme IIa and the chloroplast RNA polymerase are strongly inhibited by rifamycins AF/013 and PR/19, two derivatives of rifamycin SV. (These were kindly provided by Dr. Luigi G. Silvestri of Gruppo Lepetit.) At 100 μM-rifamycin PR/19 the enzymes were inhibited about 70%.

The chloroplast enzyme sedimented a little ahead of a 500000 dalton marker (ribulose diphosphate carboxylase) and enzyme IIb sedimented together with a 16S marker (β-galactosidase) on glycerol-density-gradient centrifugation.

All these enzymes are more active with $Mg^{2+}$ than with $Mn^{2+}$; RNA polymerases of animal origin (e.g. see urchin and rat liver as reported by Roeder & Rutter, 1970) are generally much more active with $Mn^{2+}$.

Both nuclear RNA polymerase IIa and the chloroplast enzyme appear to contain two large polypeptide chains. Fig. 3 shows that one of the large subunits of the plastid polymerase differs from that of the nuclear enzyme. Comparisons of the smaller subunits have not yet been completed. It seems very likely that enzyme IIb will resemble enzyme IIa although the former, which transcribes native DNA better, may be more complex. We have no information of this sort about maize nuclear enzyme I.

We are particularly anxious to complete our comparisons of the subunits of enzymes IIa and IIb and to compare nuclear RNA polymerase I with both of these enzymes. We hope such comparisons will reveal whether there is some factor or subunit involved in the preference for native over denatured DNA. Additional work is also needed to obtain a solubilized plastid enzyme which prefers native DNA either by a new extraction procedure or by reconstitution.

How many maize RNA polymerases are there? We have discussed three. Of these, one can be obtained in two forms (IIa and IIb). There is very likely to be an RNA polymerase in the mitochondria. Further, one interpretation of the rifamycin SV–rifampin data of Bogorad & Woodcock (1970), which show (1) that one of these compounds inhibits chlorophyll formation and plastid rRNA production whereas the other just blocks the latter and (2) that these compounds inhibit plastid RNA polymerase activity incompletely, is that there may be more than one chloroplast RNA polymerase. Beyond this there may certainly be more undetected nuclear polymerases. We also have yet to learn how the activity and specificity of each of the enzymes we know about is regulated and whether there are different regulatory mechanisms for each enzyme.

## Genes Specifying Chloroplast Ribosomal Proteins

The emerging information on RNA polymerases of nuclear and chloroplast origin reassures us, if we were uncertain before, that transcription can go on in both compartments. But what is being transcribed? Where are the genes which specify proteins and RNA species of the chloroplast? DNA–RNA hybridization techniques have been useful in searching for the genes for chloroplast rRNA species (e.g. see discussion by Woodcock & Bogorad, 1971) but the answers have not always been clear or correct. A sample of the complications is shown in the work of Kung *et al.* (1972) who found some sequence homology between chloroplast and nuclear DNA species of broad bean: this may be a clue to the basis of some of the problems. In any event, the hybridization method is limited to situations in which highly labelled direct products of gene transcription are available in adequate amounts. If specific mRNA species could be obtained in adequate quantity etc., it would probably be possible to look for the sites of genes for specific proteins. Lacking that, we have engaged in a different sort of analysis. The general approach is (*a*) to somehow mark a single chloroplast protein and (*b*) to determine whether it is inherited in a Mendelian or uniparental manner (and preferably to assign the gene to a known linkage group).

In order to carry on the genetic analyses, an organism which can easily be crossed is necessary and it would clearly be very useful if some of its genes have already been mapped. *Chlamydomonas reinhardii* fits into this category and has the further advantage that the vegetative cells are haploid. Thus mutations are expressed directly.

The protein could be marked by a mutation in the gene which specifies it. The mutation should be easy to select for and the protein should be present in relatively large amounts and be readily isolated.

Ribosomal proteins seemed to be reasonable candidates. First, as pointed out earlier, plastid and cytoplasmic ribosomes are distinctive in their susceptibility to some antibiotics, thus mutants could be screened and selected, and the ribosomes of the cytoplasm and chloroplasts differ in size and are readily separated from one another by sucrose-density-gradient centrifugation. By selecting appropriate conditions of salts and sucrose-density-gradient centrifugation, it is also possible to separate each ribosomal type into subunits and each subunit from all the others in a single centrifugation run. Further, ribosomes are present in large amounts and the proteins are easily solubilized and separated by electrophoresis on urea-containing polyacrylamide gels or by ion-exchange chromatography. Incidentally, analysis by urea–polyacrylamide-gel electrophoresis reveals that the protein complement of ribosomes from the cytoplasm, chloroplasts and mitochondria of a single species are distinctive (e.g. Vasconcelos & Bogorad, 1971) and Hoober & Blobel (1969) had compared chloroplast and cytoplasmic ribosomes of *Chlamydomonas*. Thus, some kinds of alterations in proteins might also be reasonably easily detectable by gel electrophoresis.

*Chlamydomonas* is sensitive to erythromycin but some erythromycin-resistant strains were reported by Sager & Raminis (1970) and it seemed likely from the known binding of chloramphenicol to chloroplast ribosomes that erythromycin would act similarly. It was in fact possible to show (Mets & Bogorad, 1971) that [$^{14}$C]erythromycin binds specifically to the large (52S) subunit of *Chlamydomonas* chloroplast ribosomes and to measure the binding constant of these ribosomes with erythromycin. Ribosomes adhere to Millipore filters and so erythromycin was measured by mixing ribosomes with various concentrations of [$^{14}$C]-erythromycin. The solution was poured through the filter, which was then washed and, finally, the amount of radioactivity retained on the filter by the ribosomes was measured. This is essentially the procedure of Teraoka (1970). Ribosomes from wild-type *C. reinhardii* bind erythromycin at $K^+$ concentrations of 25 mM or higher. In this and in some other respects the binding is not identical with that reported for bacteria. The affinity of *Chlamydomonas* ribosomes for erythromycin is 1 to 2 orders of magnitude below that of bacterial ribosomes. The equilibrium constant is about $8 \times 10^4 M^{-1}$ and the typical preparation of crude ribosomes contains one binding site for every $8 \times 10^6$–$9 \times 10^6$ daltons of nucleic acid (Mets & Bogorad, 1971). Growth of wild-type cells is completely blocked by 0.5 mM-erythromycin.

To obtain erythromycin-resistant strains, wild-type, mating-type$^+$ cells (from strain 137c obtained from R. P. Levine) were treated with ethylmethanesulphonate and grown with 0.5 mM-erythromycin. About 75 isolates were selected for their insensitivity to the antibiotic. All the stable resistant isolates were unaffected by the antibiotic when grown either in the light on minimal medium or in the dark on an acetate-containing medium. Nine of these strains, a representative sample, were selected for further study and have been maintained in the complete absence of erythromycin for two years now.

The first problem was to determine whether the algae had become resistant because of lowered permeability to the antibiotic, because of a newly acquired capacity to detoxify the drug, or because of an alteration in the capacity of the ribosomes to bind the antibiotic. Alterations of the first two types would not be very promising for our purposes; we were interested in mutations of the third type. That is, strains with chloroplast ribosomes that could no longer bind erythromycin. Consequently, ribosomes of the mutant strains were analysed for [$^{14}$C]erythromycin binding by the Millipore filter-retention method. It was found that the ribosomes of none of these nine mutants would bind erythromycin in 25mM-$K^+$ (Fig. 4). Thus, the resistance mutation in each case apparently led to either alterations in the erythromycin-binding site of the ribosome or the loss of chloroplast ribosomes carrying these sites from the crude preparations. The latter seems not to be the case because, first, the mole percent of chloroplast ribosomes in crude mixtures from each mutant is similar to that from wild-type except for mutant strains 1S3 and 12S3, where the content is lowered to about

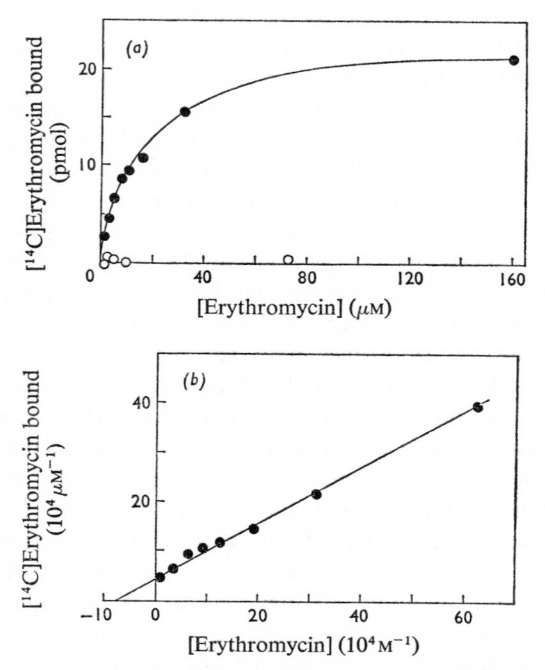


Fig. 4. *Binding of [$^{14}$C]erythromycin to crude ribosomes*

*a*) ●, Binding to wild-type ribosomes; ○, binding to ribosomes from mutant 11S7. Reaction mixtures of 0.1 ml containing buffer, [$^{14}$C]erythromycin and 0.27 mg of nucleic acid in a crude ribosome preparation were incubated for 30 min at 25°C, diluted with 3 ml of buffer, and then poured over Millipore HA filters. The filters were washed six times with 3 ml portions of buffer, dried, and the radioactivity was counted in toluene-based scintillation fluor. The [$^{14}$C]erythromycin (24c.p.m./pmol) was a generous gift from Dr. R. H. Williams (Lilly Research Laboratories). (*b*) Double-reciprocal plot of the data for wild-type ribosomes in (*a*). The binding constant is the negative of the intercept on the abscissa (from Mets & Bogorad, 1971).

one-half. Second, mutant ribosomes bound erythromycin in amounts similar to that taken up by the wild-type but only at concentrations of $K^+$ higher than needed for maximum binding by wild-type ribosomes. Some similar effects of potassium have been found in *E. coli* (Taubman *et al.*, 1966; Dekio *et al.*, 1970). There are some complications about binding in very high potassium concentration by some of the mutant strains, but this information is not crucial to our discussion of the problem.

So, the phenotype we are now following is no longer simply erythromycin resistance but binding of erythromycin to the 52S subunit of the chloroplast ribosome. Again, the implication is that there is some alteration in a component or components of these subunits.

The next problem then is to track the genes coding for these characters to one or another of the organism's genomes. Table 3 summarizes the results of genetic experiments with these nine mutants. Tetrad analyses of various crosses yielded the following results. (1) Each of the mutants, except 2L1, behaves as a single, normal Mendelian gene giving 2:2 segregation in meiosis, i.e. one-half of the offspring of a mating with wild-type is resistant and one-half is susceptible to erythromycin. (2) When a gamete of any of the mutant strains listed in Group A of Table 3 is mated with a gamete of any other of the strains in this group, all the progeny are resistant. The same is true for the strains included in Group B. These are the bases for including a strain in Group A or B. On the other hand, Groups A and B are clearly different from one another, for some of the products of a cross of mutant 12S3 with any of the Mendelian mutants in the other linkage group do not survive whereas other offspring are erythromycin-sensitive; we take these results to indicate that recombination occurs. (3) Mutant 2L1 behaves like a normal uniparental gene. When carried in a mating-type 'plus' strain, it is transmitted to all four progeny and when in a mating-type 'minus' strain, it is transmitted to none.

Dr. Sager has recently completed crosses with our mutant 2L1 and found that it maps near $sm_4$ on the uniparental linkage map. It is not unlikely that it is allelic with the erythromycin-resistant gene found by Sager & Ramanis (1970). The evidence that this linkage group is in the chloroplast of *Chlamydomonas* has been discussed extensively by Sager & Ramanis (1970) and also by Sager (1972).

Thus, the genetic analyses indicate that at least three loci in *C. reinhardii* can mutate to alter the capacity of the 52S chloroplast ribosomal subunit to bind

Table 3. *Linkage of erythromycin resistance mutations in C. reinhardii*

| | Mutant | Segregation | Locus |
|---|---|---|---|
| Group A | 1L5 | 2:2 | ery-M1 |
| | 11S7 | 2:2 | |
| | 13L1 | 2:2 | |
| | 15S4 | 2:2 | |
| Group B | 1S3 | 2:2 | ery-M2 |
| | 1S5 | 2:2 | |
| | 2L2 | 2:2 | |
| | 12S3 | 2:2 | |
| | 2L1 | 4:0 | ery-U1 |

erythromycin. Two of these loci are inherited in a Mendelian manner and are thus judged to be in the nuclear genome; one is in the uniparental genome.

Incidentally, the observation that a combination of mutant 12S3 with any of the Mendelian mutants in the other linkage group is lethal indicates that the presence of either of the mutated Mendelian genes does not interfere noticeably with ribosome function, but ribosomes with both alterations are not functional. The sensitivity of *Chlamydomonas* to erythromycin, even when it is grown on acetate, suggested that the chloroplast plays an important role in the metabolism and maintenance of the cell beyond doing photosynthesis. But, of course, erythromycin may have deleterious effects on cell metabolism besides stopping protein synthesis on chloroplast ribosomes. However, the genetic experiments described here reinforce the possibility that the chloroplast ribosomes are essential for the survival of *Chlamydomonas*.

### 52 S *subunit of the chloroplast ribosome*

We were encouraged in seeking altered ribosomal proteins in erythromycin-resistant strains by the report of Otaka *et al.* (1970) that erythromycin-resistant mutants of *E. coli* have an altered ribosomal protein. However, there was also some evidence of other possible modifications of ribosomes which might alter the affinity of ribosomes for erythromycin. Mao & Putterman (1969) concluded from the effects of various agents that alter RNA, but not proteins, that the 23 S rRNA of the 50 S ribosomal units of *E. coli* might be involved in erythromycin binding. This possibility was supported by Lai & Weisblum (1971), who reported that $N^6$-dimethyl adenine is absent from erythromycin-susceptible cells of *Staphylococcus aureus* but present in the 23 S rRNA of the erythromycin-induced erythromycin-resistant strain of this bacterium. There is no easily detectable difference in the ribosomal proteins of the sensitive and resistant strains although the methods used to detect that possibility were not exhaustive. Looking for rRNA differences is much more difficult than seeking altered proteins. So, we sought altered proteins.

Ribosomal proteins have been looked at for years by polyacrylamide-gel electrophoresis in urea. The proteins separate on the basis of charge. Alterations of proteins which do not result in a change in net charge would not show up by using this procedure, but because this method requires only small amounts of protein, is easy and fast, and is very sensitive for charge changes, this still seemed a reasonable way to begin.

The first problem was to obtain enough purified 52S subunits of *C. reinhardii* chloroplast ribosomes to be able to examine their proteins by gel electrophoresis. Cells were grown on a high-salt acetate medium (Sueoka, 1960) in equilibrium with 5% $CO_2$ in air and at a light intensity of 4300–7500 lx (400–700 ft-candles). The cells were collected by centrifugation and frozen. They were then broken by high pressure and phase transition. Ribosomes were extracted and purified first by centrifugation through a solution of 1 M-sucrose. Then, with Tris–HCl buffer, pH 7.5, a range of concentrations of $MgCl_2$ and KCl were tested to get ribosomes to separate into subunits and the subunits to separate from one another on iso-kinetic sucrose density gradients. A buffer which is 2.5 mM-$MgCl_2$–200 mM-KCl

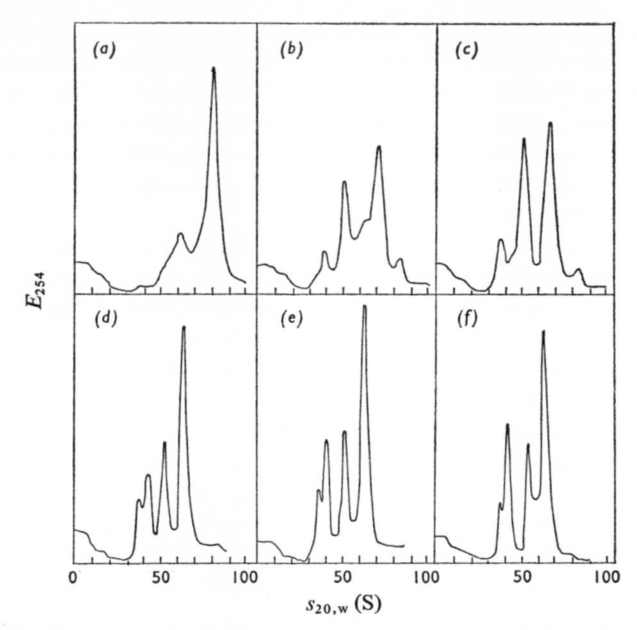


Fig. 5. *Distribution of subunits of C. reinhardii ribosomal subunits after centrifugation on a sucrose density gradient*

Wild-type cells were frozen, broken, thawed in 25mM-$MgCl_2$–25mM-KCl–25mM-Tris–HCl buffer, pH7.5, and centrifuged to remove membranes and other insoluble compounds. Ribosomes were purified as described in the text, then resuspended in (*a*) 25mM-$MgCl_2$–25mM-KCl, (*b*) 2.5mM-$MgCl_2$–25mM-KCl, (*c*) 2.5mM-$MgCl_2$–50mM-KCl; (*d*) 2.5mM-$MgCl_2$–100mM-KCl, (*e*) 2.5mM-$MgCl_2$–150mM-KCl; (*f*) 2.5mM-$MgCl_2$–200mM-KCl; all were in 25mM-Tris–HCl buffer, pH7.5. Each was centrifuged in a 13ml 10–34% (w/v) sucrose isokinetic gradient in the same $MgCl_2$–KCl–Tris solution for 4.5h at 39000rev./min in a Beckman SW-40 rotor. The $E_{254}$ shows the final positions of the subunits as monitored by an ISCO UV Analyzer during displacement of the gradient from the centrifuge tube (Mets, 1972).

was decided on and the separation on a 10–34% (w/v) isokinetic gradient run in a Spinco SW-40 rotor is shown in Fig. 5.

In order to have enough material to work with for analyses of ribosomal proteins, the whole operation was scaled up. Cultures (12 litre) were grown, and the cells were harvested in a continuous-flow centrifuge, and then processed as already described except that the ribosomal subunits were separated on a 1517ml gradient in the Beckman B-15Ti rotor with centrifugation at 30000rev./min for 23h. Fig. 6(*a*) shows the extinction profile at 254nm of the contents of the rotor after centrifugation. Ribosomes from the central portion of each peak were sedimented and rerun on the SW-40 rotor and scanned to check the homogeneity of each zone. The scans of the 'central portions' recentrifuged in 25mM-$MgCl_2$–25mM-KCl are shown in Figs. 6(*b*)–6(*e*). The 61S particles have a marked tendency to aggregate under these conditions and both the 37S and 41S particles dimerized to a certain extent. The purity and identity of the separated particles was further established by gel electrophoresis of the RNA which they contained. Samples were collected, ribosomes were pelleted, sodium dodecyl sulphate was added to dissociate the ribosomal particles, and the rRNA species were analysed by gel electrophoresis. Fig. 7 is a composite of scans of the polyacrylamide gels of the rRNA species contained in each class of particles. Some degradation of the $1.09 \times 10^6$ dalton rRNA is seen and there may be some slight contamination of subunits with the component of the next lower molecular weight. By these criteria the subunits isolated on the zonal gradient were judged to be at least 98% pure. Thus, we could obtain large amounts of highly purified 52S subunits by one run on the zonal rotor.

The next problem was to obtain and analyse the proteins of the 52S subunits. Leboy *et al.* (1964) introduced the use of LiCl–urea to simultaneously solubilize ribosomal proteins (in the urea) and precipitate the rRNA (by the $Li^+$). We used their procedure. After removing the LiCl and concentrating the proteins, the latter were analysed by polyacrylamide-gel electrophoresis at low pH in 8 M-urea (Moore *et al.*, 1968). The results of the analysis of 52S ribosomal subunits from wild-type and from several of the mutant strains are shown in Fig. 8.

First, no differences could be distinguished between the banding patterns of proteins from mutants 1L5 and 11S7 and those of the wild type. Yet we know from the erythromycin-binding data that the ribosomes of these mutant strains are clearly different from the wild-type, but such differences are not revealed by this technique which depends on net charge. Second, when compared with the wild type, one band is missing from among the proteins of the 52S subunits of 12S3, but another band which is not present in wild type does appear; it seems that the electrophoretic mobility of one of the proteins present in wild type is lowered in the 12S3 strain. Finally, there are several differences between the migration patterns of the proteins of the 52S subunits of 2L1 and wild type. These

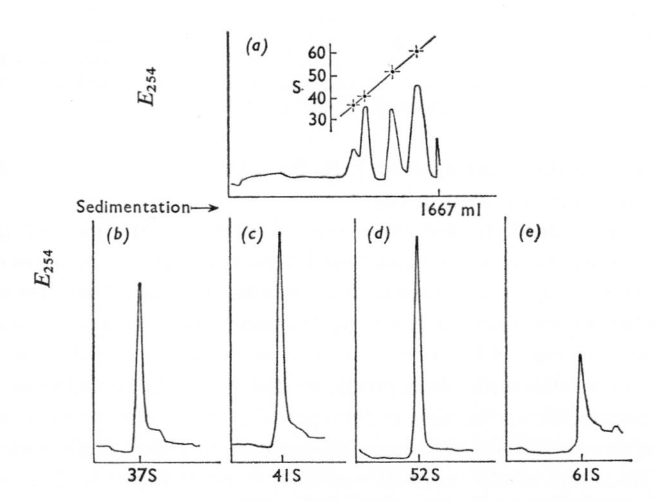


Fig. 6. *Separation of C. reinhardii ribosomal subunits by sucrose-density-gradient zonal centrifugation*

(*a*) Ribosomal subunits were prepared as described in the experiment shown in Fig. 5 but centrifugation was in a 1517 ml isokinetic gradient ranging from 40–7.25% (w/v) sucrose contained in a Beckman B-15 Ti rotor. (*b*), (*c*), (*d*) and (*e*), As described in the text, a portion from the central zone of each of the bands in (*a*) was rerun under the conditions described in the legend for Fig. 5 (Mets, 1972).

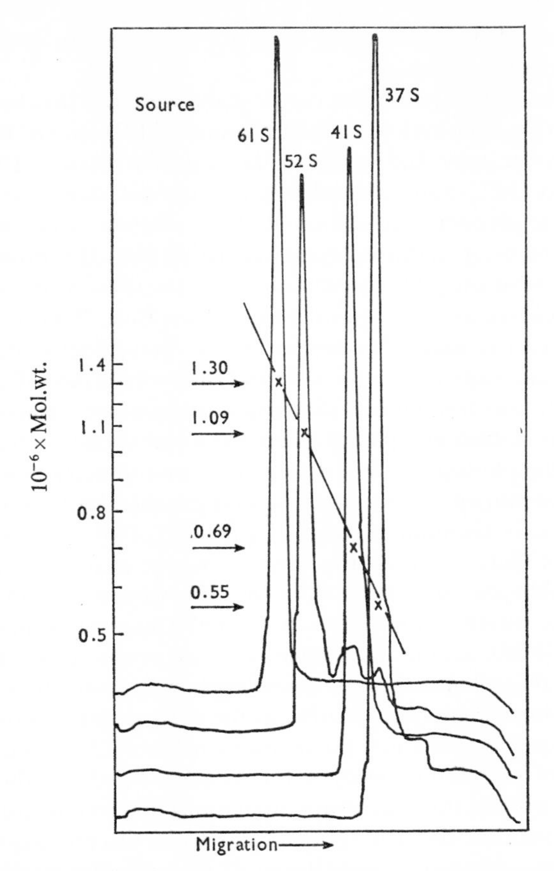


Fig. 7. *Analyses and identification by sodium dodecyl sulphate–polyacrylamide-gel electrophoresis of the major RNA species of the ribosomal subunits isolated by zonal sucrose-density-gradient centrifugation*

See Fig. 6 and the text for details (Mets, 1972).

---

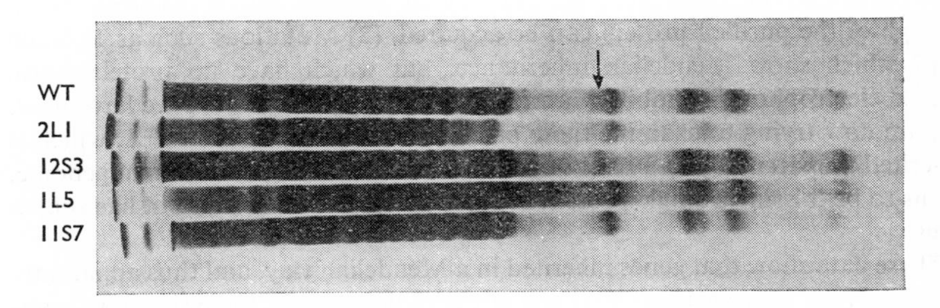


Fig. 8. *Proteins of the 52S subunits of chloroplast ribosomes of several strains of C. reinhardii separated by electrophoresis in urea-containing polyacrylamide gels* (*Mets*, 1972)

The arrow shows the position of the alteration in strain 12S3. WT, wild type.

differences are revealed more clearly when the proteins are analysed by two-dimensional gel electrophoresis.

For the two-dimensional analyses the proteins were run first under the usual conditions, i.e. in 8 M-urea and 4% polyacrylamide. At the end of this run the gel was moved from the tube and fused to the top of a 10 cm × 10 cm × 0.32 cm ($\frac{1}{8}$in) thick slab of 10% polyacrylamide gel which had been cast earlier. The upper buffer of the gel contained sodium dodecyl sulphate which moves into the gel electrophoretically. Conditions were selcted so that the proteins 'stacked' across the first-dimension gel before running into the slab. At the end of the run the slabs were removed and stained with Coomassie Blue. Plate 1 is a composite photograph. In order to make this composite, two first-dimension disc gels were run and one of them was fused to the slab and run as described. The stained slab is shown here with the original first-dimension gel removed. The duplicate first-dimension gel was stained and placed where the original had been across the top of the slab and the photograph was taken. Two first-dimension gels were done simply to show the entire process. Plate 2 is a comparable set of gels of the proteins of the 52S subunits of the mutant strain 2L1. Finally, Plate 3 shows parts of the gel slabs shown in Plates 1 and 2; the circle shows the position of a polypeptide present in the wild-type but absent from the 52S ribosomal subunits in algae carrying the uniparentally transmitted 2L1 gene. This protein has a molecular weight of about 33000. Strain 2L1 appears to have several proteins not present in wild-type 52S ribosomal subunits. These appear as small streaks with about the same electrophoretic mobility in urea as the corresponding protein found in wild-type cells. However, these new polypeptides in strain 2L1 do not migrate into the second-dimension sodium dodecyl sulphate gel as rapidly as the missing protein. The mutated protein thus has an apparent molecular weight greater than that of the wild-type one, but the new spots appear to be discrete aggregates of the missing protein since they occur at multiples of its molecular weight. The alteration of the protein appears to increase its tendency to aggregate.

We have seen three cases among *Chlamydomonas* 52S ribosomal subunits which do not bind erythromycin in the normal way. (1) Mutations such as 12S3 which are inherited in a Mendelian manner and in which a single protein is altered in its electrophoretic mobility. It seems highly likely that this alteration is in a structural gene for a chloroplast ribosomal protein. However, the final proof will require amino acid analyses and tryptic peptide 'finger printing' when enough of the purified protein can be acquired. (2) Mutations such as 1L5 and 11S7 which show Mendelian inheritance, but which have no proteins with altered electrophoretic mobility in urea gels under the conditions we have used. We are now trying to examine these proteins further. (3) Mutant 2L1, which is inherited uniparentally and has been mapped on the non-chromosomal genome, seems to be missing a single protein of the wild-type 52S subunit but has substitutes.

These data show that genes inherited in a Mendelian way, and thus apparently in the nucleus, can specify at least one of the proteins in the large ribosomal subunit of the *Chlamydomonas* chloroplast ribosome. That some co-operation of the nuclear and extranuclear genomes is involved in making the full complement of plastid ribosomal proteins is demonstrated by the alteration of a protein in 2L1

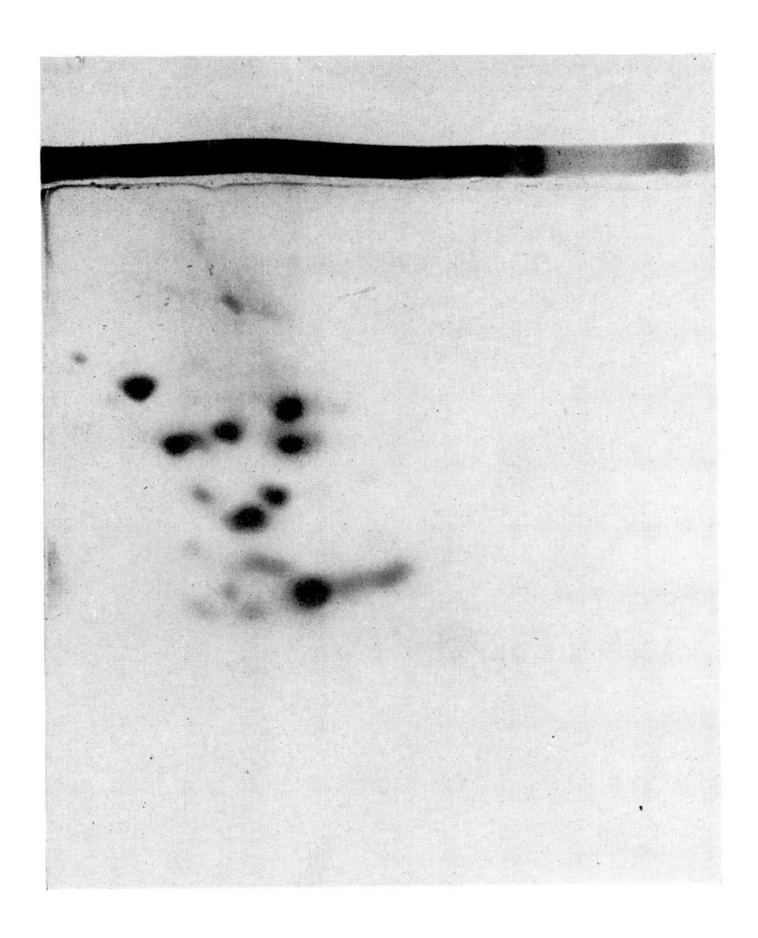

EXPLANATION OF PLATE I

*Proteins of 52S chloroplast ribosomal subunits prepared from wild-type C. reinhardii*

The proteins were separated by electrophoresis in urea–polyacrylamide-gel cylinders (left to right) and then by electrophoresis in sodium dodecyl sulphate–polyacrylamide slab (top toward bottom) (Mets, 1972).

L. BOGORAD AND OTHERS

(*Facing p.* 34)

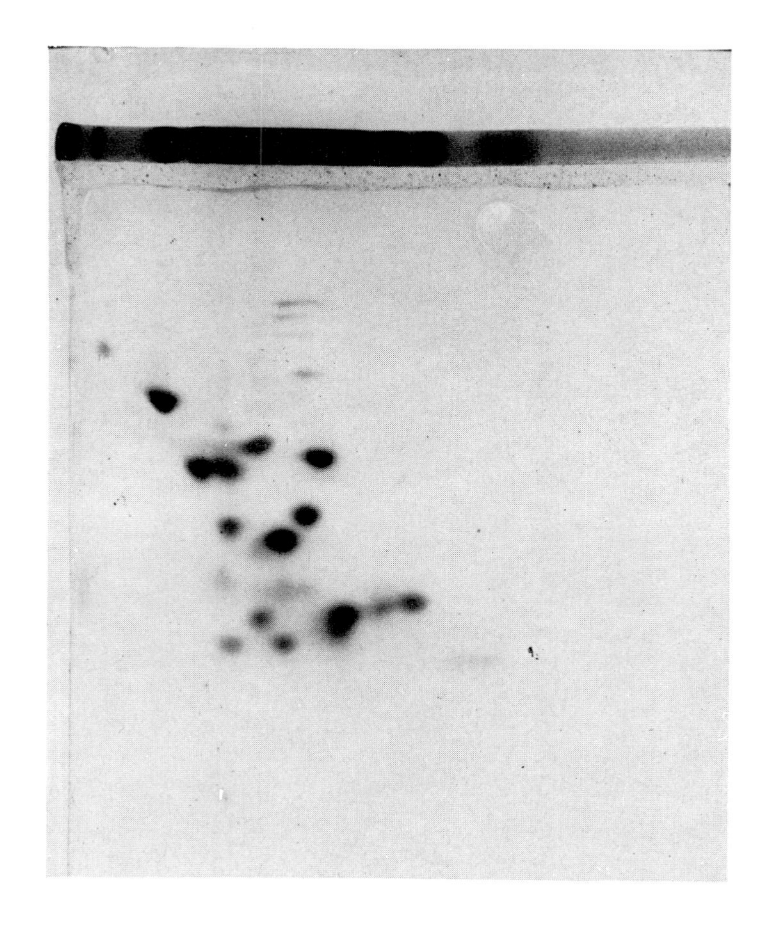

EXPLANATION OF PLATE 2

*Proteins of the* 52*S chloroplast ribosomal subunits of C. reinhardii erythromycin-resistant strain* 2*L*1

The proteins were separated as described in Plate 1. The character is transmitted uniparentally (Mets, 1972).

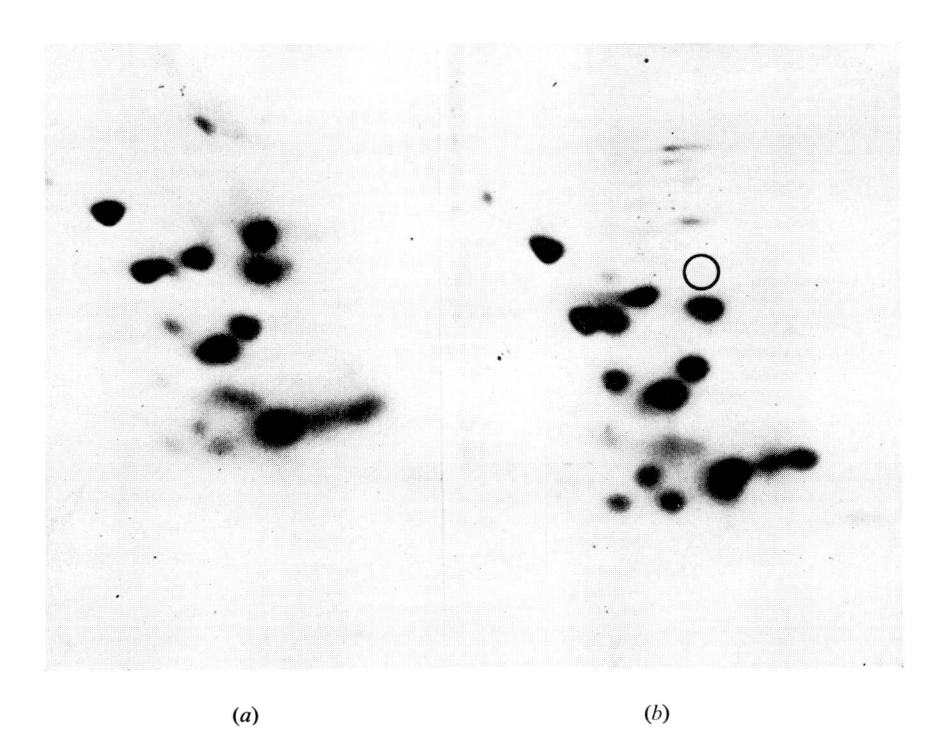

(*a*) (*b*)

EXPLANATION OF PLATE 3

*Composite of portions of the gel slabs shown in* (*a*) *Plate* 1 *and* (*b*) *Plate* 2 *to facilitate comparisons*

The circle shows the position of the polypeptide present in 52S subunits of wild-type cells but absent from those of strain 2L1. Some 'new' bands of equal electrophoretic mobility in urea gels but of higher molecular weights than the 'wild-type proteins' are seen in the 2L1 gel slab.

different from the 12S3 protein although the exact nature of this alteration is unclear at present.

**Integration of Cellular Activities: Interrelationships Between the Nuclear–Cytoplasmic System and Organelles**

Every cell of *Chlamydomonas* has only a single chloroplast. Cells of a single type in higher plants have comparatively uniform numbers of mature plastids of similar morphology and composition. Yet in other nearby cells plastids differ in morphology or have not matured. Plastids in bundle sheaths develop differently from those in mesophyll cells of tropical grasses. Chromoplasts of tomato fruit skins and chloroplasts of tomato leaves are different developmental stages of the same organelle. Amyloplasts and oil-storing plastids are still other expressions of the plastid potential in plants which bear chloroplasts that are carrying on photosynthesis in other cells. And, plastids can and are converted from one type into another during different times in the lifetime of the cell. How does a cell keep from dying of overproliferation of plastids? How is plastid development regulated? By what mechanisms does the nuclear–cytoplasmic system regulate plastid growth, multiplication and differentiation?

Regulation of the medium by the selective activities of root-cell membranes is illustrated by the maize mutation 'yellow stripe 1' described by Bell *et al.* (1958, 1962). The ability of plants with two $ys_1$ alleles to take up iron is impaired; the yellow striping is a consequence of iron deficiency and phenocopies of the wild type are very easily made by administering iron-containing solutions directly to the leaves. Changes in the medium which arrest plastid development in barley because of amino acid deficiencies have been studied by von Wettstein *et al.* (e.g. see recent review by Walles, 1971). One mutant strain fails to become green, but becomes phenotypically normal on the administration of aspartate. Another mutant behaves similarly, but needs leucine instead. Both the iron- and the amino acid-deficiency situations are extreme and finer control of the levels of such substances might permit the kinds of regulation needed without affecting processes in the cytoplasm itself. Plastid development and differentiation could be regulated by mechanisms of these sorts, but control by the nuclear–cytoplasmic system of the availability of materials needed by the organelle alone would be safer and perhaps more selective.

The two systems which have been described here, (1) the RNA polymerases of the nucleus and the chloroplast and (2) the presence of genes specifying plastid ribosomal proteins in the nuclear genome, could be significant means by which the nuclear–cytoplasmic system might control chloroplast growth and development by regulating the availability of components specifically required by the organelle for its growth and development.

Information about the various RNA polymerases is still meagre. There is as yet no evidence for RNA polymerase-mediated transcriptional control of development in eukaryotes. Nor is there any evidence in eukaryotes for small protein factors which act on or with RNA polymerases to specify the genes which are read. Yet, what could be a better sort of communication system than the controlled production by the nuclear–cytoplasmic system of all or parts of the chloroplast polymerase or of components which specify its activity? In such ways plastid activity might be altered by messages which could be obeyed only by transcription machinery of the plastids. The direct specification of plastid ribosomal proteins by nuclear genes clearly suggests one way in which the expression of a nuclear gene could affect the level of protein synthesis on chloroplast ribosomes. This is still only a possibility, for it remains to be seen whether the production of these proteins is alterable and thus does serve as a method for regulating the protein-synthetic activity of the chloroplasts.

**Origin and Evolution of Organelles**

The gene specifying the 12S3 protein is inherited in a Mendelian manner. It appears to be in the nuclear genome. How did it get there? This question can only be considered in the context of other unresolved problems such as: How did chloroplasts and mitochondria originate and what are the rules of selection and evolution within a single cell?

We can consider two hypotheses of the origin of organelles. (1) The endosymbiont hypothesis which suggests that chloroplasts and mitochondria are derived from ancestral forms of blue–green algae or photosynthetic bacteria which invaded a nucleated cell. (Or, if a separate mitochondrial progenitor is required an aerobic bacterium might have been the invader or a co-invader.) According to this view the host and the invader found it to their mutual advantage to continue the association. It is implied that they then evolved together to become an integrated system. (2) The other alternative, which is also quite apparent, we will call the cluster-clone hypothesis. According to this hypothesis, the progenitor of the modern eukaryotic cell was a single primitive prokaryotic cell. It lacked a nucleus, chloroplasts and mitochondria. Clusters of genes in such a cell could be cut off from the remainder of the cell by membranes thus making separate compartments. Each of these membrane-limited compartments would give rise to one particular clone of organelles, plastids by division of primaeval plastids, nuclei from nuclei, etc. So there are three requirements: (*a*) the clustering of genes into groups; (*b*) the formation of a membrane around these genes to form one or more gene-containing structures and a gene-free space, the cytoplasm; and (*c*) the division and faithful reproduction of each gene-containing compartment.

The endosymbiont hypothesis disregards some problems such as the origin of the nucleus but other problems are common to both of these possible mechanisms. Examples of the latter are the evolution of eukaryotic-type chromosomes and the origin of phylogenetic heterogeneity of ribosomes.

A basic premise of the endosymbiont hypothesis is that the invader brought with it information as well as the completely developed capacity for transcription and translation and that the invader and host systems have retained the marks of their origin to this date. Thus, the genes for the ribosomal proteins of the invader would have originally been in the invader alone (Scheme 1) and the evolution of ribosomes into eu- and pro-karyotic types (if this distinction will ultimately prove to be valid and useful) must have preceded the establishment of the symbiotic relationship.

I. Endosymbiont Hypothesis

Premise: the genes for the ribosomal protein of the invader are only in the invader. Then, during evolution after symbiotic relationship has been established:

Possibility A. Gene transfer.

1. Invader's gene 'transforms' host's genome.
2. Invader loses gene.

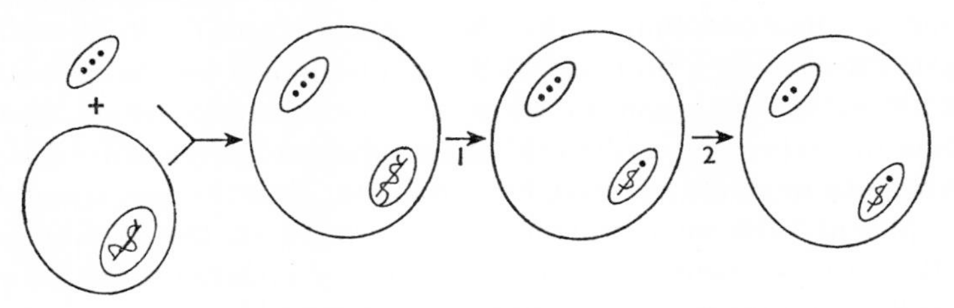


●, Organelle gene.

Possibility B. Protein and gene substitution.

1. Organelle gene mutates and organelle's ribosomal protein is lost.
2. A protein of the cytoplasmic ribosome substitutes for the lost organelle protein.
3. Organelle ribosomal protein gene in the nucleus may be retained even after the 'parent' cytoplasmic ribosomal protein changes.

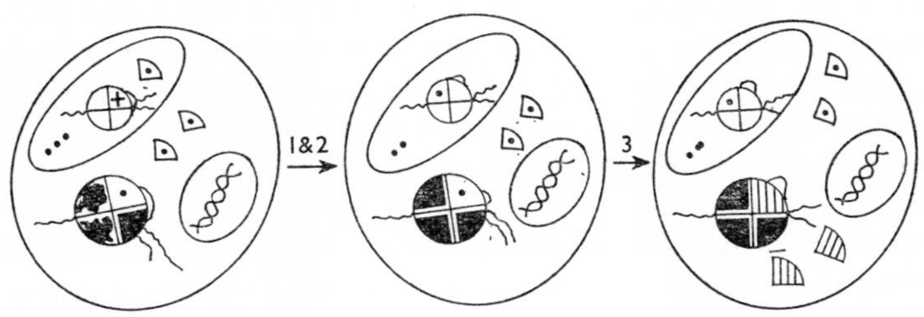


●, Organelle gene; ⊞, original organelle ribosomal protein; ⊡, 'parent' cytoplasmic ribosomal protein; ▥, modified cytoplasmic ribosomal protein.

II. Cluster-clone Hypothesis

Premise: the procaryotic progenitor had:

| | One copy of each gene or | Several copies of each gene |
|---|---|---|
| Possibilities: | A ↓ | B ↓ |
| Clustering patterns | 1. Some genes in each compartment | 1. The extreme would be that each compartment has all the genes. |
| Selection and evolution | 2. The evolutionary problem is formally equivalent to that of the Endosymbiont Hypothesis. | 2. Some genes may be lost from each compartment. Some genes may be retained in all compartments. |

Scheme 1. *Summary of the two hypotheses of the origin of organelles*

It is assumed in this discussion that ribosomes originated once and then evolved along two or more lines.

The gene for the protein 12S3, which is a component of the plastid ribosome, seems to be in the nuclear genome now; if the endosymbiont hypothesis is correct how might this have happened? Two possibilities are apparent. (*a*) Gene transfer: a copy of the gene of the invader might have migrated into the nucleus (of the host) and become incorporated into the genome of the host. The gene for this particular protein might then be present in both genomes. However, even if the invader now lost the gene specifying that protein, the organelle would survive as long as the product of the gene could enter it. (*b*) Protein and gene substitution: another possibility is that the organelle's gene for the protein might have mutated to provide a protein which could no longer be used as part of the organelle ribosome. The organism, or at least the organelle, would then perish unless some other protein in the cell could substitute; a protein of the cytoplasmic ribosome might be a reasonable candidate. Thus the gene specifying this particular plastid ribosomal protein would now be in the nuclear genome. A close alternative, of course, is that at the time of invasion some cytoplasmic and organellar ribosomal proteins were interchangeable. There would be no external selection pressure (we do not know about internal pressures) to retain the organelle's gene for that protein. Or the converse, the organelle's gene might be retained to provide information for the cytoplasmic ribosomal protein or proteins.

(In a more general way, if any of the organelle's proteins were interchangeable with those of the host, there might be no internal or external selective pressures leading to the retention of a particular gene in the nucleus or in the organelle. An organellar gene might be retained to provide information for cytoplasmic components; a nuclear gene might be retained to code for organellar components. Selections in these two directions could easily vary from one evolutionary line to another.)

If ribosomal protein substitution had occurred one should expect to find that same protein in ribosomes of both the plastids and cytoplasm. In the particular case of the 12S3 protein we do not know if this is the case but the cytoplasmic ribosomes do not bind erythromycin and it would not be surprising to find this protein only in the plastid 52S ribosomal subunits. This is troublesome for the 'substitution possibility' but suppose that there were more than one copy of the gene for a protein which is common to the plastid and cytoplasmic ribosomes. If one of these genes mutated the protein product might be compatible with the other proteins and the rRNA of, for example, the plastid but not the cytoplasmic ribosomes. Because of the complex nature of ribosomes and the accommodations required among many proteins and RNA species it is not at all unlikely that changes in a protein might render it unusable for one, but not another, type of ribosome. An important point is that intracellular evolution did not cease at the moment the first eukaryotic cell formed, no matter how that came about.

The cluster-clone hypothesis, as stated above, assumes that organelles developed within a single prokaryotic cell (a one-compartment cell containing one or several copies of each gene and chromosome) by the clustering of groups of genes, the formation of membranes around these genes, and the propagation by division, i.e. cloning, of each of the gene-containing compartments.

If the ancestral prokaryotic cell contained two or more copies of each of its genes, the possibilities of assortment would range from the equivalent of a multinucleate organism in which each cluster (=genome) contained all the genes necessary for the survival of the organism to an assortment arrangement in which some clusters included at least one copy of all the genes whereas others had an incomplete set and on to an arrangement in which each gene (and its copies) would be present in only one cluster (Scheme 1). The resulting subcellular clones of organelles would reflect these types of gene distributions. Starting with a cluster pattern of even the most redundant type, i.e. in which every gene is present in every cluster (i.e. genome), reduction at random or as a result of selection according to undetermined rules could result in the presence of each gene in only one genome, or could stop somewhere between the extremes of complete redundancy and complete reduction. In the case of the ribosomal proteins, some genes could be in one and some in another genome or in more than one.

A fundamental difference between the endosymbiont and cluster-clone hypotheses is in the time at which organelle and cytoplasmic ribosomes evolved to be different and when the distinctive outer membranes of nuclei, plastids and mitochondria evolved. According to the cluster-clone hypothesis these events occurred only after the primitive eukaryote formed; according to the endosymbiont hypothesis, on the other hand, the fundamental differences were well established before the symbiotic relationship occurred. In both cases we visualize the establishment of genomes that do not interbreed freely. The limiting membranes which persist around the organelles even during stages when the nuclear membrane breaks down and the differences in the organization of nuclear and organellar chromosomes probably all serve as barriers to free interchange of genetic material.

The evidence that nuclear genes may code for chloroplast ribosomal proteins requires that the endosymbiont hypothesis, which has been accepted frequently and uncritically in the recent past, be viewed as a possibility, not a fact. Regardless of the manner or manners in which eukaryotic cells arose, evolutionary changes followed. The rules for such 'intracellular' selection and evolution need to be ascertained. Some guidance may come as we learn the modern consequences of such evolution, the integrative mechanisms in organellar–nuclear/cytoplasmic relationships. But, just as there is no evidence available to show that direct gene transfer has occurred, there is no evidence that it has not or cannot happen.

## Summary

Three RNA polymerases have been identified in cells of maize leaves. Two of these polymerases are from maize nuclei and one is from maize chloroplasts. Each of these enzymes has been at least partly purified. The three enzymes are entirely different from one another with regard to the properties studied to date. In addition, one of the nuclear enzymes has been isolated in a form which is most effective when a template of native double-stranded DNA is provided but it can be converted into a form which prefers single-stranded DNA. This may be a favourable situation for studying this particular facet of template specificity.

Nine mutant strains of *C. reinhardii* which are resistant to the antibiotic erythromycin have been isolated. The 52S subunits of wild-type chloroplast ribosomes of this alga bind erythromycin but similar subunits from any of these mutant strains fail to do so under similar circumstances. Genetic analyses show that at least three loci may be involved.

Eight of the nine mutations are transmitted in a Mendelian manner and are judged to be in the nuclear genome. Other evidence is discussed which shows that these eight mutations fall into two categories; each category may represent one locus. The ninth mutant character is transmitted uniparentally and is thus judged to be in an extranuclear genome. Alterations in ribosomal proteins of the 52S subunit have been demonstrated for the uniparental and for one of the Mendelian mutants.

The possibility that intracellular integration (i.e. accommodation between the organelles and the nuclear–cytoplasmic system) might be effected at the level of transcription via nuclear regulation of the production of some or all of the components of chloroplast RNA polymerase and at the level of translation by nuclear regulation of the production of proteins essential for the functioning of chloroplast ribosomes was discussed.

Finally, some problems regarding the origin and evolution of eukaryotic cells were discussed. The implications of both the endosymbiont and the cluster-clone hypotheses of plastid origin were examined and some mechanisms were advanced to account for the present distribution of genes specifying organellar components among subcellular compartments.

## Notes added in proof

(1) Highly purified preparations of maize nuclear DNA-dependent RNA polymerase IIa contain polypeptides with molecular weights of 200000, 160000, 35000, 25000, 20000 and 17000 (Mullinix *et al.*, 1973).

(2) Selected aspects of the data contained in Mets (1972) were reported by Mets & Bogorad (1972).

(3) Locus ery-M1 has been mapped to the right of pf-2 on *Chlamydomonas* nuclear linkage group XI (J. N. Davidson, unpublished work). The ribosomal protein which is altered in all of the four resistant strains that map to this locus has now been identified by gel electrophoresis (J. N. Davidson, M. R. Hanson & L. Bogorad, unpublished work).

The experimental work was supported in part by a research grant from the National Institutes of General Medical Sciences, National Institutes of Health (GM 14991) in part by a research grant from the National Science Foundation and in part by the Maria Moors Cabot Foundation of Harvard University.

## References

Bell, W. D., Bogorad, L. & McIlrath, W. J. (1958) *Bot. Gaz. (Chicago)* **120**, 36–39
Bell, W. D., Bogorad, L. & McIlrath, W. J. (1962) *Bot. Gaz. (Chicago)* **124**, 1–8
Bogorad, L. (1967*a*) in *Biochemistry of Chloroplasts* (Goodwin, T. W., ed), vol. 2, pp. 615–631, Academic Press, London and New York
Bogorad, L. (1967*b*) *Symp. Soc. Develop. Biol.* **26**, 1–31
Bogorad, L. & Woodcock, C. L. F. (1970) in *Autonomy and Biogenesis of Mitochondria and Chloroplasts* (Boardman, N. K., Linnane, A. W., & Smillie, R. M., eds.), pp. 92–97, North-Holland Publishing Co., Amsterdam and London
Bottomley, W., Smith, H. J. & Bogorad, L. (1971*a*) *Proc. Nat. Acad. Sci. U.S.* **68**, 2412–2416.
Bottomley, W., Spencer, D., Wheeler, A. M. & Whitfield, P. (1971*b*) *Arch. Biochem. Biophys.* **143**, 269–275

Boulter, D., Ellis, R. J. & Yarwood, A. (1972) *Biol. Rev. Cambridge Phil. Soc.* **47**, 113–175
Burgess, R. R. (1969) *J. Biol. Chem.* **244**, 6168–6176
Burgess, R. R. (1971) *Annu. Rev. Biochem.* **40**, 711–740
Dekio, S., Takata, R., Osawa, S., Tanaka, K. & Tamaki, M. (1970) *Mol. Gen. Genet.* **107**, 39–49
Ellis, R. J. & Hartley, M. R. (1971) *Nature (London) New Biol.* **233**, 193–196
Georgopoulos, G. P. (1971) *Proc. Nat. Acad. Sci. U.S.* **68**, 2977–2981
Hoober, J. K. & Blobel, G. (1969) *J. Mol. Biol.* **41**, 121–138
Kirk, J. T. O. (1964) *Biochem. Biophys. Res. Commun.* **14**, 393–397
Kirk, J. T. O. (1970) *Annu. Rev. Plant Physiol.* **21**, 11–42
Kung, S. D., Moscarello, M. A. & Williams, J. P. (1972) *Biophys. J.* **12**, 474–483
Lai, C. & Weisblum, B. (1971) *Proc. Nat. Acad. Sci. U.S.* **68**, 856–860
Leboy, P. S., Cox, E. C. & Flaks, G. P. (1964) *Proc. Nat. Acad. Sci. U.S.* **52**, 1367–1374
Mans, R. & Novelli, G. D. (1964) *Biochim. Biophys. Acta* **91**, 186–188
Mao, J. C.-H. & Putterman, M. (1969) *J. Mol. Biol.* **44**, 347–361
Mets, L. J. (1972) Doctoral Thesis, Harvard University
Mets, L. J. & Bogorad, L. (1971) *Science* **174**, 707–709
Mets, L. J. & Bogorad, L. (1972) *Proc. Nat. Acad. Sci. U.S.* **69**, 3779–3783
Moore, P. B., Trant, R. R., Nowlar, H., Pearson, P. & Delius, H. (1968) *J. Mol. Biol.* **31**, 441–461
Mullinix, K. P., Strain, G. C. & Bogorad, L. (1973) *Proc. Nat. Acad. Sci. U.S.* **70**, 2386–2390
Otaka, E., Teraoka, H., Tamaki, M., Tanaka, K. & Osawa, S. (1970) *J. Mol. Biol.* **48**, 499–510
Roeder, R. G. & Rutter, W. J. (1969) *Nature (London)* **224**, 234–237
Roeder, R. G. & Rutter, W. J. (1970) *Proc. Nat. Acad. Sci. U.S.* **65**, 675–682
Sager, R. (1972) *Cytoplasmic Genes and Organelles*, Academic Press, New York and London
Sager, R. & Ramanis, Z. (1970) *Proc. Nat. Acad. Sci. U.S.* **65**, 593–600
Smillie, R. M. & Scott, N. S. (1969) in *Progress in Subcellular and Molecular Biology* (Hahn, F. E., Springer, F. E., Puck, T. T. & Wallenfels, K., eds.), vol. 1, pp. 136–202, Springer, Berlin, Heidelberg and New York
Stout, E. R. & Mans, R. J. (1967) *Biochim. Biophys. Acta* **134**, 327–336
Strain, G. C., Mullinix, K. P. & Bogorad, L. (1971) *Proc. Nat. Acad. Sci. U.S.* **68**, 2647–2651
Sueoka, N. (1960) *Proc. Nat. Acad. Sci. U.S.* **46**, 83–91
Taubman, S. B., Jones, N. R., Young, F. E. & Corcoran, J. W. (1966) *Biochim. Biophys. Acta* **123**, 438–440
Teraoka, H. (1970) *J. Mol. Biol.* **48**, 511–515
Vasconcelos, A. C. L. & Bogorad, L. (1971) *Biochim. Biophys. Acta* **228**, 492–502
Walles, B. (1971) in *Structure and Function of Chloroplasts* (Gibbs, M., ed.), pp. 51–88, Springer, Berlin, Heidelberg and New York
Wehrli, W. & Staehlin, M. (1971) *Bacteriol. Rev.* **35**, 290–309
Woodcock, C. L. F. & Bogorad, L. (1971) in *Structure and Function of Chloroplasts* (Gibbs, M., ed.), pp. 89–128, Springer, Berlin, Heidelberg and New York

*Biochem. Soc. Symp.* (1973) **38**, 43–56
*Printed in Great Britain*

# Transfer Ribonucleic Acid and Transfer Ribonucleic Acid-Recognizing Enzymes in Bean Cytoplasm, Chloroplasts, Etioplasts and Mitochondria

By G. BURKARD, P. GUILLEMAUT, A. STEINMETZ and J. H. WEIL

*Laboratoire de Chimie Biologique, Université Louis Pasteur, Rue Descartes, Esplanade,* 67000 *Strasbourg, France*

## Synopsis

The reverse-phase chromatographic profiles of leucyl-, methionyl- and valyl-tRNA species from bean cytoplasm, chloroplasts and mitochondria are given. The attachment of leucine and valine has been studied in homologous and heterologous systems. Less leucine or valine is attached to chloroplast tRNA when a cytoplasmic enzyme instead of a chloroplast enzyme is used, and this is due to the fact that the cytoplasmic enzyme cannot catalyse the aminoacylation of the chloroplast-specific $tRNA^{Leu}$ and $tRNA^{Val}$. These plastid-specific tRNA species seem to be preferentially synthesized on greening. Hybridization of chloroplast and cytoplasmic leucyl-tRNA species with nuclear and chloroplast DNA has been studied. Differences in the chloroplast and cytoplasmic aminoacyl-tRNA synthetases specific for the same amino acid were revealed by differences in sensitivity to various agents (cations, amino acid analogues). *N*-Formylmethionyl-tRNA and an active transformylase were characterized in chloroplasts, etioplasts and mitochondria. A tRNA adenylyl(cytidylyl)transferase activity was found in the cytoplasm and the mitochondria, and the tRNA methylases from the cytoplasm and the chloroplasts were compared.

## Introduction

In eukaryotic cells, protein synthesis takes place not only in the cytoplasm but also in cellular organelles such as mitochondria and chloroplasts. Protein synthesis in these organelles appears, by some criteria [size of the ribosomes, formylation of the initiator methionyl-tRNA, sensitivity to some antibiotics (Boulter *et al.*, 1972)] to resemble more protein synthesis in bacteria than that in the surrounding cytoplasm, and this has been considered as supporting the theory that these organelles have evolved from endosymbiotic prokaryotic cells. Barnett & Brown (1967) and Barnett *et al.* (1967) were first to report that, in *Neurospora crassa*, there are mitochondrial tRNA species and aminoacyl-tRNA synthetases which are different from their cytoplasmic counterparts. In our laboratory, we have been, in the past few years, engaged in a comparative study of the first step of protein biosynthesis, namely the reactions leading to the attachment of the amino acids to their cognate tRNA species, in the cytoplasm, the plastids and the mitochondria of the same organism, *Phaseolus vulgaris*.

## Experimental

### *Materials*

*P. vulgaris* var. Saxa was used for these studies. Cytoplasmic tRNA species and enzymes were prepared from hypocotyls grown in the dark on moist vermiculite (sterilized at 180°C) at 30°C for 4–5 days. Chloroplasts and etioplasts were obtained from green and etiolated leaves respectively, by using a non-aqueous technique described by Charlton *et al.* (1967); contaminations by micro-organisms were estimated as described by Burkard *et al.* (1970) and found to be very small. Mitochondria were obtained from dark-grown hypocotyls as described by Guillemaut *et al.* (1972).

*tRNA species and enzyme preparations.* tRNA species and enzymes were obtained from hypocotyl cytoplasm, plastids and mitochondria as described by Burkard *et al.* (1970).

*ATP–pyrophosphate exchange and aminoacylation of tRNA species.* These were performed as described previously (Burkard *et al.*, 1970).

*Fractionation of aminoacyl-tRNA species.* This was performed by reverse-phase chromatography by using the reverse-phase chromatography technique described by Weiss & Kelmers (1967) (RPC-2), or the technique described by Pearson *et al.* (1971) (RPC-5) or the rapid technique described by Kelmers & Heatherly (1971).

*Formylation of the methionyl-tRNA species.* This was performed either together with aminoacylation (Burkard *et al.*, 1969*a*) or, after re-extraction of methionyl-tRNA species, by using the conditions of Merrick & Dure (1971).

*Determinations of tRNA adenylyl(cytidylyl)transferase activity.* These were performed as described by Rether (1972) by using tRNA species deprived of their terminal pCpCpA sequence according to Zubay & Takanami (1964).

*Enzymic methylation of tRNA species.* A 30 μg sample of tRNA was incubated with Tris–HCl buffer, pH 9 (16 μmol), dithiothreitol (0.2 μmol), EDTA (0.2 μmol) ammonium acetate (50 μmol), 5-[*Me*-$^{14}$C]adenosylmethionine (3 nmol, 0.15 μCi) and 100 μg of enzymic proteins in a final volume of 120 μl for 1 h at 37°C. A 100 μl sample was then put on a disc of Whatman 3MM paper which was plunged into 5% (w/v) trichloroacetic acid, washed, and the radioactivity was counted as in the case of the aminoacylation reaction (Burkard *et al.*, 1970).

## Results

### *Comparison of the leucyl- and valyl-tRNA species and of the leucyl- and valyl-tRNA synthetases in the cytoplasm and organelles*

The leucine- and valine-specific tRNA species found in the organelles differ markedly from those found in the cytoplasm. Fig. 1 shows the elution profiles (after RPC-5 chromatography) of the leucyl-tRNA species from the cytoplasm, the chloroplasts and the mitochondria obtained in each case in the presence of the homologous enzyme. The cytoplasmic leucyl-tRNA species show two main peaks, which are also present in the mitochondria and the chloroplasts, but the organelles contain more iso-accepting peaks. There are also two valyl-tRNA peaks common to the three tRNA species, as can be seen in Fig. 2.

In addition to these studies in homologous systems, we have also compared the

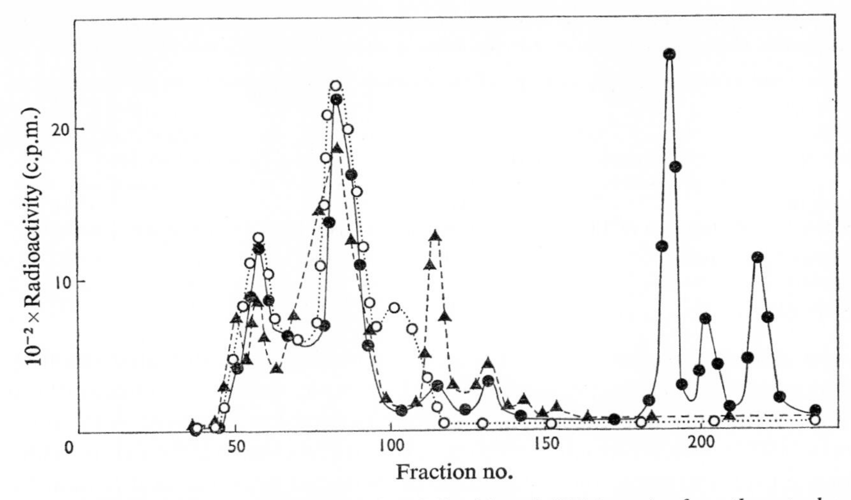


Fig. 1. *Reverse-phase chromatography (RPC-5) of leucyl-tRNA species from the cytoplasm* (○), *chloroplasts* (●) *and mitochondria* (▲), *aminoacylated by using homologous enzymes*

An NaCl gradient from 0.4 to 0.8 M in 0.01 M-acetate buffer (pH 4.7)–0.01 M-$MgCl_2$ was used.

aminoacylation in homologous and heterologous systems. In our first studies (Burkard *et al.*, 1970) we had measured the attachment of 18 amino acids to cytoplasmic and chloroplast tRNA species catalysed either by the homologous or heterologous aminoacyl-tRNA synthetase. As interesting differences in the extent of aminoacylation had been observed for several amino acids, especially leucine and valine, we have extended these comparative studies to include mitochondrial tRNA species and enzymes. The results are summarized in Table 1.

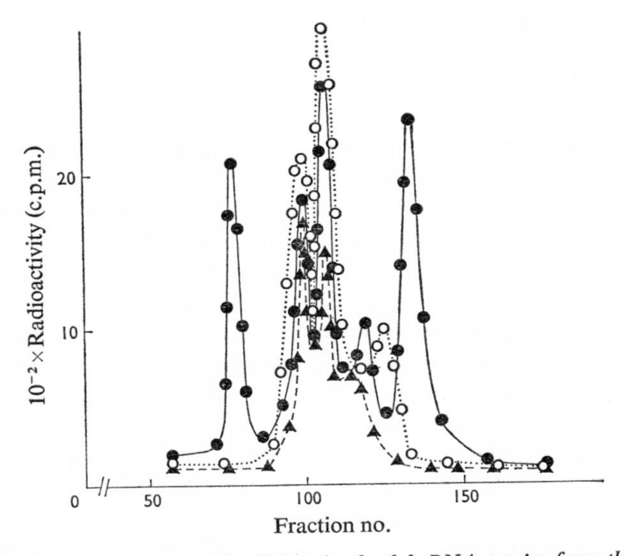


Fig. 2. *Reverse-phase chromatography (RPC-5) of valyl-tRNA species from the cytoplasm* (○), *chloroplasts* (●) *and mitochondria* (▲), *aminoacylated by using homologous enzymes*

Table 1. *Amount of attachment of leucine and valine to cytoplasmic, chloroplast and mitochondrial tRNA species catalysed by the homologous or the two heterologous enzymes*

The results are expressed as percentages of the aminoacylation obtained in the homologous reaction.

| | Leucine Source of enzyme | | | Valine Source of enzyme | | |
|---|---|---|---|---|---|---|
| Source of tRNA | Cytoplasm | Chloroplast | Mito-chondria | Cytoplasm | Chloroplast | Mito-chondria |
| Cytoplasm | 100 | 94 | 99 | 100 | 102 | 99 |
| Chloroplast | 60 | 100 | 72 | 36 | 100 | 63 |
| Mitochondria | 92 | 91 | 100 | 64 | 101 | 100 |

It appears that cytoplasmic tRNA species can be charged to the same extent by the homologous enzyme preparations and by chloroplast or mitochondrial enzymes. With only one exception, a similar situation is observed with mitochondrial tRNA species which can be charged with leucine equally well by using any of the three enzymes, and can be charged with valine to the same extent by using enzymes from the mitochondria or the chloroplasts. The chloroplast tRNA species, on the contrary, can only be fully charged by using chloroplast enzyme preparations.

If one now looks at the charging capacity of the various enzymes, it appears that chloroplast enzymes can fully charge the tRNA species from all three sources, whereas the enzymes prepared from the cytoplasm or the mitochondria charge only partly chloroplast tRNA species.

It should be pointed out that when one compares the elution profiles (after RPC-5 chromatography) of chloroplast leucyl-tRNA or valyl-tRNA species obtained in the presence of chloroplast and mitochondrial enzymes, all the peaks usually found in chloroplast tRNA species are found in both cases. In other words mitochondrial enzymes seem to recognize all chloroplast iso-accepting tRNA species (no peak missing in the profile), and the reason for the lower extent of charging (as compared with that observed with the homologous chloroplast enzymes) is not clear.

We were able, however, to explain the lower extent of charging of chloroplast tRNA species with cytoplasmic enzyme preparations, as compared with that obtained with a chloroplast enzyme. The lower values obtained with a cytoplasmic enzyme could not be raised by increasing the time of incubation or the concentration of cytoplasmic enzyme (Burkard *et al.*, 1969*b*). It turned out that this lower extent of aminoacylation is due to the fact that only some iso-accepting $tRNA^{Leu}$ and $tRNA^{Val}$ species found in the chloroplasts can be recognized and charged by the cytoplasmic enzyme preparations, whereas cytoplasmic leucyl- and valyl-tRNA species can be fully charged with the chloroplast enzymes as well as with the cytoplasmic enzymes. Figs. 3 and 4 show that the chloroplast-specific $tRNA^{Leu}$ and $tRNA^{Val}$ are only charged by the chloroplast enzymes. These profiles differ slightly from those shown in Figs. 1 and 2, as they were obtained in earlier experiments, before we started using the RPC-5 system which gives better resolution of iso-accepting tRNA species.

In order to study whether these plastid-specific tRNA species are already present in etioplasts, or if they appear only in chloroplasts after exposure of plants

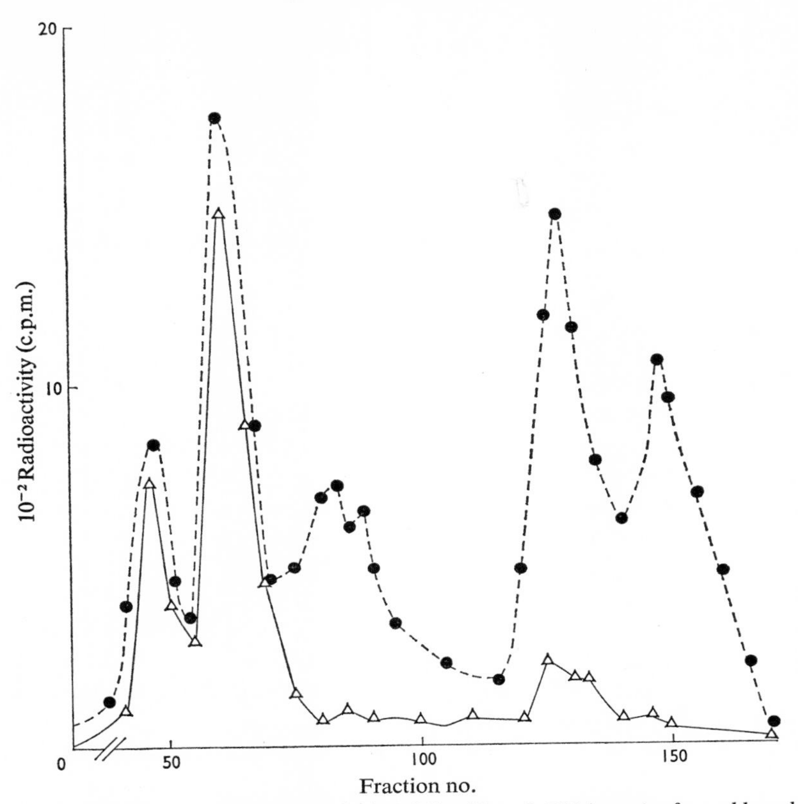


Fig. 3. *Reverse-phase chromatography (RPC-2) of leucyl-tRNA species from chloroplasts*
The tRNA was aminoacylated by using a chloroplast enzyme (●) or a cytoplasmic enzyme (△).

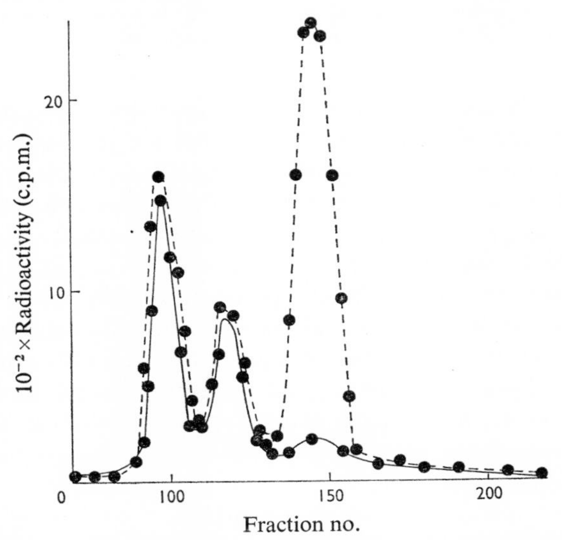


Fig. 4. *Reverse-phase chromatography (RPC-2) of valyl-tRNA species from chloroplasts*
The tRNA was aminoacylated by using a chloroplast enzyme (●---●), or a cytoplasmic enzyme (●——●).

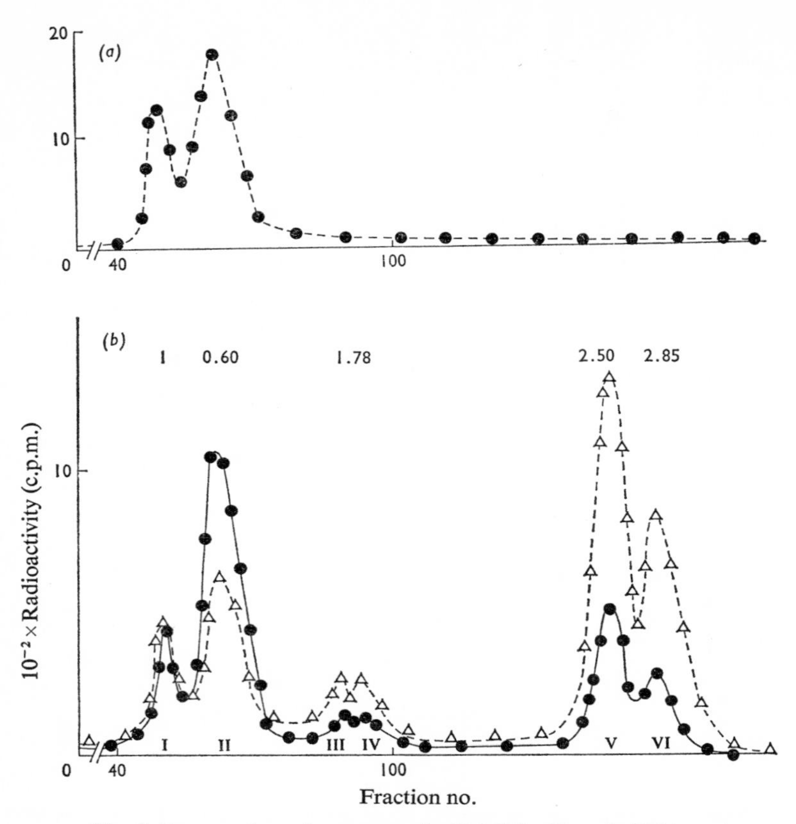


Fig. 5. *Reverse-phase chromatography* (*RPC*-2) *of leucyl-tRNA species*

The tRNA species were obtained from (*a*) the cytoplasm (●---●), and (*b*) chloroplasts (△---△) and etioplasts (●——●) and were aminoacylated by using homologous enzymes. The values represent the chloroplast tRNA/etioplast tRNA ratio of each peak.

to light, we have compared the profiles of leucyl- and valyl-tRNA species obtained from chloroplasts and etioplasts. Fig. 5 shows the two profiles after chromatography of leucyl-tRNA species on RPC-2 (together with the profile of cytoplasmic leucyl-tRNA species as a reference); if peak 1, which is found in both types of plastids and in the cytoplasm, is taken as a reference and if the ratio of chloroplast tRNA to etioplast tRNA for this peak is 1, it can be seen that this ratio is increased 2 to 3 times for the peaks of plastid-specific leucyl-tRNA species (peaks III–VI). In the case of valyl-tRNA species the ratio is increased for peaks I and V which are plastid-specific (this time the ratio for peak II, which is found in the cytoplasm and in both types of plastids, was taken as a reference) (Fig. 6). These results suggest that a preferential synthesis of the plastid-specific tRNA species takes place on transformation from etioplasts to chloroplasts (Burkard *et al.*, 1972).

In an attempt to approach the problem of the site of synthesis of these plastid-specific tRNA species, we have started to study the hybridization of chloroplast and cytoplasmic [$^{3}$H]leucyl-tRNA species to chloroplast and nuclear DNAs.

In order to prevent the hydrolysis of the ester bond between the amino acid and the tRNA, which would result in the loss of the label, these hybridization assays were performed at 40°C in the presence of 8 M-urea as described by Kourilsky *et al.* (1970). Preliminary results showed that cytoplasmic leucyl-tRNA species hybridize to nuclear DNA but not to chloroplast DNA, whereas chloroplast leucyl-tRNA species hybridize better to chloroplast DNA, but also to some extent to nuclear DNA.

The fact that the plastid-specific $tRNA^{Leu}$ and $tRNA^{Val}$ species can be aminoacylated by the chloroplast enzyme, but not by the cytoplasmic enzyme, already suggests that the chloroplast and the cytoplasmic leucyl-tRNA synthetases are different, and that this is also true for the two valyl-tRNA synthetases. In the latter case, we have additional evidence that the two enzymes are different. We have shown (Burkard *et al.*, 1970) that the cytoplasmic valyl-tRNA synthetase

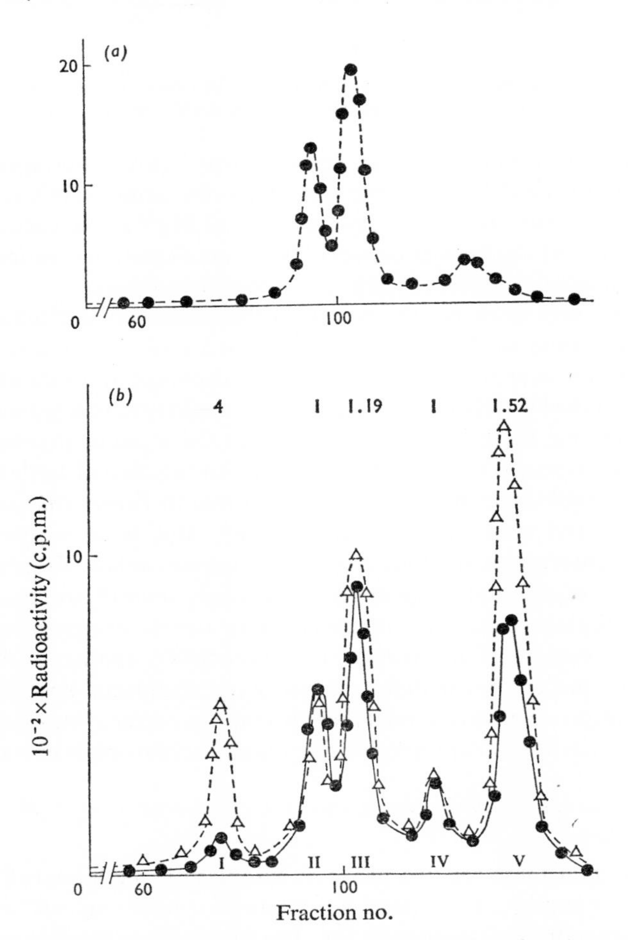


Fig. 6. *Reverse-phase chromatography (RPC-5) of valyl-tRNA species*

The tRNA species were obtained from (*a*) cytoplasm (●---●), and (*b*) chloroplasts (△---△) and etioplasts (●——●), and were aminoacylated by using homologous enzymes. The values represent the chloroplast tRNA/etioplast tRNA ratio of each peak.

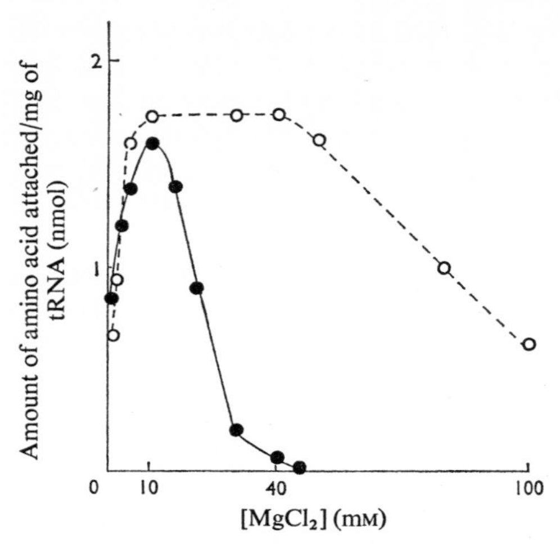


Fig. 7. *Effects of $Mg^{2+}$ concentration on the attachment of valine to cytoplasmic tRNA, with either a cytoplasmic enzyme* (●) *or a chloroplast enzyme* (○)

is more sensitive to inhibition by an excess of $Mg^{2+}$ (Fig. 7) or spermidine, or thiol groups than the chloroplast enzyme, when the same tRNA (cytoplasmic tRNA) is used as a substrate. $Co^{2+}$ is as effective as $Mg^{2+}$ in the case of the cytoplasmic enzyme, and $Ca^{2+}$ less effective (in the aminoacylation reaction), whereas the opposite situation is observed with the chloroplast enzyme.

Another approach to detect the possible differences between the chloroplast and cytoplasmic aminoacyl-tRNA synthetases for a given amino acid is to study the effects of amino acid analogue on the ATP–pyrophosphate exchange reaction. We have first studied the effects of canavanine, an analogue of arginine (Burkard *et al.*, 1970). During these studies, we found that the arginine-dependent ATP–pyrophosphate exchange is stimulated by tRNA in the case of both the chloroplast and the cytoplasmic enzymes, as also reported in *Escherichia coli* (Mehler & Mitra, 1967) and yeast (Mitra & Smith, 1969). But as far as canavanine is concerned, the chloroplast and the cytoplasmic enzymes behave quite differently: a concentration of 20 $\mu$mol of analogue/ml strongly inhibits the reaction catalysed by the cytoplasmic enzyme, whereas it stimulates the exchange catalysed by the chloroplast enzyme. The cytoplasmic leucyl-tRNA synthetase is different from the chloroplast enzyme in that it cannot charge the plastid-specific $tRNA^{Leu}$ species; in addition we have observed that the cytoplasmic enzyme is less sensitive to higher concentrations of trifluoroleucine than the chloroplast enzyme.

*Studies of the methionyl-tRNA species and the transformylases of the cytoplasm and the organelles*

Fig. 8 shows that there are two peaks of cytoplasmic methionyl-tRNA (after RPC-5 chromatography), three peaks of chloroplast methionyl-tRNA and two peaks of mitochondrial methionyl-tRNA. The first peak in the elution profile is found in the cytoplasm and in both types of organelles. In addition the last peak (peak 2) in the mitochondria seems identical with the last peak (peak 3) in the chloroplasts. The two mitochondrial $tRNA^{Met}$ species can be aminoacylated by

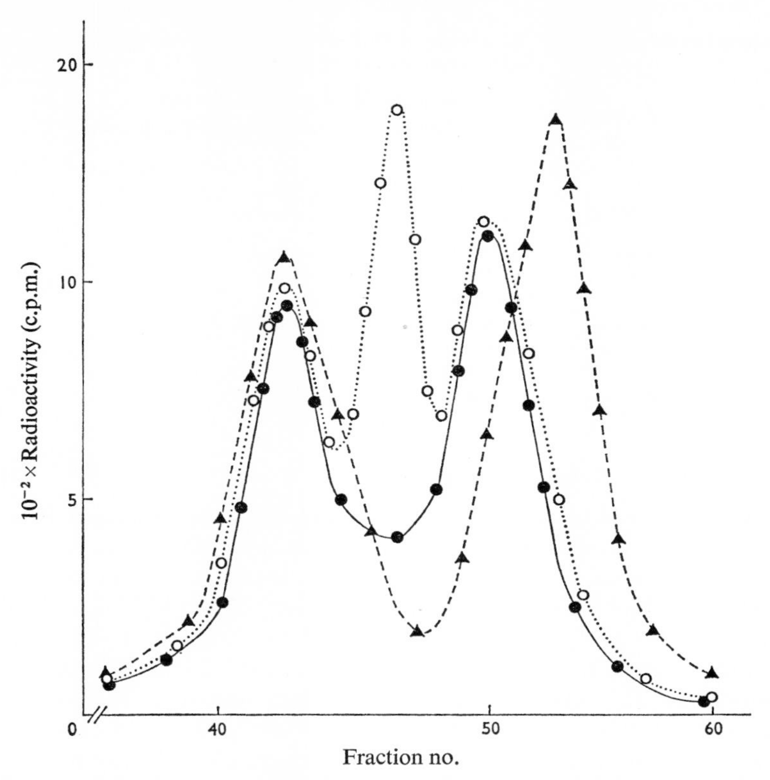


Fig. 8. *Reverse-phase chromatography (RPC-5) of methionyl-tRNA species from the cytoplasm* (▲), *chloroplasts* (○) *and mitochondria* (●), *aminoacylated by using homologous enzymes*

using a chloroplast enzyme preparation and the three chloroplast $tRNA^{Met}$ species can be charged by using a mitochondrial enzyme.

In order to determine whether *N*-formylmethionyl-tRNA is present in the cytoplasm and the organelles, tRNA species from these various sources were aminoacylated with [$^{35}$S]methionine by using the homologous enzyme and incubated in the presence of formyltetrahydrofolate with either a homologous or an *E. coli* enzyme preparation. The tRNA species were then re-isolated, treated with pancreatic ribonuclease, and the resulting methionyladenosine and formylmethionyladenosine were characterized by electrophoresis as previously described (Burkard *et al.*, 1969*a*). The results are summarized in Table 2. When tRNA species from chloroplasts, etioplasts or mitochondria are aminoacylated in the presence of the corresponding homologous enzymes, *N*-formylmethionyladenosine can be characterized and the formylation ratios are 27, 28 and 33% respectively. This ratio can be increased when a bacterial transformylase is being used. Controls have shown that cytoplasmic methionyl-tRNA species cannot be formylated, whether a chloroplast or a bacterial transformylase is used. With *E. coli* tRNA we find a formylation ratio similar to that reported by other authors (Marcker, 1965). These results show that chloroplasts, etioplasts and mitochondria all contain methionyl-tRNA that can be formylated and an active transformylase.

Table 2. *Formylation of various [$^{35}S$]methionyl-tRNA species in the presence of a homologous or heterologous transformylase*

$$\text{Formylation ratio} = \frac{\text{formylmethionyladenosine} \times 100}{\text{methionyladenosine} + \text{formylmethionyladenosine}}$$

| Source of tRNA | Source of methionyl-tRNA synthetase | Source of transformylase | Formylation ratio (%) |
|---|---|---|---|
| Chloroplasts | Chloroplasts | Chloroplasts | 27 |
| Etioplasts | Etioplasts | Etioplasts | 33 |
| Etioplasts | Etioplasts | *E. coli* | 60 |
| Mitochondria | Mitochondria | Mitochondria | 28 |
| Mitochondria | Mitochondria | *E. coli* | 56 |
| Cytoplasm | Chloroplasts | Chloroplasts | 2.5 |
| Cytoplasm | Cytoplasm | *E. coli* | 3.8 |
| *E. coli* | *E. coli* | *E. coli* | 66 |

*Presence in plant cytoplasm and organelles of other enzymes acting on tRNA species*

The tRNA adenylyl(cytidylyl)transferase has been characterized in mitochondrial extracts. In order to characterize such enzymic activity in cytoplasmic extracts, it was necessary to perform $(NH_4)_2SO_4$ fractionation and the activity was found in the fraction which precipitates between 45 and 60% saturation.

We have also studied the tRNA methylases of the cytoplasm and the chloroplasts. Table 3 shows the results obtained when cytoplasmic tRNA species are methylated either in a homologous system (with cytoplasmic enzymes) or in a heterologous system (with chloroplast enzyme), and when chloroplast tRNA species are methylated in either a homologous or heterologous system.

In addition, we have re-isolated the tRNA species at the end of the methylation reaction, submitted them to alkaline hydrolysis, fractionated the nucleotides by two-dimensional t.l.c. on cellulose plates and detected the radioactive nucleotides by radioautography. All tRNA species methylated in the four (homologous and heterologous) reactions yield three main radioactive spots corresponding to 1Me-AMP, 5Me-CMP and *N*-diMe-GMP + *N*-Me-GMP. There are some quantitative differences, but they are difficult to interpret and a further study of the methylases from the various cell compartments would require the use of purified methylases and purified tRNA species.

## Discussion

The comparison of the elution of cytoplasmic, chloroplast and mitochondrial leucyl- and valyl-tRNA species after aminoacylation (with the homologous enzyme preparation) and reverse-phase chromatography has shown that there are some tRNA species common to the cytoplasm and to both types of organelles,

Table 3. *Methylation of cytoplasmic and chloroplast tRNA species by homologous and heterologous enzymes*

The results are expressed as radioactivity (c.p.m., $^{14}CH_3$ group) incorporated per mg of tRNA.

| | Cytoplasmic enzymes | Chloroplast enzymes |
|---|---|---|
| Cytoplasmic tRNA | 5790 | 27010 |
| Chloroplast tRNA | 55160 | 46960 |

but this comparison has also revealed a number of differences due essentially to the existence of organelle-specific tRNA species. Differences between cytoplasmic and mitochondrial tRNA species have also been reported in *N. crassa* (Barnett & Brown, 1967; Epler, 1969), in yeast (Accoceberry & Stahl, 1971), in animal cells (Buck & Nass, 1968, 1969), in *Euglena* (Kislev *et al.*, 1972) and in plant cells (Anderson & Cherry, 1969; Guderian *et al.*, 1972). Differences between cytoplasmic and chloroplast tRNA species have been shown by Aliyev & Filippovich (1968) in peas, by Barnett *et al.* (1969), Reger *et al.* (1970) and Kislev *et al.* (1972) in *Euglena*, Burkard *et al.* (1969*b*, 1970) in bean and Guderian *et al.* (1972) in tobacco.

The existence of multiple iso-accepting tRNA species and aminoacyl-tRNA synthetases in the various compartments of plant cells has been taken advantage of in order to follow physiological changes such as germination, differentiation or senescence (Henshall & Goodwin, 1964; Vold & Sypherd, 1968; Anderson & Cherry, 1969; Anderson & Fowden, 1969; Brown, 1969; Bick *et al.*, 1970; Bick & Strehler, 1971). A particularly interesting case of differentiation takes place on illumination of dark-grown plants (or cells, in the case of green algae). Our results show that the transformation of etioplasts into chloroplasts is accompanied by an increase in the relative amounts of plastid-specific $tRNA^{Leu}$ and $tRNA^{Val}$. Williams & Williams (1970) have also observed that only certain bean $tRNA^{Leu}$ species are preferentially synthesized on greening, and Merrick & Dure (1971) have reported an increase in certain cotton iso-accepting $tRNA^{Met}$, $tRNA^{Val}$ and $tRNA^{Ile}$, suggesting that these species which increase in relative concentration may be localized in the chloroplasts. In *Euglena* chloroplasts Barnett *et al.* (1969) and Reger *et al.* (1970) have shown the existence of light-induced $tRNA^{Ile}$ and $tRNA^{Phe}$ species and of a light-inducible isoleucyl-tRNA synthetase.

The results of our preliminary experiments on the hybridization properties of cytoplasmic and chloroplast tRNA species have shown that cytoplasmic leucyl-tRNA species hybridize only with nuclear DNA, whereas chloroplast leucyl-tRNA species hybridize preferentially with chloroplast DNA but also with nuclear DNA. These results could be explained if the two peaks of leucyl-tRNA which are common to the cytoplasm and the chloroplasts were coded by nuclear DNA, and if the chloroplast-specific leucyl-tRNA peaks were coded by chloroplast DNA. Williams & Williams (1970) have studied the hybridization of total bean leaf leucyl-tRNA to nuclear and chloroplast DNA and have found the hybridization with chloroplast DNA to be more than twice that with the same amount of nuclear DNA. In yeast (Halbreich & Rabinowitz, 1971; Cohen *et al.*, 1972; Casey *et al.*, 1972; Reijnders & Borst, 1972) and in rat liver (Nass & Buck, 1970) it has been shown that at least some of the mitochondrial tRNA species hybridize specifically to mitochondrial DNA.

*N*-Formylmethionyl-tRNA is the initiator tRNA in bacterial protein biosynthesis. That the same initiator is also present in chloroplasts was suggested by the fact that *N*-formylmethionine is the first amino acid in polypeptides synthesized in a cell-free system prepared from *Euglena* chloroplasts and programmed by phage $f_2$ RNA as a messenger (Schwartz *et al.*, 1967). We were able to demonstrate the presence of *N*-formylmethionyl-tRNA in bean chloroplasts (Burkard *et al.*, 1969*a*) and this initiator was then also shown to be present in chloroplasts of *Acetabularia* (Bachmayer, 1970), maize (Bianchetti *et al.*, 1971), cotton (Merrick & Dure, 1971) and wheat germ (Leis & Keller, 1971). The tissues of *Vicia faba* were also shown to contain a formylatable $tRNA^{Met}$ species, to which has been ascribed the role of initiator tRNA in the organelles of this plant (Yarwood *et al.*, 1971). The same initiator tRNA is also present in the mitochondria of yeast (Smith & Marcker, 1968; Halbreich & Rabinowitz, 1971), fungi (Epler *et al.*, 1970) and mammalian cells (Smith & Marcker, 1968; Galper & Darnell, 1969). We have been able to show that this initiator is also present in plant mitochondria as well as in chloroplasts and etioplasts. In plant cytoplasm, initiation involves a non-formylated $tRNA^{Met}$ species (Leis & Keller, 1970; Marcus *et al.*, 1970; Tarrago *et al.*, 1970; Ghosh *et al.*, 1971; Yarwood *et al.*, 1971). Our results also show that the three organelles contain an active transformylase; the fact that a higher formylation ratio is obtained when an *E. coli* enzyme is used (rather than the homologous transformylase) is puzzling, but one should remember that the actual values of the formylation ratio obtained in our experiments may not necessarily reflect the relative proportions of $tRNA^{Met}_{M}$ and $tRNA^{Met}_{F}$ in plant organelles. Among possible causes of error, the fact that the ester linkage between methionine and tRNA is less stable than the bond between formylmethionine and tRNA, as already reported by Marcker (1965), could be responsible for a more rapid hydrolysis of methionyl-$tRNA^{Met}_{M}$ during the incubation *in vitro* and this would result in a higher formylation ratio.

Methylation of tRNA species occurs after transcription, at the polynucleotide level, and is catalysed by a number of specific methylases which have been found in a variety of organisms ranging from bacteria to mammals. It has been observed that the methylation pattern can change not only from one organism to the other, but also within the same organism during differentiation or on tumour formation. Only a few studies have been devoted to plant tRNA methylases, which have been characterized in spinach (Srinivasan & Borek, 1963), wheat germ (Nichols & Lane, 1969; Streeter & Lane, 1970) and in the green alga *Chlamydomonas reinhardii* (Wells & Moore, 1970). It has been recently shown that tRNA methylation can be stimulated by cytokinin in tobacco buds (Schaeffer & Sharpe, 1970) and by gibberellic acid in aleurone cells of barley seeds (Chandra & Duynstee, 1971). But no comparative study had been done on the tRNA methylases of the different compartments of the same plant cell. Our results have shown that cytoplasmic methylases catalyse only a low incorporation of methyl groups into cytoplasmic tRNA species, but catalyse a tenfold higher incorporation into chloroplast tRNA species. Cytoplasmic tRNA species are methylated in the presence of chloroplast enzymes to an extent which is 4–5 times higher than that observed in the presence of homologous (cytoplasmic) enzymes. But chloroplast tRNA species are almost as much methylated in the presence of chloroplast enzymes as in the presence of cytoplasmic enzymes, a result difficult to interpret, as it is usually assumed that tRNA species can only be over-methylated by heterologous enzymes. The actual values obtained for the methylation of chloroplast tRNA species are in both cases quite high (an incorporation of 50000 c.p.m./mg of tRNA corresponds to 1 mmol of methyl group incorporated per mmol of tRNA) and it is therefore unlikely that the values obtained for chloroplast tRNA methylation by cytoplasmic and chloroplast enzymes are equal because of the failure of the methylation reaction. One of the remaining possible explanations is that chloroplast tRNA species contain a certain proportion of methyl-deficient molecules which can be methylated by cytoplasmic as well as by chloroplast enzymes.

## References

Accoceberry, B. & Stahl, A. (1971) *Biochem. Biophys. Res. Commun.* **42**, 1235–1243
Aliyev, K. A. & Filippovich, I. I. (1968) *Mol. Biol.* (*USSR*) **2**, 364–373
Anderson, M. B. & Cherry, J. H. (1969) *Proc. Nat. Acad. Sci. U.S.* **62**, 202–209
Anderson, J. W. & Fowden, L. (1969) *Plant Physiol.* **44**, 60–68
Bachmayer, H. (1970) *Biochim. Biophys. Acta* **209**, 584–586
Barnett, W. E. & Brown, D. H. (1967) *Proc. Nat. Acad. Sci. U.S.* **57**, 452–458
Barnett, W. E., Brown, D. H. & Epler, J. L. (1967) *Proc. Nat. Acad. Sci. U.S.* **57**, 1775–1781
Barnett, W. E., Pennington, C. J. & Fairfield, S. A. (1969) *Proc. Nat. Acad. Sci. U.S.* **63**, 1261–1268
Bianchetti, R., Lucchini, G. & Sartirana, M. (1971) *Biochem. Biophys. Res. Commun.* **42**, 97–102
Bick, M. D. & Strehler, B. L. (1971) *Proc. Nat. Acad. Sci. U.S.* **68**, 224–228
Bick, M. D., Liebke, H., Cherry, J. H. & Strehler, B. L. (1970) *Biochim. Biophys. Acta* **204**, 175–182
Boulter, D., Ellis, R. J. & Yarwood, A. (1972) *Biol. Rev. Cambridge Phil. Soc.* **47**, 113–175
Brown, G. N. (1969) *Plant Physiol.* **44**, 272–276
Buck, C. A. & Nass, M. M. (1968) *Proc. Nat. Acad. Sci. U.S.* **60**, 1045–1052
Buck, C. A. & Nass, M. M. (1969) *J. Mol. Biol.* **41**, 67–82
Burkard, G., Eclancher, B. & Weil, J. H. (1969*a*) *FEBS Lett.* **4**, 285–287
Burkard, G., Guillemaut, P. & Weil, J. H. (1969*b*) *C.R. Soc. Biol.* **163**, 2731–2736
Burkard, G., Guillemaut, P. & Weil, J. H. (1970) *Biochim. Biophys. Acta* **224**, 184–198
Burkard, G., Vaultier, J. P. & Weil, J. H. (1972) *Phytochemistry* **11**, 1351–1353
Casey, J., Cohen, M., Rabinowitz, M., Fukuhara, H. & Getz, G. S. (1972) *J. Mol. Biol.* **63**, 431–440
Chandra, G. & Duynstee, E. (1971) *Biochim. Biophys. Acta* **232**, 514–523
Charlton, J. M., Treharne, K. & Goodwin, T. W. (1967) *Biochem. J.* **105**, 205–212
Cohen, M., Casey, J., Rabinowitz, M. & Getz, G. S. (1972) *J. Mol. Biol.* **63**, 441–451
Epler, J. L. (1969) *Biochemistry* **8**, 2285–2290
Epler, J. L., Shugart, L. & Barnett, T. (1970) *Biochemistry* **9**, 3575–3579
Galper, J. & Darnell, J. (1969) *Biochem. Biophys. Res. Commun.* **34**, 205–214
Ghosh, K., Grishko, A. & Ghosh, H. (1971) *Biochem. Biophys. Res. Commun.* **42**, 462–468
Guderian, R. H., Pulliam, R. L. & Gordon, M. P. (1972) *Biochim. Biophys. Acta* **262**, 50–65
Guillemaut, P., Burkard, G. & Weil, J. H. (1972) *Phytochemistry* **11**, 2217–2219
Halbreich, A. & Rabinowitz, M. (1971) *Proc. Nat. Acad. Sci. U.S.* **68**, 294–298
Henshall, J. D. & Goodwin, T. W. (1964) *Phytochemistry* **3**, 677–691
Kelmers, A. D. & Heatherly, D. E. (1971) *Anal. Biochem.* **44**, 486–495
Kislev, N., Selsky, M., Norton, C. & Eisenstadt, J. M. (1972) *Fed. Proc. Fed. Amer. Soc. Exp. Biol.* **31**, 3740
Kourilsky, Ph., Manteuil, S., Zamansky, M. H. & Gros, F. (1970) *Biochem. Biophys. Res. Commun.* **4**, 1080–1087
Leis, J. & Keller, E. (1970) *Biochem. Biophys. Res. Commun.* **49**, 416–421
Leis, J. & Keller, E. (1971) *Biochemistry* **10**, 889–894
Marcker, K. (1965) *J. Mol. Biol.* **14**, 63–70
Marcus, A., Weeks, D., Leis, J. & Keller, E. (1970) *Proc. Nat. Acad. Sci. U.S.* **67**, 1681–1687
Mehler, A. & Mitra, S. (1967) *J. Biol. Chem.* **242**, 5495–5499
Merrick, W. C. & Dure, L. S. (1971) *Proc. Nat. Acad. Sci. U.S.* **68**, 641–644
Mitra, S. & Smith, C. (1969) *Biochim. Biophys. Acta* **190**, 222–224
Nass, M. M. & Buck, C. A. (1970) *J. Mol. Biol.* **54**, 187–198
Nichols, J. & Lane, B. (1969) *Can. J. Biochem.* **47**, 863
Pearson, R. L., Weiss, J. F. & Kelmers, A. D. (1971) *Biochim. Biophys. Acta* **228**, 770–775
Reger, B. J., Fairfield, S. A., Epler, J. L. & Barnett, W. E. (1970) *Proc. Nat. Acad. Sci. U.S.* **67**, 1207–1213
Reijnders, L. & Borst, P. (1972) *Biochem. Biophys. Res. Commun.* **47**, 126–133
Rether, B. (1972) Thèse Doctorat ès-Sciences, Strasbourg

Schaeffer, G. & Sharpe, F. (1970) *Biochem. Biophys. Res. Commun.* **38**, 312–318
Schwartz, J., Meyer, R., Eisenstadt, J. & Brawerman, G. (1967) *J. Mol. Biol.* **25**, 571–574
Smith, A. & Marcker, K. (1968) *J. Mol. Biol.* **38**, 241–243
Srinivasan P. & Borek, E. (1963) *Proc. Nat. Acad. Sci. U.S.* **49**, 529–533
Streeter, D. & Lane, B. (1970) *Biochim. Biophys. Acta* **199**, 394–404
Tarrago, A., Monasterio, O. & Allende, J. (1970) *Biochem. Biophys. Res. Commun.* **41**, 765–773
Vold, B. & Sypherd, P. S. (1968) *Proc. Nat. Acad. Sci. U.S.* **59**, 453–458
Weiss, J. F. & Kelmers, A. D. (1967) *Biochemistry* **6**, 2507–2513
Wells, C. & Moore, B. (1970) *Arch. Biochem. Biophys.* **137**, 409–414
Williams, G. R. & Williams, A. S. (1970) *Biochem. Biophys. Res. Commun.* **39**, 858–863
Yarwood, A., Boulter, D. & Yarwood, J. (1971) *Biochem. Biophys. Res. Commun.* **44**, 353–361
Zubay, G. & Takanami, M. (1964) *Biochem. Biophys. Res. Commun.* **15**, 207–213

*Biochem. Soc. Symp.* (1973) **38**, 57–76
*Printed in Great Britain*

# Function of Elongation Factors in Peptide Synthesis

By A. B. LEGOCKI

*Institute of Biochemistry, College of Agriculture, Poznan, Wolynska* 35, *Poland*

## Synopsis

This paper describes a convenient method for the purification of elongation factors (EF 1 and EF 2) from a higher-plant source (wheat germ) and their use in studies on the mechanism of peptide-bond formation in a plant system. Some properties of the elongation factors are presented.

## Introduction

The elongation of the peptide chain during protein biosynthesis may be considered as a sequence of discrete steps which, after the binding of aminoacyl-tRNA and the formation of the peptide bond, end up with the translocation of the peptidyl-tRNA from the aminoacyl site (A site) to the peptidyl site (P site) of the ribosome–mRNA complex.

There is a great deal of good evidence on the presence of at least three ribosome-specific elongation factors in prokaryotic organisms, EF-Tu, EF-Ts, and EF-G* (Lucas-Lenard & Lipmann, 1966; Skoultchi *et al.*, 1968). From eukaryotic organisms two complementary factors have been isolated which have been designated as the binding enzyme (EF 1) and the translocase (EF 2) (Hardesty *et al.*, 1963; Gasior & Moldave, 1965).

Studies on bacterial systems reveal that in the first phase of the elongation process the aminoacyl-tRNA molecule is bound to the ribosomal A site. Binding reaction involves the formation of ternary complex between elongation factor Tu, GTP and aminoacyl-tRNA in a two-step manner (Scheme 1). This complex is subsequently transferred on the mRNA-programmed ribosome which contains peptidyl-tRNA bound at the P site (Brot *et al.*, 1970; Shorey *et al.*, 1969). The next phase is the formation of a new peptide bond which is preceded by the cleavage of GTP bound in the ternary complex as well as the release of both EF-Tu–GDP and $P_i$ from the ribosome (Shorey *et al.*, 1969; Waterson *et al.*, 1970). The peptide bond is then formed between the terminal carboxyl group of the peptidyl residue of the peptidyl-tRNA and the α-amino group of the aminoacyl-tRNA. This reaction is catalysed by peptidyltransferase, a component of a large ribosomal subunit (Monro *et al.*, 1969). In the last phase of the elongation (translocation) after the release of the discharged tRNA from the P site, the newly lengthened peptidyl-tRNA occupying the ribosomal A site is shifted to the P site. Translocation requires the third bacterial elongation factor G and another GTP molecule. At this stage hydrolysis of GTP takes place, as in the binding

* In this paper the revised proposal of uniform nomenclature for translation factors (Caskey *et al.*, 1972) is used.

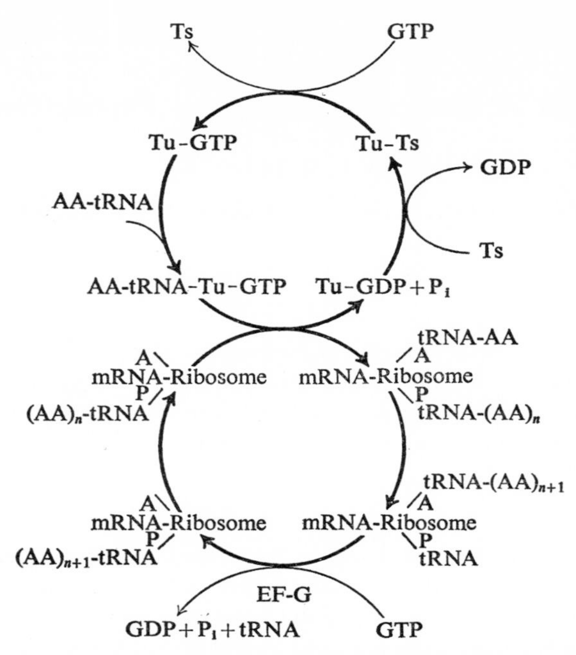


Scheme 1. *Intermediate steps of chain elongation in prokaryotic systems*

phase. At the same time the ribosome moves along the mRNA by the length of one codon in the 5′ to 3′ direction. The cycle of reactions is now repeated and a new peptide bond is synthesized.

Although eukaryotic elongation factors have been defined, little information is available about the formation of intermediates which compose the elongation process. This paper presents our studies on the intermediate steps in the course of peptide-chain elongation and our attempts at the characterization of purified factors from wheat germ.

**Purification of EF 1 and EF 2 from Wheat Germ**

It has been shown that the general mode of action of the elongation factors in plants is similar to that in bacteria and mammals (Jerez *et al.*, 1969; Ciferri & Parisi, 1970; Legocki & Marcus, 1970).

For a better understanding of the properties of the factors they were subjected to systematic purification. The details of the procedure will be presented in forthcoming papers and for the time being some topics of the isolation will be described. The results of purification of the two factors are summarized in Table 1.

The initial steps, blending of the wheat germ and centrifugation (Legocki & Marcus, 1970), were followed by $(NH_4)_2SO_4$ precipitation to 70% saturation. The precipitate was dissolved in 20mM-Tris–HCl buffer, pH 7.35, containing 10mM-KCl, 4mM-2-mercaptoethanol and 10% (v/v) glycerol. After dialysis the protein solution was passed through a column (6cm × 65cm) of Sephadex G-150 equilibrated with the above buffer. The activity of EF 1 was determined as the increase in the amount of [$^{14}$C]phenylalanyl-tRNA bound to wheat ribosomes.

Table 1. *Purification of wheat-germ elongation factors*

One unit of activity corresponds to 1 pmol of [$^{14}$C]phenylalanyl-tRNA bound (for EF 1) or 1 pmol of [$^{14}$C]phenylalanine polymerized (for EF 2) under the standard assay conditions.

| Step | Protein (mg) | Specific activity (units/mg) | Purification factor | Yield (%) |
|---|---|---|---|---|
| | | Factor EF 1 | | |
| 120000*g* supernatant | 1470 | 148 | 1 | 100 |
| $(NH_4)_2SO_4$ | 864 | 214 | 1.4 | 85 |
| Sephadex G-150 | 158 | 1093 | 7.4 | 79 |
| DEAE-cellulose | 73 | 1478 | 10.0 | 49 |
| Hydroxyapatite | 5.5 | 10598 | 71.6 | 27 |
| | | Factor EF 2 | | |
| 120000*g* supernatant | 1470 | 80 | 1 | 100 |
| $(NH_4)_2SO_4$ | 864 | 102 | 1.3 | 75 |
| Sephadex G-150 | 270 | 288 | 3.6 | 66 |
| $(NH_4)_2SO_4$ | 168 | 416 | 5.2 | 59 |
| DEAE-cellulose | 58 | 755 | 9.4 | 37 |
| $(NH_4)_2SO_4$ | 26 | 1460 | 18.3 | 32 |
| Hydroxyapatite | 2.8 | 11738 | 146.7 | 28 |
| Phospho-cellulose | 0.64 | 26784 | 334.8 | 15 |

The second factor EF 2 was assayed by testing its ability to complement EF 1 in poly(U)-directed polyphenylalanine synthesis.

Fig. 1 shows the distribution profiles of both the factors. The fractions with the highest specific activities for either factor were combined and used for further purification. In our conditions gel filtration on Sephadex G-150 was the most effective method for the resolution of the wheat elongation factors.

After resolution elongation factor 1 was purified to near homogeneity by a two-step procedure. First DEAE-cellulose chromatography was applied. Almost 90% of the activity was eluted by 20mM-Tris–HCl buffer, pH 7.35, containing 100mM-KCl. This step removed the remaining EF 2 activity. Further purification of EF 1 was obtained by fractionation on a column of hydroxyapatite with a linear gradient of 10mM-potassium phosphate buffer, pH 7.2, to 250mM-potassium phosphate buffer, pH 8.0. The final product was approx. 70-fold purified compared with the soluble supernatant fraction of wheat-germ homogenate. The final product showed as practically a single band on polyacrylamide-gel disc electrophoresis. The molecular weight of EF 1 was estimated as $180000 \pm 6000$ by gel filtration and disc electrophoresis. The purified EF 1 preparations were stored in the presence of phosphate buffer at −30°C where they were stable for at least a week.

The purification procedure applied to factor EF 2 was similar to the one outlined above. Since SH groups seem to be essential for EF 2 activity, all the steps were performed in the presence of 6mM-2-mercaptoethanol or 5mM-dithiothreitol. The pooled fractions obtained from gel filtration were precipitated with $(NH_4)_2SO_4$ to 57% saturation and after dialysis the protein was subjected to chromatography on DEAE-cellulose. The EF 2 was eluted with 300mM-KCl in 20mM-Tris–HCl buffer, pH 7.35. After further $(NH_4)_2SO_4$ fractionation (between 20 and 57% saturation) the enzyme was adsorbed on hydroxyapatite and eluted with 7.5mM-potassium phosphate buffer, pH 7.95. Final purification was

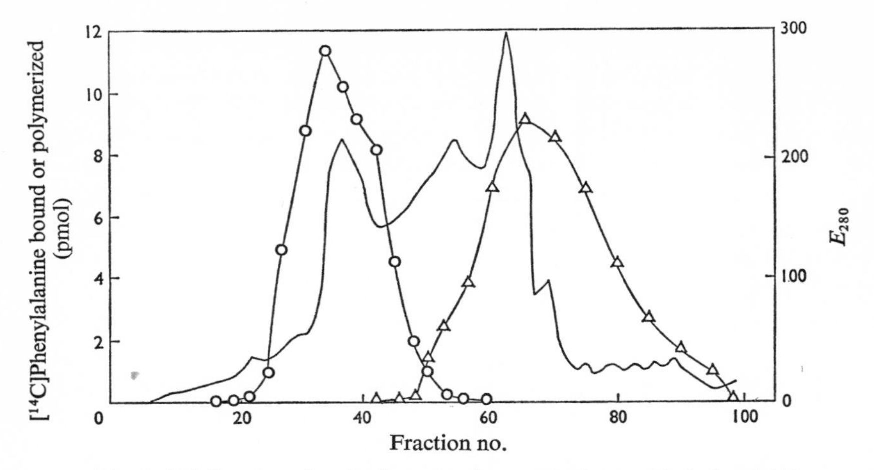


Fig. 1. *Gel filtration of crude elongation factors of wheat on Sephadex G-150*

The $(NH_4)_2SO_4$ fraction of high-speed supernatant (1.3g of protein) was applied to a column (6cm × 65cm) of Sephadex G-150 and filtration was carried out as described in the text. Fractions (10ml) were collected at a flow rate of 45ml/h. Protein concentration was determined as $E_{280}$ (——). EF 1 factor (○) was assayed by the binding of [$^{14}$C]phenylalanyl-tRNA to wheat ribosomes in a reaction mixture containing (in 0.3ml): 50mM-Tris–HCl buffer, pH7.7, 5mM-magnesium acetate, 70mM-KCl, 3mM-dithiothreitol, 30mM-GTP, 10μg of poly(U), 30pmol of [$^{14}$C]phenylalanyl-tRNA (8000c.p.m.), 130μg of ribosomes and 10μl of alternate fraction. After incubation for 10min at 30°C the reaction was stopped by the addition of 3.5ml of cold wash medium containing 10mM-Tris–HCl buffer, pH7.35, 80mM-KCl and 10mM-magnesium acetate, and passed through a cellulose nitrate filter (Sartorius, 0.45μm) previously boiled in wash medium. After washing with 14ml of the same medium and drying, filters were placed in a toluene scintillator and radioactivity was determined with 65% efficiency for $^{14}$C. The activity of EF 2 (△) was measured by phenylalanine polymerization assay in the reaction mixture containing (in 0.4ml): 50mM-Tris–HCl buffer, pH7.7, 7mM-magnesium acetate, 70mM-KCl, 5mM-dithiothreitol, 60mM-GTP, 15μg of poly(U), 30pmol of [$^{14}$C]phenylalanyl-tRNA (8000 c.p.m.), 130μg of ribosomes, 10μg of EF 1 preparation and 10μl of assayed fraction. After incubation for 10min at 30°C the reaction was stopped by the addition of 3ml of 5% (w/v) trichloroacetic acid and the amount of [$^{14}$C]phenylalanine incorporated into hot trichloroacetic acid-insoluble material was determined.

by column chromatography on phosphocellulose. The EF 2 which eluted with 5mM-potassium phosphate buffer, pH7.5, was stored as an $(NH_4)_2SO_4$ precipitate at −10°C for at least a fortnight without any significant decrease in activity. Electrophoresis on polyacrylamide gel of the freshly prepared enzyme in the presence or absence of sodium dodecyl sulphate revealed a single band. The molecular weight of EF 2 estimated by gel filtration and gel electrophoresis is 70000 ± 4000. It is worth noting that the molecular weight of wheat EF 1 corresponds to that which McKeehan & Hardesty (1969) reported for mammalian binding enzyme. The molecular weight of EF 2 from wheat is of the same order as the molecular weights of both bacterial and mammalian translocases (Parmeggiani & Gottschalk, 1969; Collins *et al.*, 1971).

## Properties of Wheat Elongation Factors

The wheat elongation factor EF 1 has been defined as a binding enzyme and as such forms specific complexes with GTP and aminoacyl-tRNA and catalyses the attachment of aminoacyl-tRNA to the ribosome. Since ribosomal binding may occur in the presence (at low $Mg^{2+}$ concentration) or absence (at high $Mg^{2+}$ concentration) of binding enzyme, it was interesting to determine whether both types of binding lead to the same product. When the products of either enzymic or non-enzymic binding were subjected to mild alkaline hydrolysis and separated by paper chromatogaphy only negligible amounts of phenylalanine peptide were detected (Fig. 2). It indicates that in both types of binding, aminoacyl-tRNA has the same affinity to ribosomal binding sites and that in both EF 1 and ribosome preparations factor EF 2 is absent. It also reflects an absolute requirement for elongation factor 2 in the amino acid polymerization under the conditions described here.

Both enzymic and non-enzymic ribosomal binding reactions are dependent on

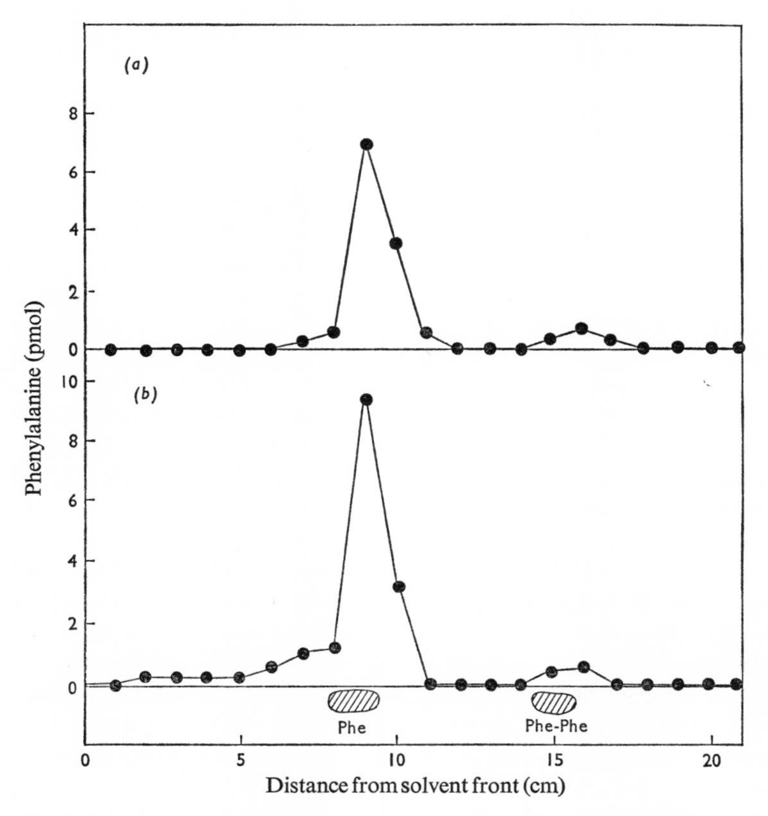


Fig. 2. *Paper chromatography of the enzymic and non-enzymic binding-reaction products after hydrolysis from the ribosomes*

The reactions were performed for 10min at 30°C in a total vol. of 0.15ml (*a*) with 22mM-magnesium acetate and no EF 1 and (*b*) with 4.6mM-magnesium acetate and 6$\mu$g of EF 1. The products were hydrolysed with 0.5M-KOH at 37°C for 3.5h and chromatographed on Whatman no. 1 paper in a manner similar to that described by Krisko *et al.* (1969) except that the solvent system was ethyl acetate–pyridine–acetic acid–water (55:25:8:10, by vol.). The paper was cut into sections (1cm × 1cm) and the radioactivity was counted by liquid scintillation. The $R_F$ values were phenylalanine, 0.42 and phenylalanylphenylalanine, 0.72.

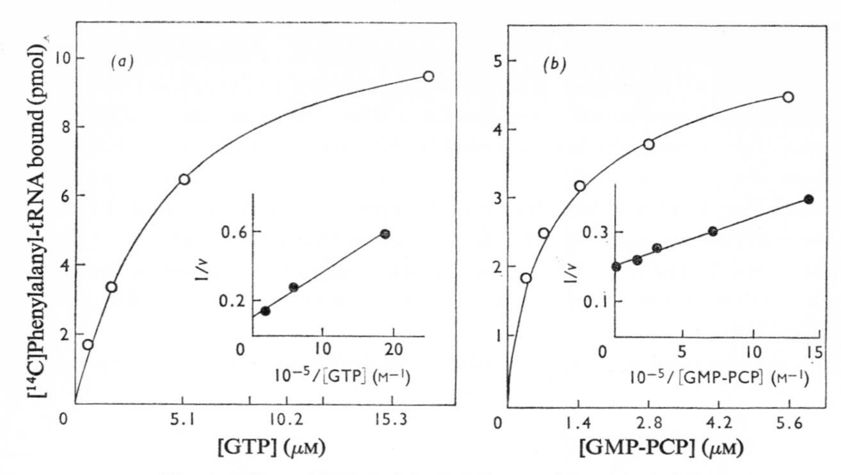


Fig. 3. *Effect of GTP and GMP-PCP on ribosomal binding*

The conditions of ribosomal binding in the presence of (*a*) GTP or (*b*) GMP-PCP were similar to those described in Fig. 1.

the presence of GTP. A number of detailed studies have been performed to determine whether, during the attachment of aminoacyl-tRNA to the ribosome, the GTP molecule is cleaved into GDP and $P_i$. For this purpose the structural analogue of GTP, guanylyl-5′-methylenediphosphonate (GMP-PCP) which cannot undergo enzymic cleavage, was substituted for GTP. It was found that in bacterial systems GMP-PCP stimulates the binding of aminoacyl-tRNA to ribosomes but does not promote formation of peptide (Haenni & Lucas-Lenard, 1968). Similar experiments with wheat EF 1 indicate that in this system GMP-PCP can mimic the GTP-dependent ribosomal binding. In the presence of the analogue about 40% of the original binding was observed (Legocki, 1972). As the kinetic patterns demonstrate (Fig. 3) the $K_m$ value for GMP-PCP is $6.7 \times 10^{-7}$M, whereas for GTP it is $2 \times 10^{-6}$M. In the wheat system, however, as in bacterial systems, phenylalanyl-tRNA bound in the presence of GMP-PCP does not incorporate phenylalanine into peptide. Recently Shorey *et al.*, (1972), contributing to the participation of GMP-PCP in ribosomal binding, indicated the inability of this analogue to replace GTP during the course of ribosomal binding at low $Mg^{2+}$ concentrations. EF-Tu–GMP-PCP complex bound to ribosomes also precludes the exchange with the GTP molecule.

Ibuki & Moldave (1968) demonstrated an interaction of GTP and aminoacyl-tRNA with rat liver factor 1 (EF 1) during the course of its thermal inactivation. In their studies GTP interacted with the factor leading to its inactivation in the absence of aminoacyl-tRNA as measured in the EF 1-dependent incorporation of amino acids into peptide. An apparently similar interaction of GTP and aminoacyl-tRNA with wheat-germ EF 1 is shown in Fig. 4. When factor 1 was incubated in buffered salts solution only, 40% of the ribosomal binding activity remained after 6 min at 38°C. The presence of aminoacyl-tRNA during pre-incubation prevents this loss of activity to a small extent. Addition of GTP caused a rapid inactivation of EF 1 so that only 10% of the initial activity remained. In

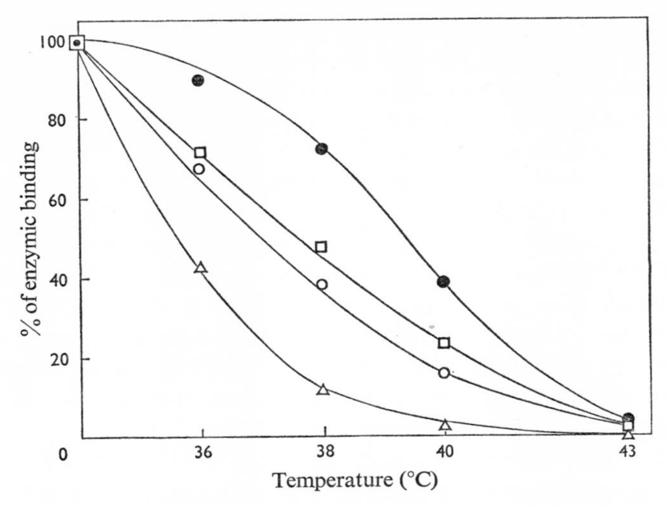


Fig. 4. *Effect of GTP and aminoacyl-tRNA on the heat stability of factor EF* 1

EF 1 (8 $\mu$g) was incubated at the indicated temperatures alone (○), or with 7 $\mu$M-GTP (△), 42 $\mu$g of [$^{14}$C]phenylalanyl-tRNA (□), or both GTP and [$^{14}$C]phenylalanyl-tRNA (●). After 6 min 7 $\mu$M-GTP and 42 $\mu$g of [$^{14}$C]phenylalanyl-tRNA were added to the tubes incubated without these components and all the tubes were incubated for a further 10 min at 30°C in the presence of poly(U) and ribosomes. Ribosomal binding was determined as described in Fig. 1. Activity was expressed as a percentage of that observed without heat inactivation of samples kept at 0°C.

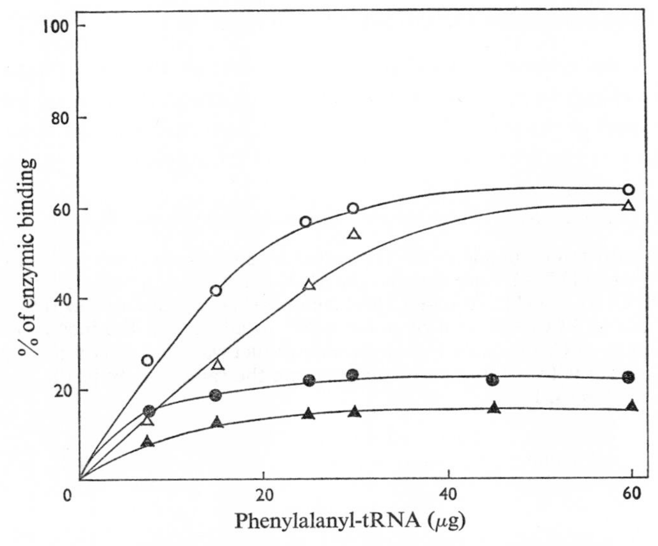


Fig. 5. *Effect of aminoacyl-tRNA on the heat stability of EF* 1

Samples (8 $\mu$g) of EF 1 were heated for 6 min at 38°C in buffered solutions containing either [$^{14}$C]phenylalanyl-tRNA from wheat germ with (○) and without (●) GTP or [$^{14}$C]phenylalanyl-tRNA from yeast with (△) and without (▲) GTP. The ribosomal binding activity was measured as described in Fig. 1.

the presence of both GTP and aminoacyl-tRNA, however, the enzyme is much more protected against heat inactivation. Thus, the heat stability of factor 1 is markedly dependent on the presence of GTP. These results also indicate that there is a direct interaction between EF 1, GTP and aminoacyl-tRNA even in the absence of EF 2 and ribosomes.

The protection of EF 1 against heat inactivation depends on aminoacyl-tRNA concentration (Fig. 5). The enzyme was heated for 6 min at 38°C at various concentrations of phenylalanyl-tRNA from wheat or yeast in the presence or absence of GTP. The enzyme was stabilized by relatively low concentrations of aminoacyl-tRNA from both sources in the presence of GTP. Yeast phenylalanyl-tRNA shows a smaller protective effect than its counterpart from wheat germ, which may be indicative of some species specificity in the interaction of EF 1 with aminoacyl-tRNA. Deacylated tRNA has practically no effect on the shift in the stability of EF 1 in the presence of GTP.

Evidence for the translocase function of wheat elongation factor EF 2 resulted from the interpretation of the puromycin reaction. The reaction with this analogue of the terminal aminoacyl-adenosine moiety of tRNA is considered as a model for functional translocation. Since puromycin reacts only with peptidyl-tRNA bound to the ribosomal P site, the enzyme that catalyses the formation of peptidyl-puromycin must be involved in the movement of peptidyl-tRNA from the puromycin unreactive site on the ribosome (A site) to where the product can be formed (P site). The ability of each of the factors to catalyse the formation of the peptidyl-puromycin is shown in Table 2. Only EF 2 is active, thus giving evidence of the translocation function of this factor.

## Formation of the Elongation Factor 1–GTP Complex

In order to understand the detailed mechanism of the first phase of polypeptide-chain elongation it was very helpful to examine the role of EF 1 factor (binding enzyme) in the individual steps of the ribosomal binding reaction.

The first step is the formation of factor 1–GTP complex. In a wheat embryo

Table 2. *Requirements for the formation of peptide–puromycin*

The standard incubation components for ribosomal binding reaction were added (see Fig. 1) except that EF 1 and GTP were omitted and the $Mg^{2+}$ concentration was 21.5 mM. After 6 min incubation at 20°C, 30 μM-GTP, 70 μM-KCl and factors as indicated were added to a volume of 0.6 ml. After 5 min at 30°C, 100 nmol of puromycin was added and the incubation was continued for 30 min at 30°C. The peptide–puromycin product was then determined by extracting the reaction mixtures with ethyl acetate and counting the radioactivity in the organic phase (from Legocki & Marcus, 1970).

| Omission from first incubation | Factor added (μg) | | Peptide–puromycin (pmol) |
|---|---|---|---|
| | EF 1 | EF 2 | |
| | 16 | | 0 |
| | 32 | | 0 |
| | | 0.75 | 2.3 |
| | | 1.5 | 2.8 |
| | | 3.0 | 3.3 |
| Poly(U) | | 1.5 | 0.1 |
| Ribosomes | | 1.5 | 0 |

Table 3. *Nucleotide specificity of binding to factor EF* 1

The incubation mixtures contained in a total volume of 200 μl: 50 mM-Tris–HCl buffer, pH 7.3, 80 mM-KCl, 10 mM-magnesium acetate, 5 mM-dithiothreitol, 5 μg of EF 1 and indicated amounts of labelled nucleotides. They were incubated for 10 min at 0°C, and the reaction was stopped by the addition of 3.5 ml of cold wash medium. The samples were filtered through cellulose nitrate filters as described in Fig. 1 and the radioactivity retained on the filters was assayed.

| Nucleotide added (nmol) | Nucleotide bound (pmol) |
|---|---|
| [$^{14}$C]GTP (0.5) | 12.9 |
| [$^{14}$C]ATP (0.6) | 2.1 |
| [$^{14}$C]UTP (0.5) | 0.4 |
| [$^{14}$C]CTP (0.6) | 0.4 |
| [$^{14}$C]GDP (0.5) | 4.4 |
| [$^{14}$C]GMP (0.6) | 0.7 |
| [$\gamma$-$^{32}$P]GTP (0.4) | 13.6 |

system Jerez *et al.* (1969) demonstrated the ability of a supernatant fraction to form a protein–GTP specific complex. This complex could interact with unblocked aminoacyl-tRNA from *Escherichia coli* and yeast as shown by a membrane-filtration technique. We would like to extend these observations and add some details about the formation of EF 1–GTP and aminoacyl-tRNA–EF 1–GTP complexes obtained by using the purified EF 1 preparation. By way of analogy with the bacterial intermediates (Shorey *et al.*, 1969) we have designated the EF 1–GTP complex complex I and the ternary complex aminoacyl-tRNA–EF 1–GTP complex II.

A very useful technique for examination of both complexes proved to be filtration through cellulose nitrate membranes. Allende *et al.* (1967) showed that the bacterial complex I is retained by the filter whereas complex II is released (Gordon, 1968). These findings were later confirmed in a number of laboratories.

Table 3 shows that factor EF 1 has a high specificity for GTP. Of the three other nucleoside triphosphates only ATP shows a certain affinity for the factor (16% of the binding capacity of GTP). GDP was about three times less effective than GTP in binding to EF 1 and GMP showed only an insignificant interaction. When [$^{14}$C]GTP was replaced with [$\gamma$-$^{32}$P]GTP a similar result was obtained, which may suggest that the hydrolysis of GTP does not occur during the reaction.

Since the commercial GTP (unlabelled or labelled) is usually contaminated with GDP, each preparation was analysed by t.l.c. and, if necessary, GDP contamination was converted into GTP by the procedure of Beaud & Lengyel (1971).

The total amount of complex I is dependent on the amount of EF 1 present in the incubation mixture. Fig. 6 indicates the amount of the complex formed. The amounts of $^{3}$H, $^{14}$C or $^{32}$P label retained by the filter are essentially the same, also suggesting that there is no GTP cleavage during formation of the complex.

Fig. 7 shows the rate of the EF 1–GTP complex formation. GTP becomes rapidly bound to the factor, with saturation yields after 6 min incubation at 0°C. In separate experiments it was found that complex I is highly unstable, and after 15 min incubation at 25°C it slowly decomposes.

GTP interacts with EF 1 even in the absence of $Mg^{2+}$ ions (Fig. 8). In addition, the univalent cation requirements were also unrestricted, but a concentration of 80 mM-KCl or -$NH_4Cl$ seemed to be optimum for the reaction. At higher con-

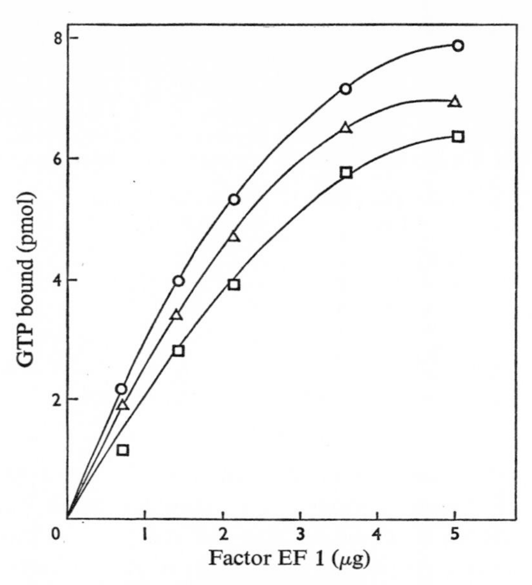


Fig. 6. *Binding of GTP by factor EF* 1

Measurement of the formation of the EF 1–GTP complex in the presence of labelled 1.2 μM-GTP was performed as described in Table 3. ○, [γ-$^{32}$P]GTP; △, [$^{14}$C]GTP; □, [$^{3}$H]GTP.

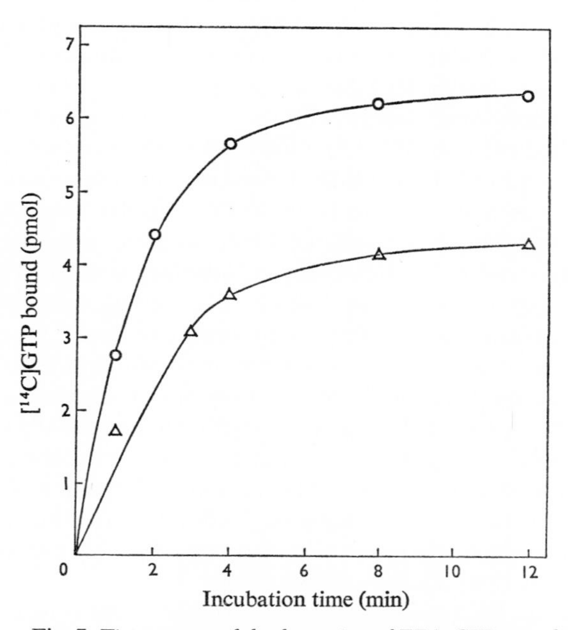


Fig. 7. *Time-course of the formation of EF* 1–*GTP complex*

The reaction was assayed as described in Table 3 with 3 μg (△) or 5 μg (○) of EF 1.

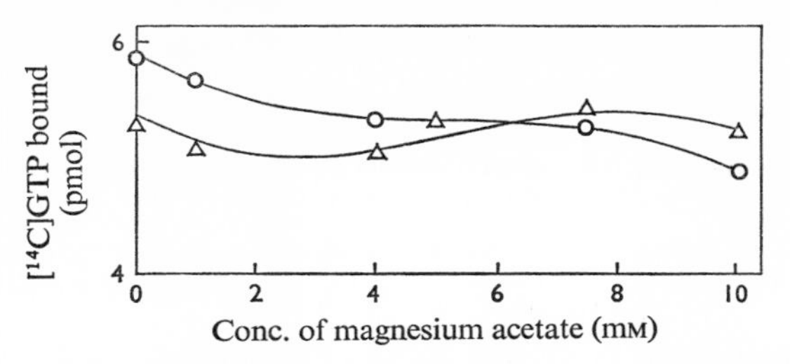


Fig. 8. *Effect of $Mg^{2+}$ on the formation of complex I at* 0°C (○) *and* 20°C (△)

The formation of the complex was determined at various magnesium acetate concentrations after 10min standard incubation.

centrations these salts slightly inhibit the formation of the complex. $NH_4Cl$ was about 20% less effective than KCl (not shown here).

As shown in the detailed studies by Cooper & Gordon (1969), the *E. coli* binding factor preferentially bound GDP rather than GTP in the absence of aminoacyl-tRNA on testing of retention on filters. Similar properties are exhibited by the binding factor from yeast, although its affinity to GDP in the presence of GTP was not as high as in the bacterial system (Richter, 1970). On the other hand, our competition studies on wheat germ revealed that the elongation factor 1 preferentially bound GTP if GDP was added simultaneously. Fig. 9 shows that GDP was ineffective in displacing GTP bound to EF 1 even when GDP was present in high excess. Instead, an eightfold excess of GTP decreased the retention of GDP on filters by about 90% as measured under the same conditions. In separate experiments the ability of either GDP or GTP to react with the

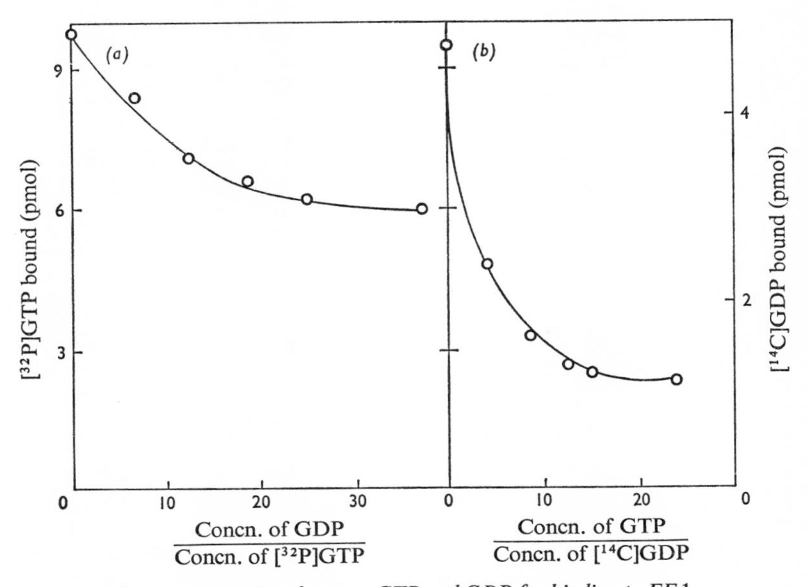


Fig. 9. *Competition between GTP and GDP for binding to EF* 1

Reaction mixtures were prepared as described in the legend to Table 3 except that the binding was carried out in the presence of various amounts of unlabelled GDP (*a*) and GTP (*b*).

Table 4. *Competition between GTP and GDP for EF* 1 *in the binding to cellulose nitrate filters*

The reaction mixtures were prepared as described in Table 3 except that various additions were made as indicated in column 1. After incubation for 10 min at 0°C [14C]GTP or [14C]GDP was added as indicated in column 2. After incubation for an additional 10 min at 0°C, the reaction mixtures were then assayed for $^{14}C$ label retained on the filters.

| First incubation (0 min) | Second incubation (10 min) | $^{14}C$ retained on filter (pmol) |
|---|---|---|
| EF 1 | [14C]GTP (0.5 nmol) | 10.2 |
| EF 1 | [14C]GDP (0.7 nmol) | 3.5 |
| EF 1 + GTP (2 nmol) | [14C]GTP (0.5 nmol) | 1.3 |
| EF 1 + GTP (2 nmol) | [14C]GDP (0.7 nmol) | 0.4 |
| EF 1 + GDP (5 nmol) | [14C]GTP (0.5 nmol) | 7.0 |
| EF 1 + GDP (5 nmol) | [14C]GDP (0.7 nmol) | 0.8 |

preformed EF 1–GTP and EF 1–GDP complexes was studied. The results (Table 4) confirm the higher affinity of EF 1 to GTP than to GDP and indicate that both GTP and GDP bound in complex I, exchange with free nucleotides. The latter property of wheat complex I is similar to that of the bacterial complex (Shorey *et al.*, 1969).

In order to determine the stoicheiometry of GTP binding to EF 1, complex I formation was followed with limited amounts of enzyme. With 2 μg (10 pmol) of EF 1 maximum binding was obtained with just over 1 μM-GTP whereas about four times greater concentration of GDP was required for maximum GDP binding (Fig. 10). The apparent Michaelis constant determined for GTP was

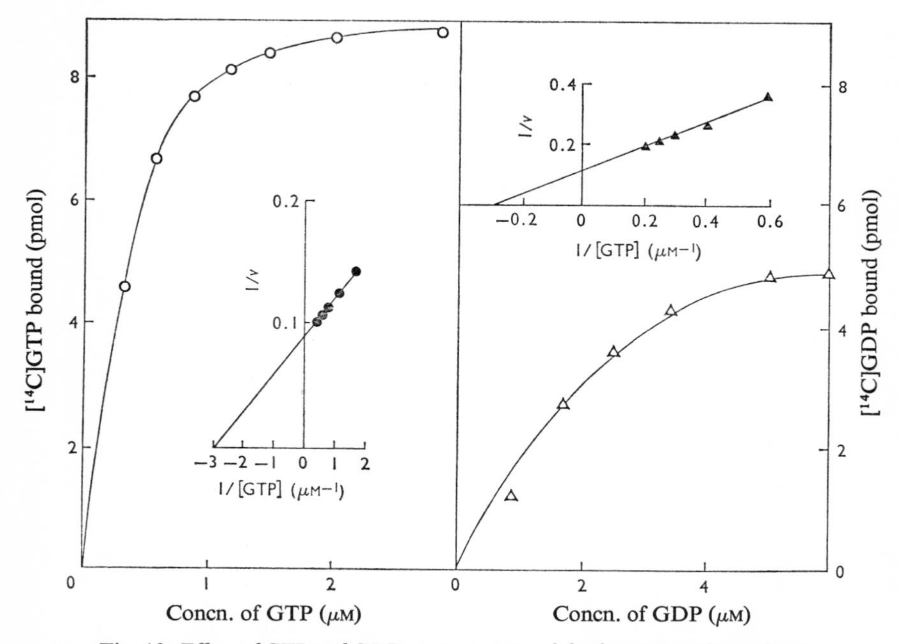


Fig. 10. *Effect of GTP and GDP concentration of the formation of complex* I

The formation of complex I was determined at various GTP (○) and GDP (△) concentrations after 10 min standard incubation. Assays were performed with 2 μg of EF 1.

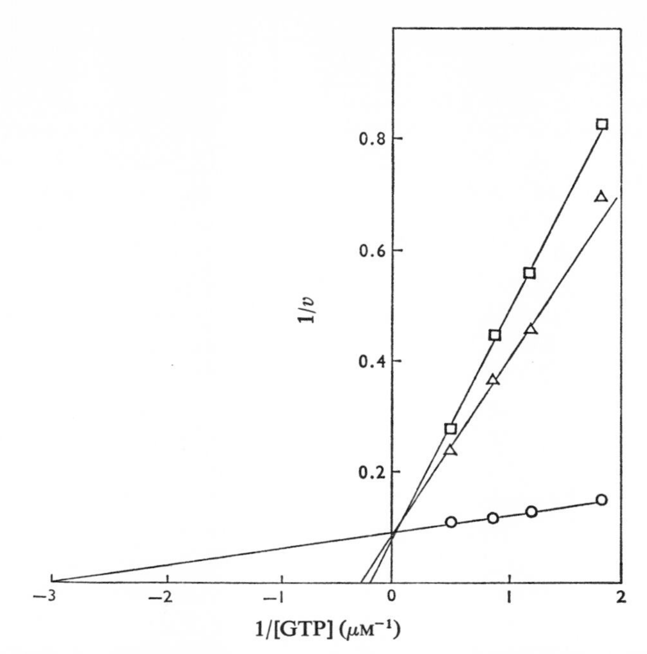


Fig. 11. *Competitive inhibition of the formation of EF 1–GTP complex by GDP and GMP-PCP*

The standard assay system was used except as follows: ○, GTP only; △, GTP + 5 μM-GDP; □, GTP + 5 μM-GMP-PCP.

$2.9 \times 10^{-7}$ M and for GDP $3.7 \times 10^{-6}$ M. The molar ratio of the maximum amount of GTP bound per mol of EF 1 was 0.9 (calculated for the saturation level of GTP). In some cases lower values were obtained when aged enzyme preparations were used. These data may suggest that the value of 1.0 is a theoretical limit of the mole ratio in complex I. The lower experimental value is perhaps caused by trace amounts of GDP still present in the GTP. It may also be related to the progressive inactivation of the relatively unstable enzyme. It was observed that GDP and GMP-PCP are effective inhibitors of complex I formation. In the conditions of assay both the compounds appear to show a simple competitive inhibition (Fig. 11). From these results the $K_i$ value for GDP is $3.3 \times 10^{-6}$ M and for GMP-PCP, $4.5 \times 10^{-6}$ M.

### Formation of the Aminoacyl-tRNA–EF 1–GTP Complex

As mentioned above, the binding of GTP by bacterial transfer factor Tu is the prerequisite step of the formation of the ternary complex, aminoacyl-tRNA–Tu–GTP. This complex serves as the direct source of aminoacyl residue for the growing polypeptide chain and thus plays the essential role in the ribosomal binding reaction. With the isolated complex it was possible to study the structural requirements of tRNA for the specific interaction with the binding factor and GTP (Krauskopf *et al.*, 1972). It was also possible to elucidate the fate of the GTP molecule while attaching aminoacyl-tRNA to ribosomes (Shorey *et al.*, 1969).

Since complex II is not retained on a filter, the disappearance of the enzyme–GTP complex resulting from the addition of aminoacyl-tRNA may be a fairly good estimation method of its formation (Fig. 12). The reaction reaches completion within 2–4 min at 30°C and within 8 min at 0°C. It was also found that aminoacyl-tRNA from different sources (phenylalanyl-tRNA from wheat germ or yeast, isoleucyl-tRNA from lupin seeds) may form ternary complexes with wheat EF 1 (Fig. 13). In this reaction the homologous tRNA seems to be the most effective, which may be another confirmation of some specificity of EF 1 with regard to aminoacyl-tRNA (see Fig. 4). Whether the tRNA samples were charged with non-radioactive amino acids to the same extent should, however, be examined more carefully. Fig. 13 shows that deacylated tRNA species were not effective in the formation of the ternary complexes.

The observed decrease in the retention of the EF 1–GTP complex due to the presence of aminoacyl-tRNA should be accompanied by the appearance of the EF 1 activity in the filtrate, as reported for the *E. coli* system (Gordon, 1968). Fig. 14 shows that an increase in factor 1 activity in the filtrate is parallel to the increased concentration of the phenylalanyl-tRNA added during the ternary complex formation. The EF 1 activity in the filtrate was measured by its ability to complement factor EF 2 and ribosomes in poly(U)-directed synthesis of polyphenylalanine. The appearance of the EF 1 activity in the filtrate also depends on the presence of GTP.

Although the filtrate contains, besides the complex, the unchanged GTP and some unbound phenylalanyl-tRNA as well, it may be used in further investigations. By using the filtrate containing complex II it was possible to follow the

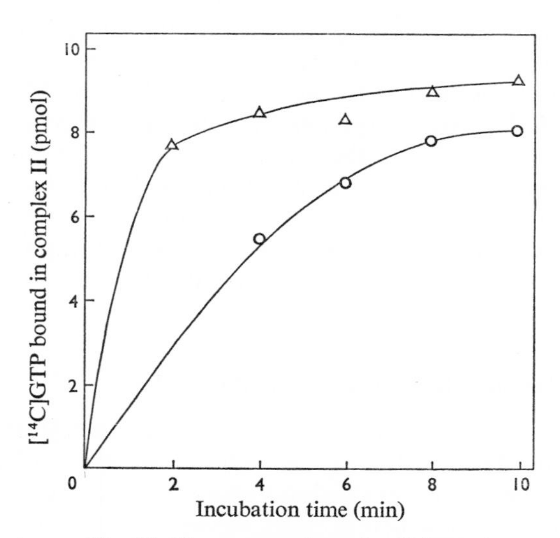


Fig. 12. *Time-course of complex II formation*

Phenylalanyl-tRNA (24 μg), [$^{14}$C]GTP (2.1 μM) and EF 1 (8 μg), in buffered solution were incubated at 0°C (○) or 30°C (△) for various times. The incubation mixtures were assayed for $^{14}$C retained on filters. In the controls, the phenylalanyl-tRNA was omitted. The decrease in the amount of radioactivity that was retained by the filter due to the presence of phenylalanyl-tRNA was used as a measure of complex formation.

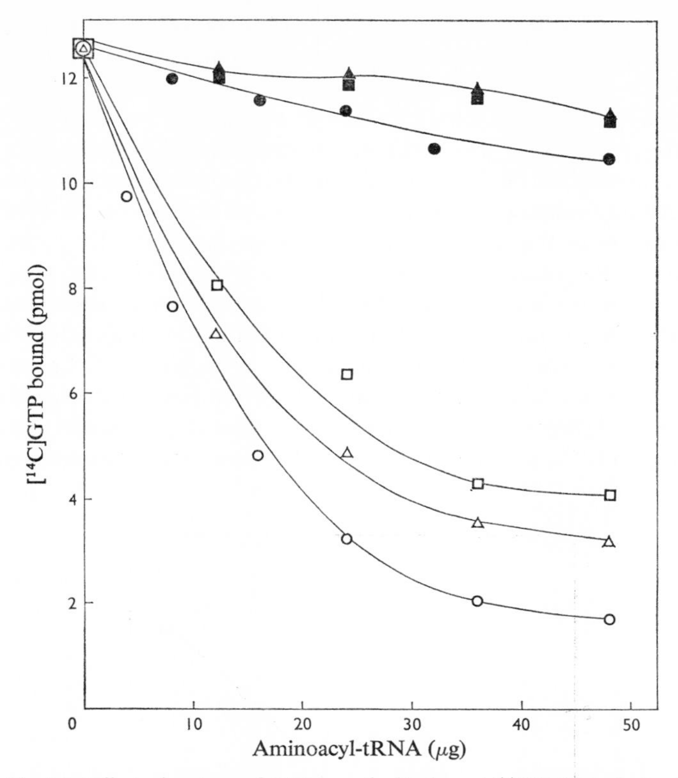


Fig. 13. *Effect of aminoacyl-tRNA on the binding of* [14C]*GTP to factor EF* 1

Assay conditions were as in Table 3 except that 8 μg of EF 1 preparation were used. Incubation was carried out in the presence of various amounts of different aminoacyl-tRNA or stripped tRNA preparations: $tRNA^{Phe}$ from wheat germ charged with phenylalanine (○) and uncharged (●); $tRNA^{Ile}$ from lupin seeds charged with isoleucine (△) and uncharged (▲); yeast $tRNA^{Phe}$ charged with phenylalanine (□) and uncharged (■).

course of ribosomal binding. The transfer of [14C]phenylalanyl-tRNA from complex II to ribosomes is strictly dependent on the presence of poly(U) and is proportional to the concentration of ribosomes (Fig. 15).

Interesting results were obtained by gel filtration of the wheat EF 1 complexes. In these studies several columns (1.1 cm × 27 cm) of Sephadex G-150 were used for each experiment. The filtrations were performed in 50 mM-Tris–HCl buffer, pH 7.35, containing 10 mM-magnesium acetate, 30 mM-KCl, 50 mM-$NH_4Cl$ and 3 mM-2-mercaptoethanol. The void volume of the columns was 8 ml. Fig. 16(*a*) shows the elution profile of factor 1 alone in buffered salts solution. As detected by phenylalanine polymerization assay the EF 1 activity appears as a single sharp peak in the fraction corresponding to the void volume. Similar EF 1 activity profiles were obtained when factor 1 was incubated with GTP or with phenyl-alanyl-tRNA before filtration (Figs. 16*b* and 16*c*). The appearance of $^{32}P$ radio-activity derived from GTP in the fractions corresponding to factor 1 (Fig. 16*b*) indicates that the EF 1–GTP complex was apparently formed. There was no radioactivity associated with factor 1, however, when it was incubated with labelled phenylalanyl-tRNA (Fig. 16*c*). This observation suggests that there is very little, if any, direct interaction between factor 1 and aminoacyl-tRNA in the absence of GTP, as shown by the filtration technique.

On the other hand, significant changes in the elution profile of EF 1 were observed after the enzyme was incubated under conditions in which a ternary complex is supposed to be formed (i.e. with both aminoacyl-tRNA and GTP) (Fig. 16*d*). An additional peak of factor 1 appeared in fraction 15. Whereas the first peak emerges in the same fraction as non-preincubated EF 1, the second peak appears at the volume corresponding to a lower-molecular-weight form. The appearance of the second peak of EF 1 is associated with the shift of [$^{14}$C]-phenylalanyl-tRNA from fraction 21 to fraction 16 (Fig. 16*e*), which indicates that the ternary complex has been formed. The explanation of the results may lead to the conclusion that there exist at least two different forms of the wheat elongation factor 1. Since the second peak of EF 1 activity appeared in the region corresponding to EF 2, its molecular weight may be roughly estimated at 70000.

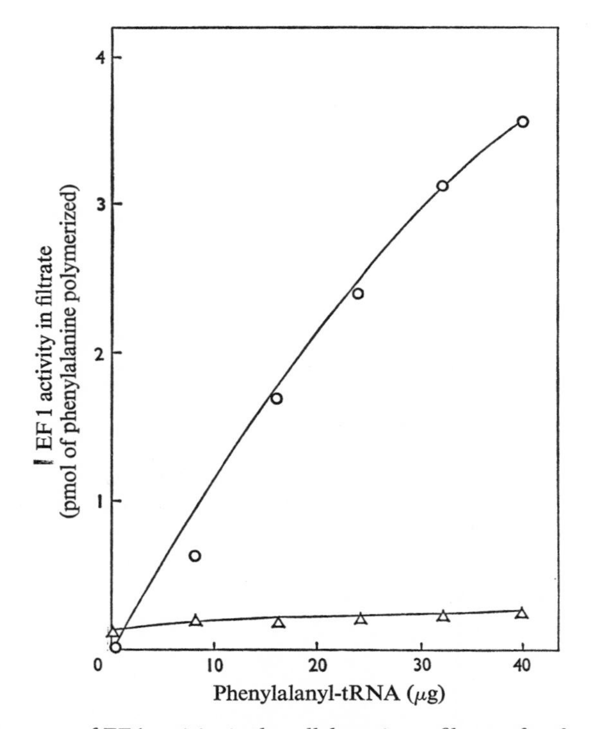


Fig. 14. *Appearance of EF 1 activity in the cellulose nitrate filtrate after formation of complex II*

Each reaction mixture contained the standard concentration of reactants except that unlabelled GTP was used with various concentrations of [$^{14}$C]phenylalanyl-tRNA. [$^{14}$C]Phenylalanyl-tRNA in the presence (○) or absence (△) of 3.4 μM-GTP were included. Incubations were performed in a total volume of 0.2 ml and terminated by addition of 0.6 ml of wash medium containing 3 mM-dithiothreitol. The diluted solution was passed through the filters which were then washed twice with 0.3 ml of wash medium. The filtrates were collected, 3.4 μM-GTP was added to those tubes in which GTP was omitted during the preincubation and samples were assayed for EF 1 activity in the phenylalanine polymerization test.

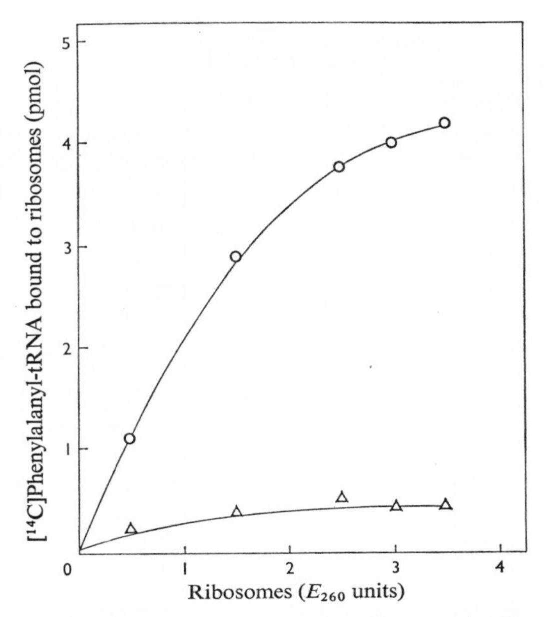


Fig. 15. *Binding of* [$^{14}$*C*]*phenylalanyl-tRNA form complex II to ribosomes*

Complex II was formed in the standard conditions and recovered in membrane filtrate as outlined in the legend to Fig. 14. A 100 $\mu$l sample of the filtrate containing 6.8 pmol of [$^{14}$C]phenylalanyl-tRNA bound in the complex was supplemented with various amounts of ribosomes in the presence (○) or absence (△) of poly(U). After 10 min incubation at 30°C, the amounts of [$^{14}$C]phenylalanyl-tRNA bound to the ribosomes were determined by the filter technique.

The second of the EF 1 forms reveals itself only as a result of the interaction with GTP and aminoacyl-tRNA. It may represent the interconversion of the native state of factor 1 into a lower-molecular-weight form due to interaction with GTP in the presence of aminoacyl-tRNA.

Moon *et al.* (1972) described similar properties of factor 1 from calf brain. They claim the existence of two separate species of elongation factor 1 which were designated as EF $1_A$ (mol.wt. > 150000) and EF $1_B$ (mol.wt. 60000–80000). Our observations on the properties of wheat elongation factor 1 suggest that their formulation of the interactions of EF 1 with GTP and aminoacyl-tRNA may be applied to higher plants as well. Thus the interactions can be illustrated as follows:

$$\text{EF } 1_A + \text{GTP} \rightleftharpoons \text{EF } 1_A\text{–GTP} \quad (1)$$

$$\text{Aminoacyl-tRNA} + \text{EF } 1_A\text{–GTP} \rightleftharpoons \text{Aminoacyl-tRNA–EF } 1_B\text{–GTP} \quad (2)$$

There are still, however, some questions left such as: what is the fate of the GTP molecule in the course of EF $1_A \rightleftharpoons$ EF $1_B$ interconversion and what is the role of GDP in this process? Another interesting problem is which of the EF 1 forms directly participates in the interaction with ribosomes.

The fact that EF 1 exists in two different forms may partially explain the sigmoidal relationship of the EF 1-dependent polymerization of phenylalanine (Fig. 17). Such a kinetic pattern was also observed in previous investigations on

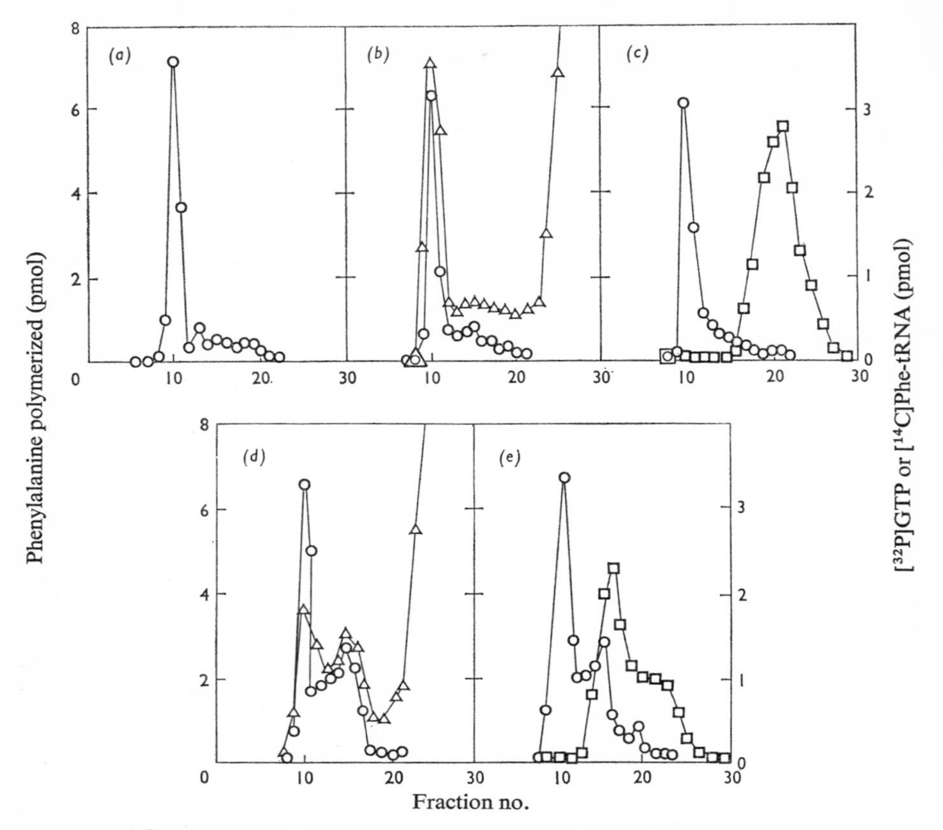


Fig. 16. *Gel-filtration patterns showing the interactions between EF*1, *GTP and phenylalanyl-tRNA*

A 20 $\mu$g sample of EF 1 was incubated for 10 min at 25°C (*a*) alone, (*b*) with 3 nmol of [$^{32}$P]GTP, (*c*) with 78 nmol of [$^{14}$C]phenylalanyl-tRNA, (*d*) with 3 nmol of [$^{32}$P]GTP + 3 nmol of phenylalanyl-tRNA, or (*e*) with 3 nmol of GTP + 3 nmol of [$^{14}$C]phenylalanyl-tRNA. After incubation samples were applied to columns (1.1 cm × 27 cm) of Sephadex G-150 equilibrated with 50 mM-Tris–HCl buffer, pH 7.35, containing 10 mM-magnesium acetate, 30 mM-KCl, 40 mM-$NH_4Cl$, 3 mM-2-mercaptoethanol and 0.8 ml fractions were collected. The EF-1 activity (○) was tested by a polymerization assay as described in Fig. 1. The concentration of [$^{32}$P]GTP (△) and [$^{14}$C]-phenylalanyl-tRNA (□) was determined in samples of each fraction.

polypeptide synthesis in the wheat-germ system (Legocki & Marcus, 1970). It may result from the fact that only one of the EF 1 forms can interact with ribosomes and the course of elongation interferes with the interconversion of factor 1. Such an explanation, however, should be regarded as tentative until more direct studies are performed.

The properties of the elongation factors described above contribute to a certain degree to current information about the elongation process in higher plants. Comparison of the particular steps to the analogous ones from bacterial or mammalian systems would be a purely formal procedure. Undoubtedly the overall mechanism of peptide-chain elongation in plants is similar to that in bacterial and mammalian cells. Hence some features of the plant system are comparable with those of the bacterial one and some others with the mammalian systems.

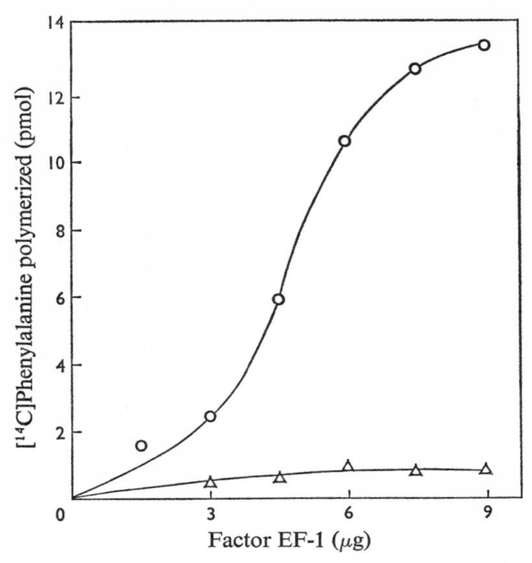


Fig. 17. *Dependence of polyphenylalanine formation on factor EF* 1

Standard incubation conditions were employed with various amounts of EF 1 in the presence (○) or absence (△) of poly(U).

These investigations, like most of the studies on peptide-chain elongation, have been performed in a simple model system in which poly(U) plays the role of template, and promotes the conversion of phenylalanyl-tRNA into polyphenyl-alanyl-tRNA. Such a system can function at the appropriate $Mg^{2+}$ concentration (in prokaryotic systems above 8mM, in eukaryotic systems above 5mM) without proper chain initiation as well as termination signals. This fact greatly facilitates the investigation. It should be kept in mind, however, that each model system is artificial and extending the conclusions drawn on its basis to the properties of a living cell may be an oversimplification.

## References

Allende, J. E., Seeds, N. W., Conway, T. W. & Weissbach, H. (1967) *Proc. Nat. Acad. Sci. U.S.* **58**, 1566–1573
Beaud, G. & Lengyel, P. (1971) *Biochemistry* **10**, 4899–4906
Brot, N., Redfield, B. & Weissbach, H. (1970) *Biochem. Biophys. Res. Commun.* **41**, 1388–1395
Caskey, T., Leder, P., Moldave, K. & Schlessinger, D. (1972) *Science* **176**, 195–197
Ciferri, O. & Parisi, B. (1970) *Progr. Nucl. Acid Res. Mol. Biol.* **10**, 121–144
Collins, J. F., Raeburn, S. & Maxwell, E. S. (1971) *J. Biol. Chem.* **246**, 1049–1054
Cooper, D. & Gordon, J. (1969) *Biochemistry* **8**, 4289–4292
Gasior, E. & Moldave, K. (1965) *J. Biol. Chem.* **240**, 3346–3352
Gordon, J. (1968) *Proc. Nat. Acad. Sci. U.S.* **59**, 179–183
Haenni, A.-L. & Lucas-Lenard, J. (1968) *Proc. Nat. Acad. Sci. U.S.* **61**, 1363–1369
Hardesty, B., Arlinghaus, R., Shaeffer, J. & Schweet, R. (1963) *Cold Spring Harbor Symp. Quant. Biol.* **28**, 215–222
Ibuki, F. & Moldave, K. (1968) *J. Biol. Chem.* **243**, 44–50
Jerez, C., Sandoval, A., Allende, J., Henes, C. & Ofengand, J. (1969) *Biochemistry* **8**, 3006–3014
Krauskopf, M., Chen, C.-M. & Ofengand, J. (1972) *J. Biol. Chem.* **247**, 842–850
Krisko, J., Gordon, J. & Lipman, F. (1969) *J. Biol. Chem.* **244**, 6117–6123

Legocki, A. B. (1972) in *Nucleic Acid and Protein Metabolism* (Farkas, G., ed.) pp. 169–173, Hungarian Academy of Sciences, Budapest
Legocki, A. B. & Marcus, A. (1970) *J. Biol. Chem.* **245**, 2814–2818
Lucas-Lenard, J. & Lipmann, F. (1966) *Proc. Nat. Acad. Sci. U.S.* **55**, 1562–1566
McKeehan, W. L. & Hardesty, B. (1969) *J. Biol. Chem.* **244**, 4330–4339
Monro, R. E., Staehelin, T., Celma, M. L. & Vazquez, D. (1969) *Cold Spring Harbor Symp. Quant. Biol.* **34**, 357–366
Moon, H.-M., Redfield, B. & Weissbach, H. (1972) *Proc. Nat. Acad. Sci. U.S.* **69**, 1249–1252
Parmeggiani, A. & Gottschalk, E. M. (1969) *Biochem. Biophys. Res. Commun.* **35**, 861–867
Richter, D. (1970) *Biochem. Biophys. Res. Commun.* **38**, 864–870
Shorey, R. L., Ravel, J. M., Garner, C. W. & Shive, W. (1969) *J. Biol. Chem.* **244**, 4555–4564
Shorey, R. L., Ravel, J. M. & Shive, W. (1972) *Arch. Biochem. Biophys.* **146**, 110–117
Skoultchi, A., Ono, Y., Moon, H.-M. & Lengyel, P. (1968) *Proc. Nat. Acad. Sci. U.S.* **60**, 675–682
Waterson, J., Beaud, G. & Lengyel, P. (1970) *Nature* (*London*) **227**, 34–38

*Biochem. Soc. Symp.* (1973) **38**, 77–96
*Printed in Great Britain*

# The Binding of Aminoacyl-Transfer Ribonucleic Acid to Wheat Ribosomes

By JORGE E. ALLENDE, ADELA TARRAGÓ, OCTAVIO MONASTERIO, SIMÓN LITVAK, MARTA GATICA, JOSÉ M. OJEDA AND MARÍA MATAMALA

*Departamento de Biología, Facultad de Ciencias, and Departamento de Bioquímica y Química, Facultad de Medicina, Universidad de Chile, Casilla* 6671, *Santiago*-4, *Chile*

## Synopsis

Wheat ribosomal particles provide a structure in which the codon and anticodon nucleotides of mRNA and tRNA, respectively, can interact with precision. In addition, wheat extracts contain a protein factor that facilitates the binding of aminoacyl-tRNA to the ribosome–mRNA complex at more physiological concentrations of $Mg^{2+}$ (4–10 mM).

The factor-mediated reaction requires GTP which interacts with the elongation factor to form a complex, EF 1–GTP, which in turn reacts with aminoacyl-tRNA to form a ternary complex, EF 1–GTP–aminoacyl-tRNA. The ternary complex, which can be isolated by gel filtration, reacts directly with the ribosome–mRNA complex transferring the aminoacyl-tRNA moiety to the particles.

In addition to the above role, GTP, and also GDP, causes a shift in the molecular form of EF 1 present in the cruder preparations of the factor. A high-molecular-weight form of EF 1 (mol.wt. 190000) is converted into a lighter form (mol.wt. 40000–50000) in the presence of GTP or GDP. The heavier form of EF 1 is more stable to heat denaturation and requires much greater concentrations of aminoacyl-tRNA to form a ternary complex.

The presence of EF 1 lends specificity to the binding of aminoacyl-tRNA to wheat ribosomes. The factor does not react with (*a*) unesterified tRNA, (*b*) aminoacyl-tRNA that has been deaminated or acylated in the amino group of the aminoacyl residue, and (*c*) aminoacyl-tRNA that has been denatured.

In addition it forms an imperfect and unstable complex with the initiator methionyl-tRNA which may explain the failure of this aminoacyl-tRNA to transfer its aminoacyl moiety to the internal positions of the peptide chain.

Wheat EF 1 reacts specifically with the RNA from plant viruses (tobacco mosaic virus and turnip yellow-mosaic virus) that have been aminoacylated enzymically. The elongation factor fails to react with the unacylated viral RNA. This finding may be related to the participation of elongation factors in viral RNA replication discovered in bacterial systems.

Wheat ribosomes in the absence of added factors show a preferential binding of the initiator species of wheat methionyl-tRNA at low $Mg^{2+}$ concentrations (4–10 mM).

## Introduction

Aminoacyl-tRNA is a key molecule in the deciphering of genetic messages in the translation process of protein biosynthesis. Fidelity of translation demands perfect specificity in the interaction of the mRNA codons with the anticodon nucleotides of the various aminoacyl-tRNA species. Since the delicate positioning that brings both nucleotide sequences properly together occurs on the ribosome, the interaction of aminoacyl-tRNA with ribosomes in the presence of mRNA is of considerable interest.

The detailed mechanism of aminoacyl-tRNA binding to ribosomes is not yet known in any system, but the work of several groups has gathered considerable information about the steps operative in bacterial systems. Legocki (1973) has summarized the scheme that has emerged from this work in his presentation before this Symposium.

We are much more ignorant about the details of the process in eukaryotic systems. It appears, however, that there are similarities and differences with the prokaryotic mechanism that make it important to investigate the reaction in plant and animal systems.

The present report outlines the work that we have done in our laboratory on the interaction of aminoacyl-tRNA and ribosomes obtained from ungerminated wheat embryos.

## Methods

Most of the methods used in the experiments that are described below have been published previously.

The wheat embryos from ungerminated seeds (*Triticum durum*) were prepared according to the procedure of Johnston & Stern (1957).

The preparation of ribosomes, supernatant fraction and tRNA and the assays for aminoacyl-tRNA binding to ribosomes and polyphenylalanine synthesis from Phe-tRNA have been described by Allende (1969).

The binding of GTP to the wheat elongation factor and the assay of aminoacyl-tRNA interaction with EF 1–GTP by the nitrocellulose membrane method have been published by Jerez *et al.* (1969).

The aminoacylation of viral RNA species from tobacco mosaic virus and turnip yellow-mosaic virus have also been described (Yot *et al.*, 1970).

Wheat elongation factor 1 was partly purified by a procedure that will be published in detail elsewhere. Essentially the procedure consists of $(NH_4)_2SO_4$ precipitation of the wheat supernatant fraction between 40 and 80% saturation of the salt. The desalted fraction is passed through a column (2.5 cm × 15 cm) of DEAE-cellulose and the activity is eluted with 0.1 M-KCl–0.02 M-Tris–HCl buffer, pH 7.5–2 mM-$MgCl_2$–5 mM-β-mercaptoethanol–10% (v/v) glycerol. The fractions from the DEAE-cellulose column are then introduced into a column (2.5 cm × 15 cm) of phospho-cellulose equilibrated with 0.2 M-KCl–0.05 M-Tris–HCl buffer, pH 8.0–2 mM-$MgCl_2$–5 mM-β-mercaptoethanol–10% (v/v) glycerol. The GTP-binding activity of EF 1 is then obtained by eluting the column with a 0.05–0.5 M gradient of KCl in the same buffer. The active fractions are pooled, precipitated with 80% saturation of $(NH_4)_2SO_4$, dissolved in a minimal amount of 0.05M-Tris–HCl buffer, pH7.5–0.05M-KCl–2mM-$MgCl_2$–1mM-β-mercapto-ethanol–10% (v/v) glycerol, and stored at −20°C.

The sucrose density gradients for ribosomal subunit preparation were from 5 to 20% sucrose in 0.01M-Tris–HCl buffer, pH7.0–0.5M-$NH_4Cl$–0.01M-$MgCl_2$–5mM-β-mercaptoethanol and were run at 20000rev./min for 12h in an SW-25 rotor.

The gradients for EF 1 analysis were done with 5 to 20% sucrose in 0.01M-Tris–HCl buffer, pH7.5–0.01M-$MgCl_2$–0.05M-$NH_4Cl$–1mM-β-mercaptoethanol. They were run for 18h at 37000rev./min in a SW-39 rotor.

## Results

*Enzymic binding of aminoacyl-tRNA to wheat ribosomes*

Messenger-directed binding of aminoacyl-tRNA to wheat ribosomes can occur in the presence or in the absence of elongation factor 1(EF 1). The key parameter that regulates the requirement for EF 1 in this reaction is the concentration of $Mg^{2+}$. Fig. 1 shows the binding of phenylalanyl-tRNA to wheat ribosomes in the presence of poly(U) at different $Mg^{2+}$ concentrations. It is clear that at concentrations of $Mg^{2+}$ greater than 20mM the binding of aminoacyl-tRNA to ribosomes is not stimulated by EF 1. At $Mg^{2+}$ concentrations of 3–10mM, the presence of the protein factor causes a very striking stimulation of the reaction.

One may assume that the physiological reaction involves the participation of the elongation factor whereas the reaction observed at high $Mg^{2+}$ concentrations constitutes only an interesting artifact which probably reflects the drastic conformational effects that this ion has on ribosomal structure.

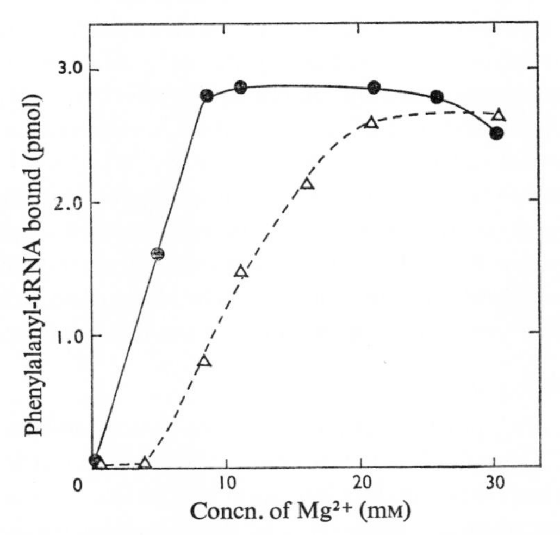


Fig. 1. *Effect of $Mg^{2+}$ concentration on enzymic and non-enzymic binding of phenylalanyl-tRNA to wheat ribosomes*

The binding reaction was carried out as described by Allende (1969). All reactions contained approx. 10pmol of [$^{14}$C]phenylalanyl-tRNA, 2.3 $E_{260}$ units of wheat ribosomes and 40μg of poly(U). ●, 10μg of EF 1 and 1mM-GTP added; △, no enzyme added.

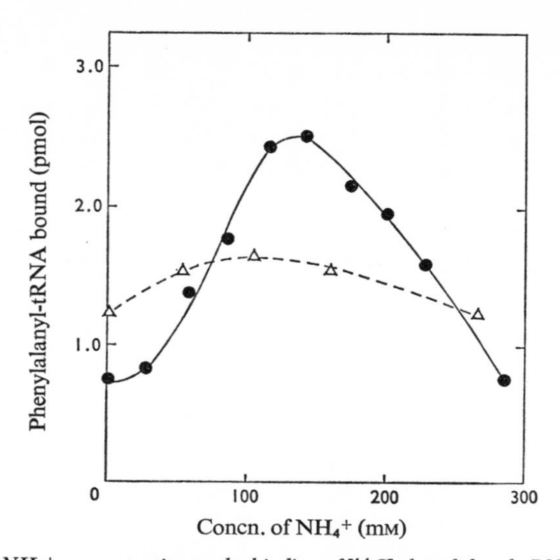


Fig. 2. *Effect of $NH_4^+$ concentration on the binding of [$^{14}C$]phenylalanyl-tRNA to wheat ribosomes*

The binding reaction was performed as reported by Allende (1969). Assays that contained 10 μg of EF 1 and 1mM-GTP (●) were carried out at 5 mM-$MgCl_2$ whereas the tubes that did not contain elongation factor or GTP (△) had 20mM-$MgCl_2$.

Codon specific non-enzymic binding also indicates that the ribosome has built into its structure the features that allow the correct interaction between mRNA and tRNA. One can then imagine that the function of the elongation factor is to make the whole process more efficient.

Another difference between enzymic and non-enzymic binding of aminoacyl-tRNA to ribosomes is found in the requirement for $NH_4^+$. EF 1-dependent binding requires the addition of $NH_4^+$ at a concentration of 120mM whereas non-enzymic binding at 20mM-$Mg^{2+}$ is not significantly affected by $NH_4^+$ (Fig. 2). Again this phenomenon is possibly related to the effect of $NH_4^+$ on ribosomal structure. High concentrations of this ion have been found to cause the dissociation of the wheat ribosomal sub-particles (Allende *et al.*, 1973).

Two other differences between enzymic and non-enzymic binding of aminoacyl-tRNA to wheat ribosomes reside in the specificity of the factor-mediated reaction and in its requirement for GTP. These characteristics are interesting because they may give us an insight into the function of the elongation factor in the overall process, and therefore we shall examine them in the following sections.

*EF* 1 *is a GTP-binding protein*

A few years ago, after participating in the work that established the binding of *Escherichia coli* EF Tu to GTP (Allende *et al.*, 1967), we set out to look for a GTP binding factor in wheat extracts. In 1969, we reported finding such a factor in the supernatant fraction of wheat embryos and described properties that indicated its participation in the elongation process of protein biosynthesis (Jerez *et al.*, 1969).

Our main conclusions relative to the properties and identity of the GTP-binding factor will be presented here in a summarized form.

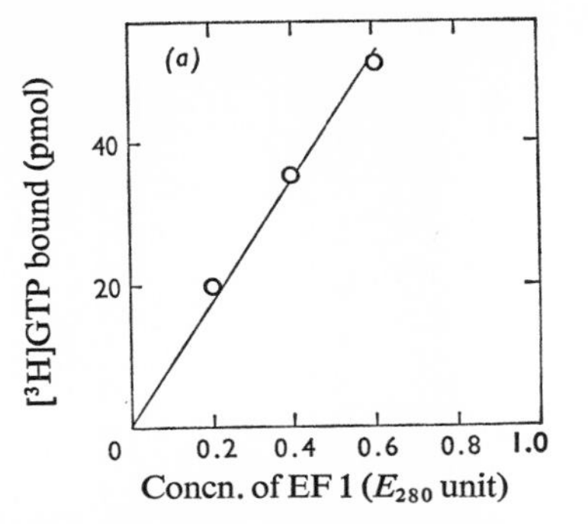


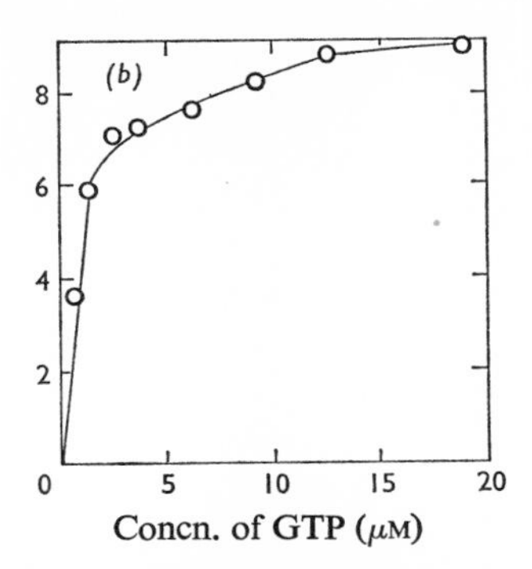


Fig. 3. *Effect of the concentration of EF* 1 *and GTP on the retention of* [$^{3}$H]*GTP on nitrocellulose membranes*

Assays were performed as described by Jerez *et al.* (1969).

The interaction between the factor found in extracts of wheat embryos and radioactive GTP can be easily measured because the factor–GTP complex is retained quantitatively on nitrocellulose filters. The formation of the factor–GTP complex depends on both the concentration of factor and the concentration of GTP (Fig. 3). The formation of this complex is extremely rapid; at 0°C the reaction is complete in less than 5 min. For this reason, our measurements always refer to a reaction that has reached equilibrium rather than to initial velocities.

The factor specifically binds GTP and GDP but does not interact with GMP or other nucleoside triphosphates. The following evidence strongly suggests that the GTP binding factor is wheat elongation factor 1.

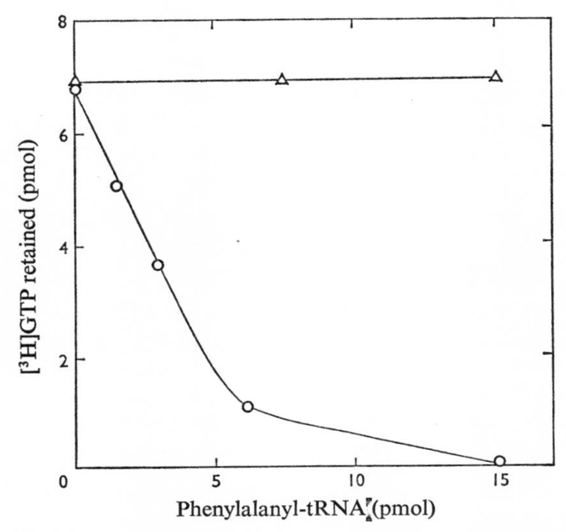


Fig. 4. *Effect of aminoacyl-tRNA and deacylated tRNA on the retention of* [$^{3}$*H*]*GTP on nitrocellulose filters in the presence of wheat EF* 1

The assays were performed as described by Jerez *et al.* (1969). ○, [$^{14}$C]Phenylalanyl-tRNA added; △, control tubes, containing the same preparation of tRNA that had been deacylated by hydrolysis at pH 10 for 30 min at 37°C.

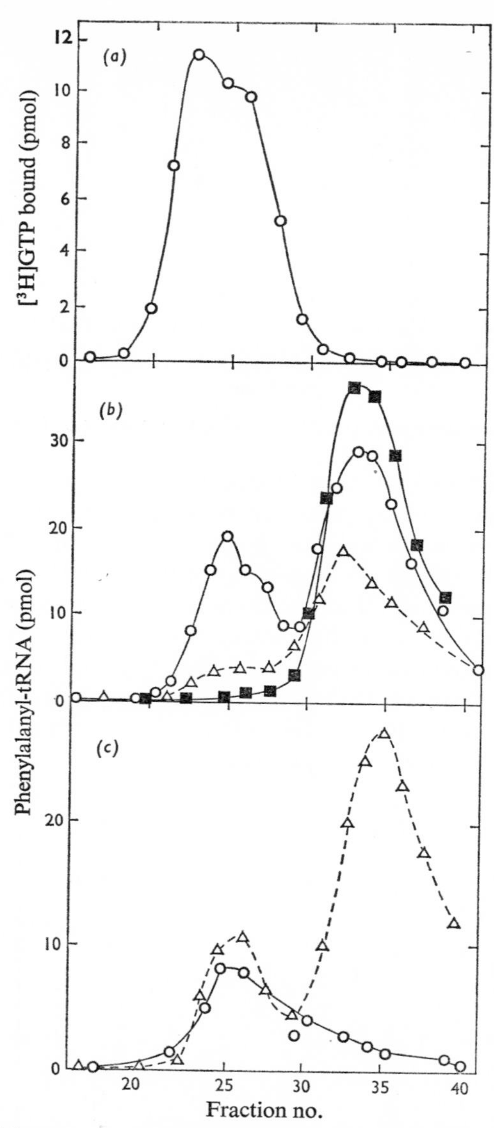


Fig. 5. *Isolation of the ternary complex, EF 1–GTP–aminoacyl-tRNA on a column of Sephadex G-75 and the binding of* [[14]C]*phenylalanyl-tRNA from the ternary complex to wheat ribosomes*

(*a*) Elution on a column (1.5 cm × 55 cm) of Sephadex G-75 equilibrated with 50 mM-cacodylate buffer, pH 7.1–120 mM-$NH_4Cl$–5 mM-$MgCl_2$–1 mM-glutathione–7.5 mM-GTP of EF 1. The factor was measured by its capacity to retain [[3]H]GTP on nitocellulose membranes. (*b*) ○, [[14]C]-Phenylalanyl-tRNA was incubated with EF 1 and GTP and subjected to column chromatography as above. Part of it elutes together with EF 1 forming the ternary complex, whereas the rest emerges as free phenylalanyl-tRNA. The measurements were done by precipitating samples with cold 5% (w/v) trichloroacetic acid and determining the amount of precipitable radioactivity. ■, Active EF 1 was replaced with EF 1 that had been heated at 42°C for 10 min. △, GTP was absent from the column buffer. (*c*) △, [[14]C]Phenylalanyl-tRNA was incubated with EF 1 and GTP as in (*b*, ○); ○, samples of the same Sephadex column fractions were incubated with 2 $E_{260}$ units of wheat ribosomes and 40 μg of poly(U) at 5 mM-$MgCl_2$ under the conditions of enzymic binding described in Fig. 1.

(*a*) The GTP-binding factor interacts specifically with aminoacyl-tRNA. This interaction can be conveniently visualized by the decrease in [$^3$H]GTP retained on the filter because the ternary complex, factor–GTP–aminoacyl-tRNA, is not bound to the nitrocellulose membranes. The titration of the binary complex, factor–GTP, by aminoacyl-tRNA gives a result as shown in Fig. 4. (*b*) The GTP-binding factor that has been partially purified by $(NH_4)_2SO_4$ precipitation, DEAE-cellulose and phospho-cellulose column chromatography stimulates the binding of the phenylalanyl-tRNA to ribosomes at low $Mg^{2+}$ concentrations and in the presence of GTP (Fig. 1).

Aminoacyl-tRNA binding to wheat ribosomes can also occur directly from the isolated ternary complex, factor–GTP–aminoacyl-tRNA. Fig. 5 shows that the ternary complex can be separated from free aminoacyl-tRNA by gel filtration on Sephadex G-100. By using [$^{14}$C]phenylalanyl-tRNA, it is apparent that in the presence of factor and GTP, part of the phenylalanyl-tRNA emerges earlier than free aminoacyl-tRNA indicating that it is complexed to a larger macromolecule.

Incubation of the fractions from such a Sephadex column with ribosomes and poly(U) at 5 mM-$Mg^{2+}$ shows that only the complexed aminoacyl-tRNA can bind the ribosomal particles. As expected, free aminoacyl-tRNA cannot bind non-enzymically at this low $Mg^{2+}$ concentration.

(*c*) The same partially purified preparations are able to complement elongation factor 2 (translocase), that has been purified by the procedure of Legocki & Marcus (1970). Both factors must be present to allow the overall elongation reaction. In this reaction, ribosomes and both supernatant factors catalyse the synthesis of polyphenylalanine starting from phenylalanyl-tRNA under the direction of poly(U). The complementary effect of EF 2 and the GTP-binding factor can be seen in Table 1.

*Effect of GTP on EF* 1

Sucrose-density-gradient analyses of the GTP-binding activity of preparations of EF 1 at different stages of purification are shown in Fig. 6. It is evident that the cruder preparations of the factor contain two species that are able to bind GTP and that these differ widely in molecular weight. There are some variations in the different density-gradient runs, but comparison of the sedimentation of known proteins under similar conditions allow us to calculate approximate molecular weights of 190000 for the heavier peak (peak II) and between 40000 and 50000 for the lighter form (peak I).

When the DEAE-cellulose EF 1 preparation is previously incubated with GTP or GDP at a concentration of 0.5 mM and sedimented in a sucrose density gradient that contains 0.05 mM-nucleotide throughout, the heavy peak dissappears. The addition of GMP or ATP under similar conditions does not have this effect (Fig. 7).

Table 1. *Complementation of EF* 1 *with wheat EF* 2 *in the synthesis of polyphenylalanine*

| Additions | Phenylalanine polymerized (pmol) |
|---|---|
| EF 1 | 0.50 |
| EF 2 | 0.58 |
| EF 1 + EF 2 | 7.90 |

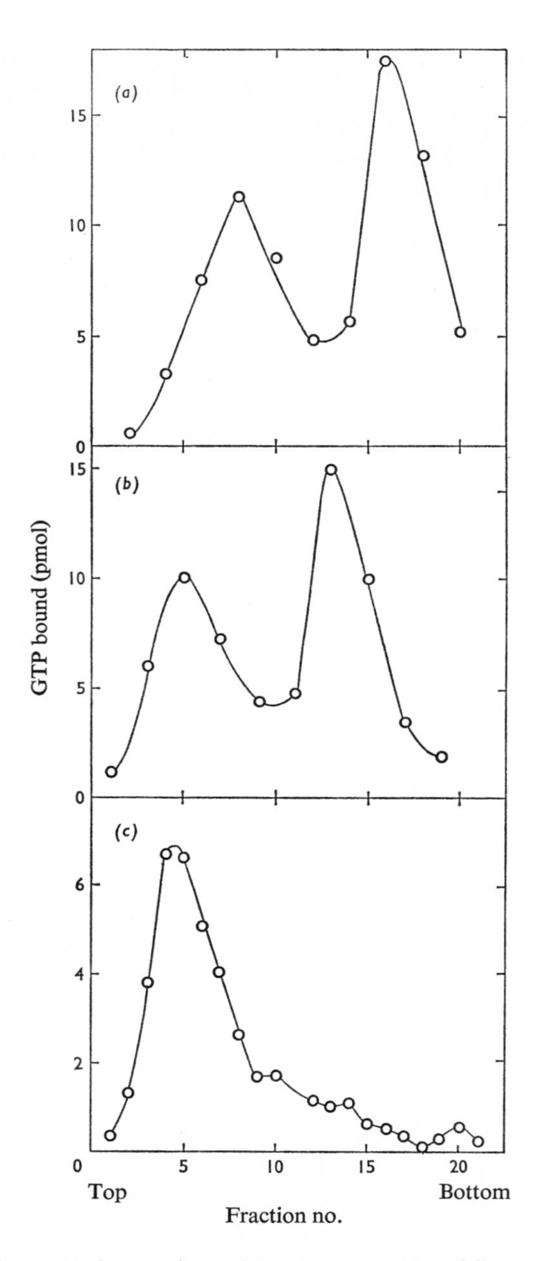


Fig. 6. *Sucrose-density-gradient analysis of EF* 1 *preparations at different stages of purification*

Density gradients of 5–20% sucrose were centrifuged for 18 h at 37000 rev./min in an SW-39 rotor. Fractions (6 drops) were collected and 50 μl samples from the different fractions were assayed for [$^3$H]GTP binding by the nitrocellulose-filter method. (*a*) Fraction (3 mg) obtained by precipitating (between 40 and 80% saturation) the high-speed supernatant of a wheat embryo extract. (*b*) EF 1 preparation (600 μg) purified by DEAE-cellulose chromatography. (*c*) EF 1 preparation (approx. 150 μg) that had been further purified by phospho-cellulose column chromatography.

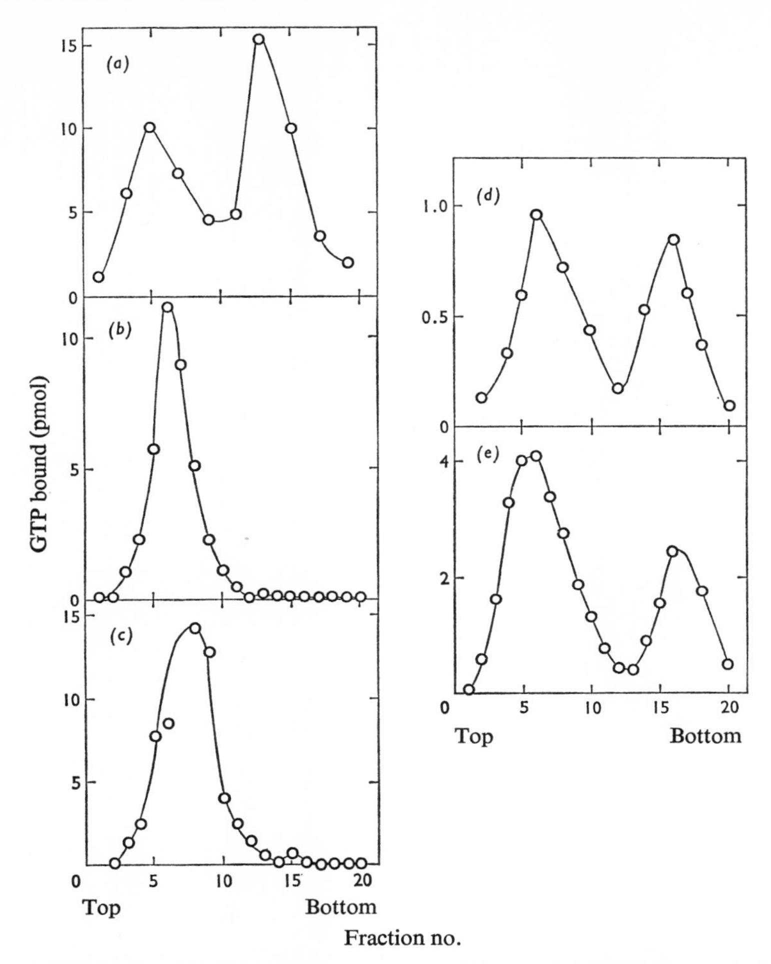


Fig. 7. *Effect of different nucleotides on the sucrose-density-gradient patterns of GTP-binding activity*

The gradients were run and the fractions were analysed as in Fig. 6. All experiments were done with approx. 600 $\mu$g of an EF 1 preparation that had been purified by DEAE- cellulose chromatography. The preparations were incubated for 10 min at 0°C with 0.5 mM-nucleotide before being introduced to the gradient. Each gradient also contained the same nucleotide (0.05 mM) throughout the tube. (*a*) Control tube, nucleotide absent; (*b*) GTP; (*c*) GDP; (*d*) GMP; (*e*) ATP.

The disappearance of peak II is apparently not due to its inactivation since resedimentation of isolated peak II in the presence of GTP causes peak I to appear whereas, in the absence of the nucleotides, that preparation maintains its heavy sedimentation characteristics (Fig. 8). Also peak II is considerably more stable to heat inactivation than peak I (Fig. 9).

The most noteworthy difference between peak I and peak II is the reactivity with aminoacyl-tRNA and GTP to form the ternary complex. By using the nitrocellulose assay described above, but expressing the ordinate of the graph as ternary complex formed instead of decrease in [$^3$H]GTP retained on the filter, Fig. 10 shows that peak I reacts very readily with aminoacyl-tRNA. Peak II,

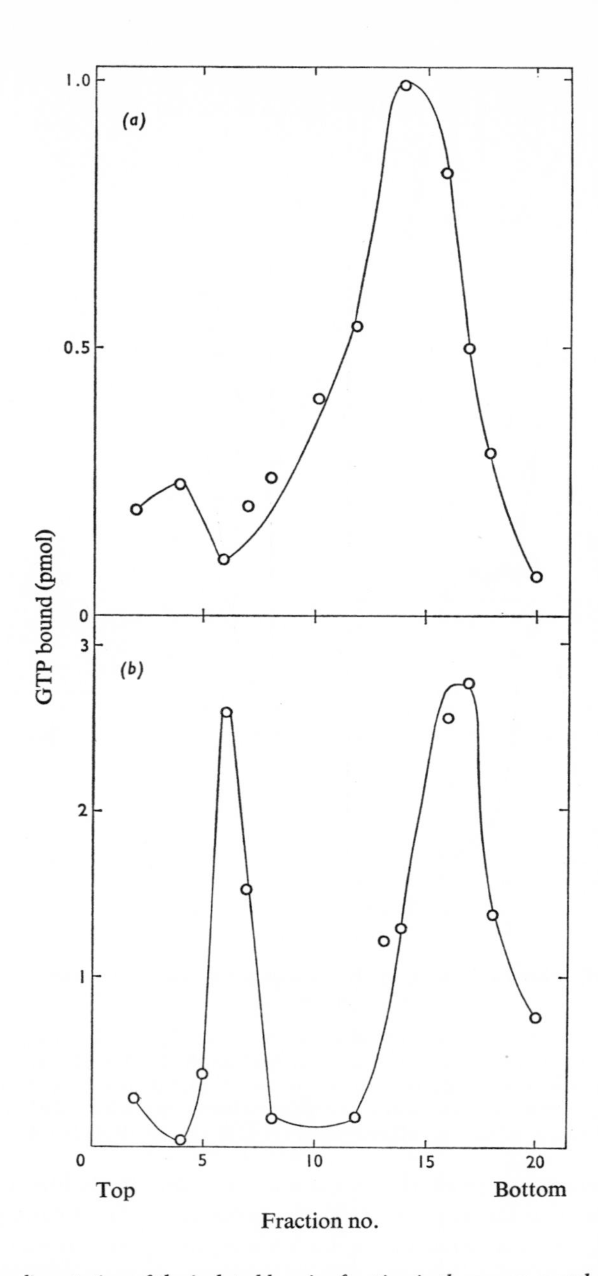


Fig. 8. *Resedimentation of the isolated heavier fraction in the presence and absence of GTP*

Three sucrose-density-gradient tubes were run in the absence of added nucleotides in order to fractionate the light and heavy peaks of GTP-binding activity present in the DEAE-cellulose preparation of EF 1. The active heavy fractions were pooled and precipitated with 80%-saturated $(NH_4)_2SO_4$. (*a*) Approx. 50 $\mu$g of the fraction obtained was added to a similar sucrose density gradient; (*b*) an identical sample was incubated with GTP and sedimented in a GTP-containing gradient as detailed in Fig. 7. Fractions were analysed as before for GTP-binding activity.

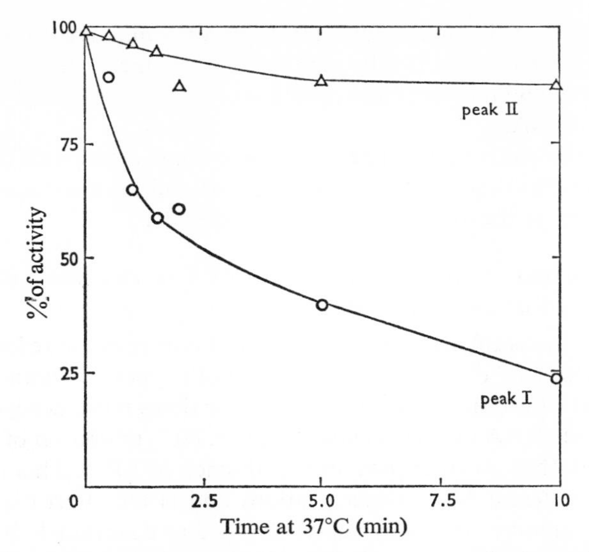


Fig. 9. *Inactivation of both GTP-binding peaks from sucrose density gradients at 37°C*

The light (peak I) and heavy (peak II) peaks were obtained by subjecting fractions, separated by DEAE-cellulose chromatography, to sucrose-density-gradient centrifugation followed by precipitation with 80%-saturated $(NH_4)_2SO_4$. Bovine serum albumin was added to bring the protein concentration of both resuspended fractions to 1 mg/ml. The fractions were dissolved in 10 mM-Tris–HCl buffer, pH 7.5–1 mM-β-mercaptoethanol. Samples were withdrawn at the times indicated and chilled to 0°C before being asssayed for GTP-binding activity.

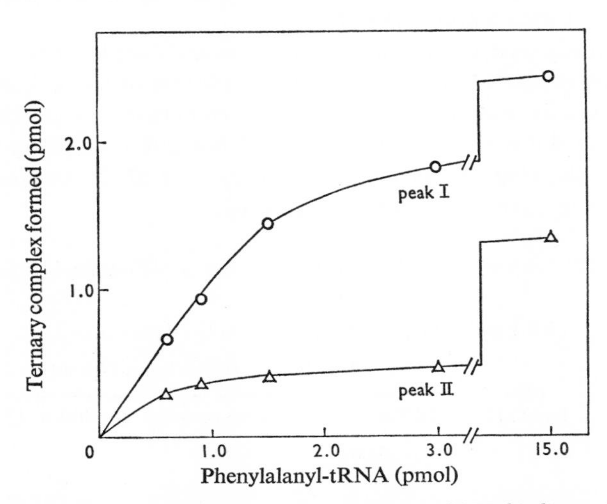


Fig. 10. *Formation of the ternary complex by both fractions of GTP-binding activity obtained from sucrose density gradients*

Peaks I and II isolated by sucrose-density-gradient centrifugation were made to react with [$^3$H]GTP and [$^{14}$C]phenylalanyl-tRNA as described in Fig. 4. Both fractions in the absence of aminoacyl-tRNA were capable of binding approx. 2.6 pmol of [$^3$H]GTP in the above assay. The decrease of [$^3$H]GTP bound caused by the addition of phenylalanyl-tRNA was plotted as amount of ternary complex formed.

however, requires much higher concentrations of aminoacyl-tRNA to react. Thus, the conversion of peak II into peak I caused by increasing concentrations of GTP or GDP would represent an activation of the capacity of EF 1 to interact with aminoacyl-tRNA.

We are presently studying this phenomenon in some detail with regard to the nature of the proteins that interact to form peak II and the mechanism by which GTP and GDP cause the dissociation of this aggregate.

*Specificity of enzymic binding of aminoacyl-tRNA to ribosomes is due to the interaction of EF* 1 *with aminoacyl-tRNA*

The difference in specificity of the enzymic and non-enzymic reactions can be shown by the effect of deacylated tRNA on both types of aminoacyl-tRNA binding to wheat ribosomes. The addition of four times more deacylated tRNA than phenylalanyl-tRNA (Table 2) causes approx. 80% inhibition of the binding of phenylalanyl-tRNA to ribosomes in the absence of EF 1. This inhibition is constant at the different $Mg^{2+}$ concentrations and is the effect expected if we assume that the affinities of phenylalanyl-tRNA and deacylated tRNA for the wheat ribosomes are similar.

The results obtained in a similar experiment done in the presence of EF 1 and GTP are quite different. At low $Mg^{2+}$ concentrations, at which the reaction is almost completely dependent on EF 1, there is minimal inhibition by deacylated tRNA. At higher $Mg^{2+}$ concentrations, as the reaction loses its enzymic character, the specificity is gradually lost and the inhibition approached the theoretical 80% value.

The following results show that the basis for this specificity resides in the interaction of EF 1 with aminoacyl-tRNA.

As we have already seen, the interaction of the binary complex, EF 1–GTP, with aminoacyl-tRNA is easily assayed by the nitrocellulose-filtration method. This method also serves to study the specificity requirements in this interaction. For instance, from Fig. 4 it is clear that unacylated tRNA fails completely to interact with EF 1, thus showing the incapacity of deacylated tRNA to compete with aminoacyl-tRNA in enzymic binding to ribosomes.

Table 2. *Effect of tRNA on enzymic and non-enzymic binding of phenylalanyl-tRNA at different* $Mg^{2+}$ *concentrations*

−, EF 1 and GTP absent; +, EF 1 and GTP present.

| $Mg^{2+}$ (mM) | EF 1 and GTP | Phenylalanyl-tRNA bound (pmol) | | |
|---|---|---|---|---|
| | | tRNA absent | tRNA present | Inhibition (%) |
| 4 | − | 0.00 | 0.00 | — |
| 7.5 | − | 0.68 | 0.11 | 84 |
| 10 | − | 1.42 | 0.36 | 75 |
| 15 | − | 2.07 | 0.59 | 73 |
| 20 | − | 2.55 | 0.60 | 76 |
| 4 | + | 1.60 | 1.49 | 7 |
| 7.5 | + | 2.76 | 1.93 | 30 |
| 10 | + | 2.77 | 1.69 | 39 |
| 15 | + | 2.81 | 1.32 | 53 |
| 20 | + | 2.83 | 1.10 | 61 |

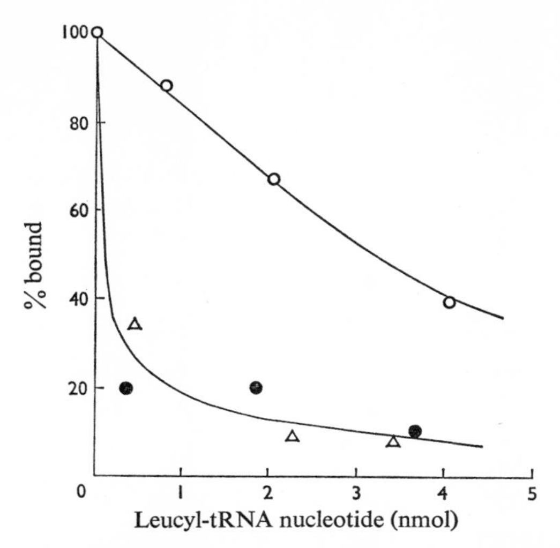


Fig. 11. *Ability of denatured and renatured leucyl-tRNA of yeast to react with EF 1–GTP*

Denatured leucyl-tRNA (○), renatured leucyl-tRNA (●) and native (△) yeast leucyl-tRNA were reacted with EF 1 and GTP as described in Fig. 4. Denatured yeast leucyl-tRNA was separated from the native species by Sephadex G-100 chromatography (Lindahl *et al.*, 1967), divided in two portions and dissolved in (*a*) 10 mM-cacodylate–1 mM-EDTA buffer, pH 7.0, and (*b*) 10 mM-cacodylate–1 mM-EDTA buffer, pH 7.0–10 mM-$MgCl_2$. The two solutions were heated at 60°C for 2 min to (*a*) denature and (*b*) renature the leucyl-tRNA. After cooling to 0°C the samples were assayed for reactivity with the EF 1–GTP complex. Leucyl-tRNA nucleotide was calculated on the basis of [$^{14}$C]leucine precipitable with cold 5% (w/v) trichloroacetic acid and by using a chain length of 85 nucleotides per $tRNA^{Leu}$ (reproduced from Jerez *et al.*, 1969).

This assay has also served to establish the importance of the amino group of the esterified amino acid in the interaction with wheat EF 1. Acetylation of the amino group of phenylalanyl-tRNA or deamination of the same aminoacyl-tRNA yielded esterified tRNA species that did not react with the wheat elongation factor (Jerez *et al.*, 1969). Recently, a noteworthy difference has been reported between the plant and the bacterial elongation factors since *E. coli* EF Tu can

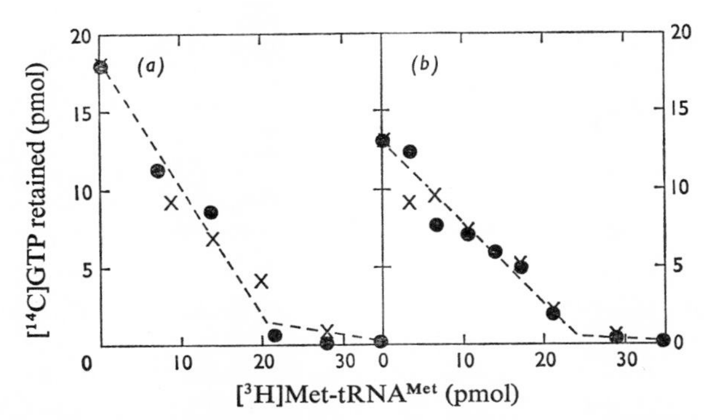


Fig. 12. *Interaction of wheat methionyl-tRNA$_m^{Met}$ (a) and methionyl-tRNA$_i^{Met}$ (b) with EF 1 and GTP measured by the nitrocellulose-filter-retention method*

The assays were done as described in Allende *et al.* (1973) (reproduced from Richter *et al.*, 1971). ● and × constitute averages of duplicates of two separate experiments.

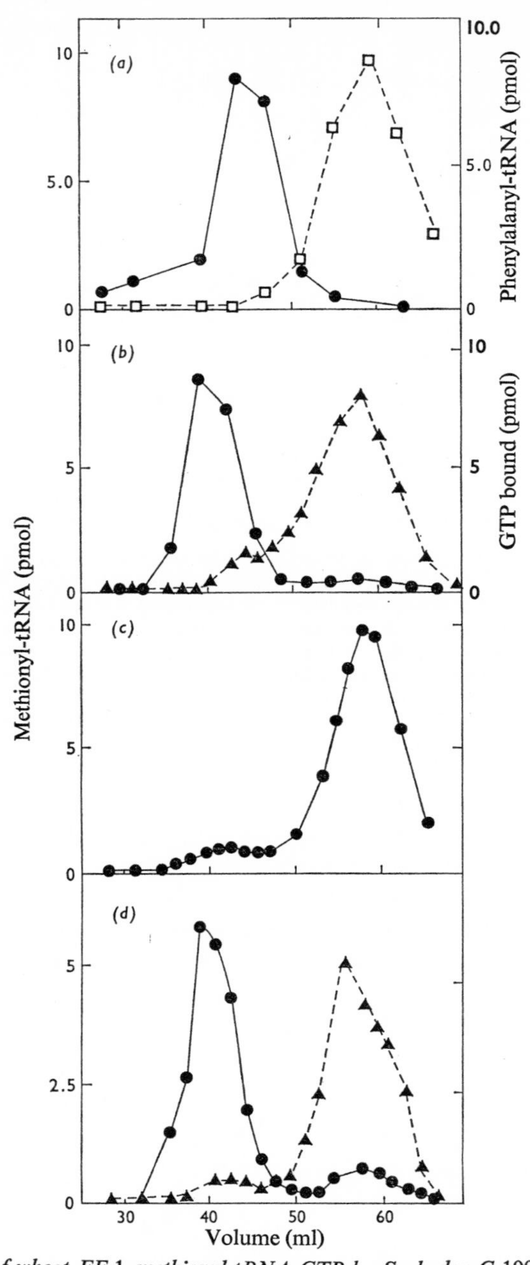


Fig. 13. *Isolation of wheat EF 1–methionyl-tRNA–GTP by Sephadex G-100 column chromatography*

(*a*) Separation of wheat EF 1, measured by its capacity to bind [$^{3}$H]GTP (●), and of [$^{14}$C]-phenylalanyl-tRNA measured by precipitation with 5% (w/v) trichloroacetic acid. (*b*) Elution of wheat [$^{14}$C]methionyl-tRNA$_m$ (or [$^{14}$C]methionyl-tRNA$_2$): ●, incubated with EF 1 and GTP; ▲, GTP omitted from the incubation mixture and elution buffer. (*c*) Elution of [$^{14}$C]-methionyl-tRNA$_i$ or ([$^{14}$C]methionyl-tRNA$_1$) incubated and eluted with the complete system. (*d*) Elution of an incubation to form ternary complex in which [$^{14}$C]methionyl-tRNA$_i$ (▲) and [$^{3}$H]methionyl-tRNA$_m$ or methionyl-tRNA$_2$ (●) were added together (reproduced from Tarragó *et al.*, 1970).

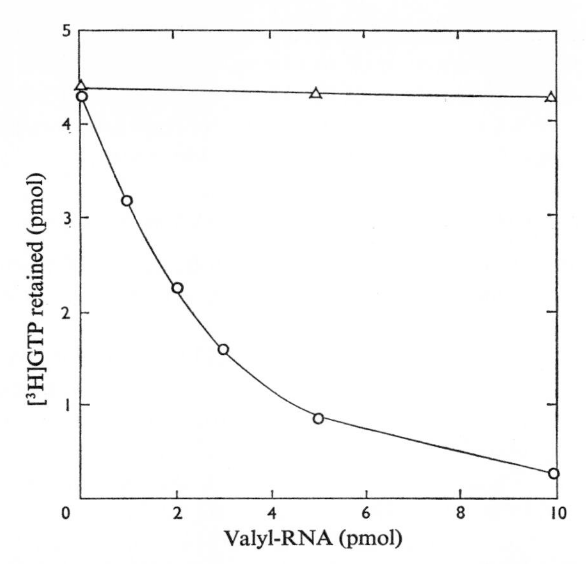


Fig. 14. *Interaction of wheat EF* 1 *with turnip yellow-mosaic virus RNA esterified with valine* (○)

The reaction was carried out as described for Fig. 4 but with valyl-RNA from turnip yellow-mosaic virus instead of phenylalanyl-tRNA. Of the molecules of viral RNA 70% were acylated with [$^{14}$C]valine by a partially purified *E. coli* valyl-tRNA synthetase. The same preparation of turnip yellow-mosaic virus RNA that had not been esterified with valine showed no formation of ternary complex (△).

react with the deaminated product of phenylalanyl-tRNA, phenyl-lactyl-tRNA, (Fahnestock *et al.*, 1972).

In addition, the specificity in the interaction between EF 1 from wheat and aminoacyl-tRNA is related to the fine structure of the tRNA involved. Denatured yeast leucyl-tRNA, for instance, reacts very poorly with wheat EF 1 as compared with the native or renatured leucyl-tRNA from the same species (Fig. 11).

Physiologically more noteworthy is the capacity of wheat EF 1 to discriminate between the two major species of $tRNA^{Met}$ found in wheat embryos. The nitrocellulose-membrane assay indicates that both methionyl-tRNA species can react with wheat EF 1 (Fig. 12).

The slopes of the curves, 0.90 for methionyl-$tRNA_m$ and 0.59 for methionyl-$tRNA_i$, however, seem to indicate a lower affinity of EF 1–GTP for methionyl-$tRNA_i$ than for methionyl-$tRNA_m$ (where the subscripts m and i indicate the tRNA species for methionine and for initiator methionine respectively).

The difference in the interaction of the two species, however, is much more evident when the Sephadex G-100 method for isolation of the ternary complex is used. Fig. 13(*b*) shows that methionyl-$tRNA_m$ can completely bind to EF 1 to form a ternary complex in the presence of GTP. Figs. 13(*c*) and 13(*d*) show that methionyl-$tRNA_i$ does not form a complex that is stable enough to be isolated by this method (Tarragó *et al.*, 1970; Richter *et al.*, 1971).

The significance of these findings lies in the fact that data from other laboratories (Leis & Keller, 1970) have indicated that the methionyl-$tRNA_i$ species acts as an initiator of protein synthesis in wheat, and that this species transfers its methionyl residue exclusively to the amino terminal positions of nascent peptide chains. The imperfect interaction of methionyl-tRNA$_i$ with EF 1 explains why this aminoacyl-tRNA cannot enter the ribosomal A site through the EF 1–GTP–aminoacyl-tRNA complex and thus why its methionyl residue cannot be transferred into the internal positions of the peptide chains.

*Interaction of EF* 1 *with aminoacylated RNA from plant viruses*

The RNA from two important plant viruses, tobacco mosaic virus and turnip yellow-mosaic virus, can be enzymically aminoacylated at its 3′ terminus (Yot *et al.*, 1970; Litvak *et al.*, 1973).

Fig. 14 shows that the binary complex, EF 1–GTP, reacts very well with valyl-RNA from turnip yellow-mosaic virus but does not react with the unacylated viral RNA. The same is true for histidyl-RNA from tobacco mosaic virus.

This finding may be of some interest in the light of the recent discovery that identified *E. coli* EF Tu as one of the host factors required for the replication of the RNA of bacteriophage Q $\beta$ (Blumenthal *et al.*, 1972).

In any case, the interaction of EF 1 with the aminoacylated RNA species of plant viruses indicates that the structures present in the 3′ terminus of the viral RNA are similar enough to tRNA species to be recognized as such by this protein

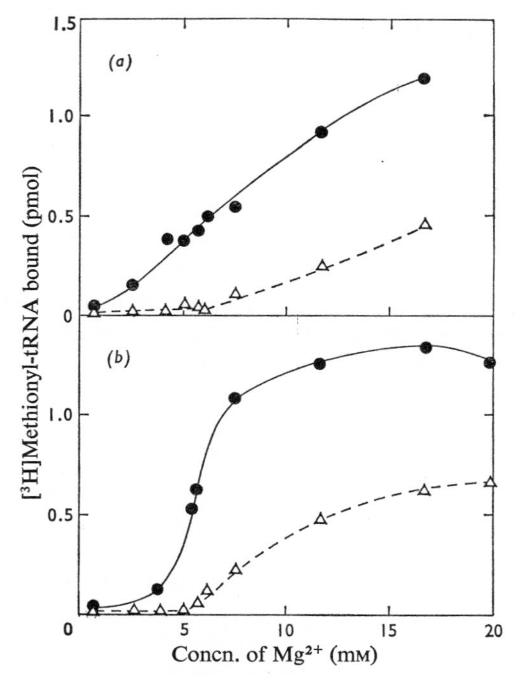


Fig. 15. *Binding of wheat methionyl-tRNA species to wheat and E. coli ribosomes at different $Mg^{2+}$ concentrations*

The binding reaction was carried out as described in Fig. 5, but in the presence of 0.18 $E_{260}$ units of ApUpG. (*a*) Washed wheat ribosomes (2 $E_{260}$ units); (*b*) *E. coli* ribosomes (2.2 $E_{260}$ units). ●, Methionyl-tRNA$_i$; △, methionyl-tRNA$_m$ (reproduced from Monasterio *et al.*, 1971).

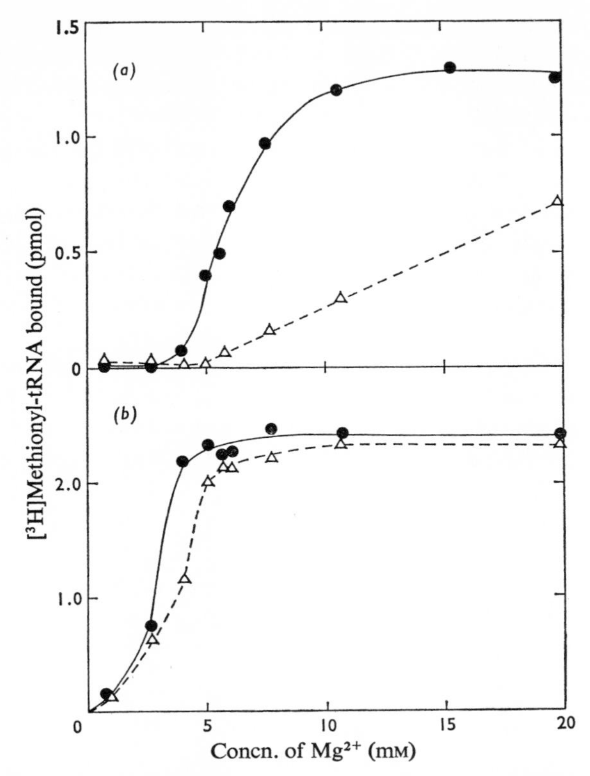


Fig. 16. *Binding of E. coli methionyl-tRNA species to wheat and E. coli ribosomes*

Procedures and reaction mixture were as described for Fig. 15. (*a*) *E. coli* ribosomes (2.2 $E_{260}$ units); (*b*) wheat ribosomes (2.0 $E_{260}$ units). ●, Methionyl-tRNA$_f$; △, methionyl-tRNA$_m$ (reproduced from Monasterio *et al.*, 1971).

factor as well as by the aminoacyl-tRNA synthetase and the nucleotidyl transferase from *E. coli* (Litvak *et al.*, 1970). It seems important to investigate the possible physiological role of the aminocylation of viral RNA species.

*Specificity of the ribosomes for the initiator methionyl-tRNA*

Although the results above have been presented to show that the specificity of binding of aminoacyl-tRNA to the ribosomal A site resides in the discriminatory capacity of elongation factor 1, we have observed an instance in which the ribosomal particle itself is able to differentiate between two apparently very similar aminoacyl-tRNA species.

By assaying the binding of the two major species of methionyl-tRNA to washed wheat ribosomes in the presence of the trinucleotide ApUpG, it was found that the particles bound preferentially the initiator species at the lower $Mg^{2+}$ concentrations (Fig. 15*a*). *E. coli* ribosomes, washed with 0.5 M-$NH_4Cl$, also bind preferentially the wheat initiator methionyl-tRNA over the non-initiating species (Fig. 15*b*).

It is noteworthy that a similar preferential binding of *E. coli* methionyl-tRNA$_f$

(where the subscript f indicates the tRNA species for initiator formyl-methionine) as compared with methionyl-$tRNA_m$ is observed with the *E. coli* ribosomes (Marcker *et al.*, 1966). Wheat particles, however, bind both bacterial species with similar efficiency (Figs. 16*a* and 16*b*).

It must be pointed out that the preferential binding of the ribosomes for the initiator methionyl-tRNA species differs from the effect caused by the initiation factors of bacteria and of wheat described in this Symposium by Marcus *et al.* (1973). The bacterial factors only catalyse the binding of the formylated methionyl-$tRNA_f$ and require GTP whereas Marcus *et al.*'s (1973) wheat initiation factors operate at a much lower $Mg^{2+}$ concentration (1–2mM) and also require GTP.

Fig. 17 shows that the 40S subunits of wheat ribosomes are responsible for the preferential binding of the methionyl-$tRNA_i$ and that the 60S subunits do not bind at all the aminoacyl-tRNA species. The wheat ribosomal subunits prepared by dissociating wheat particles with 0.5M-$NH_4Cl$–10mM-$MgCl_2$ are active in protein synthesis (Table 3).

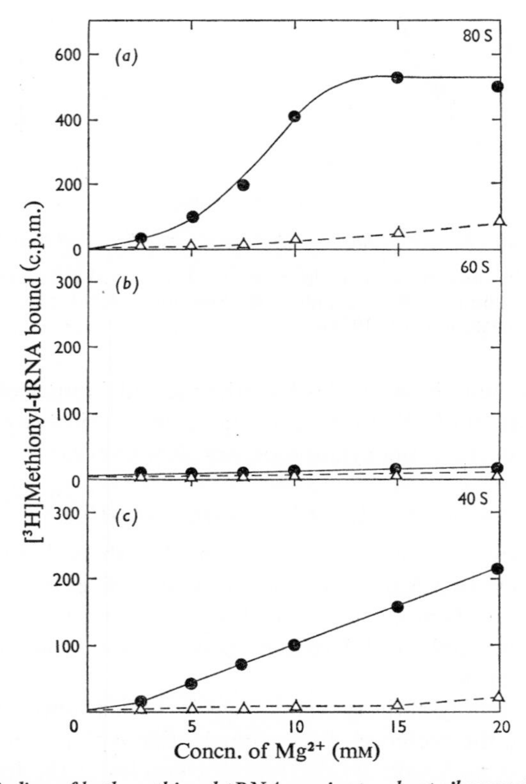


Fig. 17. *Binding of both methionyl-tRNA species to wheat ribosomal subunits*

The ribosomal subunits were prepared by sucrose-density-gradient fractionation as indicated in the Methods section. The amount of ribosomal particles used was 0.26 $E_{260}$ unit of 40S subunits, 0.33 $E_{260}$ unit of 60S subunits and 1.6 $E_{260}$ units of 80S particles. In all experiments 2.25pmol of [$^3$H]methionyl-$tRNA_i$ (●) and 3.2pmol of [$^3$H]methionyl-$tRNA_m$ (△) were used.

Table 3. *Polyphenylalanine synthesis activities of wheat ribosomal particles*

| System | Phenylalanine polymerized (pmol) |
|---|---|
| 80S wheat ribosomes (1.80 $E_{260}$) | 4.40 |
| No ribosomes | 0.00 |
| 60S subunits (0.33 $E_{260}$) | 0.19 |
| 40S subunits (0.13 $E_{263}$) | 0.12 |
| 60S + 40 S subunits (0.33 + 0.13 $E_{260}$) | 2.10 |

## Comments

Most features of the binding of aminoacyl-tRNA to wheat ribosomes are similar to those found in bacterial systems. The most noteworthy difference is the absence in the eukaryotic system of a factor with the properties of EF Ts. The finding that wheat EF 1 exists in at least two forms of different molecular weight and of different reactivity with aminoacyl-tRNA may indicate that eukaryotic EF 1 factors can interact with other proteins that regulate their activity. No evidence has been obtained, however, that indicates that any of the EF 1 forms contain a factor which, like EF Ts, is essential for the elongation process.

The specific interaction of EF 1 with aminoacyl-tRNA in the presence of GTP is of interest in several respects. Since the elongation factor recognizes all different aminoacyl-tRNA species, with the exception of the poorly interacting initiator methionyl-tRNA, the interaction between the two macromolecules can be used to study the structural characteristics common to all aminoacyl-tRNA species. Such a study can complement the widespread investigation of the specific recognition site of aminoacyl-tRNA species for their respective aminoacyl-tRNA synthetases.

Also of note in the specific interaction of EF 1 and GTP with aminoacyl-tRNA is the effect that this may have in the enzymic formation of aminoacyl-tRNA species. Since the reaction catalysed by the aminoacyl-tRNA synthetases is fully reversible and since the EF 1 and GTP react with aminoacyl-tRNA, but not with unesterified tRNA, it is conceivable that the presence of the elongation factor and GTP may serve to pull the aminoacylation of tRNA species. It seems important to start investigating the effects that sequential enzymes have on each other's efficiency since it may help us to understand the large differences that favour the physiological processes over those obtained in purified systems *in vitro*.

This work was supported in part by CONICYT, Chile, by the University of Chile Committee for Research. A. T. and O. M. are pre-doctoral fellows of CONICYT, Chile.

## References

Allende, J. E. (1969) in *Techniques in Protein Biosynthesis* (Campbell, P. N. & Sargent, J. R., eds.), vol. 2, pp. 55–100, Academic Press, London and New York

Allende, J. E., Seeds, N. W., Conway, T. W. & Weissbach, H. (1967) *Proc. Nat. Acad. Sci. U.S.* **58**, 1566–1573

Allende, J. E., Tarragó, A., Monasterio, O., Gatica, M., Ojeda, J. & Matamala, M. (1973) *Lat. Amer. Symp. Protein Syn. Nucl. Acids 11th, La Plata*; in *Basic Life Sciences* (Hollaender, A., ed.), vol. 1, pp. 411–427, Plenum Press, New York and London

Blumenthal, J., Landers, T. A. & Weber, K. (1972) *Proc. Nat. Acad. Sci. U.S.* **69**, 1313–1317

Fahnestock, S., Weissbach, H. & Rich, A. (1972) *Biochim. Biophys. Acta* **269**, 62–66

Jerez, C., Sandoval, A., Allende, J. E., Henes, C. & Ofengand, J. (1969) *Biochemistry* **8**, 3006–3014
Johnston, F. B. & Stern, H. (1957) *Nature* (*London*) **179**, 160–161
Legocki, A. B. (1973) *Biochem. Soc. Symp.* **38**, 57–76
Legocki, A. B. & Marcus, A. (1970) *J. Biol. Chem.* **245**, 2814–2818
Leis, J. P. & Keller, E. B. (1970) *Fed. Proc. Fed. Amer. Soc. Exp. Biol.* **29**, 468 (Abstr.)
Lindahl, T., Adams, A. & Fresco, J. R. (1967) *J. Biol. Chem.* **242**, 3129–3134
Litvak, S., Carré, D. S. & Chapeville, F. (1970) *FEBS Lett.* **11**, 316–319
Litvak, S., Tarragó, A., Tarragó-Litvak, L. & Allende, J. E. (1973) *Nature* (*London*) *New Biol.* **241**, 88–90
Marcker, K. A., Clark, B. F. C. & Anderson, J. S. (1966) *Cold Spring Harbor Symp. Quant. Biol.* **31**, 279–285
Marcus, A., Weeks, D. F. & Seal, S. N. (1973) *Biochem. Soc. Symp.* **38**, 97–109
Monasterio, O., Tarragó, A. & Allende, J. E. (1971) *J. Biol. Chem.* **246**, 1539–1541
Richter, D., Lipmann, F., Tarragó, A. & Allende, J. E. (1971) *Proc. Nat. Acad. Sci. U.S.* **68**, 1805–1809
Tarragó, A., Monasterio, O. & Allende, J. E. (1970) *Biochem. Biophys. Res. Commun.* **41**, 765–773
Yot, P., Pinck, M., Haenni, A. L., Duranton, H. M. & Chapeville, F. (1970) *Proc. Nat. Acad. Sci. U.S.* **67**, 1345–1352

*Biochem. Soc. Symp.* (1973) **38**, 97–109
*Printed in Great Britain*

# Protein Chain Initiation in Wheat Embryo

By ABRAHAM MARCUS, DONALD P. WEEKS and SAMARENDRA N. SEAL

*Institute for Cancer Research, Fox Chase, Philadelphia, Pa.* 19111, *U.S.A.*

## Synopsis

Ribosomes and supernatant of dry wheat embryos support protein synthesis *in vitro* catalysed by exogenous mRNA. The wheat-embryo supernatant supplies, in addition to tRNA and tRNA synthetases, two peptide-chain elongation factors (EF 1 and EF 2) and at least two initiation factors (C and D). Peptide chain initiation requires the two initiation factors, ribosomes, mRNA and both ATP and GTP. The first reaction involved in the initiation sequence occurs on the 40S ribosomal subunit and results in the formation of a 50S complex containing the ribosomal subunit and an initiator species of methionyl-tRNA. This complex subsequently combines with the 60S ribosomal subunit, forming an 80S ribosome–mRNA–methionyl-tRNA complex. Three inhibitors, aurintricarboxylic acid, pactamycin and 2-(4-methyl-2,6-dinitroanilino)-*N*-methylpropionamide, are shown to be specific for protein chain initiation. Use of these inhibitors allows the definition of several partial reactions in the initiation process.

## Introduction

The process of protein biosynthesis can be divided into three phases: chain initiation, in which the ribosome and mRNA become aligned so as to allow the correct translation of the particular messenger; chain elongation, in which the mRNA is translated; finally, chain termination in which the peptide product is released from the ribosome [see Lucas-Lenard & Lipmann (1971) and Boulter (1970) for recent reviews]. Chain elongation has been studied in a number of cell-free systems from both bacterial and eukaryotic sources. Chain initiation, however, has been described in detail only in extracts of bacteria, wheat embryos and reticulocytes. Our interest in this aspect of protein biosynthesis arose as a consequence of studies of seed embryo germination with embryos isolated from wheat seed. It could be shown that one of the earliest processes activated by water uptake was the 'turning on' of protein biosynthesis (Marcus & Feeley, 1964), and that a major aspect of this 'turn on' was the attachment of a preformed messenger (Marcus & Feeley, 1966; Weeks & Marcus, 1970) to the endogenous messenger-free ribosomes, i.e. a chain-initiation reaction. Substitution of one of several plant viral RNA species for the endogenous embryo messenger provided a convenient system for the further analysis of the initiation reaction (Marcus *et al.*, 1968).

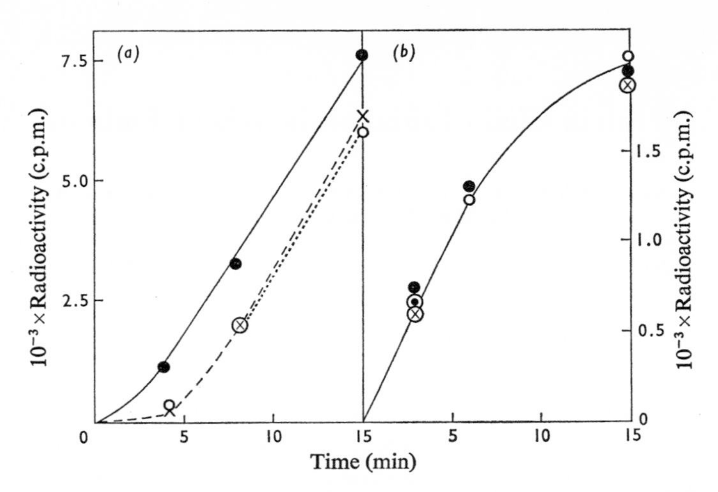


Fig. 1. *Comparison of kinetics of incorporation of* [$^{14}$*C*]*leucine by* (*a*) *the TM virus RNA and* (*b*) *the polyribosomal amino acid incorporating systems*

A 6min preliminary incubation was carried out either at 30°C with ATP present (●), at 30°C with ATP absent (×), or at 0°C with ATP present (○). (*a*) For the TM virus RNA system, the preliminary incubation contained (in 0.34ml): 10$\mu$mol of Tris–acetate buffer, pH 8.1, 0.4$\mu$mol of ATP (where applicable), 3.2$\mu$mol of creatine phosphate, 16$\mu$g of creatine phosphate kinase, 0.01$\mu$mol of GTP, ribosomes (twice-washed, 200$\mu$g of RNA; Marcus, 1972), 10$\mu$g of TM virus RNA (Marcus, 1972), 0.9$\mu$mol of dithiothreitol, 17.5$\mu$mol of KCl, 0.84$\mu$mol of magnesium acetate and 0.12ml of S-100 DEAE (Marcus *et al.*, 1968). (*b*) The preliminary incubation for the polyribosome system contained similar amounts of Tris–acetate buffer, creatine phosphate, creatine phosphate kinase, dithiothreitol, GTP, KCl and S-100 DEAE, 17$\mu$mol of $NH_4Cl$, 1.92$\mu$mol of magnesium acetate, and polyribosomes (48$\mu$g of RNA) (Weeks & Marcus, 1969). After the preliminary incubation, 16$\mu$g of tRNA and 0.125$\mu$Ci of [$^{14}$C]leucine were added to a final volume of 0.40ml to all vessels and ATP (0.4$\mu$mol) to those vessels lacking ATP during the preliminary incubation. Incubations were then continued at 30°C for the times indicated and the radioactivity incorporated into hot trichloroacetic acid-insoluble material was determined (Marcus *et al.*, 1966).

## Ribosome–Messenger Attachment Reaction

An early observation with the wheat embryo system with TM virus RNA (tobacco mosaic virus RNA) as messenger was that amino acid incorporation occurred with a 5–8min lag and that the lag could be removed by preincubation of ribosomes, supernatant and messenger in the presence of ATP (see Fig. 1). In contrast with the TM virus RNA reaction, poly(U)-dependent incorporation of phenylalanine, as well as amino acid incorporation by polyribosomes (isolated from germinated embryos), was linear from zero time and was unaffected by pre-incubation (Marcus, 1970*a*). When the respective reactions were studied by following the incorporation of amino acid from aminoacyl-tRNA, the TM virus RNA reaction required both ATP and GTP, whereas both the poly(U) and the polyribosome reactions required only GTP (Table 1). The most plausible explanation for these observations is that the TM virus RNA system represents a reaction in which ribosomes attach to a ‘natural’ messenger in a rate-limiting step. The polyribosome and poly(U) reactions, on the other hand, utilize respectively a preformed and a non-specific ribosome–messenger attachment. By implication,

Table 1. *Nucleotide requirement for incorporation of aminoacyl-tRNA*

The conditions for amino acid incorporation by the TM virus RNA system were as in Fig. 1 except that after the preliminary incubation 16 μg of a mixture of eight $^{14}C$-labelled aminoacyl-tRNA species (7894 c.p.m.) (Marcus, 1970*a*) and 400 nmol of eight unlabelled amino acids were added to start the reaction. The mixture of eight $^{14}C$ labelled aminoacyl-tRNA species and unlabelled amino acids were also added to the polyribosome system which contained components similar to the polyribosome system of Fig. 1. The components of the poly(U) assay were as for the TM virus RNA system with the following exceptions: 10 μg of poly(U) was substituted for TM virus RNA, the salt concentrations were 65 mM-KCl and 4.1 mM-magnesium acetate, and 30 μg of [$^{14}C$]phenylalanyl-tRNA (2054 c.p.m.) and 400 nmol of [$^{12}C$]phenylalanine were added. No preliminary incubation was performed in the polyribosome and poly(U) assays. All amino acid incorporation incubations were for 10 min at 30°C.

| Messenger RNA source | Addition (μmol) | Amino acid incorporated (c.p.m.) |
|---|---|---|
| Poly(U) ([$^{14}C$]phenylalanyl-tRNA) | None | 32 |
| | GTP (0.1) | 1515 |
| | ATP (0.1) | 52 |
| | ATP + GTP | 1472 |
| TM virus RNA | None | 21 |
| | GTP (0.1) | 38 |
| | ATP (0.1) | 379 |
| | ATP + GTP | 985 |
| Polyribosome | None | 169 |
| | GTP (0.1) | 1220 |
| | ATP (0.1) | 199 |
| | ATP + GTP | 1231 |

both the ATP and the supernatant components required (during preincubation) for removal of the kinetic lag are functioning in the messenger–ribosome attachment reaction.

Direct evidence that such a ribosome–mRNA complex is indeed formed in the TM virus RNA reaction, is obtained by carrying out the preincubation reaction in the presence of radioactive TM virus RNA and examining the products by sucrose-density-gradient centrifugation (Fig. 2). A new radioactive component is formed which sediments slightly faster than the 80S ribosomes. Formation of this new component requires the presence of ATP as well as the supernatant components previously shown to be necessary for removal of the kinetic lag (Marcus, 1970*b*).

## Initiation Factors

The earlier experiments provided indirect evidence suggesting that the supernatant fraction contained initiation factors. By the use of aurintricarboxylic acid a specific inhibitor of protein-chain initiation (Grollman & Stewart, 1968; Marcus *et al.*, 1970*a*) we were able to demonstrate directly the presence of at least two initiation factors in wheat embryo supernatant and to devise an assay that allowed the resolution and partial purification of these factors.

Fig. 3 compares the effect of aurintricarboxylic acid on TM virus RNA-dependent amino acid incorporation and on amino acid incorporation by polyribosomes. At 25 μM-aurintricarboxylic acid, the TM virus RNA-dependent system is completely inhibited. In contrast, polyribosome-catalysed incorporation

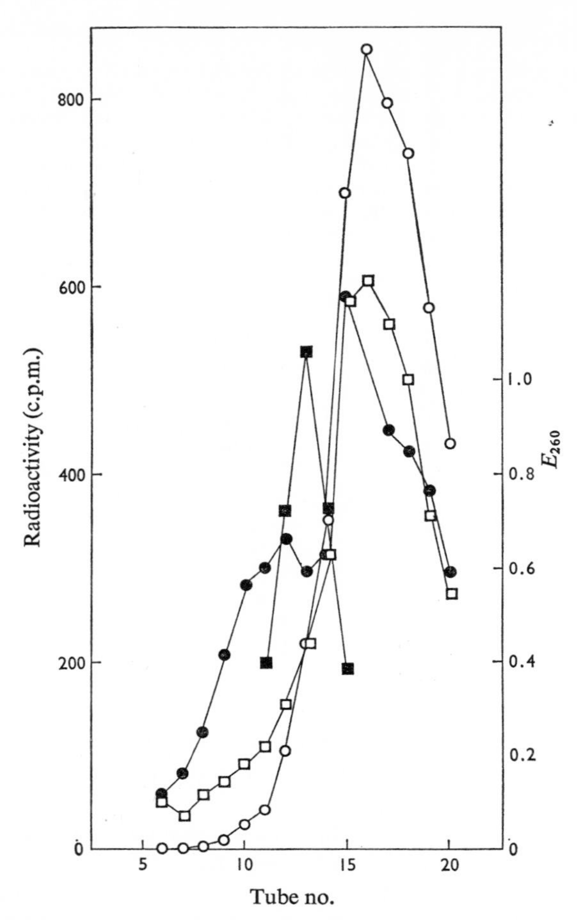


Fig. 2. *Formation of the ribosome–$^{14}$C-labelled TM virus RNA initiation complex*

After a preliminary incubation similar to that of Fig. 1, the reaction mixture was layered on a 5–25% sucrose density gradient and centrifuged for 1.5 h at 160 000 *g* (Marcus, 1970*b*). Gradient fractions were collected and $E_{260}$ (■) and trichloroacetic acid-insoluble radioactivity determined. Incubations contained 11 μg of $^{14}$C-labelled TM virus RNA (5700 c.p.m.) and were either complete (●), lacked ribosomes (○) or contained 1 mM-GTP in place of ATP (□).

is unaffected. These results establish that aurintricarboxylic acid is a specific inhibitor of the ribosome–mRNA-attachment reaction, and that it does not affect aminoacyl transfer to the growing peptide chain. This conclusion is confirmed and extended by the preincubation experiments of Table 2. When aurintricarboxylic acid is added at the outset of the reaction, it inhibits both the TM virus RNA and the poly(U)-catalysed reactions. If, however, the ribosomes and mRNA are preincubated before the addition of aurintricarboxylic acid, the reagent has little effect on either system. The conditions required for the prevention of aurintricarboxylic acid inhibition differ for the two systems. With poly(U), a preincubation at 0°C is sufficient and no soluble factors are required. With TM virus RNA, a 0°C preincubation is without effect and two soluble components

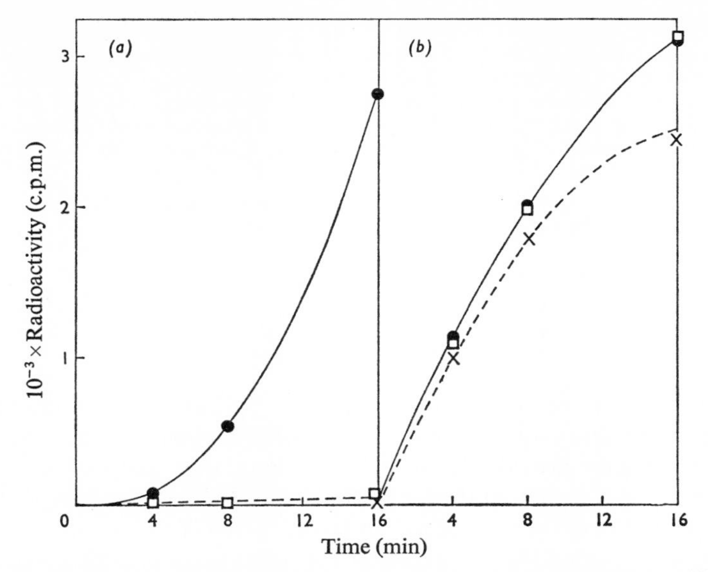


Fig. 3. *Effect of aurintricarboxylic acid on the TM virus RNA (a) and polyribosome (b) amino acid-incorporating systems*

Both incorporating systems were similar to those described in Table 1 except that all components were added in a single incubation (i.e. neither system was preincubated). The incubations were at 30°C with either no (●), 25 $\mu$M (□), or 50 $\mu$M (×) -aurintricarboxylic acid.

---

Table 2. *mRNA–ribosome complex formation and its inhibition by aurintricarboxylic acid*

The poly(U) reaction system (see legend to Table 1) was used with 70mM-$K^{+}$ and 7mM-$Mg^{2+}$ maintained both during a 6min preincubation at either 0°C or 30°C and during the subsequent 10min at 30°C amino acid incorporation incubation. In addition to aurintricarboxylic acid and [$^{14}$C]phenylalanyl-tRNA, the component omitted from the preincubation was supplemented in the incorporation incubation. The ribosomes were deoxycholate-washed (Legocki & Marcus, 1970) and the supernatant was 0.02ml of undialysed S-100. The TM virus RNA reaction system was similar to that of Table 1 except that in some incubations initiation factors [0.06ml of fraction C and/or 0.05ml of fraction D (Seal *et al.*, 1972)] were used in place of S-100 DEAE.

| Components added in preincubation | Amino acid incorporated (pmol) |
|---|---|
| Poly(U) reaction | |
| Poly(U) + ribosomes + ATA* (0°C or 30°C) | 0.1 |
| Poly(U) + ribosomes 30°C† | 5.4 |
| Poly(U) + ribosomes 0°C† | 7.7 |
| Poly(U) + supernatant 30°C† | 0.1 |
| Ribosomes + supernatant 30°C† | 0.1 |
| TM virus RNA reaction | |
| mRNA + ribosomes + S-100 DEAE + ATA 30°C | 0.25 |
| mRNA + ribosomes + S-100 DEAE 30°C† | 10.3 |
| mRNA + ribosomes + S-100 DEAE 0°C† | 0.28 |
| mRNA + ribosomes + C + D 30°C† | 4.1 |
| mRNA + ribosomes + C 30°C† | 0.15 |
| mRNA + ribosomes + D 30°C† | 0.21 |

* ATA, Aurintricarboxylic acid (25$\mu$M).
† 25$\mu$M-ATA added after the preincubation.

Table 3. *Soluble factor requirement for TM virus RNA-catalysed amino acid polymerization*

The reaction was similar to that for the TM virus RNA system in Fig. 3 except that fractionated initiation factors and elongation factors [C-final, 50 $\mu$g; D-final, 114 $\mu$g; T1(E), 29.7 $\mu$g; T2(E), 20.7 $\mu$g] (Seal *et al.*, 1972) were used in place of S-100 DEAE and eight unlabelled amino acids (1 mM) and 60 pmol of eight $^{14}$C-labelled aminoacyl-tRNA species (1 pmol = 450 c.p.m.) were added. After incubation for 15 min at 30°C, the radioactive material insoluble in hot trichloroacetic acid was determined.

| Factor omitted | Amino acid incorporated (pmol) |
|---|---|
| — | 7.3 |
| C | 0.18 |
| D | 0.21 |
| T1 | 0.16 |
| T2 | 0.14 |

are necessary. These results establish that the two soluble components are functioning in the formation of the ribosome–mRNA complex, i.e. as initiation factors. Further, they provide the basis for a quantitative assay of each of the factors. The primary feature of the assay (Seal *et al.*, 1972) is that after an appropriate preincubation in the presence of an excess of one of the factors and a limiting amount of the second factor, excess of supernatant (to provide a saturating amount of elongation factors) is added together with aurintricarboxylic acid and [$^{14}$C]aminoacyl-tRNA. The amount of amino acid polymerized is linearly proportional to the amount of the factor added in limiting quantity during the preliminary incubation. By using this initiation-factor assay, together with a poly(U) assay specific for elongation factors (Legocki & Marcus, 1970), we have resolved the wheat embryo soluble fraction into four distinct components. TM virus RNA-catalysed amino acid incorporation is completely dependent on each of these four resolved components (Table 3). Partial assays (Seal *et al.*, 1972) identify factors C and D as initiation factors and factors T1 and T2 as elongation factors. In addition to the ATP requirement already noted, careful analysis of the initiation reaction shows that GTP is clearly involved at this stage as well as in the chain-elongation reaction (Table 4).

## Methionyl-tRNA and Protein Chain Initiation

Protein synthesis in bacteria is initiated by a ribosome–mRNA-attachment reaction that requires in addition to initiation factors, a specific initiating amino-

Table 4. *Requirement for GTP in the formation of the initiation complex*

The standard TM virus RNA preincubation system similar to that of Table 1 was used with ATP at 1.2 mM and the NTP-generating system (creatine phosphate and creatine phosphate kinase) omitted. After 6 min at 30°C, 40 $\mu$M-aurintricarboxylic acid, amino[$^{14}$C]acyl-tRNA and an NTP-generating system (2 mM-phosphoenolpyruvate and 7.5 $\mu$g of phosphoenolpyruvate kinase) were added; GTP was supplemented to a final concentration of 0.5 mM, and in compensation, the $Mg^{2+}$ concentration was raised to 4.1 mM. Incorporation was measured after 8 min at 30°C.

| GTP during preincubation (mM) | Amino acid incorporation (pmol) |
|---|---|
| 0 | 3.9 |
| 0.06 | 6.4 |
| 0.6 | 6.0 |

acyl-tRNA, *N*-formylmethionyl-tRNA (Marcker & Sanger, 1964; Webster *et al.*, 1966). It had been proposed that in the cytoplasm of both mammalian and plant cells a similar function is served by unformylated methionyl-tRNA. In support of this idea, two cytoplasmic methionine tRNA species were obtained by fractionation on benzoylated DEAE-cellulose (Leis & Keller, 1970; Tarragó *et al.*, 1970) which could be distinguished by puromycin reactivity or by ribosome binding at high $Mg^{2+}$ concentration (see Allende *et al.*, 1973). A more direct test of the initiation function of methionyl-tRNA was carried out in our laboratory, by using the TM virus RNA-catalysed reaction *in vitro* with each of the two methionyl-tRNA species radioactively labelled. The peptide products were then analysed for *N*-terminal radioactivity (Table 5). Of the methionine transferred from one of the species, methionyl-$tRNA_m$, 98% was internal, whereas 57% of the methionine transferred from the other species, methionyl-$tRNA_i$, was *N*-terminal. The fluorodinitrobenzene procedure used for the *N*-terminal analysis yields only 75% in the case of methionine (Porter & Sanger, 1948), so that the actual percentage *N*-terminal for methionyl-$tRNA_i$ is considerably greater. We therefore concluded that methionyl-$tRNA_i$ is the initiator species in the wheat embryo system (Marcus *et al.*, 1970*b*). The observation that both wheat transformylase and its substrate, methionyl-$tRNA_f$, are found predominantly in the chloroplast (Leis & Keller, 1971), and the demonstration that the *N*-formylated product from *Escherichia coli* is not deformylated in the wheat embryo system (Lundquist *et al.*, 1972) provide final evidence that the cytoplasmic methionyl-$tRNA_i$ initiates without formylation.

Having ascertained the initiating function of methionyl-tRNA, we examined its ability to bind to ribosomes under initiating conditions (1.3–2.6mM-$Mg^{2+}$). Table 6 shows that such a reaction can indeed be obtained. Methionyl-tRNA binding is messenger-dependent and requires the presence of the two initiation factors, C and D, as well as ATP and GTP. The reaction is specific for the initiating species, methionyl-$tRNA_i$.

## Role of the Ribosomal Subunits in the Initiation Reaction

Having obtained evidence for the participation of soluble factors and of an initiating aminoacyl-tRNA in the formation of the ribosome–mRNA complex,

Table 5. *N-Terminal analysis of peptide products from methionyl-$tRNA_i$ and methionyl-$tRNA_m$*

After preincubation of the TM virus RNA system (see the legend to Fig. 1) [$^{14}$C]methionyl-tRNA was added. In the first experiment [$^{14}$C]methionyl-$tRNA_i$ (30pmol, 9000c.p.m.) was added after preincubation in two equal parts, half at the onset of the incubation and half 7min later; the incubation was stopped after 21min. In the second experiment [$^{14}$C]methionyl-$tRNA_m$ (23pmol, 6700c.p.m.) was added in two equal parts, half after 20 min incubation and half 7min later. The amount of each methionyl-tRNA species incorporated as *N*-terminal methionine was determined by the fluorodinitrobenzene procedure (Marcus *et al.*, 1970*b*).

| Incubation | [$^{14}$C]Methionine transferred | | |
|---|---|---|---|
| Met-tRNA | *N*-Terminal (c.p.m.) | Internal (c.p.m.) | *N*-Terminal (%) |
| [$^{14}$C]Met-$tRNA_i$ | 532 | 410 | 57 |
| [$^{14}$C]Met-$tRNA_m$ | 54 | 2889 | 1.9 |

Table 6. *Requirements for ribosomal binding of methionyl-tRNA*

The reaction mixture contained in a volume of 0.34 ml: 1.1 mM-methionine, 30 mM-Tris–acetate buffer, pH 8, 1.1 mM-ATP, 60 μM-GTP, 10 μg of TM virus RNA, 2.6 mM-dithiothreitol, 1.3 mM-magnesium acetate, 51 mM-KCl, ribosomes (220 μg of RNA), 38 pmol of unresolved [$^{14}$C]-methionyl-tRNA or 34 pmol of [$^{14}$C]methionyl-tRNA$_i$ or [$^{14}$C]methionyl-tRNA$_m$ (1 pmol = 330 c.p.m.) and initiation factors: C, 170 μg and/or D-final, 125 μg (Seal *et al.*, 1972). After being incubated for 10 min at 20°C the amount of [$^{14}$C]methionyl-tRNA retained by nitrocellulose filters (i.e. [$^{14}$C]methionyl-tRNA bound to ribosomes) was determined.

| | | Met-tRNA bound | | |
|---|---|---|---|---|
| Substrate | Conditions | + TM virus RNA (pmol) | − TM virus RNA (pmol) | Δ |
| Met-tRNA | No factors | 0.20 | | |
| | C + D | 1.44 | 0.26 | 1.18 |
| | C alone | 0.17 | | |
| | D alone | 0.20 | 0.18 | |
| | C + D (ATP omitted) | 0.35 | 0.11 | 0.24 |
| | C + D (GTP omitted) | 0.70 | 0.11 | 0.59 |
| Met-tRNA$_i$ | C + D | 1.45 | 0.18 | 1.27 |
| Met-tRNA$_m$ | C + D | 0.42 | 0.22 | 0.20 |

we began attempts to describe the sequence of reactions. In bacteria it has been shown that the initial mRNA complex is made with the 30S ribosomal subunit (Guthrie & Nomura, 1968), and a number of studies suggest a similar situation in eukaryotic systems. As a direct determination of this point, we obtained ribosomal subunits that were active in TM virus RNA-directed amino acid incorporation, and assayed these subunits for their ability to form a ribosome–mRNA complex (Weeks *et al.*, 1972). The assay involved the incubation of the particular subunit with TM virus RNA under initiation conditions and then supplementing with the complementary subunit in the presence of aurintricarboxylic acid (to block any further mRNA attachment) and measuring the capacity for amino acid incorporation. The results (Table 7) clearly indicate that only the complex formed in the presence of 40S subunits is functional and that 60S subunits are required for subsequent peptide synthesis. 60S subunits may be supplied as such, or from

Table 7. *Formation of functional initiation complexes with* 40*S ribosomal subunits*

A standard TM virus RNA preliminary incubation (see legend to Table 2) was carried out for 7 min at 20°C except that the ribosomal component was either 0.3 $E_{260}$ unit of 40S subunits or 0.6 $E_{260}$ unit of 60S subunits. Thereafter, 1.6 nmol of aurintricarboxylic acid (ATA, final concn. 40 μM), 90 pmol of eight amino[$^{14}$C]acyl-tRNA species, 0.6 μmol of magnesium acetate, 0.03 ml of subunit-free S-100 (Weeks *et al.*, 1972) and either 40S subunits (0.3 $E_{260}$ unit), 60S subunits (0.6 $E_{260}$ unit) or 80S ribosomes (5.0 $E_{260}$ units) were added to a final volume of 0.41 ml. After a 9 min incubation at 30°C the radioactive material insoluble in hot trichloroacetic acid was determined. Ribosomal subunits were prepared by low-salt dissociation of 80S ribosomes (Weeks *et al.*, 1972).

| Preliminary incubation | Incorporation incubation (with ATA) | Amino acid incorporation (pmol) |
|---|---|---|
| 40S | 60S | 2.8 |
| — | 60S | 0.2 |
| 60S | 40S | 0.2 |
| 40S | — | 0.4 |
| 40S | 80S | 5.7 |
| — | 80S | 0.4 |

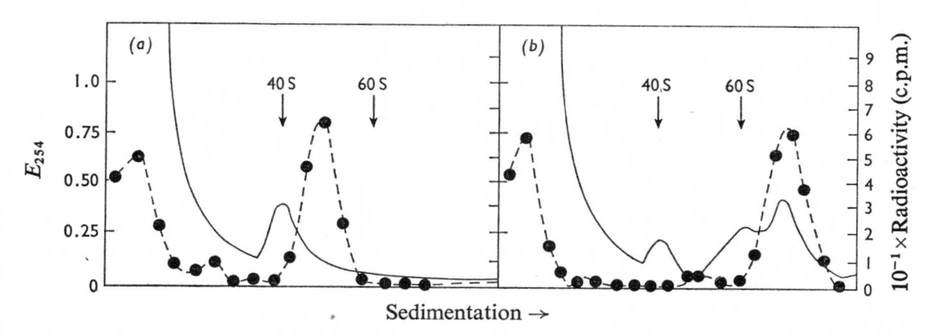


Fig. 4. *Sucrose-density-gradient analysis of the sequential formation of* (*a*) *the* 40*S ribosomal subunit–mRNA–methionyl-tRNA complex, and* (*b*) *the* 80*S ribosome–mRNA–methionyl-tRNA complex*

Methionyl-tRNA was bound to 40S subunits in incubations similar to those of the binding assay of Table 6 with the following modifications: 6 μg of STN virus RNA (the RNA of satellite tobacco necrosis virus) replaced TM virus RNA, and a subsequent incubation (3 min at 20°C) was carried out to which was added in 0.06 ml, 0.92 μmol of magnesium acetate and, in (*b*) 0.6 $E_{260}$ unit of 60S subunits. The reaction products were layered on linear 7.5–25% sucrose density gradients, and after centrifugation for 80 min at 160000*g*, the gradients were scanned at $E_{254}$ (——) and divided into 20 fractions. Each fraction was analysed for radioactivity (– – –) retained on nitrocellulose filters as in the binding assay of Table 6.

the equilibrium dissociation of 80S ribosomes. In experiments to test the ribosomal binding of methionyl-tRNA, only the 40S subunits were active and the initiating species, methionyl-$tRNA_i$, was bound in preference to the internal species.

The sequence of reactions leading to the formation of the 80S subunit–mRNA complex can be detected directly by allowing radioactive labelled methionyl-tRNA to bind to the ribosomal subunits and then examining the reaction products by sucrose-density-gradient centrifugation. Fig. 4(*a*) shows that 40S subunits, STN virus RNA (the monocistronic mRNA of satellite tobacco necrosis virus) and radioactively labelled methionyl-tRNA first react to form a complex of approx. 50S. When 60S subunits are supplied in an ensuing incubation, there is a complete shift of the methionyl-tRNA into the monoribosome region of the gradient (Fig. 4*b*). This direct approach in describing the ribosomal sequence has allowed further analysis of the partial reactions, primarily with the aid of

Table 8. *Effects of pactamycin and MDMP on the puromycin reactivity of bound methionyl-tRNA*

The methionyl-tRNA binding reactions were similar to those of Table 6 except that 0.6 mg of fraction C and 0.5 mg of fraction D (Seal *et al.*, 1972) were added and the various reaction mixtures were kept for 5 min in ice after the methionyl-tRNA binding incubation either as such or with the addition of 1.0 mM-puromycin.

| Additions | Met-tRNA bound (pmol) | | | Inhibition puromycin reaction (%) |
|---|---|---|---|---|
| | No puromycin | + Puromycin | Δ | |
| None | 1.55 | 0.65 | 0.90 | — |
| Pactamycin (0.25 μM) | 1.58 | 1.57 | 0.01 | 99 |
| None* | 1.55 | 0.90 | 0.65 | — |
| MDMP (0.1 mM)* | 1.49 | 1.40 | 0.09 | 86 |

* Binding reaction carried out at 3.6 mM-$Mg^{2+}$.

pactamycin (Cohen *et al.*, 1969; Seal & Marcus, 1972) and MDMP [2-(4-methyl-2,6-dinitroanilino)-*N*-methylpropionamide] (Weeks & Baxter, 1972), two inhibitors of protein-chain initiation. The pertinent observation with these two inhibitors is that in contrast with aurintricarboxylic acid (the initiation inhibitor described above) they have little effect on mRNA-dependent binding of methionyl-tRNA to ribosomes (Table 8). However, the methionyl-tRNA bound in the presence of either of these inhibitors is unreactive with puromycin. These results indicate that both pactamycin and MDMP, although inhibiting chain initiation, do not affect the formation of the 40S ribosome–mRNA–methionyl-tRNA complex. The possibility that these inhibitors prevent the interaction of the 40S ribosome–mRNA complex with the 60S subunit, is examined in Fig. 5. Here the

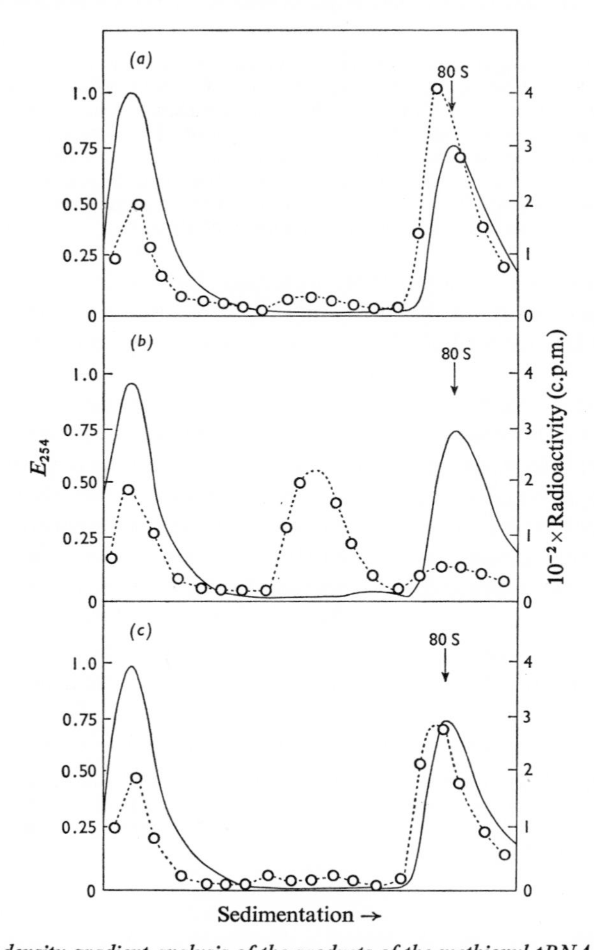


Fig. 5. *Sucrose-density-gradient analysis of the products of the methionyl-tRNA binding reaction*

Methionyl-tRNA binding was carried out as in Table 6 except that STN virus RNA was substituted for TM virus RNA and 0.6 mg of initiation factor C and 0.5mg of factor D were added. The incubations contained (*a*) no inhibitor, (*b*) 0.1 mM-MDMP and (*c*) 0.001 mM-pactamycin. The reactants were layered on sucrose density gradients and analysed as in Fig. 4. ——, $E_{254}$ ○, radioactivity.

ribosomal subunit reactions are again examined by methionyl-tRNA binding. As expected, in the control incubation methionyl-tRNA radioactivity is found predominantly at the 80S region. The presence of MDMP clearly results in retention of the radioactive methionyl-tRNA at the gradient position of the 40S ribosomal subunit complex. The reaction of the 40S subunit–mRNA complex with the 60S subunit therefore appears to be a specific interaction that can be inhibited by MDMP. In contrast with this latter inhibitor, pactamycin does not interfere with the formation of the 80S subunit initiation complex as indicated by the normal shift of the radioactive methionyl-tRNA. There is, therefore, yet another reaction, subsequent to the formation of the 80S ribosome–mRNA–methionyl-tRNA complex, which is part of the initiation process. This reaction is shown by the conversion of ribosomally bound methionyl-tRNA from a non-puromycin-reactive into a puromycin-reactive state and is prevented by the presence of pactamycin.

## Summary

The various studies of the amino acid-polymerizing system of wheat embryo *in vitro* have established that protein chain initiation proceeds via the formation of a 40S ribosomal subunit–mRNA–methionyl-tRNA initiation complex. ATP, GTP, and at least two initiation factors participate in this reaction. A 60S subunit is then added and the protein chain is then elongated. In a study with STN virus RNA, Klein *et al.* (1972) have shown that the wheat embryo system *in vitro* synthesizes the satellite virus coat protein, thereby establishing the translational fidelity of the system. The *N*-terminal amino acid of the STN virus coat is alanine (Reichmann, 1964). Thus the initial dipeptide expected with STN virus RNA is presumably methionylalanine. In preliminary experiments we have observed the formation of this dipeptide in a reaction requiring the elongation factor, T1. In continuing studies we are examining the function of both of the elongation factors in systems translating natural mRNA and are probing the mechanisms of the partial reactions.

This work was supported by Grant GM-15122 from the National Institutes of Health, and Grant GB-23041 from the National Science Foundation, by United States Public Health Service Grants CA-06927 and RR-05539 awarded to this Institute, and by an appropriation from the Commonwealth of Pennsylvania.

## References

Allende, J. E., Tarragó, A., Monasterio, O., Litvak, S., Gatica, M., Ojeda, J. M. & Matamala, M. (1973) *Biochem. Soc. Symp.* **38**, 77–96

Boulter, D. B. (1970) *Annu. Rev. Plant Physiol.* **21**, 91–114

Cohen, L. B., Herner, A. E. & Goldberg, I. H. (1969) *Biochemistry* **8**, 1312–1326

Grollman, A. P. & Stewart, M. L. (1968) *Proc. Nat. Acad. Sci. U.S.* **61**, 719–725

Guthrie, C. & Nomura, N. (1968) *Nature (London)* **219**, 232–235

Klein, W. H., Nolan, C., Lazar, J. M. & Clark, J. M., Jr. (1972) *Biochemistry* **11**, 2009–2014

Legocki, A. B. & Marcus, A. (1970) *J. Biol. Chem.* **245**, 2814–2818

Leis, J. P. & Keller, E. B. (1970) *Biochem. Biophys. Res. Commun.* **40**, 416–421

Leis, J. P. & Keller, E. B. (1971) *Biochemistry* **10**, 889–894

Lucas-Lenard, J. & Lipmann, F. (1971) *Annu. Rev. Biochem.* **40**, 409–448

Lundquist, R. E., Lazar, J. M., Klein, W. H. & Clark, J. M., Jr. (1972) *Biochemistry* **11**, 2014–2019
Marcker, K. & Sanger, F. (1964) *J. Mol. Biol.* **8**, 835–840
Marcus, A. (1970*a*) *J. Biol. Chem.* **245**, 955–961
Marcus, A. (1970*b*) *J. Biol. Chem.* **245**, 962–966
Marcus, A. (1972) *Methods Mol. Biol.* **2**, 127–145
Marcus, A. & Feeley, J. (1964) *Proc. Nat. Acad. Sci. U.S.* **51**, 1075–1079
Marcus, A. & Feeley, J. (1966) *Proc. Nat. Acad. Sci. U.S.* **56**, 1770–1777
Marcus, A., Feeley, J. & Volcani, T. (1966) *Plant Physiol.* **41**, 1167–1172
Marcus, A., Lugenbill, B. & Feeley, J. (1968) *Proc. Nat. Acad. Sci. U.S.* **59**, 1243–1250
Marcus, A., Bewley, J. D. & Weeks, D. P. (1970*a*) *Science* **167**, 1735–1736
Marcus, A., Weeks, D. P., Leis, J. P. & Keller, E. B. (1970*b*) *Proc. Nat. Acad. Sci. U.S.* **67**, 1681–1687
Porter, R. R. & Sanger, F. (1948) *Biochem. J.* **42**, 287–294
Reichmann, M. E. (1964) *Proc. Nat. Acad. Sci. U.S.* **52**, 1009–1017
Seal, S. N. & Marcus, A. (1972) *Biochem. Biophys. Res. Commun.* **46**, 1895–1902
Seal, S. N., Bewley, J. D. & Marcus, A. (1972) *J. Biol. Chem.* **247**, 2592–2597
Tarragó, H., Monasterio, O. & Allende, J. E. (1970) *Biochem. Biophys. Res. Commun.* **41**, 765–773
Webster, R. E., Englehardt, D. L. & Zinder, N. D. (1966) *Proc. Nat. Acad. Sci. U.S.* **55**, 155–161
Weeks, D. P. & Baxter, R. (1972) *Biochemistry* **11**, 3060–3064
Weeks, D. P. & Marcus, A. (1969) *Plant Physiol.* **44**, 1291–1294
Weeks, D. P. & Marcus, A. (1970) *Biochim. Biophys. Acta* **232**, 671–684
Weeks, D. P., Verma, D. S., Seal, S. N. & Marcus, A. (1972) *Nature* (*London*) **236**, 167–168

## Discussion

**J. W. Davies**: (1) You said that methionine is incorporated into the *N*-terminal position of wheat proteins, from methionyl-$tRNA_i$. However, most of your initiation experiments *in vitro* are done with TM virus RNA, and TM virus coat protein does not have an *N*-terminal methionine. Have you looked at the initial peptide sequence (e.g. tripeptides) of the product *in vitro*: do you find methionyl-serylalanine etc.? Is there evidence for removal of methionine: does wheat embryo have such an enzyme?

(2) Several laboratories have reported involvement of *N*-acetylated aminoacyl-tRNA in initiation of plant viral proteins. Would you like to comment on this? TM virus protein begins with *N*-acetylserine. Have you tried making *N*-acetyl-seryl-tRNA (by chemical acetylation) and seeing if this binds to wheat ribosome–TM virus RNA initiation complex?

**A. Marcus**: (1) In the experiments with TM virus-RNA, we looked at the total products in short-term labelling experiments and found a substantial part of the incorporated methionine at the *N*-terminus. We assume that this is due, in part to proteins where methionine naturally occurs at the *N*-terminus, in part to transient products (where the methionine hasn't been removed as yet), and in part to inability of the system to remove methionine *in vitro*. Evidence that the wheat embryo system does have capacity to remove *N*-terminal methionine is provided by the studies with STN virus RNA where the *N*-terminus of the product at the conclusion of the reaction is almost exclusively alanine. With this same viral RNA (STN virus RNA) we have data showing the formation of methionylalanine and methionylalanyl-lysine as the initial di- and tri-peptides. Presumably the methionine is removed at a later stage.

(2) The only studies where *N*-acetyl-amino acids have allowed initiation with plant mRNA have utilized bacterial incorporating systems [J. Albrecht,

W. Rozenboom, C. Vermeer & L. Bosch (1969) *FEBS Lett.* **5**, 313–315; N. J. Verhoef & L. Bosch (1971) *Virology* **45**, 75–84; N. J. Verhoef, M. C. E. Cornelissen & L. Bosch (1971) *Virology* **45**, 85–90]. We have prepared *N*-acetylseryl-tRNA and found that it is not incorporated into protein in the wheat embryo TM virus RNA-catalysed reaction. On the other hand, analysis of infected tobacco leaves *in vivo* (B. Filner & A. Marcus, unpublished work) showed the nascent coat protein is already *N*-terminally acetylated. Either the infected tobacco leaf contains a system that can utilize acetylseryl-tRNA or *N*-terminal acetylation occurs by a more complex sequence involving a displacement reaction.

It would appear that ribosome binding studies utilizing *N*-acetylseryl-tRNA with both wheat embryo and tobacco leaf ribosomes should be of interest.

*Biochem. Soc. Symp.* (1973) **38**, 111–135
*Printed in Great Britain*

# Chloroplast- and Cytoplasm-Specific Aminoacyl-Transfer Ribonucleic Acid Synthetases of *Euglena gracilis*: Separation, Characterization and Site of Synthesis

By R. KRAUSPE and B. PARTHIER

*Institute of Plant Biochemistry, Research Centre of Molecular Biology and Medicine, Academy of Sciences of the G.D.R.*, 401 *Halle* (*Saale*), *German Democratic Republic*

## Synopsis

Of 14 crude aminoacyl-tRNA synthetases prepared from greening *Euglena gracilis* 12 show a remarkable light-stimulated capacity to aminoacylate tRNA from *Anacystis*, if compared with enzymes from dark-grown cells. The light-stimulated enzyme activity is attached to chloroplast-specific synthetases; changes in the activity of the cytoplasmic enzymes are more related to the age of the cultures than to illumination. The synthetases of a chlorophyll-free *Euglena* mutant also show a limited light-stimulation of activity. For nine amino acids, the chloroplast synthetases are separable from the respective cytoplasmic enzymes by hydroxyapatite chromatography. The most striking property of the two major column-separated leucyl-tRNA synthetases (EC 6.1.1.4) is the absolute specificity toward the cellular origin of its cognate tRNA. The two enzymes differ also in their sensitivity to pretreatment at elevated temperatures and to high amounts of univalent cations present during the enzyme reaction. Both enzymes have nearly identical molecular weights of 105000 and 110000, respectively, and require the same concentration of ATP or $Mg^{2+}$ for optimum rates of aminoacylation. Experiments on greening kinetics and the effects of treatment of *E. gracilis* with chloramphenicol, nalidixic acid and cycloheximide provide evidence that the light-stimulated increase of enzyme activity can be referred to as chloroplast-specific synthetase biosynthesis. Our results suggest that the amount of the chloroplast-specific synthetases is due to an increase on illumination of the enzyme species which are already present in the proplastids. Green *Euglena* cells seem not to contain a chloroplast-specific threonyl-tRNA synthetase (EC 6.1.1.3).

## Introduction

Chloroplasts contain their own DNA and synthesize thylakoid proteins and certain enzymes involved in photosynthesis, as has been shown by inhibitor studies *in vivo* (Margulies, 1964; Kirk, 1968; Ellis, 1969; Hoober *et al.*, 1969; Bishop & Smillie, 1970; Eytan & Ohad, 1970; Surzycki *et al.*, 1970; Smillie *et al.*, 1971). Isolated chloroplasts are likewise able to incorporate labelled amino acids into polypeptides (Spencer, 1965; Boardman *et al.*, 1966; Goffeau, 1969; Smillie *et al.*, 1971). Thus the existence of a plastid-specific protein-synthesizing machinery is suggested including ribosomes and polyribosomes (Lyttleton, 1962;

Spencer, 1965; Boardman *et al.*, 1966; Stutz & Noll, 1967; Rawson & Stutz, 1969; Falk, 1969), tRNA species (Barnett *et al.*, 1969; Burkard *et al.*, 1970, 1972; Williams & Williams, 1970; Guderian *et al.*, 1972) and enzymes. By contrast with the well-investigated nucleic acid components, only little information is available about the characterization of chloroplast-specific enzymes necessary for the various steps of protein biosynthesis.

Aminoacyl-tRNA synthetases (EC 6.1.1) which catalyse the first reactions of protein biosynthesis, the activation of amino acids to acyl-adenylates and the transfer of the aminoacyl residues to cognate tRNA species, have been shown to be present in chloroplasts both of higher plants (Aliev & Filippovich, 1968; Burkard *et al.*, 1970; Guderian *et al.*, 1972) and *Euglena gracilis* (Reger *et al.*, 1970; Parthier, 1972; A. Böck, personal communication). The characterization of these enzymes, their specificity toward tRNA and their sites of biosynthesis within the green cell, however, are based on meagre information. Reger *et al.* (1970) first demonstrated that green *E. gracilis* cells contain two isoleucyl- and three phenylalanyl-tRNA synthetases. Each one of them is localized in the chloroplasts, is light-inducible and specificially aminoacylates chloroplast tRNA. The chloroplast-specific isoleucyl-tRNA synthetase was assumed to be coded by the chloroplast DNA, whereas the chloroplast-specific phenylalanyl-tRNA synthetase appeared constitutive, i.e. controlled by nuclear DNA. We have also found a light-stimulated increase of enzyme activity for most of the synthetases we have tested (Parthier *et al.*, 1972).

In the present report we give results that show the separation of nine chloroplast-specific aminoacyl-tRNA synthetases from the respective cytoplasmic enzymes by hydroxyapatite chromatography utilizing the enzymes specificities toward tRNA species of prokaryotic or eukaryotic origin. Some characteristics of the separated leucyl-tRNA synthetases are also demonstrated. Inhibition experiments were performed to determine whether the chloroplast-specific enzyme was formed inside the organelle or not.

## Materials and Methods

### *Chemicals*

L-[-$^{14}$C]Leucine (specific radioactivity 92 mCi/mmol) and the other U-$^{14}$C-labelled L-amino acids as indicated in Fig. 4 (specific radioactivity between 75 and 150 mCi/mmol) were obtained from UVVVR, Prague, Czechoslovakia; carrier-free $Na_2H^{32}PO_4$ was from Kernforschungszentrum Dresden-Rossendorf, GDR; ATP was a product of Boehringer und Soehne G.m.b.H., Mannheim, Germany. DEAE-cellulose SS (standard), benzoylated DEAE-cellulose and charcoal Norit A were purchased from Serva Feinbiochemica G.m.b.H., Heidelberg, Germany. Sephadex G materials came from Pharmacia, Uppsala, Sweden. Cycloheximide was from Fluka, Buchs, Switzerland, nalidixic acid (1-ethyl-7-methyl-1,8-naphthyridin-4-one-3-carboxylic acid) from Winthrop Products Co., London, U.K.; D-*threo*-chloramphenicol from VEB Berlin-Chemie, G.D.R. The other chemicals used had the p.a. grade.

*Organisms*

*E. gracilis*, Z strain no. 1224-5/25, was originally obtained from the algae collection of the University of Göttingen, but cultivated for several years in our laboratory. The cells were axenically grown at 25°C in Hutner's organic growth medium as described by Meissner *et al.* (1971). To maintain the cultures samples were transferred into new medium every third day. The term dark-grown cells refers to cells that were heterotrophically grown in complete darkness for many generations. These cells contained no chlorophyll. Light-induced cells were derived from dark-grown cultures, illuminated, at the time of inoculation, with artificial light of 1200lx. This was found sufficient to obtain cells with fully developed chloroplasts after 2–3 days growth. The term green cells is used for organisms permanently grown under the light conditions indicated above. The plastid mutant was obtained after u.v. irradiation of plated green cells for 15s ('u.v. mutant'), followed by 5 days growth in the dark to prevent photo-reactivation. These cells were completely free of chlorophyll, but synthesized considerable amounts of carotenoids during growth in the light.

In the light-induction experiments and for the preparation of enzymes 1 litre flasks with 250 or 500ml of growth medium were inoculated with $4 \times 10^5$ cells/ml. For the tRNA preparation the flasks were replaced by 10-litre bottles containing 6l of cell suspension which yielded approx. 150g wet wt. of material after 3 days growth.

*Anacystis nidulans*, strain Göttingen, was grown in 60cm long cylindrical 4-litre flasks at 40°C with $CO_2$+air (4:96) in Kratz & Myer's (1955) medium D. Illumination was provided by cool-white fluorescent lamps of approx. 6000 lx.

*Preparation of tRNA*

The tRNA of *Euglena* cells was prepared according to our earlier description (Meissner *et al.*, 1971). Twice-washed intact cells were suspended in 4vol. of 0.05M-Tris–HCl buffer, pH 7.8, containing 0.01M-magnesium acetate and 0.06M-KCl/g wet wt. The same volume of chloroform was added and the mixture was vigorously shaken for 30min at room temperature. The aqueous phase was separated by centrifugation at 6000*g* for 15min at 2°C, 1vol. of water-saturated phenol was added and the mixture was shaken as before. The treatment was repeated twice with 0.5vol. of phenol. The tRNA was precipitated at –30°C with 0.1vol. of 20% potassium acetate solution (pH 5) and 2.5vol. of ethanol. The precipitate was dissolved in 1vol. of 3M-potassium acetate, pH 5.0/g cell wet wt. The solution was allowed to stand overnight at 0°C to precipitate contaminations of high-molecular-weight RNA and proteins, which were then removed by centrifugation at 15000*g* for 20min. The tRNA was precipitated with 2.5vol. of ethanol as before. This step was repeated before the tRNA was dissolved in 0.5M-Tris–HCl buffer, pH 9.0, and deacylated at 37°C for 45min. The tRNA was then dissolved in three-times-distilled water and dialysed against water. This method yields tRNA preparations free of other nucleic acids, as shown by gel filtration on Sephadex and chromatography on methylated albumin kieselguhr columns (Meissner *et al.*, 1971).

The tRNA of *A. nidulans* was prepared in a similar way. Twice-washed cells were suspended in 4 vol. of buffer/g wet wt., as indicated for *Euglena*, the suspension was mixed with 1 vol. of water-saturated phenol and stirred for 1 h at room temperature. All further steps were identical with the tRNA preparation of *E. gracilis*. The crude tRNA preparation contained up to 30% ribonucleoprotein which was removed by filtration through a column (80 cm × 2 cm) of Sephadex G-100. The second u.v. light-absorbing peak is entirely tRNA, which was collected, reprecipitated, redissolved and deacylated as described.

*Chromatography of tRNA on benzoylated DEAE-cellulose columns*

Deacylated tRNA 1–2 ml (20–30 mg) was applied to a column (35 cm × 1.5 cm) of benzoylated DEAE-cellulose which had been equilibrated with 0.01 M-Tris–HCl buffer, pH 7.5, 0.01 M-magnesium acetate and 0.3 M-NaCl. The tRNA species were eluted with a 0.3–1.5 M-NaCl logarithmic gradient (Gillam *et al.*, 1967) at room temperature. The flow rate was 15 ml/h and 3 ml fractions were collected. The tRNA was precipitated with 2 vol. of ethanol, centrifuged at 8000 *g* for 20 min, dissolved in 0.1–0.3 ml of water, diluted to an appropriate concentration and examined for acceptor activity.

*Preparation of crude aminoacyl-tRNA synthetases*

Harvested cells were washed twice with 0.01 M-Tris–HCl buffer, pH 7.8, 0.01 M-magnesium acetate, 0.06 M-KCl and 0.006 M-β-mercaptoethanol (medium I) and then frozen at –30°C for several hours or days. The thawed cell paste was diluted with the same medium (1 vol./g wet wt.) and was sonicated for 3 min at 0°C (Branson sonifier, step 5, microtip). The post-ribosomal supernatant after centrifugation at 120000 *g* for 90 min at 0°C was subjected to DEAE-cellulose chromatography to remove the endogenous tRNA. The enzyme fraction was eluted with 0.3 M-KCl in medium I and dialysed over-night against several hundred volumes of medium I. Any insoluble material was removed by centrifugation at 15000 *g* for 10 min. The enzyme fraction was either used immediately or mixed with glycerol to a final concentration of 33% (v/v) glycerol for storage at –30°C.

*Separation of the synthetases by hydroxyapatite chromatography*

Hydroxyapatite was prepared as described by Tiselius *et al.* (1956). Crude enzymes were obtained as described above, but the $Mg^{2+}$ content of the disintegration buffer was decreased to 1 mM and 10% (v/v) glycerol was present from the beginning. DEAE-cellulose chromatography and dialysis were omitted. A sample (57–58 mg of protein) of the post-ribosomal supernatant was applied to a column (15 cm × 1 cm) of hydroxyapatite equilibrated with the disintegration buffer. The enzymes were eluted with a linear gradient of 0.01–0.3 M-potassium phosphate buffer, pH 7.5, containing 1 mM-$MgCl_2$, 0.5 mM-β-mercaptoethanol and 10% glycerol. The columns were run at 2°C at a flow rate of 12–18 ml/h; $E_{280}$ was continuously recorded by Uvicord II, LKB Stockholm, Sweden. Fractions (5 ml) were collected for the assay of synthetase activity.

*Preparation of Euglena chloroplast synthetases*

Freshly harvested, 3 days light-induced *Euglena* cells (100 g) were twice washed in medium I and once in the same medium containing 20% (w/v) sucrose (medium

II) before sonication in 3 vol. of medium II/g cell wet wt. for 1 min at 0°C (sonifier macrotip; step 4) was carried out. Of the cells 60–70% were destroyed by this procedure with minimum delay the homogenate was centrifuged at 600*g* for 15 min to remove whole cells and cell debris. The top three-quarters of the supernatant was centrifuged at 8000*g* for further 20 min. Care was taken that the speed was slowly and continuously increased. The dark-green upper layer of the sediment was carefully suspended in medium II, recentrifuged at 8000*g* for 20 min, the sediment carefully resuspended and layered on to a discontinuous gradient consisting of 7 ml of each of 60 (bottom), 50, 40, 30 and 25% sucrose in medium I. The whole chloroplasts were obtained in the zone of 40% sucrose after centrifugation at 8000*g* for 20 min and are well separated from remaining cells and paramylum grains and from smaller particles and cytoplasmic materials. The chloroplast fraction was then diluted to 20% sucrose with medium I and sedimented by centrifugation at 12000*g* for 15 min. The pellet was resuspended in a minimum volume of medium I, sonicated for 2 min, and the disrupted chloroplasts were dialysed overnight against medium I. Finally, the soluble fraction containing the synthetases activity was separated from the chlorophyll-containing material by centrifugation at 15000*g* for 15 min. All steps were performed at 0°C.

*ATP–pyrophosphate-exchange reaction*

Sodium [$^{32}$P]pyrophosphate was prepared from $Na_2H^{32}PO_4$ after heating at 500°C for 2h and was purified by ion-exchange chromatography with Dowex 1 (X4; 200–400 mesh; $Cl^-$ form). The peak fraction was diluted with $Na_4P_2O_7$,-10 $H_2O$ to 0.05 M final concentration. The assay was performed in 250 $\mu$l mixtures containing 100 mM-Tris–HCl buffer, pH 7.5; 5 mM-ATP; 5 mM-[$^{32}$P]pyrophosphate; 10 mM-$\beta$-mercaptoethanol; 10 mM-L-leucine, and 10–100 $\mu$g of protein of the enzyme fraction. The incubation was performed for 15 min at 30°C; stopped with 250 $\mu$l of 15% $HClO_4$ including 0.4 M-$PP_i$. A 1 ml portion of acid-activated charcoal in 50 mM-acetate buffer, pH 4.5, and 0.1 M-$PP_i$ were added and allowed to stand for 1 h at 0°C. The charcoal was washed three times with $PP_i$–acetate buffer, and the [$^{32}$P]ATP was hydrolysed with 1 ml of 2 M-HCl for 15 min at 100°C. Samples (250 $\mu$l) were diluted with 5 ml of water, and the $^{32}$P radioactivity was measured in a Tri-Carb liquid-scintillation spectrometer. The exchange rates were calculated from the d.p.m. of a $^{32}P^{32}P_i$ sample used as a reference.

*Assay of aminoacyl-tRNA synthesis*

The assay was performed for 10 min at 30°C in 250 $\mu$l mixtures containing 100 mM-Tris–HCl buffer, pH 7.5, 2–4 mM-ATP, 12 mM-$MgCl_2$, 12 mM-KCl, 1.5 mM-$\beta$-mercaptoethanol, 0.02 mM (60 $\mu$g) deacylated tRNA, a single $^{14}$C-labelled amino acid (0.004–0.01 mM), 10–40 $\mu$g of protein of the enzyme fraction. Half of the incubation mixture was used when the activities of column eluates were tested. The reaction was stopped by immediate cooling of the incubation tubes at 0°C. Samples (100 $\mu$l) were pipetted on to filter discs, and the filters were further treated as described by Mans & Novelli (1961). Radioactivity was determined in a Tri-Carb liquid-scintillation spectrometer in vials containing 13 ml of scintillation fluid [4g of 2.5-diphenyloxazole and 0.2g of 1.4-bis-(5-phenyl-oxazol-2-yl)benzene/litre in toluene]. The counting efficiency was 75%. The values of pmol of aminoacyl-tRNA formation/mg of tRNA per min as indicated in the figures and tables are calculated from the values after 10min incubation, which was still in the linear region of the kinetic curves.

*Other determinations*

Cell numbers were counted with an electronic particle counter (TuR ZG1, Transformatoren- und Roentgenwerk, Dresden) by using a 140 μm aperture. Protein was measured as described by Lowry *et al.* (1951) with bovine albumin as a standard. Chlorophyll was determined after extraction with 80% (v/v) acetone at $E_{652}$ (Arnon, 1949). RNA was measured at $E_{260}$ in three-times-distilled water; calculation was made with 24 $E_{260}$ units equivalent to 1mg of tRNA/ml.

## Results

*Aminoacyl-tRNA synthesis during the greening process*

At first we compared the activities of crude aminoacyl-tRNA synthetases from dark-grown and light-induced *E. gracilis* cells by using tRNA preparations from dark-grown, light-induced and mutant *Euglena* cells as well as from *A. nidulans*. Compared with the enzymes from dark-grown cells, it is obvious (Table 1) that leucyl- and valyl-tRNA synthesis is enhanced with enzymes from light-induced cells and tRNA from green cells. A 13-fold increase in the rate of aminoacylation is observed with *Anacystis* tRNA. The tRNA preparations from dark-grown and

Table 1. [14C]*Leucine- and* [14C]*valine-acceptor activities of various tRNA preparations during incubation with synthetases from dark-grown and light-induced E. gracilis*

The tRNA species and crude enzymes were prepared from 3-day-old cultures as described in the Material and Methods section. The $tRNA^{Leu}$ fractions 1 and 2 + 3 were separated by chromatography on benzoylated DEAE-cellulose (cf. Figs. 1 and 2). Incubation was for 10min at 30°C with the mixture indicated in the Material and Methods section. A 45 μg sample of tRNA, 30 μg of enzyme protein, 8 μM-[14C]leucine or 4 μM-[14C]valine were used.

| Source of cells | | Aminoacyl-tRNA synthesis (pmol/mg of tRNA per min) | |
|---|---|---|---|
| tRNA prep. | Enzyme prep. | [14C]Leucine | [14C]Valine |
| *E. gracilis*: | | | |
| Dark-grown | Dark-grown | 240 | 4.6 |
| | Light-induced | 224 | 4.0 |
| Light-induced | Dark-grown | 224 | 3.7 |
| | Light-induced | 285 | 5.2 |
| U.v. mutant | Dark-grown | 236 | 4.3 |
| | Light-induced | 218 | 3.0 |
| Cytoplasmic $tRNA^{Leu}$ (fraction 1) | Dark-grown | 280 | — |
| | Light-induced | 265 | — |
| Chloroplast $tRNA^{Leu}$ (fractions 2 + 3) | Dark-grown | 8 | — |
| | Light-induced | 120 | — |
| | Dark-grown | 13 | 7.0 |
| *A. nidulans* | Light-induced | 168 | 68 |

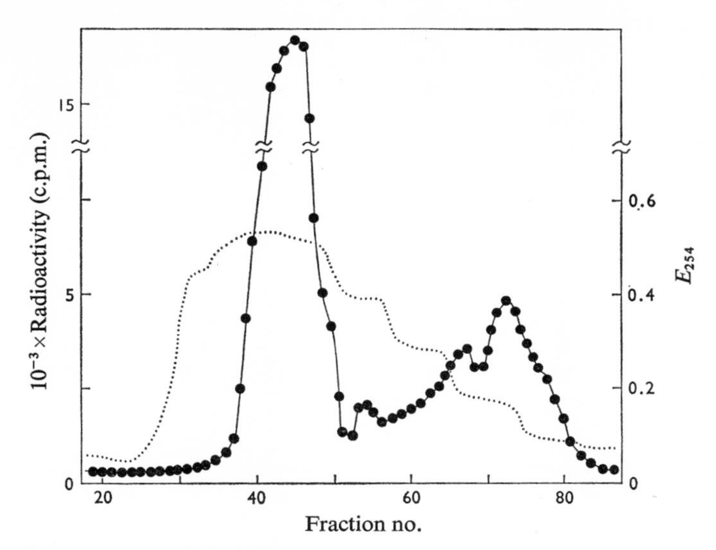


Fig. 1. *Chromatography of the tRNA preparation from light-induced E. gracilis on benzoylated DEAE-cellulose columns*

tRNA was prepared from 12 litres of 3-day-old green-cell cultures as described in the Materials and Methods section. The resulting 25mg of tRNA was applied to columns (35cm × 1.5cm) of DEAE-cellulose and eluted with a logarithmic gradient of 0.3–1.5M-NaCl in 0.01M-Tris–HCl buffer, pH7.5, and 0.01M-magnesium acetate. Fractions (3ml) were collected and pre-cipitated with 2vol. of ethanol. The tRNA was dissolved in a minimum volume of water and tested for [$^{14}$C]leucine-acceptor activity with a crude enzyme preparation from 3-day-old green *E. gracilis* cultures as described in the Materials and Methods section. ●——●, Radioactivity; ······, $E_{254}$.

mutant cells show no significant differences, if combined with the enzymes from dark-grown or light-induced cells.

Fractionation of the tRNA from light-induced cells by means of a column of benzoylated DEAE-cellulose and determination of the [$^{14}$C]leucine acceptance of the fractions with the homologous enzyme preparation, resulted in three $tRNA^{Leu}$ bands (Fig. 1) of which only the two slower-eluted, minor bands contain the $tRNA^{Leu}$ with a high acceptor specificity toward light-induced enzymes (Table 1). Since the slower-eluted tRNA fractions were exclusively aminoacylated by enzyme I fraction (chloroplast enzyme, see below), they should contain the chloroplast-specific $tRNA^{Leu}$ (Fig. 2). The faster-eluted major tRNA fractions were only charged by enzyme II (cytoplasmic enzyme) and should contain the cytoplasmic $tRNA^{Leu}$. Chromatography on benzoylated DEAE-cellulose of the tRNA prepared from the *Euglena* mutant yields cytoplasmic $tRNA^{Leu}$ only (Fig. 3).

Thus, we may tentatively suggest from these results that (i) the activity of certain synthetases specific for the charging of chloroplast tRNA and *Anacystis* tRNA is enhanced during greening, i.e. chloroplast formation, of dark-grown *E. gracilis*; (ii) no light-induced increase of aminoacyl-tRNA synthesis is observed with the u.v. mutant tRNA which corresponds to cytoplasmic tRNA. Consequently, we have replaced the chloroplast-specific tRNA by *Anacystis* tRNA and

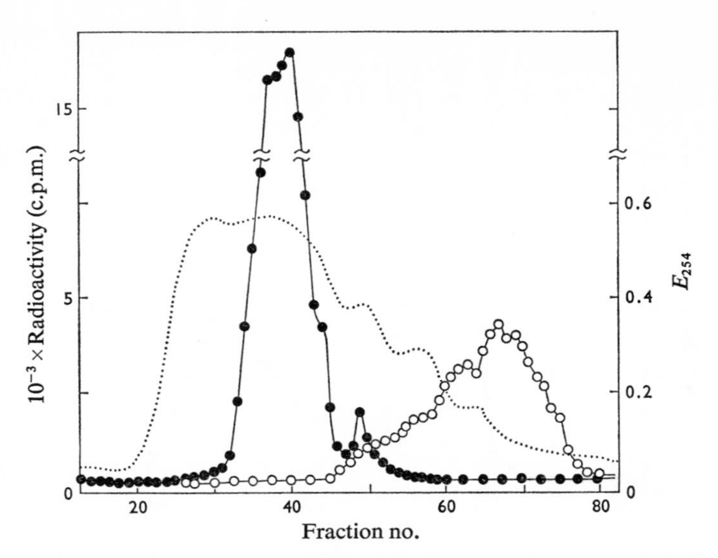


Fig. 2. *Separation of cytoplasmic and chloroplast tRNA$^{Leu}$ from green E. gracilis cells by benzoylated DEAE-cellulose chromatography*

Cells, tRNA preparation, chromatography and determination of [$^{14}$C]leucine-acceptor activity were analogous to Fig. 1, except that the leucyl-tRNA synthetases I and II (Fig. 8) were used instead of a crude enzyme preparation. ●, Enzyme II radioactivity; ○ enzyme I radioactivity; ······, $E_{254}$.

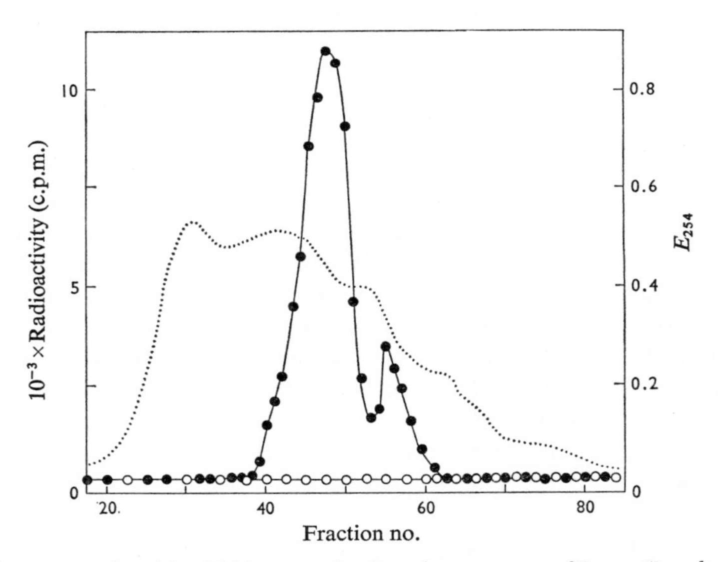


Fig. 3. *Chromatography of the tRNA preparation from the u.v. mutant of E. gracilis on benzoylated DEAE-cellulose*

The tRNA of 3-day-old illuminated cultures of the mutant was prepared as described in the Materials and Methods section and its [$^{14}$C]leucine acceptor activity was determined with separated synthetases I and II. Other details are as indicated in Fig. 1. ●, Enzyme II radioactivity; ○, enzyme I radioactivity; ······, $E_{254}$.

the cytoplasmic tRNA by the tRNA preparation from the u.v. mutant of *Euglena* in most of the experiments presented in this paper.

It is of interest to know whether the observed increase of enzyme activity after illumination of dark-grown *E. gracilis* is a general phenomenon of all synthetase species. Therefore, we tested the acceptor activity of *Anacystis* tRNA for 14 different $^{14}C$-labelled amino acids with a crude enzyme preparation from 3-day-old light-grown cultures. Recognizable differences exist between the various synthetases (Fig. 4): enzyme species showing a strong (approx. 10-fold) light-stimulation (synthetases cognate to arginine, aspartic acid, leucine, lysine, methionine, phenylalanine, tyrosine, valine); enzyme species with only a weak light-stimulated activity (alanine, glycine, isoleucine, serine); and enzymes which aminoacylate *Anacystis* tRNA at rates not significantly higher than the respective synthetases from dark-grown cells (threonine, proline). It is possible that the enzymes of the second group have lost activity during the preparation, or our incubation conditions are not optimum. However, these arguments have been excluded for the threonyl-tRNA synthetase. This enzyme species shows a high acylation rate also in dark-grown cell preparations, and this peculiarity cannot be referred to as proplastid-specific enzyme activity which is probably the

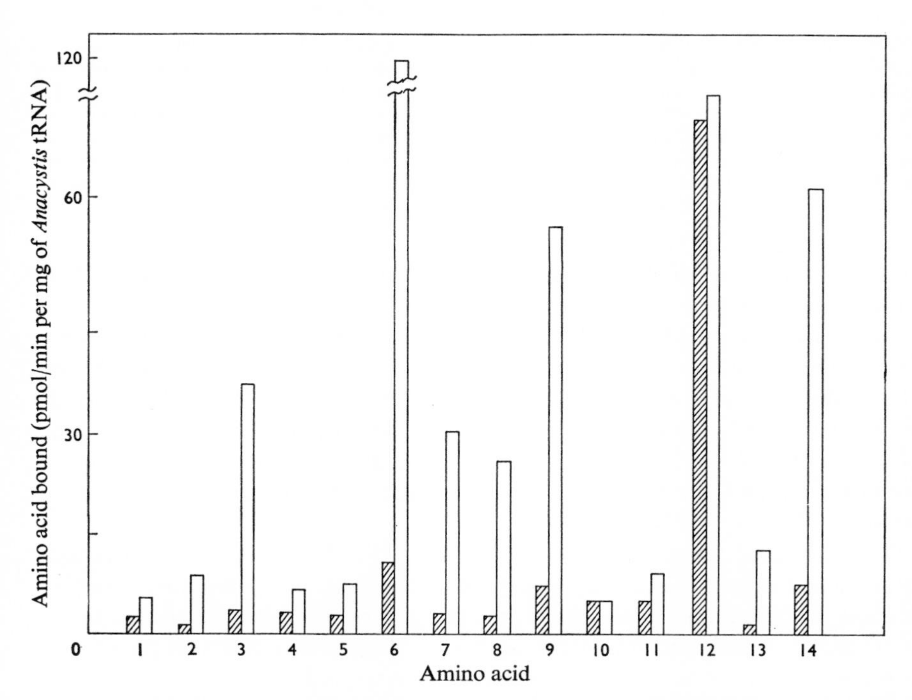


Fig. 4. *Aminoacylation of Anacystis tRNA by crude enzyme preparations from dark-grown (hatched columns) and light-induced E. gracilis (open columns)*

The preparations of tRNA and enzymes and the aminoacylation conditions were as described in the Materials and Methods section: 60 μg of *Anacystis* tRNA and 35 μg of enzyme fraction protein were used. The concentrations (nmol/ml) of the $^{14}C$-labelled L-amino acids were as follows: 1, Ala 5.2; 2, Arg 4.1; 3, Asp 9.7; 4, Gly 4.2; 5, Ile 4.0; 6, Leu 8.0; 7, Lys 6.8; 8, Met 6.6; 9, Phe 4.2; 10, Pro 3.6; 11, Ser 9.4; 12, Thr 8.0; 13, Tyr 3.0; 14, Val 4.5.

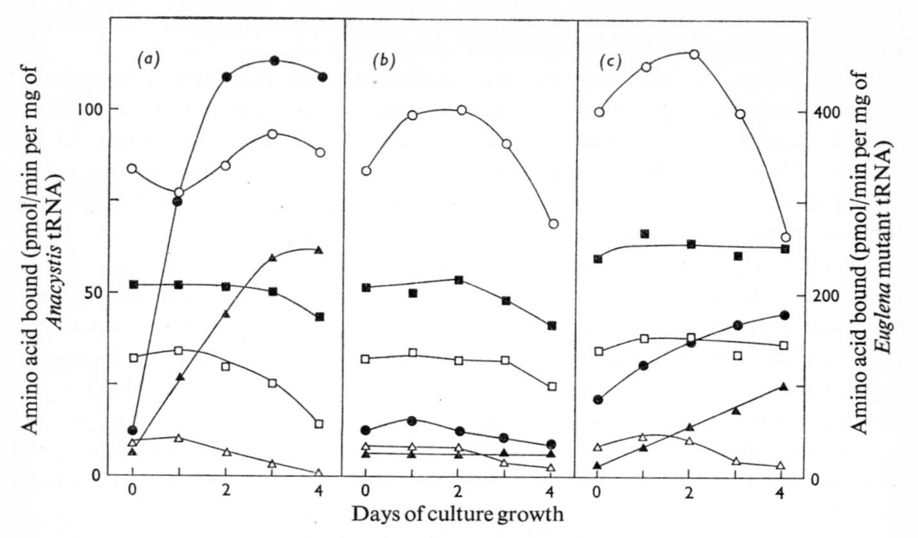


Fig. 5. *Time-course of crude (leucyl, valyl, threonyl) synthetase activities in relation to the culture conditions of heterotrophically grown E. gracilis*

(*a*) Greening cells (permanent illumination, dark-grown cells exposed at zero time); (*b*) cells growing in complete darkness; (*c*) u.v. mutant cells, illuminated at zero time. Cells were harvested at the indicated times, washed, frozen at −30°C for 1 day and thawed before sonication. The preparation of the crude enzyme fractions and tRNA and the determination of enzyme activities are described in the Materials and Methods section. The rates of aminoacyl-tRNA synthesis are calculated on protein basis. A sample of 35 $\mu$g of protein and 60 $\mu$g of tRNA were used per assay. ●, ■, ▲, *Anacystis* tRNA; ○, □, △, *Euglena* mutant tRNA. ●, ○, [$^{14}$C]Leucine; 8 nmol/ml; ▲, △, [$^{14}$C]valine, 4.5 nmol/ml; ■, □, [$^{14}$C]threonine, 8 nmol/ml.

reason for the acylation of *Anacystis* tRNA by most of the synthetases from dark-grown cells (Fig. 4).

Experiments on the kinetics of the light-induced greening of *E. gracilis* demonstrate that the increase of activity of the leucyl- and valyl-tRNA synthetases (Fig. 5*a*) is parallel to chlorophyll synthesis (cf. Fig. 16*a*). Chlorophyll accumulation is suggested as an index for chloroplast development (Kirk, 1968). Since chlorophyll shows a lag-phase in its synthesis and the synthetases do not, we suppose that the formation of these enzymes precede the formation of chloroplast lamellae (Parthier *et al.*, 1972).

The increased aminoacylation of *Anacystis* tRNA is a real light-dependent process and does not occur in dark-grown cells (Fig. 5*b*). However, if the u.v. mutant is illuminated after dark growth, a certain degree of stimulation is observed for leucyl- and valyl-tRNA synthesis (Fig. 5*c*). After 4 days growth, this stimulation is approx. 40% of the increase measured with enzymes from the green cells. No significant stimulation of binding threonine to *Anacystis* tRNA is observed. We suggest that the synthetase involved is not associated with plastid or chloroplast development.

Fig. 5 also shows that the kinetics of the formation of cytoplasmic enzymes are similar for each amino acid in dark-grown, light-induced and mutant *Euglena* cells. This is most convincing with the leucine-activating enzyme, a slight increase of the enzyme specific activity (calculated to the protein concentration of the enzyme preparation) being found during the exponential growth phase. It is followed by a decrease during the stationary phase. This alteration of activity does not support a direct correlation with the light-controlled metabolism, but may reflect the physiological changes during culture growth. However, the decline of leucyl-tRNA synthetase activity during the exponential growth phase of only greening cells (Figs. 5*a* and 16*a*) may be connected with intracellular changes of protein biosynthesis in respect to chloroplast development.

*Separation of chloroplast and cytoplasmic synthetases by hydroxyapatite chromatography*

Crude enzyme preparation from chloroplast-containing material can be separated by means of hydroxyapatite chromatography into distinct peaks of synthetase activity (Figs. 6, 7, 8; cf. Reger *et al.*, 1970; Kanabus & Cherry, 1971; Parthier, 1972; Parthier *et al.*, 1972). In view of the differences of the light-stimulation of the synthetase species (Fig. 4) it is also interesting to know whether these enzymes show a different chromatographic behaviour. The elution pattern and aminoacylation specificity of ten synthetase species are shown in Figs. 6 and 7. The separation into two enzyme activities (enzymes I and II) with different tRNA acceptor specificities was obtained very clearly for the synthetases activating leucine, valine, lysine and serine and less clearly (overlapping separation) for the enzymes specific to arginine, glycine and methionine. These findings improve and correct some of our preliminary results (Parthier, 1972). A separation of isoleucyl- and phenylalanyl-tRNA synthetases into two tRNA-specific

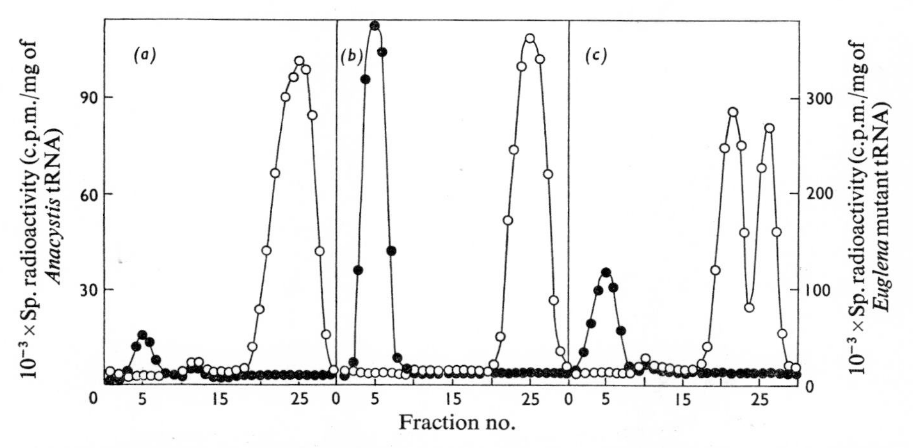


Fig. 6. *Hydroxyapatite chromatography of leucyl-tRNA synthetases from* (*a*) *dark-grown,* (*b*) *light-induced and* (*c*) *light-grown mutant cells of E. gracilis*

The crude enzymes from 3-day-old cultures were prepared as described, applied to columns (15 cm × 1 cm) of hydroxyapatite and eluted with a gradient of 0.01–0.3 M-potassium phosphate buffer, pH 7.5, containing 1 mM-$MgCl_2$, 0.5 mM-$\beta$-mercaptoethanol and 10% (v/v) glycerol. Fractions (5 ml) were collected at 0°C and then tested without dilution for enzyme activity in 125 $\mu$l mixtures containing 100 mM-Tris–HCl buffer, pH 7.5, 3 mM-ATP, 12 mM-$MgCl_2$, 12 mM-KCl, 1.5 mM-$\beta$-mercaptoethanol, 50 $\mu$g of either *Anacystis* tRNA (●) or *Euglena* mutant tRNA (○) and 0.01 mM-[$^{14}$C]leucine.

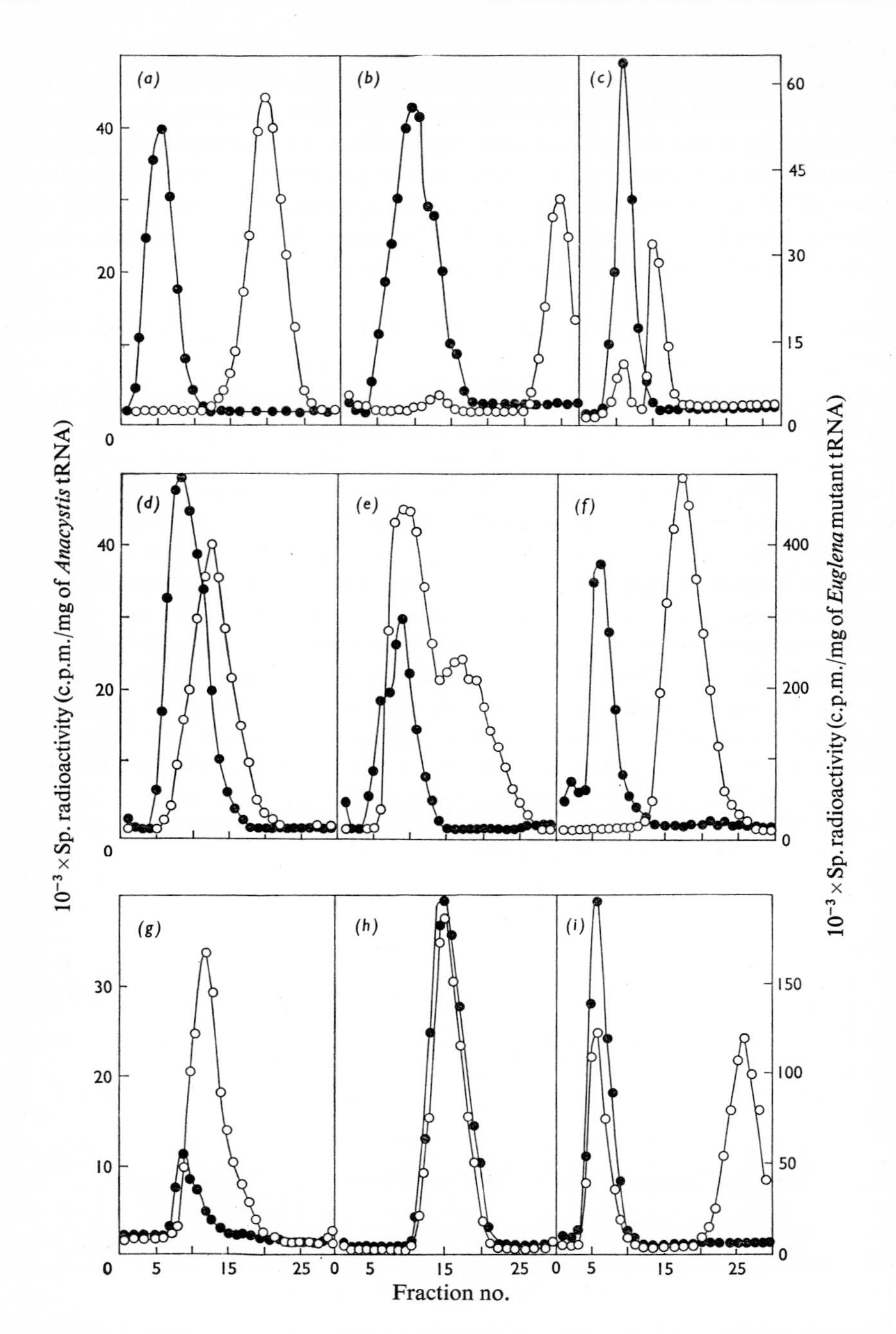


Fig. 7. *Hydroxyapatite chromatography of several aminoacyl-tRNA synthetases from 3-day-old light-induced E. gracilis cells*

The methods were as described in Fig. 6. The enzymes were cognate to the following $^{14}C$-labelled amino acids: (*a*) valine; (*b*) lysine; (*c*) methionine; (*d*) arginine; (*e*) isoleucine; (*f*) serine; (*g*) glycine; (*h*) threonine; (*i*) phenylalanine. ●, *Anacystis* tRNA; ○, *Euglena* mutant tRNA.

fractions was also observed (Figs. 7*e* and 7*i*), however; the enzymes I were likewise able to aminoacylate the tRNA of the u.v. mutant in an appreciable rate.

Threonyl-tRNA synthetase differs from the other species in its chromatographic properties. It is eluted as a single peak with a congruent charging activity for both *Anacystis* tRNA and u.v. mutant tRNA (Fig. 7*h*). This confirms our assumption that *E. gracilis* possesses a single threonine-activating enzyme which charges tRNA of prokaryotic as well as of cytoplasmic origin.

As a common feature of the separation of *Euglena* synthetases on hydroxyapatite it can be stated that the chloroplast enzymes are eluted at lower phosphate concentrations than the respective cytoplasmic enzymes. The use of the terms 'chloroplast enzyme' for enzyme I and 'cytoplasmic enzyme' for enzyme II is justified at least for the leucyl-tRNA synthetases by the finding that enzyme I exclusively charges chloroplast tRNA and enzyme II cytoplasmic tRNA (Fig. 2). It is further shown that the activities of enzyme I in dark-grown and mutant cells is much lower than in green cells (Fig. 6). Finally, the activity of both the crude enzyme (Fig. 5*a*) and the column-separated enzyme (Fig. 14) corresponds to the amount of chlorophyll and thus chloroplast formation.

Table 2 shows that isolated *Euglena* chloroplasts contain synthetase species which charge *Anacystis* tRNA at reasonable rates. It is also evident that the values obtained with cytoplasmic tRNA indicate the contamination of our chloroplast preparation with cytoplasmic synthetases (if one does not assume the presence of cytoplasm-specific enzymes inside the organelles). On the other hand, the amount of leakage of chloroplast-specific enzymes during the preparation of the chloroplast fraction can be estimated from the level of $^{14}$C-labelled amino acids bound to *Anacystis* tRNA in the presence of the soluble (cytoplasmic) fraction. For leucine, valine and serine this attachment is approximately three times higher than with the chloroplast fraction (calculated on protein basis). However, the aminoacylation by the cytoplasmic fraction of the same amount of cytoplasmic tRNA is 24 times higher compared with the aminoacylation of cytoplasmic tRNA catalysed by the chloroplast fraction. This value is still higher for the attachment of serine and valine. The tRNA of the two sources are similarly charged with [$^{14}$C]threonine, if either the chloroplast or the cytoplasm fractions are tested (Table 2). This result may also support the presence of a single threonyl-tRNA

Table 2. *Aminoacyl-tRNA synthetase activity and specificity in isolated chloroplasts*

The chloroplast fraction was prepared from 3-day-old light-grown *E. gracilis* cultures as described in the Material and Methods section. The cytoplasmic fraction was from the supernatant after chloroplast sedimentation and was centrifuged at 120000*g* for 90 min and then dialysed. Aminoacylation rates and acceptor specificity were determined with tRNA from *A. nidulans* and from the u.v. mutant of *E. gracilis*. The incubation conditions were as indicated in Table 1: 20 $\mu$g samples of protein from the fractions were used for the incubations.

| Source | | Amino[$^{14}$C]acyl-tRNA synthesis (pmol/min per mg of tRNA) | | | |
|---|---|---|---|---|---|
| Enzyme | tRNA | Leucine | Serine | Valine | Threonine |
| Chloroplast fraction | *A. nidulans* | 26 | 0.3 | 9 | 3 |
| | *E gracilis* mutant | 9 | 0.01 | 0.1 | 3 |
| Cytoplasmic fraction | *A. nidulans* | 84 | 0.7 | 29 | 28 |
| | *E. gracilis* mutant | 235 | 3.1 | 8 | 38 |

synthetase in *E. gracilis*. We suggest it is localized in the cytoplasm since the low enzyme activity in the chloroplast fraction may be due to cytoplasmic contamination.

*Some characteristics of chloroplast and cytoplasmic leucyl-tRNA synthetases*

Having established a simple method for the separation of chloroplast- and cytoplasm-specific synthetases, some aspects of their properties were examined in order to support further evidence to prove the existence of two distinctly different enzymes. The leucyl-tRNA synthetases were chosen because they are very stable, show a marked light-stimulated increase of enzyme I activity and, on the basis of the high aminoacylation rates *in vitro*, they should occur in the *Euglena* cell in higher concentrations than each of the other enzyme species.

A comparison between the two enzyme activities in the ATP–$PP_i$-exchange reaction and in the [$^{14}$C]leucyl-tRNA synthesis is shown in Fig. 8. A complete chromatographic coincidence of the activities of the two reactions was found, but different reaction rates exist between enzymes I and II. Enzyme I showed a higher ATP–$PP_i$ exchange than enzyme II, but in terms of specific enzyme activity (nmol of [$^{32}$P]ATP formed/min per mg of protein) it was similar for both enzymes (approx. 70 nmol). In contrast, the [$^{14}$C]leucyl-tRNA formation was always found to be 2–5 times higher for enzyme II than for enzyme I. The difference is not due to rate-limiting substrate concentration (e.g. enzyme I-specific $tRNA^{Leu}$ in the tRNA preparations) as we have shown with experiments on enzyme kinetics (R. Krauspe & B. Parthier, unpublished work).

Our results agree with the observation that the activation reaction is three orders of magnitude higher than aminoacyl-tRNA synthesis *in vitro* (but not *in*

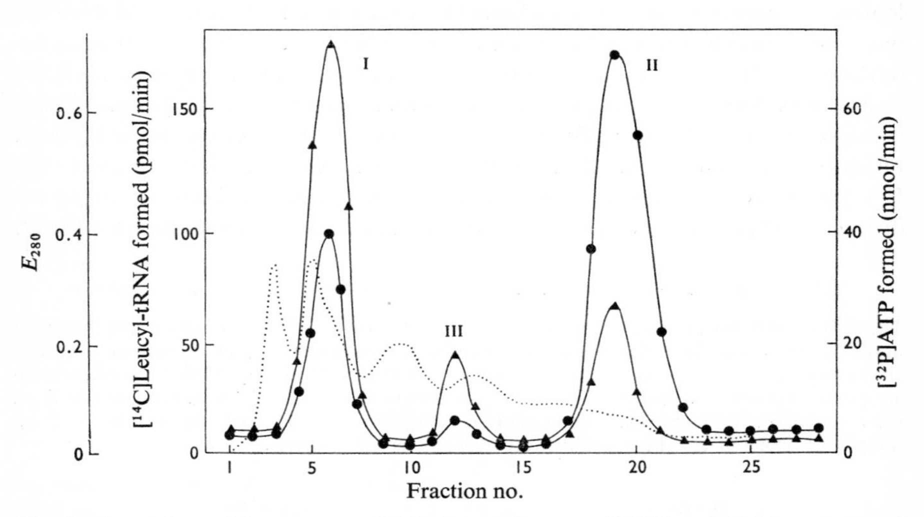


Fig. 8. *Comparison of the activities of ATP–$PP_i$ exchange and [$^{14}$C]leucyl-tRNA synthesis with leucyl-tRNA synthetases separated by hydroxyapatite chromatography*

Hydroxyapatite chromatography and determination of activity were performed as described in the Materials and Methods section. Green cultures (3-day-old) were used for the preparation of enzymes and tRNA. (In some cases a mixture of *Anacystis* tRNA and cytoplasmic $tRNA^{Leu}$ was also used.) ●, [$^{14}$C]Leucyl-tRNA synthesis; ▲, ATP–$PP_i$-exchange reaction. ······, $E_{280}$.

*vivo*; cf. Hall & Tao, 1970). From Fig. 8 we can estimate a 700-fold higher rate in the exchange reaction compared with aminoacylation for the chloroplast enzyme, but the rate is only 120-fold higher for the cytoplasmic enzyme. For bean leaf synthetase preparations (pH 5 enzymes) Hall & Tao (1970) measured 0.5 pmol of amino acid incorporated/min per mg of tRNA, which means a decrease of activity in the order of $10^4$ after destruction of the tissue. We have not determined the aminoacylation rates in *E. gracilis in vivo*, but the values observed *in vitro* were at least two orders of magnitude higher than those reported for the enzyme preparation from bean leaves.

Fig. 8 also indicates a minor third leucyl-tRNA synthetase activity eluted at 0.09 M-phosphate (enzyme III). It seems to occur only in the cells of the stationary growth phase. Its appearance is not dependent on the illumination of the culture and it acylates the tRNA of any source tested. Likewise, we can find a cleavage of the enzyme II activity into two peaks in the u.v. mutant (Fig. 6*c*) and occasionally in dark-grown cells. The two peaks show the same rate of leucyl-tRNA synthesis.

The leucyl-tRNA synthetases I and II exhibit only small differences in their molecular weights. After gel filtration on Sephadex G-200 with marker proteins a

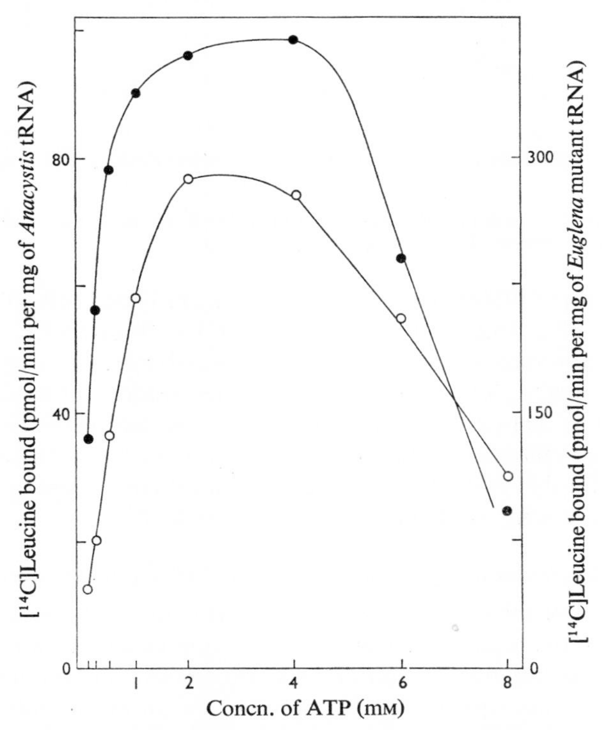


Fig. 9. *Activity of the chloroplast and cytoplasmic leucyl-tRNA synthetases in relation to ATP concentration*

Enzymes I and II of 3-day-old light-induced *E. gracilis* cultures were prepared and separated on hydroxyapatite columns as described in the text. The incubation mixture contained 12 mM-$MgCl_2$, 0.008 mM-[$^{14}$C]leucine, 60 μg of tRNA and 25 μg of enzyme protein and incubation was carried out for 10 min at 30°C. ●, Enzyme I and *Anacystis* tRNA; ○, enzyme II and *Euglena* mutant tRNA.

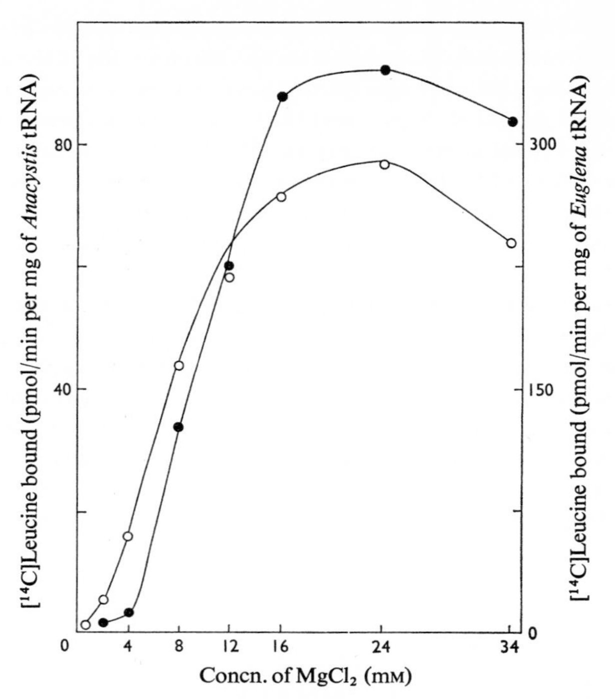


Fig. 10. *Activity of the chloroplast and cytoplasmic leucyl-tRNA synthetases in relation to $Mg^{2+}$ concentration*

The incubation mixtures contained 4mM-ATP. Other details are as in Fig. 9. ●, Enzyme I and *Anacystis* tRNA; ○, enzyme II and *Euglena* mutant tRNA.

molecular weight of 105000 was calculated for enzyme I and 110000 for enzyme II. Requirements of 2–4mM-ATP (Fig. 9) and of 16–20mM-$MgCl_2$ (Fig. 10) for optimum reaction rates are also similar for both enzymes. However, the chloroplast enzyme appears less sensitive to high concentrations of univalent cations such as $K^+$ (Fig. 11) or $NH_4^+$. It is also more resistant than the cytoplasmic enzyme to a pretreatment at elevated temperatures (Fig. 12). These findings are likewise valid for the ATP–$PP_i$-exchange reaction with the exception that here the optimum requirements are 10mM-$Mg^{2+}$ and 5mM-ATP.

## *Effects of antibiotics on the synthesis of aminoacyl-tRNA synthetases*

The existence of the synthetases in chloroplasts does not *a priori* mean their synthesis inside the organelles. Our light-induction experiments have indicated that the increased aminoacylation of *Anacystis* tRNA can be referred to as *de novo* enzyme synthesis paralleling the development of proplastids into chloroplasts. The use of certain inhibitors of protein synthesis is necessary to show whether the chloroplast-specific enzymes are synthesized on the 70S ribosomes of the plastids or on 80S ribosomes in the cytoplasm.

Dark-grown *Euglena* cells were inoculated into growth medium containing the inhibitors chloramphenicol, cycloheximide or nalidixic acid at the concentrations indicated in Fig. 13. The cycloheximide-treated cells and an untreated control were immediately illuminated, the cells treated with chloramphenicol or nalidixic acid and an untreated control were kept in darkness for 8–12 h before illumination. Fig. 13 shows a 20–50% inhibition of most of the enzyme activities from chloramphenicol-treated cultures as compared with untreated cultures. Although nalidixic acid was used in lower concentrations, 20–80% inhibition was observed. In no case was there a total suppression of the light-stimulated aminoacyl-tRNA synthesis. It is obvious from Fig. 13 and Fig. 4 that the inhibition is stronger the higher the light-stimulated enzyme activity. A very low inhibition of threonyl- and prolyl-tRNA synthesis was observed.

In contrast, the aminoacylation of *Anacystis* tRNA appears to be stimulated after the cultures have been grown in the presence of 4 $\mu$g of cycloheximide/ml (14 $\mu$M) for three days. The increase of activity amounts to 200% compared with the enzyme activity of untreated cells (Fig. 13). We may be measuring a spurious stimulation since the values can reflect a higher proportion of chloroplast-specific synthetases in the enzyme preparation of the drug-treated cells, compared with the enzyme fraction of untreated cultures. But the result indicates that the formation of enzymes aminoacylating *Anacystis* tRNA is not inhibited by a drug which blocks protein synthesis on cytoplasmic ribosomes. However, the inhibition of acylating u.v. mutant tRNA by cytoplasmic enzymes from cultures treated with cycloheximide for three days is less than we expected (10–20%). Thus, treatment of *Euglena* cultures with cycloheximide can lead to misinterpretations as the possibility of a rapid inactivation of the drug action cannot be excluded. Shorter periods of exposure have to be chosen.

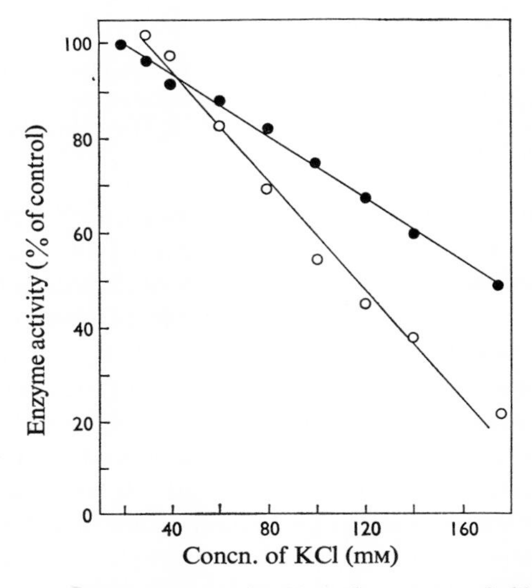


Fig. 11. *Effect of various concentrations of KCl on the activity of chloroplast and cytoplasmic leucyl-tRNA synthetases*

The incubation mixture contained 3 mM-ATP, 12 mM-$MgCl_2$ and various concentrations of KCl: incubation was for 10 min at 30°C. To balance protein concentration and $K^+$ concentration in the elution buffer, enzyme I (fraction no. 5 in Fig. 8) was mixed with fraction no. 16, and enzyme II (fraction no. 18 in Fig. 8) was mixed with fraction no. 3. Other details are as in Fig. 9. ●, Enzyme and *Anacystis* tRNA; ○, enzyme II and *Euglena* mutant tRNA.

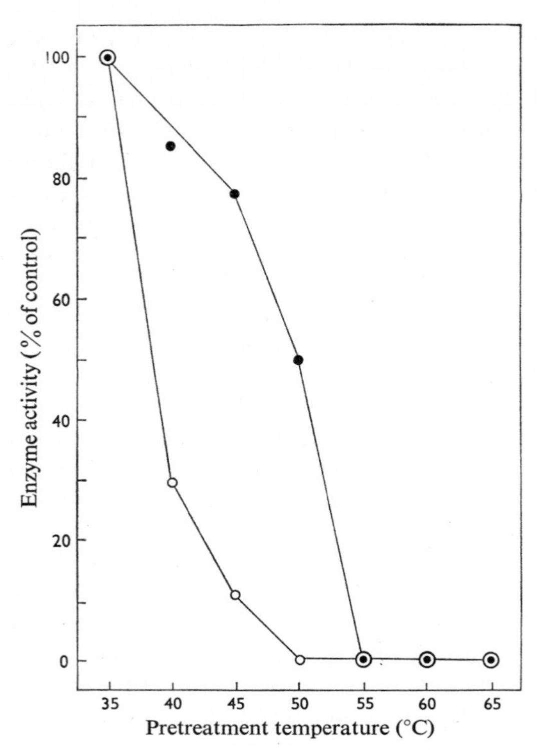


Fig. 12. *Effect of preincubation temperature on the activity of chloroplast and cytoplasmic leucyl-tRNA synthetases*

The separated enzymes were diluted as described in Fig. 11. Samples were incubated for 15 min at the temperatures indicated, rapidly cooled to 0°C, and the synthesis of [$^{14}$C]leucyl-tRNA was determined as described in Fig. 9 (3 mM-ATP, 12 mM-$MgCl_2$). The values are given as percentages of the activity of non-preincubated enzymes.

The effect of various cycloheximide concentrations after 29 h treatment is shown in Table 3. Illumination of the treated cultures caused an enhancement of both chlorophyll and chloroplast-specific leucyl-tRNA synthetase activity with a drug concentration as low as 2 μg/ml. Cycloheximide (4 μg/ml; 14 μM) inhibited chlorophyll formation by 94% and synthetase activity by 85% (the values of the dark-grown cells subtracted); at 8 μg/ml (28 μM) both processes were completely stopped. It is surprising that the cytoplasmic enzymes are less severely affected. A stimulated leucyl-tRNA synthetase activity was found after treatment of the culture with the lowest cycloheximide concentrations, which almost completely blocks cell division (Table 3). Even at the highest drug concentration cytoplasmic synthetase activity is retained. Although cell multiplication is completely prevented the cells were still alive since they move very slowly.

The probable loss of the drug action in *E. gracilis* enables the cells to survive and to recover. If in kinetic experiments, dark-grown cells are illuminated and treated with 4 μg of cycloheximide/ml, a decrease of the cytoplasmic leucyl-tRNA synthetase activity is found until a plateau is reached (Fig. 14*c*). The result can be interpreted as a disturbed turnover of the enzyme, i.e. the normal breakdown rate

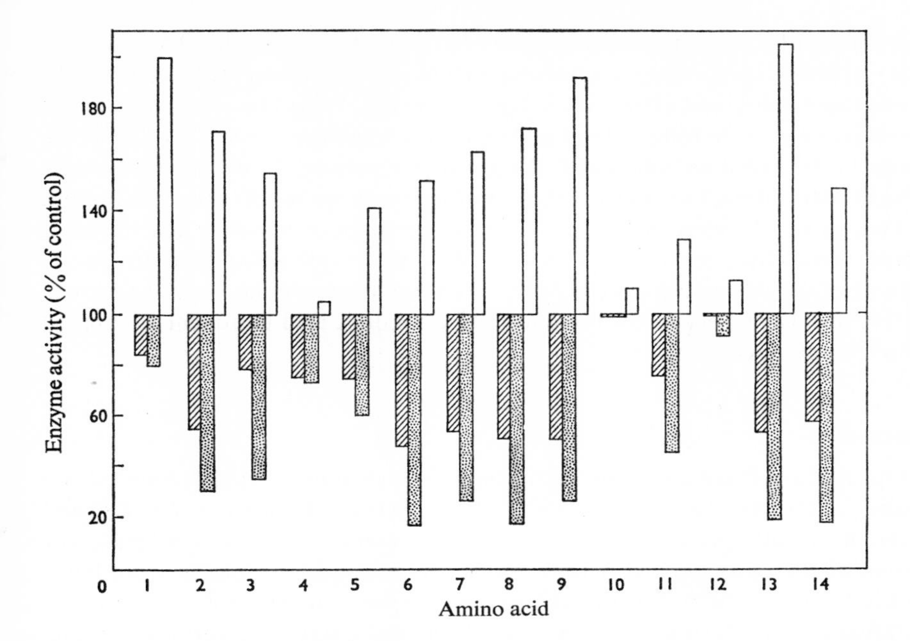


Fig. 13. *Inhibition of aminoacyl-tRNA synthesis after* 3 *days treatment of light-induced E. gracilis with chloramphenicol, nalidixic acid and cycloheximide*

Dark-grown cultures ($4 \times 10^5$ cells/ml) were treated with 3mM-chloramphenicol (hatched columns), 0.2mM-nalidixic acid (dotted columns) or 14$\mu$M-cycloheximide (open columns) and allowed to grow in light for 3 days. Chloramphenicol and nalidixic acid were added 12h before illumination, cycloheximide at the start of illumination. Crude enzymes were prepared and tested for $^{14}$C-labelled aminoacyl-tRNA synthesis with *Anacystis* tRNA. The rates of enzyme activity are expressed as percentages of the rates obtained with enzymes from untreated cells. The numbers and concentrations of the $^{14}$C-labelled amino acids used are shown in Fig. 4.

Table 3. *Influence of the cycloheximide concentration on the leucyl-tRNA synthetase activity in light-induced E. gracilis*

The drug at various concentrations was applied to dark-grown cells ($4 \times 10^5$ cells/ml) at the time of illumination which was maintained for 29h. Then the crude enzyme fractions are prepared and tested for [$^{14}$C]leucyl-tRNA synthesis with tRNA from *A. nidulans* or from the u.v. mutant of *E. gracilis*. Each incubation mixture contained 60$\mu$g of tRNA, 30$\mu$g of protein of the enzyme preparation, 0.01mM-[$^{14}$C]leucine as well as the other components indicated in the Material and Methods section. Cell number and chlorophyll contents were also measured after 29h illumination.

| Cells | Final concn. of cycloheximide (M) | $10^{-6} \times$ Cell no. (per ml) | Chlorophyll ($\mu$g/$10^6$ cells) | [$^{14}$C]Leu-tRNA synthesis (pmol/min per mg of tRNA) | |
|---|---|---|---|---|---|
| | | | | *Anacystis* tRNA | *Euglena* mutant tRNA |
| Dark-grown | 0 | 1.30 | 0.00 | 10 | 205 |
| Light-induced | 0 | 1.75 | 3.45 | 58 | 187 |
| Light-induced | $7 \times 10^{6}$ | 0.43 | 4.00 | 69 | 220 |
| Light-induced | $1.4 \times 10^{5}$ | 0.40 | 0.22 | 18 | 59 |
| Light-induced | $2.8 \times 10^{5}$ | 0.40 | 0.00 | 10 | 40 |

interferes with an arrested biosynthesis of the synthetase molecules. This turnover rate proceeds until a rapid synthesis of chlorophyll and chloroplast-specific synthetase is observed after a short lag period (Fig. 14*c*). The resynthesis of the cytoplasmic enzyme follows later and can soon reach the enzyme activity of the control cells which decreases with the age of the culture. In dependence on the drug concentration, this suggestion is consistent with the results shown in Fig. 13.

The growth kinetics of *E. gracilis* in the presence of nalidixic acid reveal a similar parallelism between the block of chlorophyll synthesis and the suppressed activity of the chloroplast-specific leucyl-tRNA synthetase (Fig. 14*b*). The activity of the cytoplasmic enzyme is not affected by nalidixic acid treatment, compared with the untreated cells.

## Discussion

Green *E. gracilis* cells contain two or three types of aminoacyl-tRNA synthetase species, as shown by Reger *et al.* (1970), Parthier (1972), Parthier *et al.* (1972) and from the results presented here. (Exceptions seem to be the synthetase for threonine and probably also for proline.) Those enzyme species from light-induced (green) cells that considerably stimulate the aminoacylation of *Anacystis* tRNA are separable from the respective cytoplasmic enzymes by chromatography on hydroxyapatite columns. The most striking property of the separated enzymes is their high specificity to the subcellular origin of their cognate tRNA species. Thus, a leucyl-tRNA synthetase eluted with 0.05M-phosphate (enzyme I)

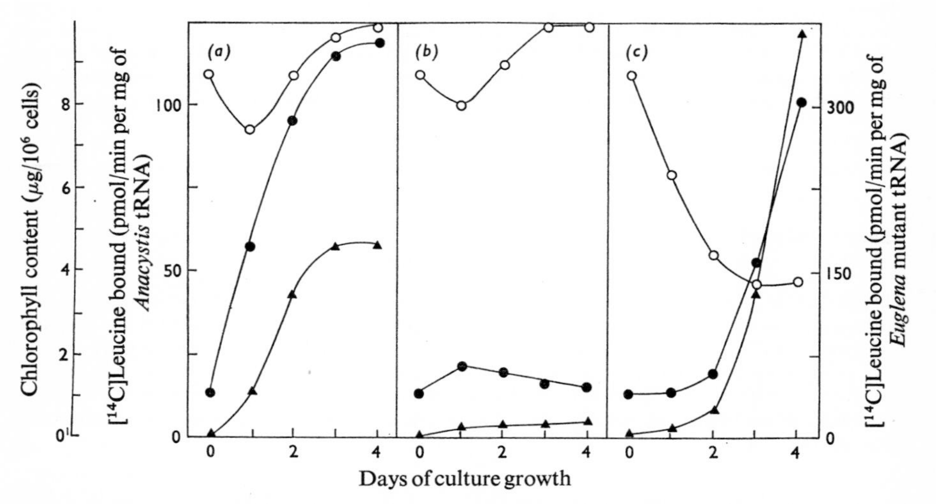


Fig. 14. *Time-course of chlorophyll synthesis and activities of chloroplast and cytoplasmic leucyl-tRNA synthetases after treatment of light-induced E. gracilis with nalidixic acid and cycloheximide*

Dark-grown cultures ($4 \times 10^5$ cells/ml) were treated as described in Fig. 13. The enzymes were prepared at various times, the leucyl-tRNA synthetases were separated by hydroxyapatite chromatography and tested for [$^{14}$C]leucyl-tRNA synthesis with *Anacystis* tRNA (●) or *Euglena* mutant tRNA (○) as described in Fig. 6. Chlorophyll (▲) was measured as described by Arnon (1949). (*a*) Untreated cells; (*b*) cells treated with 0.2mM-nalidixic acid; (*c*) cells treated with 14μM-cycloheximide.

exclusively acylates chloroplast $tRNA^{Leu}$ or *Anacystis* tRNA, but not cytoplasmic $tRNA^{Leu}$ which is, however, specifically charged by an enzyme eluted between 0.17 and 0.20 M-phosphate (enzyme II). In agreement with the reports of Barnett *et al.* (1969), Burkard *et al.* (1970, 1972), Reger *et al.* (1970) and Guderian *et al.* (1972), we can ascribe the two types of synthetases as enzymes localized in the chloroplasts (enzyme I) and in the cytoplasm (enzyme II). This is also likely for the separated leucyl-tRNA synthetases from soybean material (Kanabus & Cherry, 1971), although it has not been explicitly proved.

We have investigated the enzymic attachment of 14 different $^{14}C$-labelled amino acids to tRNA from *A. nidulans* or from a u.v. mutant of *E. gracilis* (analogous to cytoplasmic tRNA), in order to find out whether we may generalize this suggestion. Twelve of the tested synthetases acylating *Anacystis* tRNA showed an increasing activity after illumination of dark-grown *E. gracilis* cultures. A stimulation of the acylation of cytoplasmic tRNA was also observed (after a transient decrease in the period after illumination), but this phenomenon seems to be less connected with the light conditions than with other physiological alterations during the growth of the cultures. The increase of the leucyl-tRNA synthetase activity during the exponential and the early stationary growth phases as well as a remarkable decrease in the late stationary growth phase was also observed with cytoplasmic enzymes prepared from dark-grown cells or from the illuminated chlorophyll-free mutant of *E. gracilis.* Bick & Strehler (1971) have reported a decrease of leucyl-tRNA synthetase activity in aging soybean cotyledons. Our observations may be based on similar physiological changes. The activity of the chloroplast-specific enzyme was not decreased in the same period; it appears metabolically more stable than the cytoplasmic enzyme.

The two leucyl-tRNA synthetases, after separation on hydroxyapatite, showed a number of differences, e.g. in the behaviour to elevated preincubation temperatures or to a high concentration of univalent cations (Anderson & Fowden, 1970) during the ATP–$PP_i$-exchange or aminoacylation reactions. The cytoplasmic enzyme was more sensitive than the chloroplast enzyme. Burkard *et al.* (1970) have also revealed that the cytoplasmic valyl-tRNA synthetase from bean hypocotyls is more sensitive to excess of $Mg^{2+}$ or spermidine than the chloroplast enzyme. However, these differences are restricted to the transfer reaction, the ATP–$PP_i$ exchange was not affected. Both the chromatographic properties and the sensitivity to a physicochemical stress may be ascribed to differences in the amino acid sequence of the polypeptide chains of the two leucyl-tRNA synthetases. The catalytic site of the amino acid-activation step seems not to be involved, since the optimum requirements for $Mg^{2+}$ and ATP are identical for both enzymes. However, the strict tRNA specificity indicates differences of amino acid–nucleotide interactions during the transfer step. Information about the tertiary structures of the cognate tRNA species are required before speculations on the molecular differences in the catalytic sites of the two synthetases can be made.

The increase of the light-stimulated synthetase activities is connected with chloroplast development, as measured by chlorophyll accumulation. In contrast with chlorophyll synthesis, the enzyme formation shows no lag phase. Our inhibitor studies and the experiments on induction kinetics support the idea that the increase of enzyme activity can be referred to as an increase of enzyme concentration in the cell and is not due to an activation of pre-existing enzyme molecules by other light-related processes.

The light-dependent stimulation in aminoacylation of prokaryotic tRNA is neither dependent on the synthesis nor connected with the presence of chlorophyll, although the increase is higher in the presence of chlorophyll. This is shown after illumination of a dark-grown culture of the *E. gracilis* u.v. mutant which lacks the ability to synthesize chlorophyll. Since this mutant accumulates carotenoids during culture growth, it remains to be investigated whether the stage of development of the chromatophores is a prerequisite for the stimulated synthesis of plastid-specific synthetases (analogous to chloroplast development in wild-type cells). Electron-microscopic observations indicate that the mutant blocked in chloroplast development is capable of forming few rudimentary plastids (Neumann & Parthier, unpublished work). Chlorophyll synthesis as an index for chloroplast development is coupled with the formation of certain protein constituents of chloroplast lamellae (Hoober *et al.*, 1969; Eytan & Ohad, 1970; Kirk, 1968, 1970; Smillie *et al.*, 1971). Since we have observed an increase of plastid-specific synthetase activity before chlorophyll can be measured, we suggest that the biosynthesis of these synthetases precedes also the formation of thylakoids necessary for the installation of the pigment molecules. As soluble stroma enzymes the synthetases do not need thylakoid structures for their function, indeed, the multiplication of their molecules may be a prerequisite for a rapid synthesis of thylakoid proteins during chloroplast development.

Thus, the light-stimulated increase of synthetase activities during the greening of *Euglena* cells may represent a massive *de novo* synthesis of plastid-specific enzyme species which are already present in much lower amounts in the proplastides of dark-grown cells. This conclusion is consistent with the low rates of aminoacylation of *Anacystis* tRNA by enzymes of the dark-grown cells. There is evidence that proplastids or etioplasts contain 70S ribosomes (Boardman, 1966; Hirvonen & Price, 1971) able to synthesize chloroplast-specific enzymes in the dark. Consequently, we would prefer the term 'light-stimulation' instead of 'light-induction' of synthetases (Reger *et al.*, 1970) because enzymes of a new type do not appear. Similar considerations can be made for $tRNA^{Val}$ and $tRNA^{Leu}$ in etioplasts of *Phaseolus* (Burkard *et al.*, 1972).

The trigger mechanism of the light-stimulated synthetase multiplication is unknown. The increase in the chlorophyll-free mutant argues against the possibility that photosynthesis or photosynthetic products are primarily involved. It has been known for a long time that during chloroplast development both vesicle fusion to discs or thylakoids, lamellae formation and aggregation (in higher plant chloroplasts) to grana require light (von Wettstein, 1961; Schiff & Zeldin, 1968). The phytochrome system probably participates in these processes in higher plants, but not in the development of *Euglena* chloroplasts (Holowinsky & Schiff, 1970).

Contrary to the findings of Reger *et al.* (1970) we were able to show in dark-grown cells an isoleucyl-tRNA synthetase which charges *Anacystis* tRNA. There are several possibilities to explain the reasons for this disagreement, e.g. the different strain of *E. gracilis* used; the different method of chromatographic separation of the crude enzyme; the lability of the enzyme activity we have observed; finally, the enzyme may be unable to charge chloroplast $tRNA^{Ile}$, but can charge (or even mischarge) *Anacystis* tRNA.

The probability exists, however, that the enzymes from dark-grown cells specific for *Anacystis* tRNA are derived from mitochondria. This problem needs further investigation. We are unable to differentiate between plastid- and mitochondria-specific leucyl-tRNA synthetases of *E. gracilis* by hydroxyapatite rechromatography of enzyme I with a less-steep gradient. Mitochondria-specific tRNA and synthetase species have been demonstrated in *Neurospora crassa* (Barnett *et al.*, 1967) and animal cells (Buck & Nass, 1969). In tobacco leaf cells an apparently mitochondria-bound leucyl-tRNA synthetase can acylate both mitochondrial and cytoplasmic $tRNA^{Leu}$ but not the tRNA from chloroplasts (Guderian *et al.*, 1972).

Our experiments with inhibitors confirm the suggestion that the light-stimulated synthetases charging *Anacystis* tRNA are at least partially synthesized within the plastids and probably also coded by plastid DNA (Parthier *et al.*, 1972). The results are consistent with the inhibition of synthesis of chlorophyll and ribulose-diphosphate carboxylase, another light-stimulated soluble chloroplast enzyme (Smillie *et al.*, 1971; Parthier, 1972). Chloramphenicol as an inhibitor of protein biosynthesis on 70S chloroplast ribosomes suppresses the light-stimulated synthetase activity markedly although not completely. A similar effect is obtained with nalidixic acid, a drug which inhibits chloroplast replication in *E. gracilis* (Lyman, 1967; Ebringer, 1971), probably via prevention of plastid DNA synthesis.

The extent of synthetase inhibition is dependent on the drug concentration and on the time elapsed between drug application and illumination of the culture. The necessity of 6–8 h pretreatment in darkness could be explained for nalidixic acid by the mode of its action, since this time corresponds to one division per cell. It is coupled also with one replication of plastid DNA in *E. gracilis* (Schiff & Zeldin, 1968). Pretreatment with chloramphenicol for 8 h, however, yields less than 60% inhibition of synthetase activities even in concentrations as high as 1 or 2 mg/ml. After treatment with higher doses of nalidixic acid the activity never falls below the values of untreated dark-grown cultures. A selective uptake barrier for the drugs into proplastids is worthy of consideration, and we cannot exclude the participation of mitochondrial synthetases at this time. The biosynthesis of mitochondrial enzymes in *E. gracilis* is not influenced by chloramphenicol (Smillie *et al.*, 1971), moreover, these organelles appear to be increased in size and number after growth of the cells in the presence of chloramphenicol (D. Neumann & B. Parthier, unpublished work). On the other hand, we have observed that chloramphenicol-treated, illuminated cells contain plastids in certainly a higher developmental stage than the proplastids of dark-grown cells; the corresponding higher rate of aminoacyl-tRNA synthesis may reflect this situation.

The effect of cycloheximide on synthesis or activity of the plastid-specific leucyl-tRNA synthetases is more complicated. Light-induced *Euglena* cells treated for 3 days with low concentrations of cycloheximide (2–4 μg/ml) contain not only a higher content of chlorophyll and a higher activity of synthetases charging tRNA of prokaryotic origin than the untreated cells, but also a higher ribulose diphosphate carboxylase activity (Parthier, 1972) and larger chloroplasts with a higher number of lamellae. Growth in the presence of higher but sublethal doses of cycloheximide (8–10 $\mu$g/ml) for a shorter period (29 h) results in a total inhibition of the light-stimulated enzyme activity; it is retained at the value of that in dark-grown cells. This experiment reveals no direct effect of the drug on the activity of the plastid-specific enzyme; such action of cycloheximide cannot be excluded for the lowered activity of the cytoplasmic synthetase. However, the result appears better interpreted in terms of a difference in the turnover rates of the two enzymes. If their biosyntheses are equally affected by cycloheximide, the lowered activity of the cytoplasmic enzyme would then reflect a more rapid turnover of this enzyme and the higher metabolic stability of the plastid-specific synthetase. This alternative is not experimentally answered at this time. Further, a possible inactivation of cycloheximide in the *Euglena* cell could invalidate our suggestions. This process may be responsible for the rapid recovery of plastid-specific syntheses followed by the recovery of cell multiplication after a concentration-dependent time of the period of inhibition.

In this connexion we especially consider the blocked synthesis of certain cytoplasmic proteins necessary for chloroplast division (e.g. chloroplast DNA polymerase; Richards *et al.*, 1971) or for the biogenesis of the thylakoid structures (Hoober *et al.*, 1969; Ben-Shaul & Ophir, 1970; Eytan & Ohad, 1970; Kirk, 1968, 1970).

Our suggestions may be applicable to most of the light-stimulated plastid-specific synthetase species, although we have made extensive studies only with the leucyl-tRNA synthetase. The threonine-activating enzyme of *E. gracilis* obviously represents another type. It differs from the other synthetase species we have tested in a number of aspects, e.g. high activity of charging *Anacystis* tRNA in enzyme preparations of dark-grown cells; no significant stimulation of the activity during the greening process; very low, if any, inhibition after treatment with chloramphenicol or nalidixic acid; no separation by hydroxyapatite chromatography into tRNA-specific fractions. We suggest that *E. gracilis* does not contain a chloroplast-specific threonyl-tRNA synthetase. The threonyl-tRNA necessary for polypeptide synthesis on the chloroplast ribosomes may be synthesized by the enzyme localized in the cytoplasm.

There is now agreement among the laboratories engaged in the complicated story of chloroplast development that this light-induced multistep construction of morphological elements is controlled by both chloroplast and nuclear DNA (Schiff & Zeldin, 1968; Surzycki *et al.*, 1970; Kirk, 1970). The interdependence of the two major genetic systems in the green plant cell represents the degree of reproductive autonomy of the chloroplast, but we have likewise to assume a mutualism in the nutritional or metabolic autonomy of the organelles (Givan & Leech, 1971). Together with tRNA species the synthetases may play a key role in the interrelationship between metabolism and reproduction of the chloroplasts.

We thank Mrs. D. Rümpler and Mrs. Ch. Schimpf for technical assistance.

## References

Aliev, K. A. & Filippovich, I. I. (1968) *Mol. Biol.* **2**, 364–373
Anderson, J. W. & Fowden, L. (1970) *Biochem. J.* **119**, 677–690

Arnon, D. I. (1949) *Plant Physiol.* **24**, 1–15
Barnett, W. E., Brown, D. H. & Epler, J. L. (1967) *Proc. Nat. Acad. Sci. U.S.* **57**, 1775–1782
Barnett, W. E., Pennington, C. J. & Fairfield, S. A. (1969) *Proc. Nat. Acad. Sci. U.S.* **63**, 1261–1268
Ben-Shaul, Y. & Ophir, I. (1970) *Can. J. Bot.* **48**, 929–934
Bick, M. D. & Strehler, B. L. (1971) *Proc. Nat. Acad. Sci. U.S.* **68**, 224–228
Bishop, D. G. & Smillie, R. M. (1970) *Arch. Biochem. Biophys.* **137**, 179–189
Boardman, N. K. (1966) *Exp. Cell Res.* **43**, 474–482
Boardman, N. K., Francki, R. J. & Wildman, S. G. (1966) *J. Mol. Biol.* **17**, 470–489
Buck, C. A. & Nass, M. M. K. (1969) *J. Mol. Biol.* **41**, 67–82
Burkard, G., Guillemaut, P. & Weil, J. H. (1970) *Biochim. Biophys. Acta* **224**, 184–198
Burkard, G., Vaultier, J. P. & Weil, J. H. (1972) *Phytochemistry* **11**, 1351–1353
Ebringer, L. (1971) *Experientia* **27**, 586–587
Ellis, R. J. (1969) *Science* **148**, 477–478
Eytan, G. & Ohad, I. (1970) *J. Biol. Chem.* **245**, 4297–4307
Falk, H. (1969) *J. Cell Biol.* **42**, 582–585
Gillam, I., Milward, S., Blew, D., von Tigerstrom, M., Wimmer, E. & Tener, G. M. (1967) *Biochemistry* **6**, 3043–3056
Givan, C. V. & Leech, R. M. (1971) *Biol. Rev. Cambridge Phil. Soc.* **46**, 409–428
Goffeau, A. (1969) *Biochim. Biophys. Acta* **174**, 340–350
Guderian, R. H., Pulliam, R. L. & Gordon, M. P. (1972) *Biochim. Biophys. Acta* **262**, 50–65
Hall, T. C. & Tao, K. L. (1970) *Biochem. J.* **117**, 853–859
Hirvonen, A. P. & Price, C. A. (1971) *Biochim. Biophys. Acta* **232**, 696–704
Holowinsky, A. W. & Schiff, J. A. (1970) *Plant Physiol.* **45**, 339–347
Hoober, J. K., Siekevitz, P. & Palade, G. E. (1969) *J. Biol. Chem.* **244**, 2621–2631
Kanabus, J. & Cherry, J. H. (1971) *Proc. Nat. Acad. Sci. U.S.* **68**, 873–876
Kirk, J. T. O. (1968) *Planta* **78**, 200–207
Kirk, J. T. O. (1970) *Annu. Rev. Plant Physiol.* **21**, 11–42
Kratz, W. A. & Myers, J. (1955) *Amer. J. Bot.* **42**, 282–287
Lowry, O. H., Rosebrough, N. J., Farr, A. L. & Randall, R. J. (1951) *J. Biol. Chem.* **193**, 265–275
Lyman, H. (1967) *J. Cell Biol.* **35**, 726–730
Lyttleton, J. W. (1962) *Exp. Cell Res.* **26**, 312–317
Mans, R. J. & Novelli, G. D. (1961) *Arch. Biochem. Biophys.* **94**, 48–53
Margulies, M. M. (1964) *Plant Physiol.* **37**, 579–585
Meissner, L., Samtleben, S., Vogt, M. & Parthier, B. (1971) *Biochem. Physiol. Pflanz.* **162**, 526–537
Parthier, B. (1972) *Symp. Biol. Hung.* **13**, 235–248
Parthier, B., Krauspe, R. & Samtleben, S. (1972) *Biochim. Biophys. Acta* **277**, 335–341
Rawson, J. R. & Stutz, E. (1969) *Biochim. Biophys. Acta* **196**, 368–380
Reger, B. J., Fairfield, S. A., Epler, J. L. & Barnett, W. E. (1970) *Proc. Nat. Acad. Sci. U.S.* **67**, 1207–1213
Richards, O. C., Ryan, R. S. & Manning, J. E. (1971) *Biochim. Biophys. Acta* **238**, 196–201
Schiff, J. A. & Zeldin, M. H. (1968) *J. Cell Physiol.* **72**, *Suppl.* **1**, pp. 103–127
Smillie, R. M., Bishop, D. G., Gibbons, G. C., Graham, D., Grieve, A. M., Raison, J. K. & Reger, R. J. (1971) in *Autonomy and Biogenesis of Mitochondria and Chloroplasts* (Boardman, N. K., Linnane, A. W. & Smillie, R. M., eds.), pp. 422–433, North-Holland Publishing Co., Amsterdam and London
Spencer, D. (1965) *Arch. Biochem. Biophys.* **111**, 381–390
Stutz, E. & Noll, H. (1967) *Proc. Nat. Acad. Sci. U.S.* **57**, 774–781
Surzycki, S. J., Goodenough, U. W., Levine, R. P. & Armstrong, J. J. (1970) *Symp. Soc. Exp. Biol.* **24**, 13–37
Tiselius, A., Hjertén, S. & Levin, Ö. (1956) *Arch. Biochem. Biophys.* **65**, 132–141
von Wettstein, D. (1961) *Can. J. Bot.* **39**, 1537–1554
Williams, G. R. & Williams, A. S. (1970) *Biochem. Biophys. Res. Commun.* **39**, 858–863

*Biochem. Soc. Symp.* (1973) **38**, 137–162
*Printed in Great Britain*

# The Nature and Function of Chloroplast Protein Synthesis

By R. J. ELLIS, G. E. BLAIR and M. R. HARTLEY

*Division of Biological Sciences, University of Warwick, Coventry CV4 7AL, U.K.*

## Synopsis

Chloroplasts contain DNA, RNA polymerase and a protein-synthesizing system that are distinct from those outside the organelle. The DNA occurs as many copies of a 40 $\mu$m circular molecule of unique base sequence; this DNA contains genes for chloroplast rRNA, tRNA and the large subunit of Fraction I protein (ribulose diphosphate carboxylase). The chloroplast ribosomes are of the prokaryotic type, and represent up to 50% of the total ribosomes in a leaf. Isolated chloroplasts use light energy to synthesize the large subunit of Fraction I protein as the sole detectable soluble product, and several membrane-bound proteins. The chloroplast ribosomes represent a high proportion of total leaf ribosomes because one of their products is a component of probably the most abundant protein in Nature. The small subunit of Fraction I protein is encoded in the nucleus, and is probably synthesized by cytoplasmic ribosomes, as are most of the Calvin-cycle enzymes, and chloroplast RNA polymerase. The formation of chloroplasts thus results from an interplay between nuclear and plastid genomes that involves a massive transport of proteins across the chloroplast envelope.

## Introduction

Chloroplasts are the major type of plastid that distinguishes plant cells from animal cells. Life on this planet is maintained largely by the solar energy trapped by chloroplasts in the process of photosynthesis, and hence it is not surprising that a huge literature exists on the mechanism of photosynthesis. However, chloroplasts are interesting for reasons other than their possession of the photosynthetic machinery. The whole concept of the extranuclear inheritance of some cell organelles arose from the classical studies of chloroplast behaviour. An important and expanding field of modern plant biochemistry is concerned with the development and function of chloroplasts as discrete entities within the plant cell; thus, photosynthesis is now viewed as only one aspect of chloroplasts, albeit a most important one.

Chloroplasts in higher plants are lens-shaped green bodies, usually 4–6 $\mu$m in diameter; in the algae they can assume all manner of shapes and sizes. Each chloroplast is delimited by a double outer membrane which surrounds a matrix or stroma. There are no pores in this double membrane, so far as can be judged by electron microscopy. In intact cells, the stroma appears to be in constant motion (Wildman *et al.*, 1962). It contains DNA, ribosomes, osmiophilic globules, amino acids, nucleotides, organic acids, saccharides, other intermediates, inorganic ions and at least fifty enzymes (Smillie & Scott, 1970). The protein complement is unusual in that one protein is present in much larger amounts than any other; this so-called Fraction I protein can account for up to 50% of the total soluble protein in a leaf extract (Kawashima & Wildman, 1970). It may well be the most abundant protein in nature. Fraction I protein has been purified to electrophoretical and ultracentrifugal homogeneity, when it possesses only one enzymic activity, that of ribulose diphosphate carboxylase (EC 4.1.1.39). This protein has a molecular weight of just over 500000 and consists of large and small subunits (Rutner & Lane, 1967). Fraction I protein thus possesses two special features, it is both the major protein component of leaves and the enzyme which carries out the vital $CO_2$-fixing step in photosynthesis.

Embedded in the stroma is a series of membranes or lamellae which contain DNA, DNA polymerase, RNA polymerase, ribosomes, chlorophylls, carotenoids, quinones, phospholipids, glycolipids, and the proteins associated with photon capture, electron transport and photophosphorylation (Kirk & Tilney-Bassett, 1967; Boardman, 1968). The lamellae are composed of membrane-bounded sacs which are flattened so that the opposite sides are close together. Each flattened sac is termed a thylakoid (Menke, 1962). Thylakoids often occur in regular closely packed stacks, or grana. Grana are not always present in chloroplasts, and the extent of their formation can be altered by varying the conditions of growth (Ballantine & Forde, 1970). There is even a viable mutant of *Chlamydomonas* which totally lacks grana; it has single thylakoids (Goodenough *et al.*, 1969). Although the significance of grana is not clear, their regular structure is the most striking feature of chloroplasts as revealed by the electron microscope.

From this brief sketch of chloroplasts, it is possible to outline the major questions that must be answered if the intracellular differentiation of these organelles is to be understood.

(1) What are the precise biosynthetic capacities of chloroplasts? Which precursors are made in the cytoplasm, and how is their entry into the chloroplast controlled?

(2) Which proteins do chloroplast ribosomes make? Do they make all the proteins of the chloroplast, and if not, which proteins do they make? Where do mRNA species, translated by chloroplast ribosomes, come from, the chloroplast DNA, or the nuclear DNA, or both? If some chloroplast proteins are made on cytoplasmic ribosomes, how do they enter the chloroplast?

(3) What are the functions of chloroplast DNA? Are all mRNA species that are transcribed from chloroplast DNA, subsequently translated on the chloroplast ribosomes, or do some leave the chloroplast to be translated on cytoplasmic ribosomes?

(4) How are the membranes of the thylakoids assembled from their components? What controls the formation of grana?

(5) What is the mechanism ensuring the replication of chloroplasts and their transmission to the next generation?

The state of our present knowledge can be summarized by saying that only very tentative answers to a few of these questions can yet be given (Kirk, 1970, 1971*a*,*b*; Miller, 1970; Levine & Goodenough, 1970; Boardman *et al.*, 1971; Givan & Leech, 1971). In this review, the nature and function of protein synthesis by chloroplasts will be considered in the light of the concept of chloroplast autonomy.

**Concept of Chloroplast Autonomy**

The concept of chloroplast autonomy was founded on the observation that, in the algae, chloroplasts can be seen to divide and to be passed to the new cells in cell division (Strasburger, 1882; Green, 1964). Chloroplast division in the cells of higher plants has been observed very infrequently; indeed those investigators who probably have the most experience in the observation of chloroplasts (Honda *et al.*, 1971) have never directly observed it. However, from a study of the size distributions of chloroplasts in the cells of developing leaves, these workers have concluded that new chloroplasts are produced by fission of a small sub-population of constricted mature chloroplasts. These mature chloroplasts develop in the first instance from proplastids present in the undifferentiated cells in the leaf primordia. Studies of the inheritance of abnormal chloroplasts have shown that chloroplasts can be inherited via the maternal parent in some higher plants; the physical basis for maternal inheritance could be the occurrence of proplastids in the egg cell, but not in the pollen tube (Kirk & Tilney-Bassett, 1967). The cytological and genetic evidence thus supports the view, first proposed by Schimper (1885) and Meyer (1883), that plastids do not arise *de novo*, but are formed by the division of pre-existing plastids.

The term 'autonomy' means literally 'self-government'; our current dogma maintains that for any biological system to be autonomous it must contain at least four components: (*a*) DNA to code for its entire structure; (*b*) DNA polymerase to replicate the DNA; (*c*) RNA polymerase to transcribe the DNA; (*d*) protein-synthesizing apparatus to translate the mRNA species into all the necessary proteins.

In principle, an autonomous system has no necessary requirement for intermediary metabolism, since a supply of nucleotides, amino acids and inorganic ions could be obtained by uptake from the environment, and hence, in such an environment autonomous bodies could replicate. There is a large literature attesting to the fact that chloroplasts contain DNA, DNA polymerase, RNA polymerase and a protein-synthesizing apparatus. However, there is also abundant evidence that many genes concerned with chloroplast structure and function in higher plants and some algae are inherited in a Mendelian fashion and are, therefore, located in the nucleus (Kirk & Tilney-Bassett, 1967; Levine, 1969; Apel & Schweiger, 1972). The same may be true for *Chlamydomonas*, but the conclusion is less certain for this organism because reproduction is isogametic and involves chloroplast fusion (Levine & Goodenough, 1970; Cavalier-Smith, 1970). There is increasing evidence that many chloroplast enzymes, including RNA polymerase, are synthesized on cytoplasmic ribosomes (see below). It is important to realize that the demonstration of the cytological continuity of plastids is necessary, but not sufficient, to establish that plastids replicate independently of nuclear control. No successful attempts to culture chloroplasts *in vitro* have been substantiated, although isolated chloroplasts have been reported to undergo splitting (Ridley & Leech, 1970).

How then are we to regard the concept of chloroplast autonomy? It is our contention that chloroplasts are not autonomous in any meaningful sense; the term is useful only as a quick way to describe the fact that chloroplasts contain some genes and make some proteins, but no more than this. The formation of chloroplasts results from a complex interplay between plastid and nuclear genomes (Surzycki *et al.*, 1970).

The year 1962 marks a turning point in the study of chloroplast development. In that year Ris & Plaut (1962) reported deoxyribonuclease (DNAase)-sensitive fibrils lying in areas of low electron density in chloroplasts from many plants, and Lyttleton (1962) showed that chloroplasts contain 70S ribosomes which are distinct from the 80S ribosomes found in the cytoplasm of the same cells. In the subsequent decade much information has been obtained about the nature of the DNA and protein-synthesizing machinery found in chloroplasts; this information will now be considered.

## Chloroplast DNA

The existence of DNA in chloroplasts has been established by both cytological and biochemical criteria. Subsequent work to the report of Ris & Plaut (1962) showed that DNA was associated with chloroplast preparations isolated both from algae and from higher plants (Kirk & Tilney-Bassett, 1967; Smillie & Scott, 1970; Kirk, 1971*a*), and the bulk of the work on chloroplast DNA has been done with these preparations. In these preparations the DNA is attached to the membranes of the chloroplast; for example, Woodcock & Fernandez-Moran (1968) observed DNA fibrils associated with granal and intergranal membranes of lysed spinach chloroplasts. It is not clear whether this association is real or reflects some artifact of preparation. It is difficult to be sure that purified chloroplasts are not contaminated with nuclear fragments (e.g. Kung & Williams, 1969), and it is not surprising that controversy is still rampant in this field (e.g. Bard & Gordon, 1969; Kirk, 1971*a*). There are four characteristics which may be used, but not all are applicable in every case.

(1) Buoyant density: depending on the organism, chloroplast DNA can have a lower density, the same density, or a higher density than nuclear DNA. In *Euglena* and *Chlamydomonas*, chloroplast DNA has a sufficiently lower density to be resolved as a satellite band from nuclear DNA by analytical centrifugation (Sager & Ishida, 1963; Edelman *et al.*, 1964; Chiang & Sueoka, 1967). By contrast, chloroplast DNA from a wide range of higher plants has a very constant density of $1.697 \pm 0.001$ which is too close to that of nuclear DNA in species such as spinach and tobacco to permit a clear separation by analytical centrifugation (Whitfeld & Spencer, 1968; Ingle *et al.*, 1970; Kirk, 1971*a*). It must be emphasized that some of the earlier studies which claimed that chloroplast DNA from tobacco and spinach can be distinguished readily from nuclear DNA on a density basis could not be substantiated by Whitfeld & Spencer (1968), but Tewari & Wildman (1970) have reiterated that separation of the two in tobacco has been achieved in their laboratory.

(2) Ease of renaturation: a more useful criterion than the buoyant density for detecting chloroplast DNA from higher plants is the ease with which it will renature after heat denaturation (Tewari & Wildman, 1966; Whitfeld & Spencer, 1968; Wells & Birnstiel, 1969; Kung & Williams, 1969; Wells & Sager, 1971;

Bastia *et al.*, 1971). Nuclear DNA renatures to only a slight extent (Kung & Williams, 1969; Wells & Birnstiel, 1969).

(3) Absence of 5-methylcytosine: nuclear DNA usually contains 5-methylcytosine as a minor component, but chloroplast DNA from both algae and higher plants contains no detectable amount (Brawerman & Eisenstadt, 1964; Granick & Gibor, 1967). Whitfeld & Spencer (1968) regard the absence of this base as the most reliable criterion for establishing the purity of chloroplast DNA; a divergent opinion is held by Bard & Gordon (1969), who think there are two distinct chloroplast DNA species, one of which contains a small amount of 5-methylcytosine.

(4) Absence of histone: purified chloroplast DNA is not complexed with basic proteins, and in this respect stands in contrast with nuclear DNA (Tewari & Wildman, 1969). The demonstration of the absence of histone-like proteins in isolated chloroplast DNA confirms the microscopic observations by Ris & Plaut (1962), that chloroplasts contain DNA fibrils in the same form as they appear in the nucleoplasms of bacteria.

It is clear that although there is reasonable evidence for the occurrence of DNA in chloroplasts, there is still much dispute about its nature (Kirk, 1971*a*). The problems of contamination by nuclear, mitochondrial, and possibly bacterial DNA are very real. Usable criteria for chloroplast DNA must be established, but there is always a danger that some of the chloroplast DNA will not meet these criteria. For instance, if chloroplasts contain a minor DNA component which is a nuclear transcript, contains 5-methylcytosine and renatures poorly, it is doubtful whether present techniques could identify it as chloroplast in location.

The amount of DNA per chloroplast varies over a wide range, from 0.1 fg in *Acetabularia* to 40 fg in spinach; most values are in the range of 1–10 fg (Smillie & Scott, 1970). It is suspicious that by far the lowest value is for *Acetabularia* where the problem of nuclear contamination can be eliminated. The coding potential of 1 fg of DNA is about 1600 proteins each of molecular weight 27000 (Kirk & Tilney-Bassett, 1967). Such estimates suggest that chloroplast DNA has at

Table 1. *Analytical and kinetic complexities of chloroplast DNA*

| Organism | Analytical complexity (amount of DNA per chloroplast or organism in daltons) | Kinetic complexity (daltons) | Reference |
|---|---|---|---|
| *Euglena gracilis** | $5.8 \times 10^9$ | $0.9 \times 10^8$ | Stutz (1970), Manning *et al.* (1971) |
| *Chlamydomonas reinhardii* (gamete) | $5.16 \times 10^9$ | $1.94 \times 10^8$ | Bastia *et al.* (1971) |
| *Chlamydomonas reinhardii* (gamete) | $4.3 \times 10^9$ | $2 \times 10^8$ | Wells & Sager (1971) |
| *Lactuca sativa* | $2 \times 10^9$ | $1.2 \times 10^8$ | Wells & Birnstiel (1969) |
| *Nicotiana tabacum* | $3 \times 10^9$ | $1.1 \times 10^8$ | Tewari & Wildman (1970) |
| *Escherichia coli* | $2.5 \times 10^9$ | $2.5 \times 10^9$ | Watson (1965) |
| Bacteriophage T4 | $1.3 \times 10^8$ | $2.1 \times 10^8$ | Abelson & Thomas (1966) |

* Values corrected for the low G + C content of *Euglena* chloroplast DNA.

least the potential for coding for a substantial amount of genetic information, and this has bolstered the view that chloroplasts are autonomous. However, recent work suggests that each chloroplast contains multiple copies of DNA, so that the information content is less than appears from measurements of the total amount of DNA. Table 1 lists the kinetic complexities of chloroplast DNA from several algae and higher plants. The kinetic complexity is the size of the unique nucleotide sequence of a given DNA sample; this can be estimated from the rate of renaturation of the DNA (Wetmur & Davidson, 1968). It is striking that the kinetic complexities of chloroplast DNA from both algae and higher plants are all in the range $1 \times 10^8$–$2 \times 10^8$ daltons. This may be coincidental, or it could indicate that the information content necessary for chloroplast DNA is basically similar throughout the plant kingdom. In any event, these results clearly suggest that chloroplasts contain multiple copies of DNA, since the kinetic complexities are always much fewer than the analytical complexities. These data do not rule out the possibility that microheterogeneity in nucleotide sequence exists between the copies. The chloroplast DNA from *Euglena* has recently been shown to be circular with a contour length of $40 \mu m$ (Manning *et al.*, 1971). A molecule of this length has a molecular weight of $83 \times 10^6$ which is very similar to the kinetic complexity of *Euglena* DNA derived from renaturation studies. This correlation suggests that the total genetic information carried by chloroplast DNA is accommodated by the length of the circular molecule. Chloroplast DNA from other organisms has not been shown to be circular; this may be so for technical reasons, or there may be variation among different species in this respect. Woodcock & Fernandez-Moran (1968) found DNA fibrils up to $150 \mu m$ long in lysates of spinach chloroplasts. Although the existence of multiple copies lowers the coding potential of chloroplast DNA, this decreased potential is still greater than that found for mammalian mitochondrial DNA. The kinetic complexity of chloroplast DNA is similar to that of bacteriophage $T_4$ (Table 1), which is known to contain at least 100 different genes.

Although it has been known for nearly 10 years that chloroplasts contain DNA, we could not, until very recently, point to a single protein which this DNA specifies. The first demonstration that chloroplast DNA probably does have a function was provided by Edelman *et al.* (1965). They found that irradiation of the cytoplasm, but not of the nucleus, of *Euglena* with a u.v. microbeam, gave rise to a high percentage of bleached cells which could be grown heterotrophically. These growing cells contained no detectable amount of chloroplast DNA. Thus, the loss of the ability to make functional chloroplasts is correlated with the absence of chloroplast DNA. Unfortunately, such experiments have not been incisive enough to indicate what is encoded in this DNA. Many mutations which are inherited in a non-Mendelian fashion are known in higher plants and algae (Kirk & Tilney-Bassett, 1967; Levine & Goodenough, 1970). These mutations lead to noticeable changes in chloroplast structure or in the amounts of photosynthetic pigments; none of them has so far been traced to a change in any specific protein. However, Wildman reported at a recent meeting that the mutants of the large subunit of Fraction I protein are inherited in a maternal fashion (Chan & Wildman, 1972), whereas the small subunit is coded for in the nucleus (Kawashima & Wildman, 1972).

Molecular hybridization experiments suggest that one function of chloroplast DNA is to code for the RNA of the chloroplast ribosomes. In different organisms 0.5–1.5% of chloroplast DNA can be saturated with chloroplast ribosomal RNA (Scott & Smillie, 1967; Tewari & Wildman, 1968; Ingle *et al.*, 1970). However, chloroplast ribosomal RNA also hybridizes to a significant extent with nuclear DNA, and Ingle *et al.* (1970) suggest that caution should be exercised in interpreting these hybridization experiments. An intriguing possibility is that there are two types of chloroplast ribosomal RNA, one encoded in the nuclear DNA, and the other encoded in the chloroplast DNA. If this is so, it should be possible to distinguish between them by performing hybridization experiments with ribosomal RNA, which has already been hybridized with nuclear or chloroplast DNA, and then reisolated from the hybrid state.

The most direct way to demonstrate that chloroplast DNA codes for at least some of the chloroplast ribosomal RNA is to persuade isolated chloroplasts to synthesize ribosomal RNA. Berger (1967) demonstrated that chloroplasts isolated from enucleated *Acetabularia* cells incorporated [$^{14}$C]uracil or [$^{14}$C]-nucleoside triphosphates into RNA species which co-sedimented on sucrose density gradients with ribosomal and transfer RNA from *Escherichia coli*. In view of the controversy as to the precise sizes of *Acetabularia* chloroplast ribosomal RNA (e.g. Woodcock & Bogorad, 1970) and the recent finding that isolated 'chloroplasts' may in fact consist of chloroplasts associated with variable amounts of cytoplasm surrounded by a membrane (Bidwell, 1972), caution should be exercised in the interpretation of these results. Analysis of the RNA

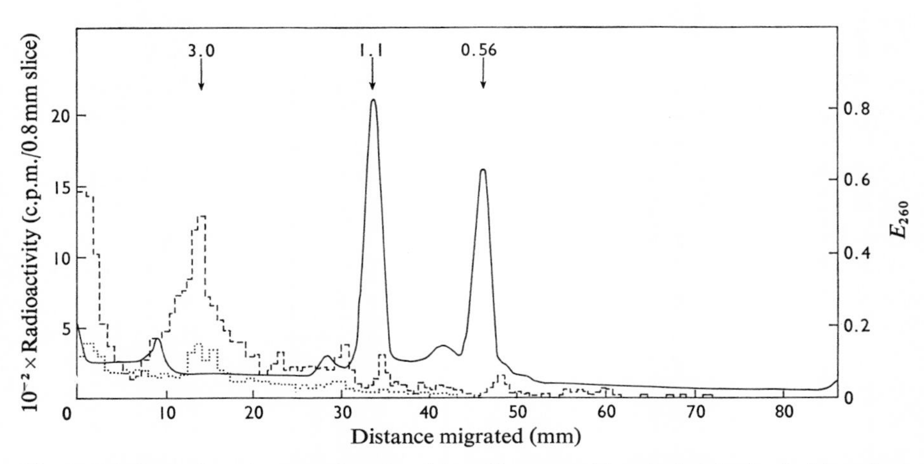


Fig. 1. *Gel-electrophoretic separation of nucleic acid extracted from isolated spinach chloroplasts incubated with [$^{3}$H]uridine*

Chloroplasts were isolated from young spinach leaves by the method described in Fig. 2. Chloroplast suspensions (1.0 ml) containing 250 $\mu$g of chlorophyll were incubated with 10 $\mu$Ci of [$^{3}$H]uridine at 20°C and 15000 lx of red light for 45 min. Nucleic acid was extracted by the phenol–detergent method (Leaver & Ingle, 1971) and 30 $\mu$g was fractionated on 2.2% polyacrylamide gels in the EDTA-containing buffer system described by Loening (1969). The values above the peaks represent their molecular weights $\times 10^{-6}$. ——, $E_{260}$; ---, radioactivity, chloroplasts incubated with [$^{3}$H]uridine; ······, radioactivity, chloroplasts incubated with [$^{3}$H]uridine plus actinomycin D (10 $\mu$g/ml).

Table 2. *Protein synthesis by isolated chloroplasts*

| Species | Characteristics of protein synthesis | Remarks | Authors |
|---|---|---|---|
| *Nicotiana tabacum* | Not dependent on added ATP. Not inhibited by ribonuclease. Stimulated twofold by both light and oxygen | Incorporation probably due largely to bacteria | Stephenson *et al.* (1956) |
| *Spinacia oleracea* | Not dependent on added ATP. Stimulated by light, unaffected by chloramphenicol. Isolated ribosomes show some dependence on added ATP for incorporation | Results with chloroplasts attributed to bacterial contamination, but results with isolated ribosomes thought not to be due to this | App & Jagendorf (1963, 1964) |
| *Euglena gracilis* | Highly dependent on added ATP and tRNA. Inhibited by ribonuclease, puromycin and chloramphenicol. Light stimulates incorporation twofold in the presence of NADP, ADP and $P_i$ | Chloroplasts broken in assay. Cytoplasmic ribosomes not inhibited by chloramphenicol | Eisenstadt & Brawerman (1964), Avadhani & Buetow (1972) |
| *Nicotiana tabacum* | Highly dependent on added ATP. Inhibited by ribonuclease, puromycin and the D-*threo* isomer of chloramphenicol. Deoxyribonuclease and actinomycin D inhibit not more than 25%. Cycloheximide does not inhibit. Ribosomes present in both stroma and thylakoid fractions | Chloroplasts probably broken in assay despite use of Honda medium. Cytoplasmic ribosomes not inhibited by chloramphenicol. No discrete products identifiable | Spencer & Wildman (1964), Francki *et al.* (1965), Boardman *et al.* (1965, 1966), Ellis (1969, 1970), Chen & Wildman (1970) |
| *Spinacia oleracea* | Highly dependent on added ATP. Ribonuclease, puromycin and chloramphenicol are strongly inhibitory. Light stimulates tenfold in presence of ADP, pyocyanine and $P_i$. Deoxyribonuclease and actinomycin D inhibit not more than 20% | Activity largely remained in chloroplasts washed with hypo-osmotic media suggesting a high proportion of membrane-bound ribosomes | Spencer (1965) |
| *A. mediterranea* | Not dependent on added ATP or light. Inhibited by puromycin and chloramphenicol, but not by ribonuclease or deoxyribonuclease. Actinomycin D inhibits at high concentration. Chloroplasts not lysed by hypo-osmotic media or Triton X-100 | Nature of energy source obscure. Characteristics very different from those found for other chloroplasts, perhaps due to an outer membrane with unusual intactness or to occurrence of 'cytoplasts' | Goffeau & Brachet (1965), Goffeau (1969), Woodcock & Bogorad (1971), Bidwell (1972) |

Table 2.—*continued*

| Species | Characteristics of protein synthesis | Remarks | Authors |
|---|---|---|---|
| *Triticum vulgare* | Highly dependent on added ATP, GTP and $Mg^{2+}$. Inhibited by puromycin, ribonuclease and chloramphenicol. Activity present in both stroma and thylakoid fractions | Chloroplasts probably broken in assay despite use of Honda medium | Bamji & Jagendorf (1966), Ranaletti *et al.* (1969) |
| *Lycopersicum esculentum* | Partially dependent on added ATP and $Mg^{2+}$. Puromycin, chloramphenicol, ribonuclease and actinomycin D inhibit not more than 50% | Precautions taken to ensure aseptic conditions | Hall & Cocking (1966), Davies & Cocking (1967) |
| *Phaseolus vulgaris* | Highly dependent on added ATP. Inhibited by ribonuclease only if chloroplasts are broken. Inhibited by chloramphenicol, spectinomycin and linomycin | Chloroplasts break during assay in Honda medium | Parenti & Margulies (1967), Margulies *et al.* (1968), Margulies & Parenti (1968), Drumm & Margulies (1970), Margulies (1970), Ellis (1970) |
| *Spinacia oleracea* | Stimulated tenfold by added ATP and 20-fold by light in absence of added catalysts. Not inhibited by 3-(3,4-dichlorophenyl)-1,1-dimethylurea | Precautions taken to isolate intact chloroplasts | Ramirez *et al.* (1968) |
| *Pisum sativum* | Stimulated tenfold by added ATP and 20-fold by light in absence of added catalysts, provided chloroplasts are kept intact. Inhibited by chloramphenicol, spectinomycin, low concentrations of lincomycin and 3-(3,4-dichlorophenyl)-1,1-dimethylurea. Not inhibited by ribonuclease or actinomycin D, if chloroplasts are kept intact | Incorporation is due to intact chloroplasts only. Products identified as large subunit of Fraction I protein and a membrane-bound protein | Ellis & Hartley (1971), Blair & Ellis (1972), Ellis & Forrester (1972) |

synthesized by isolated chloroplasts from higher plants has, in all published reports, shown it to be heterogenous in size with no discrete RNA species recognizable (Tewari & Wildman, 1969; Spencer *et al.*, 1971). However, all these experiments were carried out under conditions where the incorporation was due to lysed chloroplasts, and hence where controlling factors may have been lost by dilution. An additional problem is the ease of degradation of chloroplast rRNA (Leaver & Ingle, 1971). Recent work in our laboratory has revealed that intact chloroplasts isolated from young spinach leaves carry out a light-dependent incorporation of $[^3H]$uridine into a discrete RNA species with a high molecular weight of about $3 \times 10^6$ (Fig. 1). An RNA species of similar characteristics has been demonstrated in chloroplasts isolated from excised leaves labelled with $[^{32}P]P_i$, and kinetic studies suggest that this may be a large precursor to chloroplast rRNA (Hartley & Ellis, 1972). The fact that the incorporation *in vitro* is dependent on light, and is insensitive to added ribonuclease which cannot penetrate intact chloroplasts, indicates that this high-molecular-weight RNA species is transcribed from chloroplast DNA.

Since chloroplast ribosomes appear to be synthesized inside the chloroplast (see below), a plausible guess would be that the ribosomal proteins are encoded in the chloroplast DNA. The genetic analysis of mutants resistant to antibiotics which inhibit protein synthesis by chloroplast ribosomes may provide evidence for this, but only preliminary studies have been reported as yet (Gillham *et al.*, 1970).

## Chloroplast Protein-Synthesizing Apparatus

Chloroplast preparations capable of incorporating labelled amino acids into protein have been isolated from several higher plants and algae (Table 2). It is important to recognize that a considerable problem of interpretation always exists when the sterility of the preparation is not established. Indeed, App & Jagendorf (1964) demonstrated that protein synthesis by the spinach chloroplast preparations found in a preceding study (App & Jagendorf, 1963) could be accounted for by bacterial contamination. Another possibility that must be considered is that the incorporation is due to intact leaf cells which may be present in the chloroplast preparation. The best criterion for establishing that the incorporation is due to the chloroplasts is the dependence on an added energy source, either ATP, or light in the case of intact chloroplasts. Stimulation by light must be demonstrated in aerobic conditions, since under conditions of low oxygen tension the stimulation might result from the enhancement of bacterial metabolism by oxygen released photochemically (App & Jagendorf, 1964). Other criteria, such as sensitivity to added ribonuclease and the $Mg^{2+}$ concentration, which are useful for isolated ribosomes (Boulter, 1970), may not be suitable in cases where the incorporation of amino acids is due to intact chloroplasts. Incorporation by chloroplasts from *Acetabularia* does not meet the criterion of dependence on an added energy source but, on the other hand, no evidence for significant bacterial incorporation could be found (Goffeau & Brachet, 1965). Chloroplasts isolated from this alga differ in several other respects from those found in higher plants, e.g. in their resistance to lysis by Triton X-100 or hypo-osmotic conditions, and thus caution must be exercised in extrapolating their characteristics to other organisms. Woodcock & Bogorad (1971) have pointed out that with the type of internal morphology found in *Acetabularia*, it is to be expected that mild homogenization results in the formation of 'cytoplasts' consisting of a number of organelles and some cytoplasm surrounded by a tonoplast membrane; such structures were reported in chloroplast preparations of this alga by Bidwell (1972). Other criteria for ruling out bacteria as the agents of incorporation are insensitivity to anaerobiosis and to respiratory inhibitors (Gnanam *et al.*, 1969) and solubility of the labelled protein in Triton X-100 (Bamji & Jagendorf, 1966; Parenti & Margulies, 1967).

It can be concluded from the information summarized on Table 2 that isolated chloroplasts have some capacity for incorporating amino acids into protein, but two major questions remain to be answered: (1) what is the detailed mechanism of protein synthesis by chloroplasts, and (2) what are the functions of chloroplast ribosomes? These will now be considered.

## Mechanism of Chloroplast Protein Synthesis

### *Ribosomes*

The existence of two classes of ribosomes in green cells was first shown by Lyttleton (1962). Ribosomes of the 70S class have been isolated from chloroplasts from several algae and higher plants (Table 3). Chloroplast ribosomes thus resemble those from prokaryote cells in their S value. Their RNA components are also of the same size as those found in prokaryote ribosomes, being smaller than those found in 80S ribosomes (Table 3). However, these similarities are of size only, not in the sequence of nucleotides. Thus, rRNA from *E. coli* does not compete with chloroplast rRNA from *Euglena* for hybridization to chloroplast DNA (Scott *et al.*, 1971). The protein complement of *E. coli* ribosomes is also quite different from that of chloroplast ribosomes, as judged by immunological tests (Wittmann, 1970) and gel electrophoresis (Hoober & Blobel, 1969; Odintsova & Yurina, 1969; Vasconcelos & Bogorad, 1971). Similarly, the proteins of chloroplast and cytoplasmic ribosomes of the same plant differ distinctly in their

Table 3. *Comparison of S values of chloroplast and cytoplasmic ribosomes and rRNA*

| Organism | Chloroplast ribosomes | | Cytoplasmic ribosomes | | Reference |
|---|---|---|---|---|---|
| | Whole | RNA | Whole | RNA | |
| *E. coli* | — | — | 70 | 23, 16 | |
| *Oscillatoria* | — | — | 70 | 23, 16 | Loening & Ingle (1967) |
| *C. reinhardii* | 69 | — | 79 | — | Sager & Hamilton (1967) |
| *C. reinhardii* | 68 | 22, 16 | 80 | 25, 18 | Hoober & Blobel (1969) |
| *E. gracilis* | 70 | 23, 16 | 88 | 26, 22 | Scott *et al.* (1971) |
| *P. sativum* | 70 | 23, 16 | 80 | — | Svetailo *et al.* (1967), Leaver & Ingle (1971) |
| *S. oleracea* | 67 | 23, 16 | 78 | — | Lyttleton (1962), Gualerzi & Commarano (1969), Leaver & Ingle (1971) |
| *N. tabacum* | 70 | — | 80 | — | Boardman *et al.* (1965) |

gel-electrophoresis patterns (Lyttleton, 1968; Janda & Wittmann, 1968; Hoober & Blobel, 1969; Arglebe & Hall, 1969; Odintsova & Yurina, 1969; Gualerzi & Cammarano, 1969, 1970).

The heavy component of chloroplast rRNA is unstable, and may break down during isolation. The stability varies with the species of plant as does the size and number of fragments produced, but this component may be stabilized by bivalent cations (Leaver & Ingle, 1971). Chloroplast ribosomes also contain a 5S RNA component, which is not identical with the 5S component of cytoplasmic ribosomes from the same plant (Payne & Dyer, 1971). On the other hand, the 5.8S RNA component found in 80S ribosomes is absent from chloroplast ribosomes (Payne & Dyer, 1972).

Some of the ribosomes in chloroplasts are bound to the membranes. Falk (1969) showed by electron microscopy that whorl-like polyribosomes are attached to the thylakoid membranes of *Phaseolus vulgaris* chloroplasts. These polyribosomal whorls occur on the outermost thylakoid membranes of the grana stacks as well as on the single thylakoids running through the stroma. Chen & Wildman (1970) found that both the supernatant and membrane fractions prepared by lysing tobacco chloroplasts were capable of carrying out protein synthesis *in vitro*. They estimate that at least 50% of the ribosomes in tobacco chloroplasts are tightly associated with the thylakoids and can be removed only by treatment with detergent. If the chloroplasts are washed before osmotic lysis, then a significant proportion of the free ribosomes released exist as polyribosomes which have a higher protein-synthetic capacity than the monoribosomes (Chen & Wildman, 1967). Chloroplast polyribosomes active in protein synthesis have also been isolated from *Euglena gracilis* (Avadhani & Buetow, 1972). The formation of chloroplast polyribosomes is stimulated by light both in etiolated leaves (Brown & Gunning, 1966) and in fully greened leaves (Clark, 1964).

### *Amino acid activation*

The occurrence of specific aminoacyl-tRNA species and of the corresponding synthetases has been demonstrated in chloroplasts isolated from *E. gracilis* (Reger *et al.*, 1970), *P. vulgaris* (Burkard *et al.*, 1970), *Triticum vulgare* (Leis & Keller, 1970, 1971), cotton (Merrick & Dure, 1971) and *Nicotiana tabacum* (Guderian *et al.*, 1972). The synthesis of some of the enzymes and tRNA species is induced by light in *Euglena*. Thus, light-grown cells contain two aminoacyl-tRNA synthetases for both isoleucine and phenylalanine. Only one of the two synthetases for each amino acid is found in isolated chloroplasts, as are the light-induced phenylalanine and isoleucine tRNA species. The light-induced chloroplast tRNA species can be acylated only by the chloroplast synthetases in each case. The chloroplast isoleucyl-tRNA is light-inducible and cannot be detected in dark-grown cells or in cells of a u.v.-bleached mutant which lacks chloroplast DNA. However, the chloroplast phenylalanyl-tRNA synthetase occurs both in dark-grown cells and in the mutant, which suggests that it is coded for in the nuclear DNA, and may be synthesized on cytoplasmic ribosomes.

A study of the synthetases found in the chloroplasts and cytoplasm of *Phaseolus* showed that the synthetases for the same amino acid were distinctly different in such properties as sensitivity to inhibitors. However, it was found that for several amino acids both cytoplasmic and chloroplast enzymes are equally effective in attaching amino acids to their cognate RNA species. Specificity in heterologous systems, e.g. cytoplasmic enzyme plus chloroplast tRNA, ranged from none to absolute. Thus the cytoplasm has a leucyl-tRNA synthetase which is able to add leucine only to the two leucine tRNA species that are common to the cytoplasm and the chloroplasts, but is not able to recognize the three leucine tRNA species found only in the chloroplasts. The chloroplast enzymes, on the other hand, can aminoacylate all five leucine tRNA species. It is difficult to attach any biological significance to these variations in specificity.

It has often been suggested that there may be a control of cellular differentiation at the level of translation, as well as at the level of transcription. A possibly fruitful area for more research is the study of the changes in tRNA species and aminoacyl-tRNA synthetases during the formation of chloroplasts.

*Initiation*

Protein synthesis in bacteria is initiated by *N*-formylmethionyl-tRNA; there is evidence that *N*-formylmethionyl-tRNA also functions in chain initiation in mitochondria, but not in the cytoplasm of eukaryotic cells. The first report that the same is true for chloroplasts was provided by Schwartz *et al.* (1967), who found that RNA from bacteriophage f2 directs the synthesis of viral-coat protein by chloroplast ribosomes from *Euglena*; the *N*-terminal peptide contained *N*-formylmethionine. More detailed studies of initiating methionine tRNA species have since been reported for wheat leaves (Leis & Keller, 1970, 1971), bean chloroplasts (Burkard *et al.*, 1969) and cotton seedlings (Merrick & Dure, 1971). There are five methionine tRNA species detectable in extracts of wheat leaves. Two are localized in the chloroplasts; one of these, when charged with methionine, can be formylated both by an endogenous transformylase and by *E. coli* transformylase, whereas the other cannot be formylated by the endogenous transformylase and may serve to direct methionine into internal positions in the polypeptide chain. The two chloroplast methionine tRNA species are distinct from the two that were isolated from the cytoplasmic fraction; neither of the latter can be formylated either by endogenous transformylase or by *E. coli* transformylase. The fifth methionine tRNA species was present in small amounts and may be mitochondrial in origin. The transformylase from wheat leaves has been purified (Leis & Keller, 1971). This enzyme uses $N^{10}$-formyltetrahydrofolate as formyl donor, and is present in chloroplasts at a higher specific activity than in the other subcellular fractions.

It is probable from this type of evidence that the initiation of protein synthesis by chloroplasts is similar to that in mitochondria and *E. coli* in that it uses a formylated methionyl-tRNA, but distinct from that in the cytoplasm, which uses an unformylated methionyl-tRNA. None of the experiments reported so far indicate whether mRNA with an initiator codon for *N*-formylmethionine is transcribed from nuclear or chloroplast DNA. If mRNA can cross the chloroplast membrane, then the site of translation could be specified by the requirements for initiation and not by the site of transcription.

*Energy source*

In most studies of protein synthesis by isolated chloroplasts, ATP and an ATP-generating system have been added as an energy source (see Table 2). Since isolated chloroplasts can generate ATP by both cyclic and non-cyclic photophosphorylation, it could be possible to demonstrate that protein synthesis by chloroplasts can be driven by light. Spencer (1965) found that spinach chloroplasts would incorporate amino acids into protein in the absence of added ATP, provided that ADP, $P_i$, pyocyanine and light were supplied. The necessity for pyocyanine as catalyst indicates that the chloroplasts were broken and had lost their natural catalyst, ferredoxin. Similar results have been obtained with pea chloroplasts (Griffiths & Lozano, 1970). Much more striking results were reported by Ramirez *et al.* (1968), who found that if precautions were taken to isolate spinach chloroplasts that were largely intact, protein synthesis was stimulated 20-fold by light in the absence of added cofactors or catalysts of photophosphorylation; the rates of protein synthesis by these chloroplasts are the highest yet recorded. Protein synthesis was insensitive to 3-(3,4-dichlorophenyl)-1,1-dimethylurea; this was interpreted to mean that cyclic photophosphorylation alone was providing the ATP required for protein synthesis. These large stimulations of protein synthesis by light have been confirmed for intact pea chloroplasts (Blair & Ellis, 1972), but in this case 3-(3,4-dichlorophenyl)-1,1-dimethylurea was inhibitory. We suggest that the use of intact chloroplasts for studies of protein synthesis has a major advantage over the use of broken chloroplasts, in that conditions around the polyribosomes will be more normal with regard to controlling factors. It was by ensuring that protein synthesis occurred only in intact chloroplasts that two of the proteins made by chloroplast ribosomes have been identified (see below). There is evidence that amino acids can readily cross the outer membranes of chloroplasts (Nobel & Wang, 1970). Exogenous ATP, on the other hand, crosses at a rate which is very slow compared with its rate of synthesis by photophosphorylation (Heldt, 1969; Heber & Santarius, 1970). It is possible that protein synthesis by mature chloroplasts *in vivo* uses ATP largely provided by the chloroplast itself. However, this is not the case in developing chloroplasts, since many organisms, e.g. *Chlorella*, *Chlamydomonas*, *Pinus*, do not require light to form chloroplasts, and mutants in which the photosynthetic pathway is blocked, form otherwise normal chloroplasts (Bishop, 1966). In *Euglena*, the formation of chloroplasts is not prevented by a concentration of 3-(3,4-dichlorophenyl)-1,1-dimethylurea that inhibits photosynthesis completely (Schiff, 1971). The development of chloroplasts must therefore depend on ATP and reducing power supplied by the rest of the cell.

*Inhibitors*

It is well established that protein synthesis by isolated chloroplast ribosomes is inhibited by several antibiotics, which also inhibit protein synthesis by prokaryote ribosomes. The best studied example is chloramphenicol which was found to inhibit protein synthesis by chloroplasts from spinach (App & Jagendorf. 1963), tobacco (Spencer & Wildman, 1964), *Euglena* (Eisenstadt & Brawerman, 1964) and *Acetabularia* (Goffeau & Brachet, 1965). Inhibition of protein synthesis in chloroplasts by chloramphenicol has since been demonstrated in every case where it has been tested (Table 2). It is important to realize that inhibition is meaningful only in those studies where protein synthesis has been shown to be due to chloroplast ribosomes and not to contaminating bacteria. Protein synthesis in chloroplasts is inhibited by only the D-*threo* isomer of chloramphenicol, and in this respect resembles bacterial protein synthesis (Ellis, 1969). Protein synthesis by cytoplasmic ribosomes isolated from both green cells (Eisenstadt & Brawerman, 1964; Ellis, 1969) and non-green cells (Marcus & Feeley, 1965; Ellis & MacDonald, 1967) is unaffected by chloramphenicol. Chloroplast ribosomes are insensitive to cycloheximide, which is regarded as specific for 80S ribosomes, but there are several cases known in which 80S ribosomes from plants also appear to be insensitive to this antibiotic (Ellis & MacDonald, 1970). This insensitivity may be due to the presence of thiol compounds in the assay, rather than an intrinsic feature of the 80S ribosomes (Baliga *et al.*, 1969).

A peculiarity of chloramphenicol is that no 70S ribosomes are known from either bacteria or chloroplasts which are resistant to its action in protein synthesis. Presumably the integrity of the chloramphenicol-binding site is essential for the functioning of the 70S ribosome. The similarity between chloroplast and bacterial ribosomes as regards chloramphenicol could, therefore, reflect merely that both are of the 70S class and does not arise because they are related in evolution. However, three other unrelated antibiotics, spectinomycin, lincomycin and erythromycin, also inhibit protein synthesis by chloroplast and bacterial ribosomes, but not by cytoplasmic ribosomes (Ellis, 1970). Light-driven protein synthesis by pea chloroplasts is especially sensitive to lincomycin, 50% inhibition being given by 0.2 $\mu$g/ml (Ellis & Hartley, 1971). Although chloroplast ribosomes are sensitive to the same set of antibiotics as bacterial ribosomes, the protein complements of the ribosomes are completely different. The similarity between the ribosomes must therefore reside in the antibiotic-binding sites, and not in the immunological properties or molecular weights of the protein constituents. It must also be emphasized that there is no evidence, as yet, that the mechanism of action of bacterial antibiotics on chloroplast ribosomes is similar to that on bacterial ribosomes.

## Function of Chloroplast Ribosomes

The problem of identifying which proteins are synthesized by chloroplast ribosomes has been tackled in two ways: (*a*) by supplying inhibitors of 70S ribosomes to cells making chloroplasts, and determining which proteins are no longer synthesized, and (*b*) by identifying the products of protein synthesis by isolated chloroplasts. Both these approaches have their difficulties, but the picture that is emerging suggests strongly that many of the chloroplast proteins are synthesized by cytoplasmic ribosomes, and only relatively few by the chloroplast ribosomes.

### *Evidence from studies of inhibitors in vivo*

The validity of results from experiments with any inhibitor depends absolutely on the specificity of its action in intact cells, and there is evidence that both chloramphenicol and cycloheximide have effects on systems other than protein synthesis in some higher plants (Ellis, 1963; Hanson & Krueger, 1966; Ellis & MacDonald, 1970). Systems such as ion uptake and oxidative phosphorylation are inhibited by all the isomers of chloramphenicol, whereas the inhibition of protein synthesis by isolated chloroplast ribosomes is specific for the D-*threo*-isomer (Ellis, 1969; Wilson & Moore, 1973). This stereospecificity provides a means of establishing, for any particular tissue, whether chloramphenicol is inhibiting protein synthesis directly at the ribosomal level, or, in addition, is affecting the energy supply. It is recommended that only if an inhibition is produced specifically by the D-*threo* isomer should an interpretation directly involving protein synthesis be invoked. An additional problem is that inhibitors of 70S ribosomes also inhibit protein synthesis by mitochondria, and this factor may be important in some experiments. The use of cycloheximide can also give misleading results. For example, Pitt (1971) found that although cycloheximide prevented the increase in ribonuclease activity extracted from damaged potato tubers, no net synthesis of this protein was occurring as judged by density labelling and immunochemical assay.

Table 4 lists the suggested sites of synthesis of some chloroplast proteins; these suggestions have been derived from studies in which several 70S ribosomal inhibitors have been applied to greening cells of several algae and higher plants. It might be argued that it is unlikely that the same set of chloroplast proteins is made by chloroplast ribosomes in organisms as different as *Euglena* and *Pisum*, but, with certain reservations, the data suggest that this may be true. Thus, in all the studies, the synthesis of Fraction I protein was found to be inhibited by 70S ribosomal inhibitors. On the other hand, the enzymes of the Calvin cycle appear to be synthesized on cytoplasmic ribosomes in all the studies except those on *Zea mays* (Smillie *et al.*, 1967) and *Euglena* (Smillie & Scott, 1970). In the case of *Z. mays*, the inhibition by chloramphenicol was, however, not shown to be stereospecific, and in that of *Euglena* independent work by Schiff (1971) did not confirm the result for the triose phosphate dehydrogenase (NADP). Besides Fraction I protein, the only other proteins that appear to be synthesized by chloroplast ribosomes are some of the ribosomal and lamellar proteins, including the cytochromes.

It is noteworthy that for Fraction I protein and the lamellar proteins, the evidence from *Chlamydomonas* suggests that both 70S and 80S ribosomes are required for their synthesis. One interpretation is that the apoenzyme is synthesized on the cytoplasmic ribosomes, and requires for integration into the chloroplast in an active state, an additional protein which is synthesized on chloroplast ribosomes. It has been shown by a double-labelling experiment with barley leaves that the synthesis of the large subunit of Fraction I protein is preferentially inhibited by chloramphenicol, whereas that of the small subunit is preferentially inhibited by cycloheximide (Criddle *et al.*, 1970). The suggestion that the large subunit only is made by chloroplast ribosomes is fully supported by the data (see below) obtained *in vitro*.

*Products of protein synthesis in vitro*

The most direct way to determine the function of chloroplast ribosomes is to identify the proteins made by isolated chloroplasts. This approach has hitherto

Table 4. *Suggested sites of synthesis of chloroplast proteins*

References: [1]Ellis & Hartley (1971); [2]Thomson & Ellis (1972); [3]Ireland & Bradbeer (1971); [4]Graham *et al.* (1970); [5]Smillie *et al.* (1967); [6]Smillie & Scott (1970); [7]Schiff (1971); [8]Armstrong *et al.* (1971); [9] Hoober *et al.* (1969); [10]Hoober (1970); [11]Eytan & Ohad (1970).

| | Sites of synthesis | | | | | |
|---|---|---|---|---|---|---|
| Species .. | *P. sativum*[1,2] | *Ph. vulgaris*[3] | *Z. mays*[4] | *E. gracilis*[5,6] | *E. gracilis*[7] | *C. reinhardii*[8,9,10,11] |
| Ribosomal inhibitor used .. | Lincomycin | D-*threo*- and L-*threo*-Chloramphenicol | D-*threo*-Chloramphenicol, cycloheximide | D-*threo*- and L-*threo*-Chloramphenicol, cycloheximide | Streptomycin | D-*threo*-Chloramphenicol, cycloheximide, spectinomycin |
| Ribose phosphate isomerase | 80S | 80S | — | — | — | — |
| Phosphoribulokinase | 80S | 70S | — | — | — | 80S |
| Ribulose diphosphate carboxylase (Fraction I) | 70S | 70S | 70S | 70S | 70S | 70S, 80S |
| Phosphoglycerate kinase | 80S | 80S | — | — | — | — |
| Triose phosphate dehydrogenase (NADP) | 80S | 80S | 70S | 70S | 80S | — |
| Triose phosphate isomerase | — | 80S | — | — | — | — |
| Fructose diphosphate aldolase | — | 80S | — | 70S | — | — |
| Transketolase | — | 80S | — | — | — | — |
| Pyruvate, $P_i$ kinase | — | — | 70S | — | — | — |
| Ferredoxin | — | 80S | — | — | — | 80S |
| Ferredoxin NADP reductase | — | 80S | — | 70S | — | 80S |
| Cytochrome 552 | — | — | — | 70S | 70S | 70S, 80S |
| Cytochrome 561 | — | — | — | 70S | — | 70S, 80S |
| RNA polymerase | 80S | — | — | 70S | — | — |
| Ribosomal proteins | 70S | — | — | — | — | — |
| Membrane proteins | 70S | — | — | — | — | 70S, 80S |

given inconclusive results (Kirk, 1970; Woodcock & Bogorad, 1971). However, in our laboratory, recent work has shown that isolated pea chloroplasts can synthesize two discrete proteins, namely the large subunit of Fraction I protein (but not the small subunit) and a component of the chloroplast membranes (Blair & Ellis, 1973; Ellis & Forrester, 1972).

The rationale of our approach is that to obtain an identifiable product we must use conditions in which correct elongation, termination and release of the polypeptide chains occur in isolated chloroplasts; initiation would be desirable but is not essential for the purposes of identification. Such conditions seemed to us to be more likely met in intact chloroplasts rather than in the lysed preparations that are commonly used. We therefore used techniques which were developed originally to prepare intact chloroplasts capable of high rates of photosynthesis. It is diagnostic of such chloroplasts that they can carry out photophosphorylation in the absence of added cofactors and catalysts. By using light as the source of energy for protein synthesis, it is possible to ensure that amino acid incorporation is due solely to the intact chloroplasts in the preparation, since broken chloroplasts are unable to synthesize ATP in the absence of added cofactors.

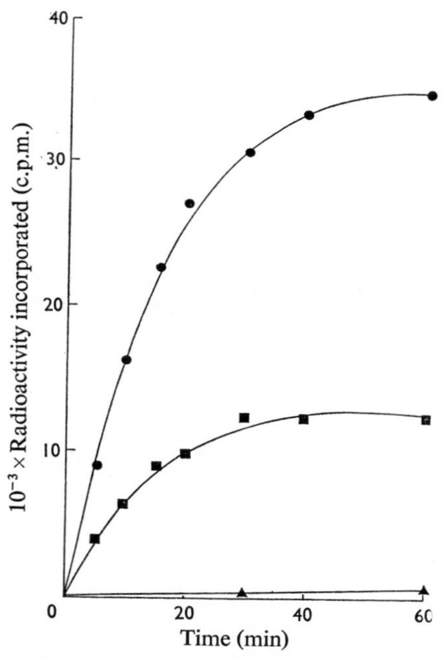


Fig. 2. *Time-course of light-driven protein synthesis*

Pea apices (15 g) were homogenized for 4 s in a Polytron homogenizer in 100 ml of sterile ice-cold 0.35 M-sucrose, 0.025 M-Hepes–NaOH buffer (pH 7.6), 2 mM-EDTA, 2 mM-sodium isoascorbate. The homogenate was strained through eight layers of muslin and centrifuged at 2500 ***g*** for 1 min. The supernatant was decanted at once and the pellet resuspended in 5 ml of sterile 0.2 M-KCl, 0.06 M-Tricine–KOH buffer (pH 8.3), 6.6 mM-$MgCl_2$; 300 $\mu$l samples were incubated in a final volume of 500 $\mu$l with 0.5 $\mu$Ci of [$^{14}$C]leucine or [$^{35}$S]methionine. The tubes were illuminated at 20°C with 25000 lx from a photoflood. Protein was extracted and the radioactivity counted at 80% efficiency as described by Ellis (1970). The rates of incorporation were in the range 0.5–1.0 nmol of [$^{14}$C]leucine/h per mg of chlorophyll. ▲, Dark; ■, 100 $\mu$g of chlorophyll; ●, 400 $\mu$g of chlorophyll.

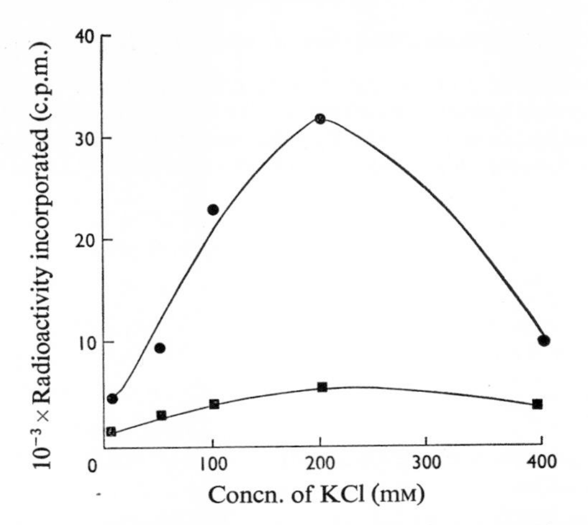


Fig. 3. *Dependence of light-driven protein synthesis on KCl concentration*

The chloroplast pellets prepared as in Fig. 2 were resuspended in media of various KCl and sucrose concentrations. Light-driven incorporation of [$^{14}$C]leucine into protein was assayed as described as in Fig. 2. ●, Sucrose absent; ■, +300mM-sucrose.

Pea seeds (*Pisum sativum* var. Meteor or Feltham First) were grown in compost for 7–10 days under a 12h photoperiod of 2000 lx provided by white fluorescent tubes. Chloroplasts were isolated from the apices by using the method described in Fig. 2. This method was derived from a technique reported by Ramirez *et al.* (1968) and produces preparations containing 40–50% intact chloroplasts, as judged by phase microscopy and the rate of ferricyanide reduction. These preparations incorporated labelled amino acids into protein when illuminated (Fig. 2). The rate of incorporation decreased to zero after about 20min; this falling rate was not accompanied by lysis of the chloroplasts. A vital component of the incubation medium was the high concentration of KCl. When KCl was replaced by sucrose, protein synthesis was much decreased (Fig. 3). If the chloroplasts were lysed by resuspension in media containing no KCl, subsequent restoration of the KCl to 0.2M did not restore the ability to incorporate amino acids into protein (Table 5). We therefore believe that the KCl is acting both as an osmoticum and as a cofactor for protein synthesis.

Some characteristics of this chloroplast system are shown in Table 5. Light can only be partially replaced as an energy source by added ATP and an ATP-generating system, and addition of ATP as well as light produces only a slight stimulation. Lysed chloroplasts show very low incorporation even when supplied with ATP. Inhibitors of photophosphorylation such as carbonyl cyanide *m*-chlorophenylhydrazone and 3-(3,4-dichlorophenyl)-1,1-dimethylurea inhibit protein synthesis, as do D-*threo*-chloramphenicol and lincomycin. The incorporation is not inhibited by added ribonuclease; this enzyme is not inactivated because analyses of the RNA in these preparations show that RNA is hydrolysed to a percentage equal to the percentage of broken chloroplasts. Actinomycin D does not inhibit protein synthesis at 10$\mu$g/ml; this concentration inhibits light-dependent incorporation of uridine into RNA by the same chloroplast preparation by about 85%. We conclude from all these findings that protein synthesis

Table 5. *Characteristics of light-driven protein synthesis by isolated pea chloroplasts*

Pea chloroplasts were isolated and incubated as described in Fig. 2. Incorporation by the complete light-driven system is designated 100. The ATP and ATP-generating system consisted of 2mM-ATP, 5mM-creatine phosphate and 100 $\mu$g of creatine phosphokinase/ml. Lysed chloroplasts were prepared by resuspending in incubation medium without KCl, and then restoring the KCl to 0.2M.

| Energy source | Treatment | Incorporation |
|---|---|---|
| Light | Complete | 100 |
| None | Zero time | 0.5 |
| None | Complete | 3.0 |
| ATP + ATP-generating system | Complete | 50 |
| Light + ATP + ATP-generating system | Complete | 125 |
| Light | + Ribonuclease (30 $\mu$g/ml) | 95 |
| None | Lysed | 5 |
| ATP + ATP-generating system | Lysed | 7.5 |
| Light | + Carbonyl cyanide *m*-chlorophenylhydrazone (5 $\mu$M) | 6 |
| Light | + 3-(3,4-Dichlorophenyl)-1,1-dimethylurea (1 $\mu$M) | 38 |
| Light | + D-*threo*-Chloramphenicol (150 $\mu$M) | 5 |
| Light | + Lincomycin (5 $\mu$M) | 25 |
| Light | + Actinomycin D (10 $\mu$g/ml) | 100 |

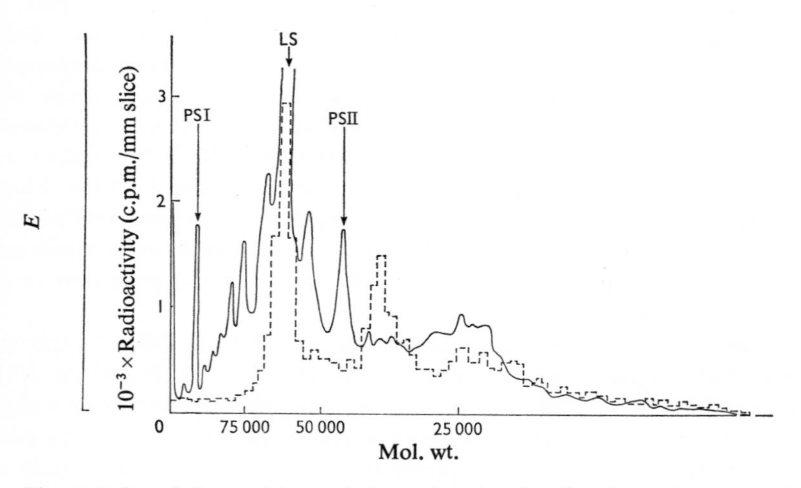


Fig. 4. *Sodium dodecyl sulphate-gel electrophoresis of labelled chloroplast preparations*

Chloroplasts were isolated from 8-day-old pea plants as described in Fig. 2 and incubated at a chlorophyll concentration of 60 $\mu$g/ml with 20 $\mu$Ci of [$^{35}$S]methionine (1 nmol) in a final volume of 0.5 ml at 20°C for 40 min with light as the energy source. The preparation was dialysed against 2.5mM-Tris–glycine buffer, pH 8.5, containing 0.1M-2-mercaptoethanol for 18h at 4°C; 10mg of sodium dodecyl sulphate was added and 100 $\mu$l samples were run on 10% (w/v) sodium dodecyl sulphate–polyacrylamide gels in borate buffer, pH 8.5, at 100V for 2h. The gels were stained in Amido Black, scanned and sliced into 1 mm slices; the slices were solubilized and the radioactivity was counted at 80% efficiency. PS, photosystem; ——, $E_{620}$; ----, radioactivity; LS, large subunit of Fraction I protein.

proceeds in intact chloroplasts only, and probably uses mRNA synthesized before the chloroplasts were isolated.

The products of protein synthesis by this system have been analysed by electrophoresis in sodium dodecyl sulphate–polyacrylamide gels. After incubation the chloroplast preparation was dialysed against 2.5mM-Tris–glycine buffer, pH 8.5, containing 0.1M-2-mercaptoethanol to remove unchanged labelled amino acid and to lyse the chloroplasts. The dialysed preparation was made 1% with respect to sodium dodecyl sulphate and subjected to electrophoresis on sodium dodecyl sulphate gels in borate or Tris–glycine buffer, pH 8.5. These gels showed two green bands which are known to be chlorophyll–protein complexes derived from photosystems I and II (Thornber *et al.*, 1967). When stained with Amido Black many protein bands became visible; the large subunit of Fraction I protein was by far the predominant band (Fig. 4). When the gels were sliced and the radioactivity measured, two major labelled peaks were found (Fig. 4). These peaks were still present when the chloroplast preparations were boiled in sodium dodecyl sulphate immediately after incubation to inactivate proteases. Neither peak was found in chloroplasts incubated in the dark or in the presence of chloramphenicol. The peaks do not coincide with the complexes derived from either photosystem I or II. If the dialysed extracts were first centrifuged at 150000$g$ for 1 h the slower-moving peak of molecular weight about 60000 remains in the supernatant; this peak runs exactly with the large subunit of Fraction I protein on 8%, 10% and 12% sodium dodecyl sulphate gels (Fig. 5). No other discrete radioactive peak was detected in the supernatant fraction. The faster-moving peak of molecular weight about 40000 could be sedimented with the chloroplast membranes at 10000$g$ for 10 min; this peak was not removed if the membrane pellet was washed several times in hypo-osmotic buffer and water (Fig. 6). This membrane-associated peak was present when the chloroplasts were treated with 10mM-puromycin near the

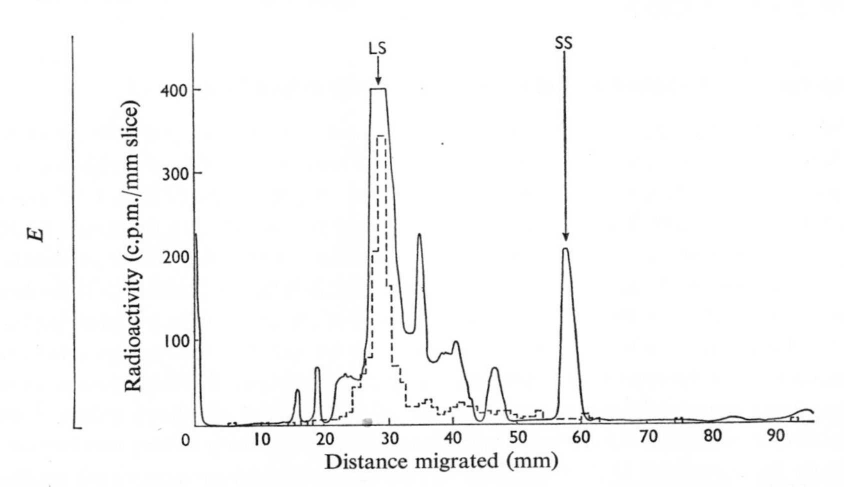


Fig. 5. *Sodium dodecyl sulphate-gel electrophoresis of soluble fraction from labelled chloroplasts*

Chloroplasts were treated as in Fig. 4 except that the amino acid was 1 μCi of [$^{14}$C]leucine (3 nmol). After dialysis the preparation was centrifuged at 150000$g$ for 1 h and the supernatant run on 12% sodium dodecyl sulphate gels. ——, $E_{620}$; ----, radioactivity; LS, large subunit; SS, small subunit of Fraction I protein.

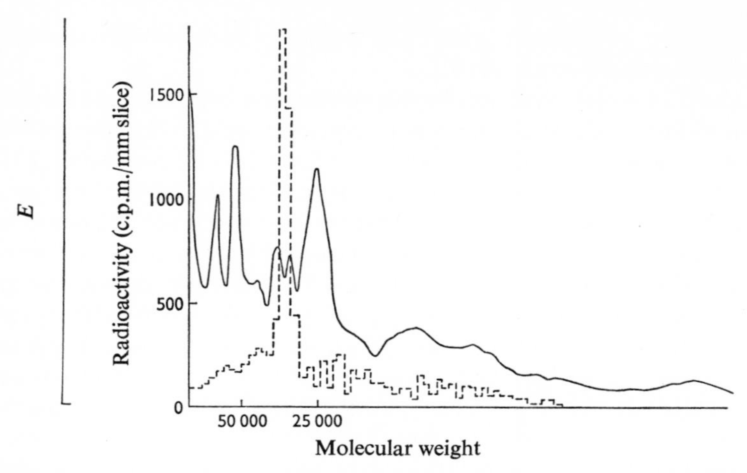


**Fig. 6.** *Sodium dodecyl sulphate-gel electrophoresis of membrane fraction from labelled chloroplasts*

Chloroplasts were incubated with [$^{35}$S]methionine as in Fig. 4, mixed with 30ml of 8mM-sodium borate buffer (pH8.9) containing 5mM-2-mercaptoethanol and centrifuged at 10000$g$ for 10min. The green pellet was washed twice more by the same procedure and boiled in sodium dodecyl sulphate for 2min before analysis on 15% sodium dodecyl sulphate gels in borate buffer. ——, $E_{620}$; ----, radioactivity.

end of the incubation; this fact plus the discrete nature of the peak (Fig. 6) suggests that this is a completed protein and not incomplete polypeptide chains. Treatment of the membrane preparation with 0.1% Triton X-100 and subsequent centrifugation at 150000$g$ for 3h resulted in 70% of the radioactivity remaining in the supernatant; this suggests that the membrane peak is not derived from the membrane-bound ribosomes.

## Co-operation of Chloroplast and Cytoplasmic Protein Synthesis

The results derived by both the inhibitor method *in vivo* and the protein-synthesis method *in vitro* both point to the same conclusion, that the chloroplast ribosomes are required to synthesize only a few of the many different proteins found in chloroplasts. The reason why chloroplast ribosomes can represent up to 50% of the total ribosomes in a leaf would appear to be that one of their products, the large subunit of Fraction I protein, occurs in much larger quantities than any other protein. Most of the other soluble proteins of the chloroplast, including the small subunit of Fraction I protein, appear to be synthesized by cytoplasmic ribosomes. An inference must therefore be drawn that specific mechanisms exist to transport across the outer membranes of the plastid all those chloroplast proteins that are made on cytoplasmic ribosomes. The nature of this mechanism can only be surmised. One possibility is that a membrane protein exists in the outer envelope which recognizes a site common to those proteins that are destined for the plastid. We are tempted to suggest that the membrane protein synthesized *in vitro* may have this function. Supporting evidence for this idea could come from the demonstration that this protein is localized in the envelope of the plastid; we

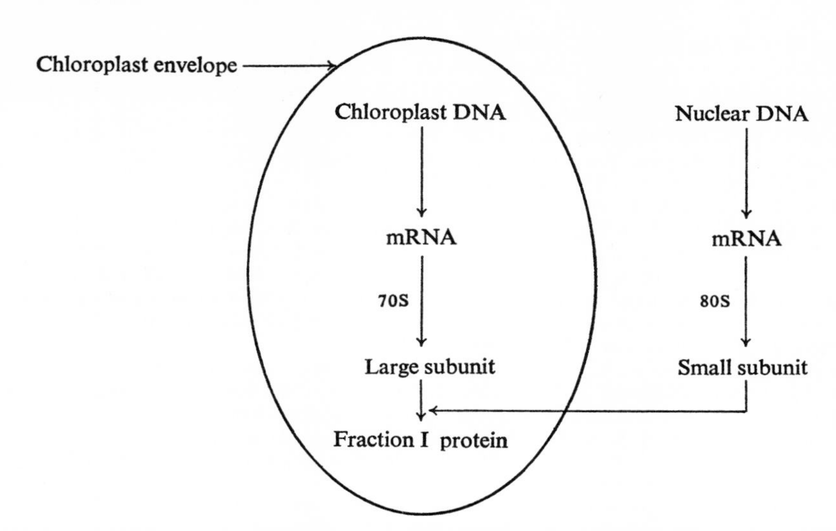


Scheme 1. *Model of co-operation between plastid and nuclear genomes in synthesis of Fraction I protein (after Kawashima & Wildman,* 1972)

are attempting to resolve this point by using the technique of envelope isolation developed by Mackender & Leech (1970).

A model which describes our current view of the co-operation between plastid and nuclear genomes in the synthesis of Fraction I protein is shown in Scheme 1. The large subunit synthesized by isolated chloroplasts does not exchange with preformed Fraction I protein, so we suggest that there is little or no pool of small subunits in the chloroplasts. The nature of the mechanism which regulates the relative rates of synthesis of the subunits in the two different cellular compartments is unknown.

## Prospects

We suggest that the following areas are ripe for development:

(*a*) the detailed mechanism of protein synthesis by chloroplast ribosomes;

(*b*) the elucidation of the mode of action of the bacterial antibiotics on chloroplast ribosomes;

(*c*) the isolation of the mRNA for the large subunit of Fraction I protein from chloroplast polyribosomes;

(*d*) the uptake of chloroplast proteins by isolated developing plastids.

## References

Abelson, J. & Thomas, C. A. (1966) *J. Mol. Biol.* **18**, 262–291
Apel, K. & Schweiger, H. G. (1972) *Eur. J. Biochem.* **25**, 229–238
App, A. A. & Jagendorf, A. T. (1963) *Biochim. Biophys. Acta* **76**, 286–292
App, A. A. & Jagendorf, A. T. (1964) *Plant Physiol.* **39**, 772–776
Arglebe, C. & Hall, T. C. (1969) *Plant Cell Physiol.* **10**, 171–182
Armstrong, J. J., Surzycki, S. J., Moll, B. & Levine, R. P. (1971) *Biochemistry* **10**, 692–701
Avadhani, N. G. & Buetow, D. E. (1972) *Biochem. J.* **128**, 353–365
Baliga, B. S., Pronczuk, A. W. & Munro, H. N. (1969) *J. Biol. Chem.* **244**, 4480–4489
Ballantine, J. E. & Forde, B. J. (1970) *Amer. J. Bot.* **57**, 1150–1159

Bamji, M. S. & Jagendorf, A. T. (1966) *Plant Physiol.* **41**, 764–770
Bard, S. A. & Gordon, M. P. (1969) *Plant Physiol.* **44**, 377–384
Bastia, D., Chiang, K., Swift, H. & Siersma, P. (1971) *Proc. Nat. Acad. Sci. U.S.* **68**, 1157–1161
Berger, A. (1967) *Protoplasma* **64**, 13–25
Bidwell, R. G. S. (1972) *Nature (London)* **237**, 169
Bishop, N. I. (1966) *Annu. Rev. Plant Physiol.* **17**, 185–208
Blair, G. E. & Ellis, R. J. (1972) *Biochem. J.* **127**, 42P
Blair, G. E. & Ellis, R. J. (1973) *Biochim. Biophys. Acta* **319**, 223–234
Boardman, N. K. (1968) *Advan. Enzymol. Relat. Areas Mol. Biol.* **30**, 1–79
Boardman, N. K., Francki, R. I. B. & Wildman, S. G. (1965) *Biochemistry* **4**, 872–876
Boardman, N. K., Francki, R. I. B. & Wildman, S. G. (1966) *J. Mol. Biol.* **17**, 470–489
Boardman, N. K., Linnane, A. W. & Smillie, R. M. (eds.) (1971) *Autonomy and Biogenesis of Chloroplasts and Mitochondria*, North-Holland Publishing Co., Amsterdam
Boulter, D. (1970) *Annu. Rev. Plant Physiol.* **21**, 91–114
Brawerman, G. & Eisenstadt, J. M. (1964) *Biochim. Biophys. Acta* **91**, 477–485
Brown, F. A. M. & Gunning, B. E. S. (1966) in *Biochemistry of Chloroplasts* (Goodwin, T. W., ed.), vol. 1, pp. 365–373, Academic Press, London and New York
Burkard, G., Echlancher, B. & Weil, J. H. (1969) *FEBS Lett.* **4**, 285–287
Burkard, G., Guillemaut, P. & Weil, J. H. (1970) *Biochim. Biophys. Acta* **224**, 184–198
Cavalier-Smith, T. (1970) *Nature (London)* **228**, 333–335
Chan, P. H. & Wildman, S. G. (1972) *Biochim. Biophys. Acta* **277**, 677–680
Chen, J. L. & Wildman, S. G. (1967) *Science* **155**, 1271–1273
Chen, J. L. & Wildman, S. G. (1970) *Biochim. Biophys. Acta* **209**, 207–219
Chiang, K. S. & Sueoka, N. (1967) *Proc. Nat. Acad. Sci. U.S.* **57**, 1506–1513
Clark, M. F. (1964) *Biochim. Biophys. Acta* **91**, 671–674
Criddle, R. S., Dau, B., Kleinkopf, G. E. & Huffaker, R. C. (1970) *Biochem. Biophys. Res. Commun.* **41**, 621–627
Davies, J. W. & Cocking, E. C. (1967) *Biochem. J.* **104**, 23–33
Drumm, H. E. & Margulies, M. M. (1970) *Plant Physiol.* **45**, 435–442
Edelman, M., Cowan, C. A., Epstein, H. T. & Schiff, J. A. (1964) *Proc. Nat. Acad. Sci. U.S.* **52**, 1214–1219
Edelman, M., Schiff, J. A. & Epstein, H. T. (1965) *J. Mol. Biol.* **11**, 769–774
Eisenstadt, J. M. & Brawerman, G. (1964) *J. Mol. Biol.* **10**, 392–402
Ellis, R. J. (1963) *Nature (London)* **200**, 596
Ellis, R. J. (1969) *Science* **163**, 477–478
Ellis, R. J. (1970) *Planta* **91**, 329–335
Ellis, R. J. & Forrester, E. E. (1972) *Biochem. J.* **130**, 28P
Ellis, R. J. & Hartley, M. R. (1971) *Nature (London) New Biol.* **233**, 193–196
Ellis, R. J. & MacDonald, I. R. (1967) *Plant Physiol.* **42**, 1297–1302
Ellis, R. J. & MacDonald, I. R. (1970) *Plant Physiol* **46**, 227–232
Eytan, G. & Ohad, I. (1970) *J. Biol. Chem.* **245**, 4297–4307
Falk, H. (1969) *J. Cell Biol.* **42**, 582–587
Francki, R. I. B., Boardman, N. K. & Wildman, S. G. (1965) *Biochemistry* **4**, 865–872
Gillham, N. W., Boynton, J. E. & Burkholder, B. (1970) *Proc. Nat. Acad. Sci. U.S.* **67**, 1026–1033
Givan, C. V. & Leech, R. M. (1971) *Biol. Rev. Cambridge Phil. Soc.* **46**, 409–428
Gnanam, A., Jagendorf, A. T. & Ranaletti, M. L. (1969) *Biochim. Biophys. Acta* **186**, 205–213
Goffeau, A. (1969) *Biochim. Biophys. Acta* **174**, 340–350
Goffeau, A. & Brachet, J. (1965) *Biochim. Biophys. Acta* **95**, 302–313
Goodenough, U. W., Armstrong, J. J. & Levine, R. P. (1969) *Plant Physiol.* **44**, 1001–1012
Graham, D., Hatch, M. D., Slack, C. R. & Smillie, R. M. (1970) *Phytochemistry* **9**, 521–610
Granick, S. & Gibor, A. (1967) *Progr. Nucl. Acid. Res. Mol. Biol.* **6**, 143–186
Green, P. B. (1964) *Amer. J. Bot.* **51**, 334–342
Griffiths, D. E. & Lozano, J. A. (1970) *Rev Espan. Fisiol.* **26**, 83–88
Gualerzi, C. & Cammarano, P. (1969) *Biochim. Biophys. Acta* **190**, 170–186
Gualerzi, C. & Cammarano, P. (1970) *Biochim. Biophys. Acta* **199**, 203–213
Guderian, R. H., Pulliam, R. L. & Gordon, M. P. (1972) *Biochim. Biophys. Acta* **262**, 50–65
Hall, T. C. & Cocking, E. C. (1966) *Biochim. Biophys. Acta* **123**, 163–171
Hanson, J. B. & Krueger, W. A. (1966) *Nature (London)* **211**, 1322
Hartley, M. R. & Ellis, R. J. (1972) *Biochem. J.* **130**, 27P–28P
Heber, V. & Santarius, K. A. (1970) *Z. Naturforsch. B* **25**, 718–728
Heldt, H. W. (1969) *FEBS Lett.* **5**, 11–14
Honda, S. I., Hongladarom, T. & Laties, G. G. (1966) *J. Exp. Bot.* **17**, 460–472

Honda, S. I., Hongladarom-Honda, T., Kwanyuen, P. & Wildman, S. G. (1971) *Planta* **97**, 1–15
Hoober, J. K. (1970) *J. Biol. Chem.* **245**, 4327–4334
Hoober, J. K. & Blobel, G. (1969) *J. Mol. Biol.* **41**, 121–138
Hoober, J. K., Siekevitz, P. & Palade, G. E. (1969) *J. Biol. Chem.* **244**, 2621–2631
Ingle, J., Possingham, J. V., Wells, R., Leaver, C. J. & Loening, U. E. (1970) *Symp. Soc. Exp. Biol.* **24**, 303–325
Ireland, H. M. M. & Bradbeer, J. W. (1971) *Planta* **96**, 254–261
Janda, H. G. & Wittman, H. G. (1968) *Mol. Gen. Genet.* **103**, 238–243
Kawashima, N. & Wildman, S. G. (1970) *Annu. Rev. Plant Physiol.* **21**, 325–358
Kawashima, N. & Wildman, S. G. (1972) *Biochim. Biophys. Acta* **262**, 42–49
Kirk, J. T. O. (1970) *Annu. Rev. Plant Physiol.* **21**, 11–42
Kirk, J. T. O. (1971*a*) in *Autonomy and Biogenesis of Mitochondria and Chloroplasts* (Boardman N. K., Linnane, A. W. & Smillie, R. M., eds.), pp. 267–276, North-Holland Publishing Co., Amsterdam
Kirk, J. T. O. (1971*b*) *Annu. Rev. Biochem.* **40**, 161–196
Kirk, J. T. O. & Tilney-Bassett, R. A. E. (1967) *The Plastids: their Chemistry, Structure, Growth and Inheritance*, W. H. Freeman, London and San Francisco
Kung, S. D. & Williams, J. P. (1969) *Biochim. Biophys. Acta* **195**, 434–445
Leaver, C. J. & Ingle, J. (1971) *Biochem. J.* **123**, 235–243
Leis, J. P. & Keller, E. B. (1970) *Proc. Nat. Acad. Sci. U.S.* **67**, 1593–1599
Leis, J. P. & Keller, E. B. (1971) *Biochemistry* **10**, 889–894
Levine, R. P. (1969) *Annu. Rev. Plant Physiol.* **20**, 523–540
Levine, R. P. & Goodenough, U. W. (1970) *Annu. Rev. Genet.* **4**, 397–408
Loening, U. E. (1969) *Biochem. J.* **113**, 131–138
Loening, U. E. & Ingle, J. (1967) *Nature (London)* **215**, 363–367
Lyttleton, J. W. (1962) *Exp. Cell Res.* **26**, 312–317
Lyttleton, J. W. (1968) *Biochim. Biophys. Acta* **154**, 145–149
Mackender, R. O. & Leech, R. M. (1970) *Nature (London)* **228**, 1347–1348
Manning, J. E., Wolstenholme, D. R., Ryan, R. S., Hunter, J. A. & Richards, O. C. (1971) *Proc. Nat. Acad. Sci. U.S.* **68**, 1169–1173
Marcus, A. & Feeley, J. (1965) *J. Biol. Chem.* **160**, 1675–1680
Margulies, M. M. (1970) *Plant Physiol.* **46**, 136–141
Margulies, M. M. & Parenti, F. (1968) *Plant Physiol.* **43**, 504–514
Margulies, M. M., Gantt, E. & Parenti, F. (1968) *Plant Physiol.* **43**, 495–503
Menke, W. (1962) *Annu. Rev. Plant Physiol.* **13**, 27–44
Merrick, W. C. & Dure, L. S., III (1971) *Proc. Nat. Acad. Sci. U.S.* **68**, 641–644
Meyer, A. (1883) *Bot. Ztg.* **41**, 489–498
Miller, P. L. (1970) *Symp. Soc. Exp. Biol.* **24**, 1–501
Nobel, P. S. & Wang, C. T. (1970) *Biochim. Biophys. Acta* **211**, 79–87
Odintsova, M. S. & Yurina, N. P. (1969) *J. Mol. Biol.* **40**, 503–506
Parenti, F. & Margulies, M. (1967) *Plant Physiol.* **42**, 1179–1186
Payne, P. I. & Dyer, T. A. (1971) *Biochem. J.* **124**, 83–89
Payne, P. I. & Dyer, T. A. (1972) *Nature (London) New Biol.* **235**, 145–147
Pitt, D. (1971) *Planta* **101**, 333–351
Ramirez, J. M., Del Campo, F. F. & Arnon, D. I. (1968) *Proc. Nat. Acad. Sci. U.S.* **59**, 606–611
Ranaletti, M., Gnanam, A. & Jagendorf, A. T. (1969) *Biochim. Biophys. Acta* **186**, 192–204
Reger, B. J., Fairfield, S. A., Epler, J. L. & Barnett, W. W. (1970) *Proc. Nat. Acad. Sci. U.S.* **67**, 1207–1213
Ridley, S. M. & Leech, R. M. (1970) *Nature (London)* **227**, 463–465
Ris, H. & Plaut, W. (1962) *J. Cell Biol.* **13**, 383–391
Rutner, A. C. & Lane, D. M. (1967) *Biochem. Biophys. Res. Commun.* **28**, 531–537
Sager, R. & Hamilton, M. G. (1967) *Science* **157**, 709–711
Sager, R. & Ishida, M. R. (1963) *Proc. Nat. Acad. Sci. U.S.* **50**, 725–730
Schiff, J. A. (1971) in *Autonomy and Biogenesis of Mitochondria and Chloroplasts* (Boardman, N. K., Linnane, A. W. & Smillie, R. W., eds.), pp. 98–118, North-Holland Publishing Co., Amsterdam
Schimper, A. F. W. (1885) *Jahrb. Wiss. Bot.* **16**, 1–247
Schwartz, J. H., Meyer, R., Eisenstadt, J. M. & Brawerman, G. (1967) *J. Mol. Biol.* **25**, 571–574
Scott, N. S. & Smillie, R. M. (1967) *Biochem. Biophys. Res. Commun.* **28**, 598–603
Scott, N. S., Munns, R., Graham, D. & Smillie, R. M. (1971) in *Autonomy and Biogenesis of Mitochondria and Chloroplasts* (Boardman, N. K., Linnane, A. W. & Smillie, R. M., eds.), pp. 383–392, North-Holland Publishing Co., Amsterdam

Smillie, R. M. & Scott, N. S. (1970) *Progr. Mol. Subcell. Bot.* **1**, 136–202
Smillie, R. M., Graham, D., Dwyer, M. R., Grieve, A. & Tobin, N. F. (1967) *Biochem. Biophys. Res. Commum.* **28**, 604–610
Spencer, D. (1965) *Arch. Biochem. Biophys.* **111**, 381–390
Spencer, D. & Wildman, S. G. (1964) *Biochemistry* **3**, 954–959
Spencer, D., Whitfield, P. R., Bottomley, W. & Wheeler, A. M. (1971) in *Autonomy and Biogenesis of Mitochondria and Chloroplasts* (Boardman, N. K., Linnane, A. W. & Smillie, R. M., eds.), pp. 372–382, North-Holland Publishing Co., Amsterdam
Stephenson, M. L., Thimann, K. V. & Zamecnik, P. C. (1956) *Arch. Biochem. Biophys.* **65**, 194–209
Strasburger, E. (1882) *Arch. Mikrosk. Anat. Entwicklungsmech.* **21**, 476–590
Stutz, E. (1970) *FEBS Lett.* **8**, 25–28
Surzycki, S. J., Goodenough, U. W., Levine, R. P. & Armstrong, J. J. (1970) *Symp. Soc. Exp. Biol.* **24**, 13–37
Svetailo, E. N., Phillppovich, I. I. & Sissakian, N. M. (1967) *J. Mol. Biol.* **24**, 405–415
Tewari, K. K. & Wildman, S. G. (1966) *Science* **153**, 1269–1271
Tewari, K. K. & Wildman, S. G. (1968) *Proc. Nat. Acad. Sci. U.S.* **59**, 569–576
Tewari, K. K. & Wildman, S. G. (1969) *Biochim. Biophys. Acta* **186**, 358–372
Tewari, K. K. & Wildman, S. G. (1970) *Symp. Soc. Exp. Biol.* **24**, 147–179
Thomson, W. W. & Ellis, R. J. (1972) *Planta* **108**, 89–92
Thornber, J. P., Gregory, R. P. F., Smith, C. A. & Bailey, J. L. (1967) *Biochemistry* **6**, 391–396
Vasconcelos, A. C. L. & Bogorad, L. (1971) *Biochim. Biophys. Acta* **228**, 492–502
Watson, J. D. (1965) *Molecular Biology of the Gene*, W. A. Benjamin, New York and Amsterdam
Wells, R. & Birnstiel, M. (1969) *Biochem. J.* **112**, 777–786
Wells, R. & Sager, R. (1971) *J. Mol. Biol.* **58**, 611–622
Wetmur, J. G. & Davidson, N. (1968) *J. Mol. Biol.* **31**, 349–370
Whitfeld, P. R. & Spencer, D. (1968) *Biochim. Biophys. Acta* **157**, 333–343
Wildman, S. G., Hongladarom, T. & Honda, S. J. (1962) *Science* **138**, 434–436
Wilson, S. B. & Moore, A. L. (1973) *Biochim. Biophys. Acta* **292**, 603–610
Wittmann, H. G. (1970) *Symp. Soc. Gen. Microbiol.* **20**, 55–76
Woodcock, C. L. F. & Bogorad, L. (1970) *Biochim. Biophys. Acta* **224**, 639–643
Woodcock, C. L. F. & Bogorad, L. (1971) in *Structure and Function of Chloroplasts* (Gibbs, M., ed.), pp. 89–128, Springer-Verlag, Berlin, Heidelberg and New York
Woodcock, C. L. F. & Fernandez-Moran, H. (1968) *J. Mol. Biol.* **31**, 627–631

*Biochem. Soc. Symp.* (1973) **38**, 163–174
*Printed in Great Britain*

# Isolation and Characterization of Chloroplast 70S and of 80S Ribosomes from *Chlamydomonas reinhardii*: Protein Synthesis *in vitro*

By NAM-HAI CHUA, GÜNTER BLOBEL and PHILIP SIEKEVITZ
*The Rockefeller University, New York, N.Y.* 10021, *U.S.A.*

## Synopsis

The 70S chloroplast ribosomes and 80S cytoplasmic ribosomes were isolated, purified and dissociated into their respective large and small subunits. The monoribosomes and the reassociated subunits are active *in vitro* in the incorporation of phenylalanine in the presence of poly(U).

## Introduction

There is no doubt by now that isolated chloroplasts have the machinery necessary for protein synthesis (Boulter *et al.*, 1972), including ribosomes which in many biochemical aspects are akin to bacterial ribosomes (cf. Boulter *et al.*, 1972). Many laboratories have attempted to isolate these ribosomes, separate from the 70S mitochondrial ribosomes as well as from the 80S extra-chloroplastic ribosomes (henceforth called cytoplasmic ribosomes), to characterize them (cf. Boulter *et al.*, 1972), and to study *in vitro* their protein-synthetic capacities and properties. However, the early work suffered from a lack of our current knowledge of the properties of these ribosomes; thus the early studies on *Euglena* ribosomes (Eisenstadt & Brawerman, 1964*a*,*b*; Schwartz *et al.*, 1965) characterized the cytoplasmic and chloroplast ribosomes as having sedimentation coefficients of 70S and 60S, whereas we now know these to be 86S and 70S, respectively (cf. Boulter *et al.*, 1972). Moreover, it is evident that the extensive work *in vitro* by Boardman *et al.* (1965, 1966) dealt with mixtures of 70S and 80S ribosomes from tobacco leaves, as did the earlier study of App & Jagendorf (1963). In some cases, either no attempts were made to characterize the ribosomes in the protein-synthesizing system (Sager *et al.*, 1963; Spencer & Wildman, 1964; Chen & Wildman, 1967; Van Kammen, 1967) or no attempt was made to use purified ribosomes for studies *in vitro* (Sager & Hamilton, 1967). Thus, as far as we know, no papers have appeared dealing with the purification of chloroplast and cytoplasmic ribosomes from the same plant tissue and dealing with the protein-synthetic properties of these purified ribosomes *in vitro*.

We would like here to report on the isolation of purified chloroplast 70S and cytoplasmic 80S ribosomes from *Chlamydomonas reinhardii* and on the synthesis of polypeptides by these particles *in vitro*. The choice of *C. reinhardii* is a fortunate one, for its photosynthetic apparatus, both morphologically (Ohad *et al.*, 1967*a*) and biochemically (Levine, 1968), is like that of higher plants; it has one single giant chloroplast, about one-half of the volume of the cell, which contains many ribosomes (Ohad *et al.*, 1967*a*,*b*), and further, the ratio of chloroplast material to mitochondrial material is much higher than in other plants and plant tissues (Ohad *et al.*, 1967*a*). This last point is important, since the sedimentation characteristics of mitochondrial ribosomes (Borst & Grivell, 1971) are similar to those of chloroplast ribosomes, and thus it becomes necessary to ensure that any 70S ribosomes are really chloroplast and not mitochondrial in origin, particularly since separation of intact chloroplasts from intact plant mitochondria is not always easy. Thus, it has been shown that the amount of ribosomal material sedimenting at 70S is proportional to the amount of chloroplast ribosomes which can be seen with the electron microscope, when wild-type *C. reinhardii* cells are compared with mutants having much fewer chloroplast ribosomes (Goodenough & Levine, 1970; Bourque *et al.*, 1971).

## Materials and Methods

Cells of the wild-type strain (137C, mating type plus) of *C. reinhardii* were grown in liquid Tris–acetate–phosphate medium at 25°C under conditions described by Gorman & Levine (1965). The culture was harvested at exponential phase of growth ($3\times10^6$–$4\times10^6$ cells/ml) by centrifugation at 2500***g*** for 5min at 0°C.

Ribosomes were prepared by a modification and extension of the procedure of Hoober & Blobel (1969). The cell pellet was washed once in a high-$Mg^{2+}$ TKMD buffer (25mM-Tris–HCl buffer, pH 7.5–25mM-KCl–25mM-$MgCl_2$–5mM-dithiothreitol) and resuspended in the same buffer to a cell concentration of approx. $5\times10^8$ cells/ml. The suspension was forced through a chilled French pressure cell maintained at a constant pressure of 26.2 MPa (3800 lb/in$^2$). All subsequent operations were carried out at 4°C. The homogenate was centrifuged at 12000 rev./min for 10min in a Sorvall SS-34 rotor, and 7.0ml of the resulting post-mitochondrial supernatant was layered over 2.5ml of 1.87M-sucrose containing the high-$Mg^{2+}$ TKMD buffer. After centrifugation at 40000 rev./min for 16h in a Spinco no. 40 rotor, the ribosomes were sedimented to the bottom of the tube and the photosynthetic membranes were retained at the 1.87M-sucrose interface. The surface of the pellet was washed twice with deionized water, and either used directly or stored at −80°C for up to several weeks.

Ribosomal pellets were resuspended in water ($E_{260}/E_{280}$ 1.84–1.88) and approx. 0.5ml (25 $E_{260}$ units) of the suspension was layered on a 12.5ml linear 10–40% sucrose density gradient containing a low-$Mg^{2+}$ TKMD buffer (25mM-Tris–HCl buffer, pH 7.5–25mM-KCl–5mM-$MgCl_2$–5mM-dithiothreitol). After centrifugation at 22500 rev./min for 15.5h at 4°C in an SB283 rotor of the IEC centrifuge, the $E_{254}$ was recorded with an ISCO Model D fractionator and analyser (Fig. 1). Fractions containing the 70S and the 80S ribosomes (hatched areas in Fig. 1) were collected separately, the $MgCl_2$ concentration in the pooled fractions was adjusted to 25mM and the ribosomes were sedimented by centrifugation at 40000 rev./min for 12–14h in a Spinco no. 40 rotor. The pellets were stored at −80°, and when used for the analytical and incorporation experiments were resuspended in ice-cold deionized water, unless otherwise indicated.

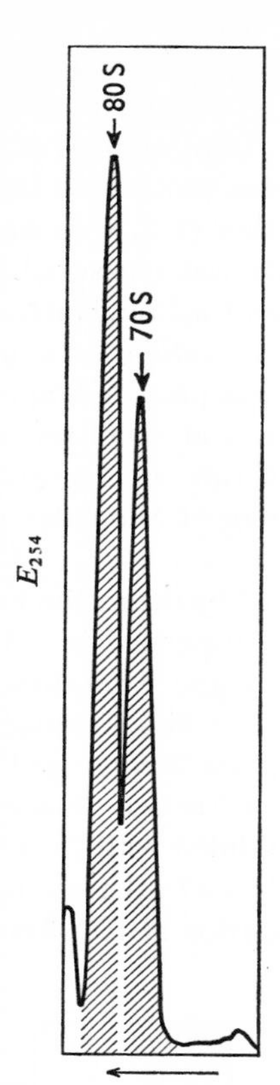


Fig. 1. *Preparative sucrose-density-gradient separation of* 70*S and* 80*S ribosomes*
The direction of sedimentation is indicated by the arrow. A sample of 25 $E_{260}$ units was layered on the gradient as described in the text.

The 70S and the 80S ribosomes were dissociated into their subunits by centrifugation of the respective suspensions in a high-salt, linear 5–20% sucrose density gradient (cf. Blobel & Sabatini, 1971). For the 70S ribosomes, the gradient contained 50mM-Tris–HCl buffer (pH7.5), 400mM-KCl, 25mM-$MgCl_2$ and 5mM-dithiothreitol. The gradients were centrifuged at 39000 rev./min for 3h at 18°C in the SB 283 rotor. The separated subunits (cf. Fig. 3) were collected and sedimented by centrifugation at 40000 rev./min for 16h at 0°C in a Spinco no. 40 rotor. The 80S ribosomes were treated similarly except that the KCl concentration was raised to 500mM. Subunits from both the 70S and the 80S ribosomes were kept at −80°C in the form of pellets for future use.

The reaction mixture for the incorporation into polyphenylalanine contained the following components: Tris–HCl buffer, pH 7.5, 25μmol; KCl, 50μmol; $MgCl_2$, 6.25μmol (for the 80S ribosomes or their subunits) or 11.25μmol (for the 70S ribosomes or their subunits); dithiothreitol, 2.5μmol; high-speed supernatant, 125μl (6–8mg of protein/ml); pH 5 enzyme, 125μl (10mg of protein/ml); creatine phosphokinase, 25μg (1.87 units); creatine phosphate, 5μmol; GTP, 0.25μmol; ATP, 0.5μmol; [$^{14}$C]phenylalanine (383μCi/mol), 0.25μCi; ribosomes, approx. 2.0 $E_{260}$ units; poly(U), 200μg, all in a final volume of 0.5ml. All the components except poly(U) were mixed together at 0°C; the reaction was started by the addition of poly(U) and the transfer of the mixture to a 37°C water bath. At 10, 20, 30 and 60min, 100$\mu$l samples of the mixture were pipetted on to Whatman 3MM filter-paper discs which were then extracted with hot 5% (w/v) trichloroacetic acid, ethanol–ether (1:1, v/v), and ether as described by Mans & Novelli (1961). Radioactivity was measured in 10ml of Liquifluor–toluene mixture (40ml of Liquifluor, New England Nuclear Corp., plus 960ml of toluene) with a Nuclear–Chicago Mark I scintillation counter.

The high-speed supernatant was prepared by centrifugation of the post-mitochondrial supernatant (see above) at 40000rev./min for 4h in a Spinco no. 40 rotor. The upper two-thirds of this supernatant was removed with a Pasteur pipette, passed through a column of Sephadex G-25 and stored at −196°C in 1.0ml samples.

The pH5 enzyme was prepared by the method of Falvey & Staehelin (1970). A 1g sample of packed cells was suspended in 2.0ml of 10mM-Tris–HCl buffer, pH 7.5, containing 5mM-$MgCl_2$ and 5mM-dithiothreitol, and disrupted by a French pressure cell at 3800lb/in$^2$. This homogenate was centrifuged at 17000 rev./min for 15min in a Sorvall SS-34 rotor and the supernatant from this was centrifuged at 40000rev./min for 3h in a Spinco no. 40 rotor. The upper two-thirds of this supernatant was diluted with 2vol. of ice-cold 5mM-dithiothreitol and the pH was adjusted to about 5.1 by the addition of 1M-acetic acid. The precipitate was collected by centrifugation at 10000rev./min in a Sorvall SS-34 rotor

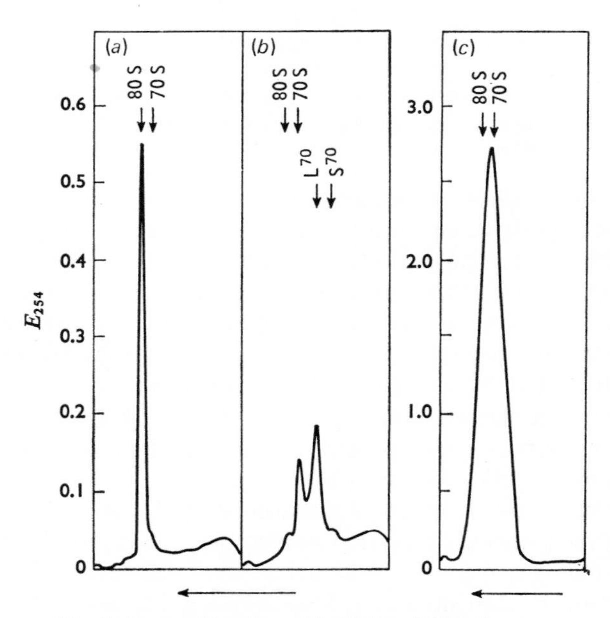


Fig. 2. *Recentrifugation of the* 70*S and* 80*S ribosomes*

The two pooled fractions, indicated by hatched areas in Fig. 1, were sedimented, resuspended in high-$Mg^{2+}$ TKMD buffer, and centrifuged for 15.5h as described in the Materials and Methods section, on 10–40% sucrose density gradients containing the same buffer. (*a*) 80S ribosomes, 0.5 $E_{260}$ units; (*b*) 70S ribosomes, 0.5 $E_{260}$ units; (*c*) 70S ribosomes, 11.4 $E_{260}$ units.

for 10min and redissolved in the low-$Mg^{2+}$ TKMD buffer to a protein concentration of about 10mg/ml. The pH of the suspension was adjusted to 7.5 by the addition of 1 M-KOH and the suspension was stored in 1 ml portions in liquid $N_2$.

## Results and Discussion

Fig. 1 shows the distribution on a preparative sucrose density gradient of the ribosome population of *C. reinhardii*. The areas under the peaks cannot be quantitatively compared with each other, since at high concentrations extinction readings traced by the ISCO analyser are not linear. However, from earlier work (Hoober & Blobel, 1969) we know that about one-third of the total population of *C. reinhardii* ribosomes is of the 70S variety. Other experiments (not shown) indicate that less than 5% of material sediments in the polyribosome region of the gradient. The volumes included within the hatched areas in Fig. 1 were always collected for the incorporation experiments described below. Fig. 2 shows that the 80S and 70S ribosomal suspensions, so separated were almost free from

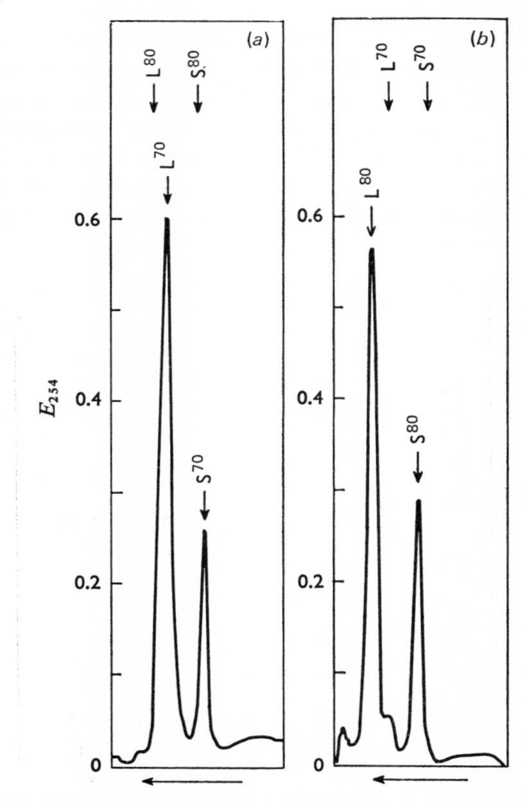


Fig. 3. *Dissociation of* 70 *S and* 80 *S ribosomes into subunits*

Ribosomes were dissociated in a high-salt sucrose density gradient as described in the text. Amounts of ribosomes used were: (*a*) 70S, 1.24 $E_{260}$ units; (*b*) 80S, 1.29 $E_{260}$ units. L and S refer to the large and small subunits, and the superscripts 80 and 70 refer to the monomers from which they are derived.

cross-contamination. There are very few 70S ribosomes in the 80S ribosomal preparation (Fig. 2*a*). In the case of the 70S ribosomes, the distribution of the purified preparation varies with load put on the gradients; when a small load is put on (Fig. 2*b*), about two-thirds of the ribosomes become dissociated, but when a larger load is placed on the gradient (Fig. 2*c*), most of the 70S ribosomes remain undissociated. However, in both cases, it can be seen that very few 80S ribosomes appear in the 70S ribosomal preparation. Control experiments showed that the extinction trace near the top of the gradient (Figs. 2*a* and 2*b*) reflects the presence there of oxidized dithiothreitol.

Both the 80S and 70S ribosomes are dissociated into their respective large (L) and small (S) subunits when they are submitted to sucrose-density-gradient centrifugation in a high-salt medium, as was found for liver ribosomes (Blobel & Sabatini, 1971), and as can be seen for *C. reinhardii* ribosomes in Fig. 3. When shorter periods of centrifugation were used, to detect undissociated ribosomes, it was found that the dissociation of both 70S and 80S *C. reinhardii* ribosomes was almost complete. By using this procedure, the four subunits can be isolated in a very pure state and are active biologically (see below). The high degree of purity of the initial 80S and 70S monomers, before dissociation, can also be estimated by the relative lack of subunit cross contamination; the subunit profile for the 70S ribosomes shows very little contamination by 80S subunits and vice versa.

Fig. 4 and Table 1 show that the purified ribosomes, collected as shown in Fig. 1, are biologically active in polypeptide synthesis. In the absence of poly(U), there was no incorporation of [$^{14}$C]phenylalanine into polypeptides, but with saturating amounts of poly(U), the incorporation was linear for the first 30 min, for both the 80S and 70S ribosomes (Fig. 4). The results in Table 1 indicate that the co-

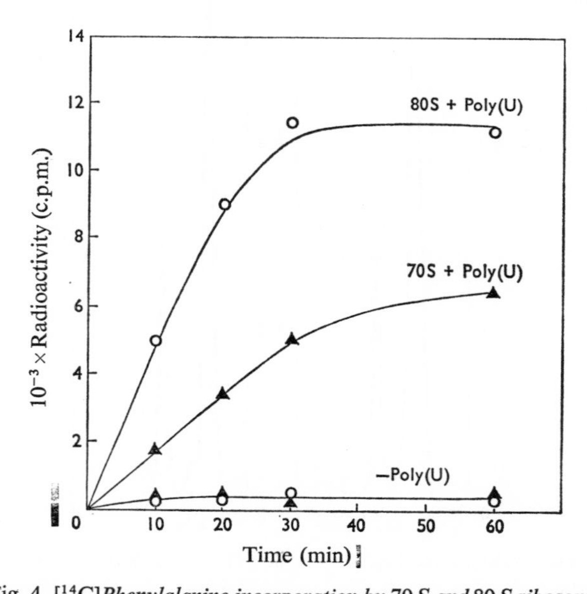


Fig. 4. [$^{14}$C]*Phenylalanine incorporation by* 70S *and* 80S *ribosomes*

The incubation conditions are described in the text. Amounts of ribosomes used were: 80S, 2.04 $E_{260}$ units (○); 70S, 1.86 $E_{260}$ units (▲).

Table 1. *Requirements for [$^{14}$C]phenylalanine incorporation by* 70 *S and* 80 *S ribosomes*

Experiments were performed as described in the text. The amounts of ribosomes used were: 80S, 2.04 $E_{260}$ units; 70S, 1.86 $E_{260}$ units.

| Conditions | Radioactivity (c.p.m.) | |
|---|---|---|
| | 70S | 80S |
| Complete | 6436 | 11181 |
| –Ribosomes | 276 | 380 |
| –Poly(U) | 366 | 374 |
| –High-speed supernatant | 8727 | 13115 |
| –pH5 enzyme | 1514 | 2683 |
| –High-speed supernatant and pH5 enzyme | 152 | 144 |
| –GTP | 1065 | 7879 |
| –ATP | 2074 | 3793 |
| –Creatine phosphate and creatine phosphokinase | 2068 | 8255 |

factor requirements, including energy sources, are the same as for other systems which have been examined *in vitro*. However, in some cases, the requirements for GTP, ATP, an ATP-generating system, and the pH5 enzyme fraction, were only partial, reflecting perhaps that the ribosomes as isolated contained some bound factors. In the case of the high-speed supernatant, there was no requirement at all, indicating that all the protein factors and tRNA required for [$^{14}$C]phenylalanine incorporation were present in the pH5 fraction; in later experiments the addition of this high-speed supernatant was omitted. On the basis of specific radioactivity calculations (c.p.m./$E$ unit), the 80S ribosomes were about 1.5 times more active than the 70S ribosomes in their incorporation of phenylalanine into polyphenylalanine.

The formation of nascent polypeptide is shown in Fig. 5 for the 70S ribosomes and in Fig. 6 for the 80S ribosomes. After a 90min incubation of the ribosomes under the incorporation conditions, they were analysed on a high-salt sucrose density gradient to dissociate these particles, as described above. It was found that, first of all, in both cases the radioactivity appeared only in the monoribosome region, coincident with the extinction profile in that region (Figs. 5*a* and 6*a*) Secondly, Fig. 5(*a*) and Fig. 6(*a*) show that more monoribosome formation occurred with the 80S preparation (Fig. 6*a*) than with the 70S preparation (Fig. 5*a*). It appears that the presence of the nascent chain stabilizes the monoribosome in the high-salt gradient; this is similar to the finding with liver ribosomes (Blobel & Sabatini, 1971; Falvey & Staehelin, 1970). In the presence of a high-salt concentration, puromycin has been found to release the nascent chain from the rat liver 80S monoribosomes, at the same time causing a dissociation into subunits (Blobel & Sabatini, 1971). The same thing happens in the case both of the 70S (Fig. 5*b*) and 80S (Fig. 6*b*) monoribosomes, indicating again the nascent nature of the incorporated radioactivity. The end result is a complete dissociation of the ribosome (Figs. 5*b* and 6*b*). The finding that puromycin plus high-salt is necessary to effect this dissociation after incorporation *in vitro*, although only high-salt conditions are necessary with ribosomes initially isolated, is an indication that, as isolated, no nascent chains are bound to the ribosomes. This absence of nascent chains, and the concomitant absence of mRNA, is a probable explanation for the lack of endogenous incorporation. Finally, it was found that apparently not all the ribosomes are active *in vitro*; whereas a large proportion is active in the case of the 80S ribosome preparation (Fig. 6*a*), only a small number is active in the case of the 70S ribosome preparation (Fig. 5*a*). Indeed, the quantitative difference in the incorporating ability of the 80S and 70S ribosome preparations (Fig. 4 and Table 1) may be due to a higher ratio of active ribosomes/total ribosomes in the 80S population than in the 70S population, and not due to any intrinsic difference in the peptide-synthesizing capability of 70S and 80S ribosomes.

After dissociation by a high-salt concentration, the subunits are still biologically active (Table 2). Subunits can be obtained from both the 80S and 70S ribosomes (cf. Fig. 3), and can be separated and purified from each other (Figs. 7*a* and 7*b* for the 70S ribosomes, and Figs. 8*a* and 8*b* for the 80S ribosomes). When the salt content is lowered and the $Mg^{2+}$ concentration is raised, the large and small

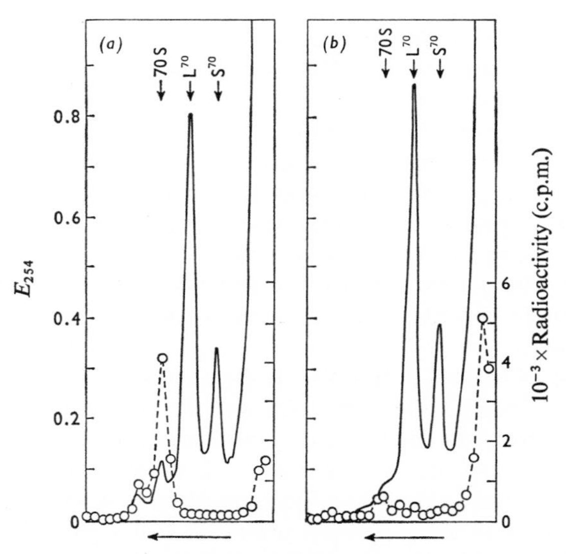


Fig. 5. *Sucrose-density-gradient analyses of* 70*S ribosomes after incorporation of* [$^{14}$*C*]-*phenylalanine*

The [$^{14}$C]phenylalanine incorporation mixture (1 ml) containing 7.76 $E_{260}$ units of 70S ribosomes was incubated at 37°C for 90 min and then chilled on ice to stop the reaction. (*b*) A 350 $\mu$l sample of the solution was mixed with 50 $\mu$l of 10 mM-puromycin (pH 7.0) followed by 400 $\mu$l of compensating buffer such that the final ionic composition was 50 mM-Tris–HCl buffer (pH 7.5), 500 mM-KCl, 25 mM-$MgCl_2$ and 5 mM-dithiothreitol (Blobel & Sabatini, 1971). After an incubation period of 10 min at 0°C, 400 $\mu$l (1.36 $E_{260}$ units) of the reaction mixture was layered on a high-salt 5–20% sucrose density gradient containing 500 mM-KCl, and centrifuged at 39000 rev./min for 1.8 h at 18°C in a SB 283 rotor. (*a*) A control sample was processed in the same way except that 50 $\mu$l of deionized water was used in place of puromycin. After centrifugation, the gradient was fractionated into 0.5 ml portions with an ISCO model D fractionator and analyzer. A 0.1 ml sample of 1% bovine serum albumin was added to each fraction as a carrier, followed by 0.2 ml of 2 M-KOH. After incubation at 37°C for 30 min, 0.2 ml of 2 M-HCl was added to neutralize the base. The proteins were precipitated with 4 ml of ice-cold 10% (w/v) trichloracetic acid. After being kept at 4°C overnight, the precipitate was washed with 5 ml of ice-cold 5% (w/v) trichloroacetic acid, dissolved in 0.5 ml of Protosol (New England Nuclear Corp.), and the radioactivity measured in 8 ml of the Liquifluor–toluene mixture with a Nuclear–Chicago Mark I scintillation counter. ——, $E_{254}$; ○, radioactivity.

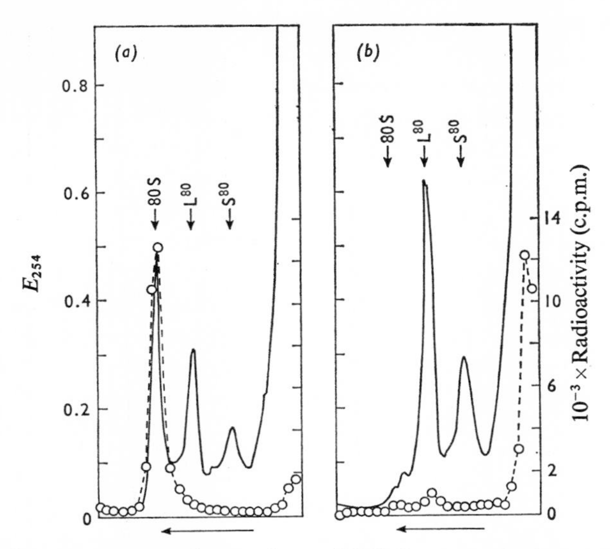


Fig. 6. *Sucrose-density-gradient analyses of* 80S *ribosomes after incorporation of* [$^{14}$*C*]*phenylalanine*

Conditions were as described in the legend to Fig. 5, except that a total of 1.33 $E_{260}$ units was initially layered on the gradient. (*a*) Control, 80S ribosomes; (*b*) 80S ribosomes + puromycin. ——, $E_{254}$; ○, radioactivity.

subunits of the 80S ribosomes can be almost completely reassociated into the monomer (Fig. 8*c*). This reassociation results in a biologically active ribosome; if the mixture is analysed by high-salt sucrose-density-gradient centrifugation after the incorporation, the radioactivity remains with the 80S monoribosome (Fig. 8*d*), and the separation on the sucrose density gradient after incorporation is the same whether the monomers are incubated (Fig. 6*a*) or whether the separated large and small subunits are incubated as a mixture (Fig. 8*d*). Indeed the specific radioactivity (c.p.m./*E* unit) of the 80S ribosomes after incubation of the large and small subunits is about 85% of that of the ribosomes after incubation of the monomers (Table 2).

In the case of the large and small subunits of the 70S ribosome, there is no reassociation (Figs. 7*a*, 7*b* and 7*c*) unless the subunits are incubated in the incorporation mixture (Fig. 7*d*), and even here the reassociation is less than in the case of the 80S ribosomes (cf. Figs. 7*d* and 8*d*). However, once again, the reassociated subunits are as active as the initially isolated monomers in phenylalanine incorporation, for after centrifugation of the incubated mixture in a high-salt medium, the same patterns emerge with the incubated subunits (Fig. 7*d*) as with the incubated monomer (Fig. 5*a*). Indeed, the specific radioactivity of the ribosomes after incorporation by isolated subunits is about 20% higher than in the case of the monomer (Table 2).

In conclusion, we have shown that it is possible to purify, for preparative purposes, both 70S and 80S ribosomes from the same plant, *C. reinhardii*, and that the former are most likely derived from the chloroplasts. These ribosomes can be disassociated into subunits and reassociated to various extents from their purified

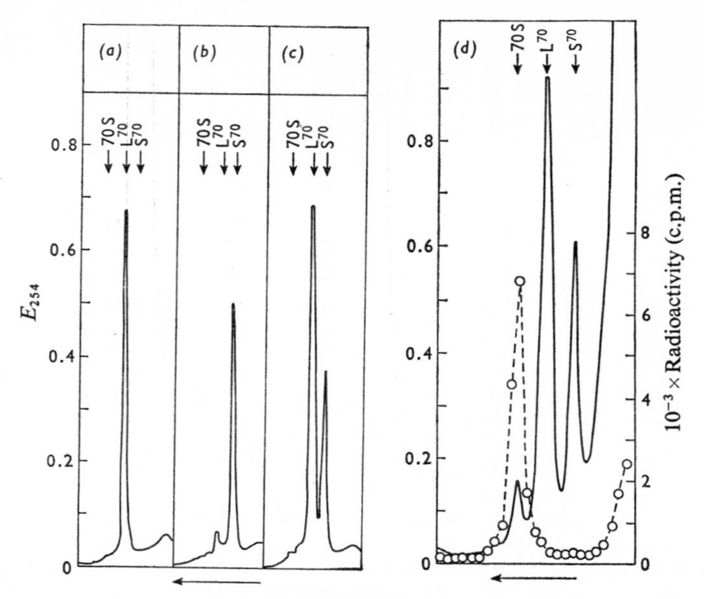


Fig. 7. *Reassociation of the large and the small subunits into* 70*S ribosomes*

(*a*) A total of 0.9 $E_{260}$ unit of $L^{70}$ (cf. Fig. 3) was suspended in the high-$Mg^{2+}$ TKMD buffer and centrifuged on a 10–40% sucrose density gradient containing the same buffer for 15.5h at 22500 rev./min in an SB 283 rotor at 4°C; (*b*) a total of 0.6 $E_{260}$ unit of $S^{70}$ (cf. Fig. 3) was treated as in (*a*); (*c*) a total of 0.9 $E_{260}$ unit of $L^{70}$ was mixed with 0.45 $E_{260}$ unit of $S^{70}$ and incubated at 0°C for 10min; the mixture was then centrifuged as in (*a*); (*d*) a total of 3.08 $E_{260}$ units of $L^{70}$ and 1.52 $E_{260}$ units of $S^{70}$ was incubated in the [$^{14}$C]phenylalanine incorporation medium (0.5ml) at 37°C for 90min and then chilled on ice to stop the reaction. A 350 μl of the solution was mixed with 450 μl of compensating buffer such that the final ionic composition was 50mM-Tris–HCl buffer, pH 7.5, 500mM-KCl, 25mM-$MgCl_2$ and 5mM-dithiothreitol. After an incubation period of 10min at 0°C, 400 μl (containing 1.08 $E_{260}$ units of $L^{70}$ and 0.53 $E_{260}$ unit of $S^{70}$) was centrifuged on a high-salt 5–20% sucrose density gradient as described in the legend to Fig. 5. The gradient was fractionated and the fractions were processed and counted for radioactivity as in Fig. 5. ——, $E_{264}$; o, radioactivity.

---

Table 2. *Reconstitution of active* 70*S and* 80*S monomers from their respective subunits*

Preparations were made of 70S and 80S ribosomes; a sample was used for incorporation, and from the remainder large and small subunits were isolated as described in the text and incubated for 60min at 37°C in the incorporation medium (as described in the Materials and Methods section) separately or immediately after mixing. For the 70S ribosomes the amounts used were as follows: monomer, 2.26 $E_{260}$ units; large subunit, 1.60 $E_{260}$ units; small subunit, 0.52 $E_{260}$ unit. For the 80S ribosomes the amounts added were: monomer, 1.80 $E_{260}$ units; large subunit, 1.40 $E_{260}$ units; small subunit, 0.66 $E_{260}$ unit. In the absence of ribosomes or subunits, the incorporation was 235c.p.m.; this value was not subtracted.

| | Radioactivity (c.p.m.) | |
|---|---|---|
| | 70S | 80S |
| Monomer | 5324 | 9079 |
| Large subunit only | 770 | 1225 |
| Small subunit only | 221 | 368 |
| Large subunit plus small subunit | 5919 | 9010 |

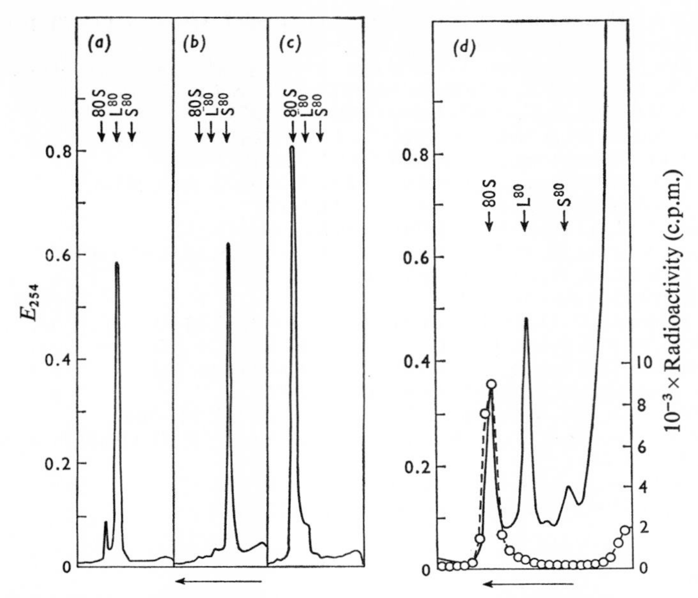


Fig. 8. *Reassociation of the large and the small subunits into* 80*S ribosomes*

(*a*) A total of 0.8 $E_{260}$ unit of $L^{80}$ (cf. Fig. 3) was suspended in high-$Mg^{2+}$ TKMD buffer and centrifuged as given in the legend to Fig. 7 (*a*); (*b*) a total of 0.8 $E_{260}$ unit of $S^{80}$ (cf. Fig. 3) was treated as in (*a*); (*c*) a total of 0.8 $E_{260}$ unit of $L^{80}$ was mixed with 0.4 unit of $S^{80}$ and incubated at 0°C for 10min, the mixture was then centrifuged as in (*a*); (*d*) experiments were performed as described in the legend to Fig. 7 (*d*) except that 0.89 $E_{260}$ unit of $L^{80}$ and 0.44 $E_{260}$ unit of $S^{80}$ were layered on the high-salt 5–20% sucrose density gradient. ——, $E_{254}$; ○, radioactivity.

subunits. Both the initial 70S and 80S ribosomes and the reassociated ribosomes are active in polypeptide synthesis *in vitro*, but only with exogenous messenger [in our case, poly(U) and phenylalanine]. The requirements for the incorporation of phenylalanine are the same as has been shown for animal and bacterial systems. Work is in progress to obtain polyribosome preparations containing either 70S or 80S ribosomes, which could be active in endogenous incorporation, indicating a retention of native mRNA and bound nascent chains. Preparations such as these would be most useful in the examination of the differences between these two populations of ribosomes, particularly with regard to the kinds of proteins each is capable of synthesizing.

## Note Added in Proof

A full paper on this topic has been published (Chua *et al.*, 1973).

The authors are grateful to G. E. Palade for his advice. This project was supported by Grant H-01689 from the National Institutes of Health.

## References

App, A. A. & Jagendorf, A. T. (1963) *Biochim. Biophys. Acta* **76**, 286–292
Blobel, G. & Sabatini, D., (1971) *Proc. Nat. Acad. Sci. U.S.* **68**, 390–394
Boardman, N. K., Francki, R. I. B. & Wildman, S. G. (1965) *Biochemistry* **4**, 872–876

Boardman, N. K., Francki, R. I. B. & Wildman, S. G. (1966) *J. Mol. Biol.* **17**, 470–489
Borst, P. & Grivell, L. A. (1971) *FEBS Lett.* **13**, 73–88
Boulter, I., Ellis, R. J. & Yarwood, A. (1972) *Biol. Rev. Cambridge Phil. Soc.* **47**, 113–175
Bourque, D. P., Boynton, J. E. & Gillham, N. (1971) *J. Cell Sci.* **8**, 153–183
Chen, J. L. & Wildman, S. G. (1967) *Science* **155**, 1271–1273
Chua, N.-H., Blobel, G. & Siekevitz, P. (1973) *J. Cell Biol.* **57**, 798–814
Eisenstadt, J. M. & Brawerman, G. (1964*a*) *J. Mol. Biol.* **10**, 392–402
Eisenstadt, J. M. & Brawerman, G. (1964*b*) *Biochim. Biophys. Acta* **80**, 463–472
Falvey, A. K. & Staehelin, T. (1970) *J. Mol. Biol.* **53**, 1–19
Goodenough, U. W. & Levine, R. O. (1970) *J. Cell Biol.* **44**, 547–562
Gorman, D. S. & Levine, R. P. (1965) *Proc. Nat. Acad. Sci. U.S.* **54**, 1665–1669
Hoober, J. K. & Blobel, G. (1969) *J. Mol. Biol.* **41**, 121–138
Levine, R. P. (1968) *Science* **162**, 768– 771
Mans, R. J. & Novelli, G. D. (1961) *Arch. Biochem. Biophys.* **94**, 48–53
Ohad, I., Siekevitz, P. & Palade, G. E. (1967*a*) *J. Cell Biol.* **35**, 521–552
Ohad, I., Siekevitz, P. & Palade, G. E. (1967*b*) *J. Cell Biol.* **35**, 553–584
Sager, R. & Hamilton, M. G. (1967) *Science* **157**, 709–711
Sager, R., Weinstein, I. B. & Ashkenazi, Y. (1963) *Science* **140**, 304–306
Schwartz, J. H., Eisenstadt, J. M., Brawerman, G. & Zinder, N. D. (1965) *Proc. Nat. Acad. Sci. U.S.* **53**, 195–200
Spencer, D. & Wildman, S. G. (1964) *Biochemistry* **3**, 954–959
Van Kammen, A. (1967) *Arch. Biochem. Biophys.* **118**, 517–524

## Discussion

**J. W. Davies**: Only about 10% or less, of your 70S ribosomes are active: do you think this is a true representation, or may be the conditions are not quite right? For instance, ideally translation should require free subunits, to get 30S ribosome, mRNA initiation. Natural dissociation mediated by dissociation factor requires low magnesium yet your extraction method uses 25mM-$Mg^{2+}$, and of course poly(U)-directed polyphenylalanine synthesis requires high $Mg^{2+}$. Have you tried natural messengers?

Also, sucrose (used in the gradients) is a notorious source of ribonuclease and may damage your ribosomes, in addition to ribonucleases from the cell extract. Have you used ribonuclease-free sucrose, and ribonuclease inhibitors, which might make a difference to the proportions of protein synthesis of 70S and 80S ribosomes *in vitro*?

**P. Siekevitz**: No, I do not think the 10% value is a true one regarding the number of active ribosomes in the cell at any one time. I do not think we have ideal conditions for getting out the ribosomes with all the factors necessary for protein synthesis; that is the reason we are trying to obtain active polyribosomes instead of monoribosomes.

In answer to the other question: We have not tried natural messenger, and though we do use ribonuclease-free sucrose, we have not added any ribonuclease inhibitor to the homogenization medium.

*Biochem. Soc. Symp.* (1973) **38**, 175–193
*Printed in Great Britain*

# Plant Mitochondrial Nucleic Acids

by C. J. LEAVER and M. A. HARMEY*

*Department of Botany, University of Edinburgh, Edinburgh EH9 3JH, U.K.*

## Synopsis

Plant mitochondria contain mitochondrial ribosomes sedimenting at 77–78 S which can be dissociated into large and small subunits. These subunits contain discrete ribosomal RNA species with estimated molecular weights of $1.12\times10^6$–$1.18\times10^6$ and $0.69\times10^6$–$0.78\times10^6$, depending on the plant species. The mitochondria also contain 4 S RNA and a 5 S ribosomal RNA component.

## Introduction

It is now generally accepted that all eukaryote cells contain interdependent genetic and protein-synthesizing systems in the nucleus and cytoplasm and in mitochondria. Green plants are unique in that they contain in addition a third such system in the plastids. Several recent reviews have considered the structure and function of the DNA, RNA and protein-synthesizing systems, of the chloroplasts of the green plant (Boulter *et al.*, 1972; Smillie & Scott, 1970) and of the mitochondria from animals and ascomycetes (Ashwell & Work, 1970; Borst & Grivell, 1971; Beattie, 1971; Kroon, 1971). The characterization and function of plant mitochondrial genetic and protein-synthetic systems has, with few exceptions, been neglected, although some studies have shown (Wolstenholme & Gross, 1968; Wells & Birnstiel, 1969; Kolodner & Tewari, 1972) that the molecular size of higher-plant mitochondrial DNA exceeds that of animal mitochondrial DNA by a factor of about 7 and that of those ascomycetes studies by a factor of about 2. Yet as far as is known all mitochondria perform the same function, namely that of respiration and the generation of energy.

Before we can ask the obvious questions as to the function of plant mitochondrial nucleic acids and the relationships between these functions and their contribution to plant growth and development, it is necessary to characterize the plant mitochondrial nucleic acids and ribosomes and produce convincing evidence that they are real mitochondrial components. In addition some knowledge of the evolutionary relationship between mitochondria and their origins may become apparent, as well as the functional and developmental relationships between the three integrated genetic systems found in the plant cell.

In this paper we shall briefly review the pertinent details of the structure and function of the mitochondrial nucleic acids and ribosomes of animals and ascomycetes and then discuss the results of our work on higher-plant mitochondria.

* Present address: Department of Botany, University College, Dublin, Irish Republic.

**Mitochondrial DNA**

The mitochondria of all organisms so far studied have been shown by a variety of techniques to contain DNA distinguishable from nuclear DNA and other satellite DNA species. Table 1 shows some of the characteristics of this mit-DNA (mitochondrial DNA) from animals, fungi and higher plants.

The buoyant densities of animal mit-DNA in CsCl vary from 1.686 g/cm$^3$ in *Drosophila* to 1.711 g/cm$^3$ in duck, equivalent to a difference in base composition of 25 mole per cent G + C. In a range of higher plants the buoyant density of mit-DNA is remarkably constant at about 1.706 g/cm$^3$ (Wells & Ingle, 1970; Suyama & Bonner, 1966) in contrast with chloroplast DNA with a constant density of about 1.697 g/cm$^3$ and the nuclear DNA species with a much more variable density of from 1.691 to 1.702 g/cm$^3$.

It has been established that mit-DNA from animal tissues exists in the form of closed circles with a contour length of about 5 $\mu$m and a molecular weight of $1.0 \times 10^7$ (Borst, 1970). Circular mit-DNA has also been isolated from yeast with a contour length of 25 $\mu$m corresponding to a molecular weight of $5.0 \times 10^7$. Mit-DNA species from higher plants have until recently always been isolated in linear molecules with mean lengths ranging from 10 to 20 $\mu$m (Wolstenholme & Gross, 1968). Kolodner & Tewari (1972) have isolated mit-DNA from pea leaves in a circular conformation with an average contour length of 30 $\mu$m. These workers, using both electron microscopy and renaturation kinetics, have shown that the molecular weight of mit-DNA from a range of higher plants is about $7.0 \times 10^7$. Wells & Birnstiel (1969) have reported that the kinetic complexity of mit-DNA from lettuce plants is equivalent to a molecular weight of $>10^8$, Kolodner & Tewari (1972) failed to show any intermolecular heterogeneity in the base compositions of different molecules or any evidence for repeating sequences in the mit-DNA. Their evidence would appear to preclude the presence of mit-DNA molecules containing different base sequences. These results agree with Borst's (1970) earlier conclusion that the potential genetic information available in all normal mit-DNA is roughly equivalent to its genome size and that large-scale gene repetitions are absent. Since the genetic complexity of mit-DNA from animals is of the order of $1.0 \times 10^7$, the total information content corresponds to about 15000 base pairs, which could be equivalent to 5000 amino acids or 20 small proteins. Evidence from hybridization studies with mitochondrial RNA and homologous mit-DNA (Dawid, 1970; Attardi *et al.*, 1970) would suggest that mitochondrial rRNA species, as well as some of the mitochondrial tRNA species are coded for by mit-DNA. To date there is no strong evidence that mit-DNA codes for its own mRNA and thus for some mitochondrial proteins. At best the amount of mit-DNA available is only sufficient to code for about 20 small proteins, which is equivalent to about 5% of the constituent proteins of the mitochondrion (Ashwell & Work, 1970). Most of the evidence to date suggests that the enzymes of the mitochondrion are coded for by nuclear genes (Ashwell & Work, 1970; Boulter *et al.*, 1972) and that the function of mit-DNA is restricted, at least in animals, to the elaboration of a special protein-synthesizing system (Dawid, 1970). This implies that all the

Table 1. *Some characteristics of mit-DNA from a number of organisms*

| Source of mitochondria | Conformation | Size ($\mu$m) | Genetic complexity based on quantitative renaturation experiments | Buoyant density in CsCl (g cm$^3$) | | References |
|---|---|---|---|---|---|---|
| | | | | Mit-DNA | Nuclear DNA | |
| Animal tissues | Circular | 4.6–5.9 | $1.0 \times 10^7$ | 1.686–1.711 | 1.694–1.707 | Borst (1970) |
| Protozoa | | | | | | |
| *Tetrahymena pyriformis* | Linear | 17.6 | $3.0 \times 10^7$–$4.0 \times 10^7$ | 1.684 | 1.688 | Suyama & Miura (1968) |
| Fungi | | | | | | |
| *Neurospora crassa* | Linear | 26 | $6.6 \times 10^7$ | 1.701 | 1.712 | Luck & Reich (1964) |
| *Saccharomyces cerevisiae* | Circular | 25 | $5 \times 10^7$ | 1.679 | 1.693 | Hollenberg *et al.* (1970)<br>Tewari *et al.* (1966) |
| Higher plants | | | | | | |
| *Phaseolus vulgaris* | Linear | 19.5 | $1.0 \times 10^8$ | 1.707 | 1.693 | Wolstenholme & Gross (1968)<br>Suyama & Bonner (1966) |
| *Lactuca* sp. | Linear | — | $1.4 \times 10^8$ | 1.706 | 1.694 | Wells & Birnstiel (1969) |
| *Pisum* sp. | Circular | 30 | $7.4 \times 10^7$ | 1.706 | 1.698 | Kolodner & Tewari (1972) |

information specifying the sequence of product proteins would be provided by the nucleus.

Higher-plant mit-DNA has a potential genetic information content equivalent to its genome size, which could be as much as 7 times the size of the mitochondrial genome of animal tissues and as with ascomycete mit-DNA (Borst, 1970) we must be careful in the indiscriminate extrapolation of results obtained with animal mit-DNA to higher-plant mit-DNA, because about 90% of the genes present in the latter are absent in the former.

## Mitochondrial Ribosomes and RNA

The decrease in information content of mit-DNA from higher plants to animals appears to be associated with or involve the ribosomal cistrons, which have been decreased in animal mitochondria to an unprecedented small size. A further complication in identification of mitochondria-specific RNA species has been the problem of contamination of preparations with a variety of extramitochondrial RNA species which are often found in larger amounts than the mit-RNA (mitochondrial RNA) which is present in amounts of the order of 1–20 $\mu$g/mg of mitochondrial protein. This contamination problem is less serious with lower eukaryotes such as the ascomycetes, which do not have well-developed endoplasmic membrane systems and associated bound cytoplasmic ribosomes. The identification of mitochondrial ribosomes and RNA has been controversial and there is still some measure of disagreement about the precise sizes. However, Table 2 gives the sedimentation characteristics of mitochondrial ribosomes and their RNA components as the situation stands at present. The mitochondrial ribosomes of the ascomycetes so far studied sediment at about '73S', a rate intermediate between its cytoplasmic counterparts and the ribosomes of *Escherichia coli*. Animal mitochondria contain so-called 'miniribosomes' which sediment between 55 and 60S. Evidence to date suggests that whereas mitochondrial ribosomes have sedimentation values distinct from the bacterial 70S ribosomes they are functionally similar in their response to a wide variety of antibiotics, and

Table 2. *Sedimentation characteristics (S values) of mitochondrial ribosomes and their RNA components compared with E. coli*

| Organism | Mitochondrial ribosomes | Mitochondrial ribosomal RNA species | References |
|---|---|---|---|
| Animals | | | |
| *Xenopus laevis* | 60 | 21/13 | Dawid (1970) |
| HeLa cells | 60 | 16/12 | Attardi *et al.* (1970) |
| Rat (liver) | 55 | 16/13 | O'Brien & Kalf (1967)<br>Ashwell & Work (1970)<br>Groot *et al.* (1970) |
| Fungi | | | |
| *Neurospora crassa* | 73 | 23/16 | Küntzel & Noll (1967) |
| *Aspergillus nidulans* | 67 | 23.5/15.5 | Edelman *et al.* (1970) |
| *Saccharomyces cerevisiae* | 74 ± 1 | 21–22/14–15 | Stegeman *et al.* (1970)<br>Grivell *et al.* (1971) |
| Protozoa | | | |
| *Tetrahymena* | '80' | 21/14 | Chi & Suyama (1970) |

also in their ability to interchange initiation and elongation factors (see Borst & Grivell, 1971; Boulter *et al.*, 1972).

The size of the component mit-RNA species from both ascomycete and animal mitochondrial ribosomes has also been the subject of some disagreement, in part owing to technical problems of comparing results from different laboratories and in part owing to the unusual structure of the mit-RNA species, which have a low mole percent G+C content. The mit-rRNA species seem to unfold easily and as a consequence their molecular weights cannot be reliably inferred from their sedimentation behaviour or electrophoretic mobility. Several workers using animal (Groot *et al.*, 1970) and ascomycete (Edelman *et al.*, 1971) mit-RNA species have now shown that the electrophoretic mobility of mit-RNA relative to homologous cytoplasmic rRNA and *E. coli* rRNA is dependent on temperature and ionic strength, so that their electrophoretic mobility in polyacrylamide gels is lower than would be expected from their sedimentation behaviour in sucrose density gradients. The most striking feature of animal mit-rRNA species is their small size which makes them little more than half the size of the bacterial species.

Various transfer RNA species have been isolated from yeast and *Neurospora* mitochondria (Epler & Barnett, 1967; Epler *et al.*, 1970) and also a '4S' RNA component has been identified in a range of animal mitochondria.

The presence of a 5S RNA component, which is found in all bacteria and eukaryotic cytoplasmic ribosomes, has not been demonstrated in mitochondria, either because mitochondrial ribosomes do not contain a low-molecular-weight RNA or because it is smaller than other 5S RNA species and sediments or moves on electrophoresis together with mitochondrial tRNA.

Electron microscopy of plant mitochondria has shown the presence of ribosome-like particles (Wilson *et al.*, 1968; Kislev *et al.*, 1965). However, characterization of mit-RNA components is limited to the results by Baxter & Bishop (1968) who in a short communication, reported that soya-bean mitochondria contained two high-molecular-weight species intermediate in size between the two major cytoplasmic rRNA components.

## Extraction and Properties of Plant Mitochondria

One of the major problems in the characterization of mitochondria-specific nucleic acids is contamination with extramitochondrial nucleic acids, membrane fragments, bacteria and in the case of higher plants, the presence of plastid material. Injury of mitochondria during isolation, coupled with the greater permeability of plant mitochondria can also result in partial degradation of the mitochondrial RNA. It is evident, therefore, that the isolation of intact mitochondria, free from other cell components, is a prerequisite to a study of their nucleic acids and protein-synthesizing systems.

We have purposely chosen as tissue sources plant material obtainable in large quantities, containing as homogeneous a population of cells as possible and with relatively small amounts of potentially contaminating cell organelles such as plastids. In addition all solutions, glassware and plant material were sterilized as a routine and bacterial counts were made of the mitochondrial preparations.

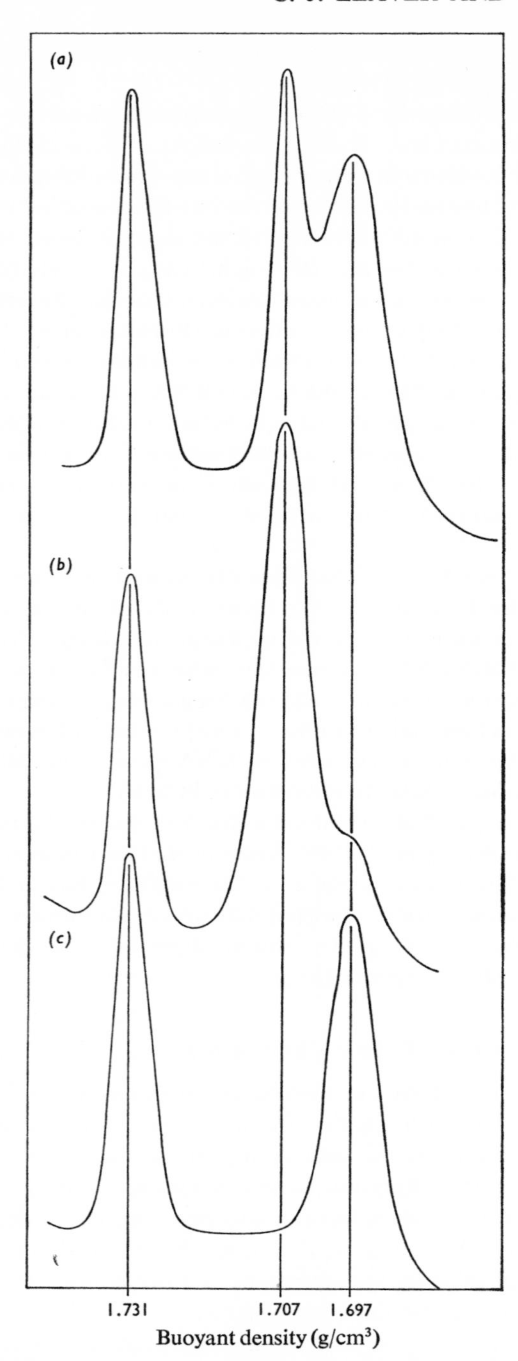


Fig. 1. *Microdensitometer tracings of u.v. photographs of mitochondrial and total cellular DNA species from turnip, obtained after analytical CsCl-density-gradient centrifugation*

The DNA samples were adjusted to a density of 1.720 g/cm³ and centrifuged at 44000 rev./min for 20 h at 25°C. The marker *Micrococcus lysodeikticus* DNA had a density of 1.731 g/cm³. (*a*) 'Crude' mitochondrial DNA; (*b*) 'purified' mitochondrial DNA; (*c*) total cellular DNA, the major band being the nuclear DNA, 1.697 g/cm³.

To date we have been successful in isolating essentially pure mitochondria from the etiolated hypocotyls of mung bean and french bean, pea shoots and maize coleoptiles, the storage roots of turnip, the tubers of potato and the inflorescence of cauliflower.

Many studies on plant mitochondria have been carried out with mitochondria-enriched fractions and we therefore decided that all our preparations should be checked for purity and intactness by electron microscopy, oxygen-electrode polarography and spectrophotometry as a routine.

It is beyond the scope of this paper to go into a detailed description of our isolation procedure and this will be published in due course.

Mitochondria were extracted from kilogram quantities of plant material by a modification of the technique described by Bonner (1967). Electron micrographs of this 'crude' preparation showed the presence of a small amount of contaminating membrane, nuclear and plastid material. Buoyant-density analysis of the DNA extracted from such preparations (Fig. 1) also showed some evidence of contamination by nuclear and/or plastid DNA. Further purification was effected by sucrose-density-gradient centrifugation. The crude mitochondria were layered on a six-step gradient, 0.9–1.8 M-sucrose in 10 mM-Tris–HCl buffer, pH 7.5–50 mM-KCl–1 mM-EDTA and centrifuged for 2.5 h at 25000 rev./min in the Spinco SW 25.1 rotor at 0°C. The mitochondria band at the 1.35–1.5 M-sucrose interface was isolated, diluted to 0.25 M-sucrose with 10 mM-Tris–HCl buffer, pH 7.5–50 mM-KCl,–10 mM-$MgCl_2$ and recovered by centrifugation.

Electron micrographs of these 'pure' mitochondria showed no evidence of contaminants and this was further borne out by buoyant-density analysis of the extracted DNA (Fig. 1) which showed a single main band at 1.706–1.707 g/cm$^3$ corresponding to mit-DNA and in some cases a slight shoulder (less than 5% of the main band) at buoyant densities between 1.692 and 1.701 g/cm$^3$ depending on the plant species.

In this manner we obtained yields of 35–45 mg of mitochondrial protein/kg, fresh weight of mung-bean hypocotyl or turnip root.

Oxygen-electrode polarography of our preparations allowed us to measure

Table 3. *Comparison of the activities of crude and sucrose-density-gradient-purified mitochondria from mung-bean hypocotyls and turnip root*

Oxygen uptake was measured polarographically with a Beckman 160 Physiological Gas Analyser at room temperature (24° ± 1 °C). Oxygen consumption with malate as a substrate is expressed as μl of $O_2$/mg of protein per h. The ADP/O ratio is expressed as phosphate esterified per μatom of oxygen consumed. The respiratory control ratio is the ratio:

$$\frac{\text{oxygen uptake in respiratory state 3}}{\text{oxygen uptake in respiratory state 4}}$$

| Source of mitochondria | $Q_{O_2}$ (μl/h per mg of protein) | | ADP/O ratio | Respiratory control ratio |
|---|---|---|---|---|
| | State 3 | State 4 | | |
| Mung bean | | | | |
| Crude | 352 | 88 | 2.4 | 4.0 |
| Purified | 426 | 159 | 2.1 | 2.7 |
| Turnip | | | | |
| Crude | 861 | 246 | 2.5 | 3.5 |
| Purified | 1036 | 576 | 2.2 | 1.8 |

oxidative phosphorylation, respiratory control ratios and values for $Q_{O_2}$ (N), thus giving some measure of the intactness and physiological competence of the mitochondria at different stages during the isolation procedure. Table 3 gives typical values for 'crude' and 'pure' mitochondria from mung bean and turnip and shows that both preparations are coupled, although after purification by density-gradient centrifugation lower respiratory control ratios are obtained, a fact which can be attributed to an increased proportion of uncoupled and therefore probably damaged mitochondria. This assumption was further confirmed by difference-spectra analysis of both types of preparation which showed that in purified mitochondria the cytochrome *c* relative to cytochrome *b* content is decreased.

## Mitochondrial Nucleic Acids

Total nucleic acid was extracted from mitochondria by the method of Parish & Kirby (1966) and precipitated with ethanol.

The total amount of RNA in turnip and mung-bean mitochondria was about 10–20$\mu$g/mg of protein, whereas the DNA content was approx. 1$\mu$g/mg of protein, a value in agreement with that reported by Suyama & Bonner (1966) who obtained a value of 0.8$\mu$g of DNA/mg of protein in mitochondria from mung bean, turnip and sweet potato. A 1 kg sample of etiolated mung-bean hypocotyls yielded approx. 500–700$\mu$g of mitochondrial nucleic acid, which corresponds to between 0.3 and 0.5% of the total cellular nucleic acids.

## Characterization of rRNA

Polycrylamide-gel electrophoresis of the mit-RNA species was carried out by the method of Loening (1969) and further details are given in the figure legends.

Fig. 2 shows the fractionation by gel electrophoresis of the total mitochondrial nucleic acid from mung-bean and turnip mitochondria. The RNA components are referred to in terms of their molecular weights since the relationship between electrophoretic mobility and log (molecular weight) has been shown to be linear with a range of viral (Bishop *et al.*, 1967) and rRNA species (Loening, 1968). *E. coli* ($1.1 \times 10^6$ and $0.56 \times 10^6$ daltons) rRNA species were co-electrophoresed with mitochondrial nucleic acids as a standard (see Fig. 3).

The two major high-molecular-weight RNA species from both mung-bean and turnip mitochondria have molecular weights of approx. $1.15 \times 10^6$ and $0.7–0.78 \times 10^6$. In addition to these major components, minor RNA species at $1.08 \times 10^6$, $0.60 \times 10^6$, $0.54 \times 10^6$ and $0.42 \times 10^6$ were found in such preparations. In some cases these may have arisen from residual cytoplasmic contamination or maybe as yet unidentified mitochondrial components, which, in the case of the lower molecular weights, arise by cleavage of larger molecules. Table 4 lists the major high-molecular-weight and presumed rRNA species, isolated from a range of plant mitochondria. The molecular weights of the homologous cytoplasmic rRNA species isolated under similar conditions are included for comparison. Chloroplast rRNA species isolated from the green leaves of the plants listed had molecular weights of $1.1 \times 10^6$ and $0.56 \times 10^6$.

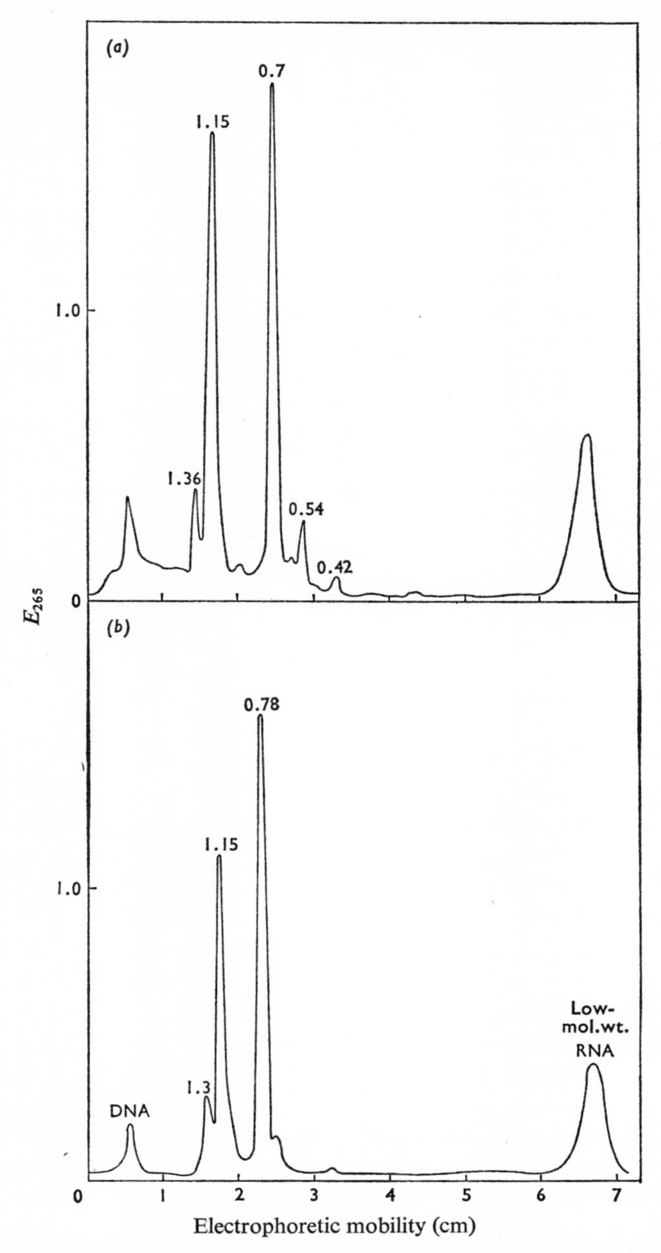


Fig. 2. *Polyacrylamide-gel fractionation of nucleic acids from purified mitochondria of* (*a*) *turnip*, (*b*) *mung bean*

Nucleic acids were prepared by the detergent–phenol procedure and fractionated on 2.4% polyacrylamide gels for 2h at 50V (6mA/8cm gel) and then scanned at 265nm in a Joyce–Loebl Chromoscan (see Leaver & Ingle, 1971). RNA components are referred to as their molecular weight in millions.

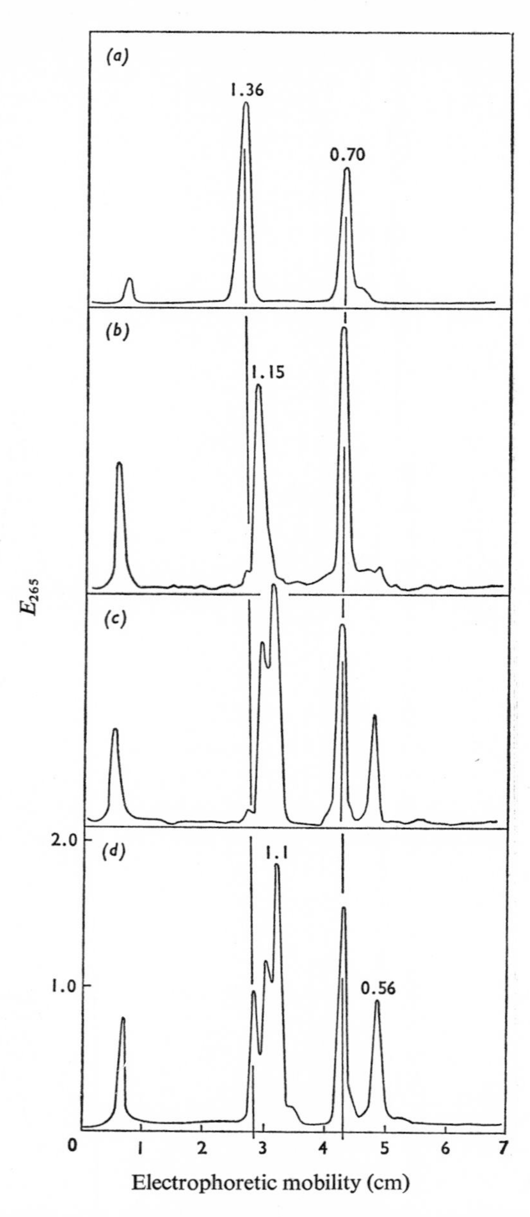


Fig. 3. *Polyacrylamide-gel electrophoresis of* (*a*) *turnip cytoplasmic nucleic acids*, (*b*) *turnip mitochondrial nucleic acids*, (*c*) *mitochondrial nucleic acids plus E. coli nucleic acids and* (*d*) *mitochondrial, cytoplasmic and E. coli nucleic acids*

Preparations were fractionated by electrophoresis on 2.4% (w/v) polyacrylamide gels for 3.5h. Low-molecular-weight RNA species have run off the gels.

Table 4. *Molecular weights of mitochondrial and cytoplasmic ribosomal RNA species from a range of tissues*

The molecular weights were determined by polyacrylamide-gel electrophoresis by using *E. coli* rRNA of mol.wt. $1.10 \times 10^6$ and $0.56 \times 10^6$ as standards.

| Tissue | Mitochondrial | | Cytoplasmic | |
|---|---|---|---|---|
| | Heavy | Light | Heavy | Light |
| Turnip root | 1.15 | 0.70 | 1.36 | 0.70 |
| Mung-bean hypocotyl | 1.15 | 0.78 | 1.30 | 0.70 |
| Potato tuber | 1.12 | 0.70 | 1.30 | 0.70 |
| Cauliflower inflorescence | 1.18 | 0.69 | 1.30 | 0.69 |
| Pea stem | 1.15 | 0.75 | 1.30 | 0.69 |

The two cytoplasmic high-molecular-weight rRNA species, $1.3 \times 10^6$ and $0.7 \times 10^6$, are found in amounts of approx. 2:1 (Fig. 3). In the case of the mit-RNA species studied the amounts of the two mit-rRNA species is less than 1.6:1, the value expected from a ribosome containing equimolar proportions of the two rRNA species. Further, in all these fractionations additional peaks were present. Even allowing for a low level of contamination of the light mit-rRNA by cytoplasmic light rRNA ($0.7 \times 10^6$ daltons) the ratio does not approach the theoretical value. The easiest way to account for the observed distributions is to assume the breakdown of certain components, and the obvious candidate for this is the $1.15 \times 10^6$-dalton mit-rRNA. Attempts to prevent the assumed breakdown by inclusion of known ribonuclease inhibitors, such as bentonite and diethyl pyrocarbonate in the extraction medium, were without success. However, previous experience (Leaver & Ingle, 1971) had shown that the heavy chloroplast rRNA ($1.1 \times 10^6$ daltons) from a range of plants was very unstable and tended to be cleaved into lower-molecular-weight products. The $1.1 \times 10^6$-molecular-weight rRNA may be stabilized by the inclusion of $Mg^{2+}$ during the extraction and electrophoresis of the RNA. When RNA is prepared from turnip mitochondria in the presence of $Mg^{2+}$ and fractionated in an $Mg^{2+}$-containing electrophoresis buffer (Fig. 4) the two mit-rRNA species are nicely separated with a ratio of 1.7:1. When the same RNA is fractionated in the normal EDTA buffer, the heavy mit-rRNA is decreased markedly relative to the light mit-rRNA and there are increases in the amounts of several of the minor lower-molecular-weight components. The ratio of heavy to light mit-rRNA species is decreased to 1.0:1.

There is some doubt as to whether we can assign an accurate molecular weight to mit-RNA by using results derived from $Mg^{2+}$-containing gels, because, as has been suggested by Loening (1969) and Grivell *et al.* (1971), the electrophoretic mobility of mit-rRNA species in $Mg^{2+}$ buffer may not reflect the true molecular weight because it remains to be proven that the conformation of the mit-rRNA species is really equivalent to that of the marker rRNA species of *E. coli*.

It should be noted that under all normal conditions of extraction the heavy cytoplasmic rRNA ($1.3 \times 10^6$ daltons) of the tissue studied was very stable and at no time was evidence of degradation of this component detected. A variable cleavage of the light cytoplasmic rRNA ($0.7 \times 10^6$ daltons) to molecules of between $0.58 \times 10^6$ and $0.65 \times 10^6$ daltons, depending on the tissue, only occurred when RNA was extracted from isolated ribosomes or subunits. In view of the reported sensitivity of the electrophoretic mobility, of ribosomal-type RNA from animal and ascomycete mitochondria, to ionic conditions and temperature we have fractionated plant mit-tRNA species under the ionic and temperature conditions similar to those used by Grivell *et al.* (1971) in studying these effects on yeast mit-RNA. In preliminary experiments with turnip mit-rRNA we could find no dramatic effect on electrophoretic mobility relative to standard rRNA species by the variety of conditions used. Preliminary base composition analysis of the turnip mit-rRNA gave a mole percent G + C of 44 compared with the value of 52 mole percent G + C for the corresponding cytoplasmic rRNA.

## Sedimentation Properties of Mitochondrial RNA

Many rRNA species show an inverse relationship between sedimentation coefficient and electrophoretic mobility (Loening, 1968); several authors, however, have reported that mit-rRNA species show exceptions to this relationship (Edelman *et al.*, 1971; Dawid, 1970). It was therefore of interest to examine

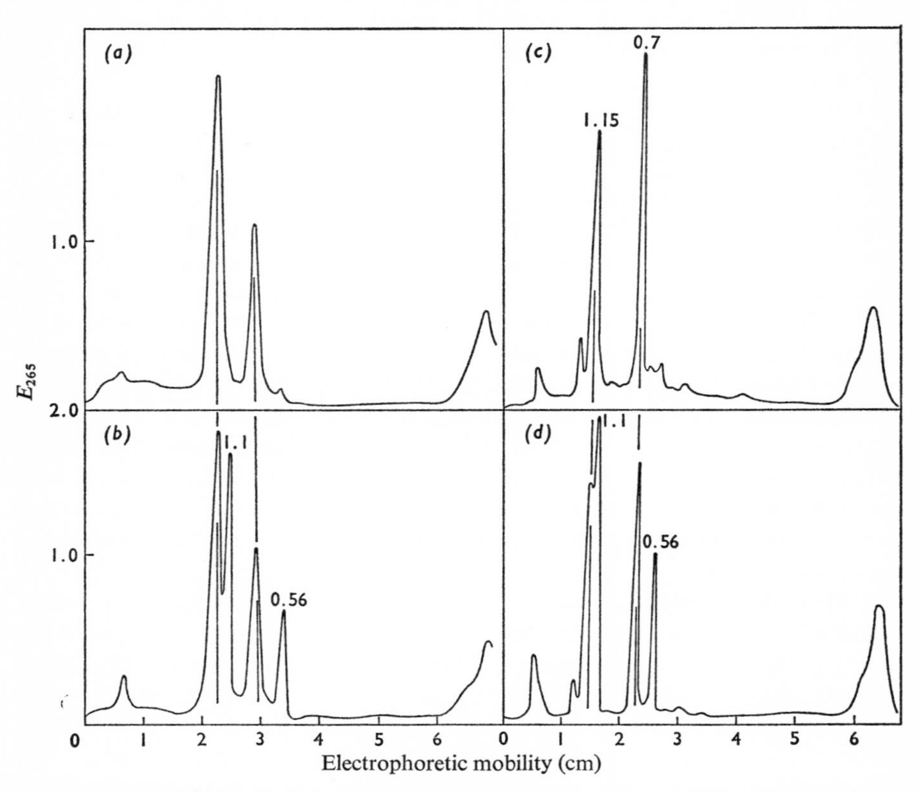


Fig. 4. *Effect of $Mg^{2+}$ and EDTA on the preparation and fractionation of turnip mitochondrial RNA*
Nucleic acids were prepared from turnip mitochondria in the presence of 10mM-$MgCl_2$ and fractionated (*a*) in a $Mg^{2+}$-containing buffer (see Leaver & Ingle, 1971), (*b*) in a $Mg^{2+}$-containing buffer + *E. coli* rRNA ($1.1 \times 10^6$ and $0.56 \times 10^6$ daltons), (*c*) in the normal EDTA-containing buffer, (*d*) in the normal EDTA-containg buffer + *E. coli* rRNA. The $Mg^{2+}$-containing gels were subjected to electrophoresis for 3h at 3mA/8cm gel and the normal EDTA-containing gels for 2h at 6mA/8cm gel.

whether plant mit-rRNA species behaved in a similar manner. In addition, the effect of $Mg^{2+}$ on the proportion of light and heavy rRNA species could be due to the electropotential applied during electrophoresis, which would appear to be sufficient to remove $Mg^{2+}$ from the RNA, resulting in breakdown.

To study these problems further we have used sucrose-density-gradient centrifugation, where the RNA is not subjected to an electropotential force.

Sedimentation profiles for the rRNA from turnip mitochondria, turnip cytoplasm and *E. coli* were determined at 5°C in a buffer containing 0.15 M-lithium acetate, pH 6.0, and 0.5% sodium lauryl sulphate. Resolution between the different RNA species was achieved by the use of rRNA species labelled with [$^{32}$P]-Pi and linear sucrose density gradients in the Spinco SW 25 rotor. Representative gradients are shown in Fig. 5. The mit-rRNA components sediment at about 24S and 18.5S (*E. coli* 23S and 16S; cytoplasmic rRNA species 25S and 18S). Gel electrophoresis of the 24S rRNA recovered from the sucrose density gradient showed the presence of a $1.15 \times 10^6$-dalton rRNA component, whereas the RNA recovered from the 18.5 S rRNA contained $0.7 \times 10^6$-dalton rRNA.

What is of further interest is that the ratio of heavy to light mit-rRNA species approaches the theoretical value expected if the mitochondrial ribosome contains equimolar amounts of the two rRNA components.

### Mitochondrial Low-Molecular-Weight RNA species

In addition to rRNA the mitochondria contain a relatively large amount of low-molecular-weight RNA with sedimentation values of 4–5S (Fig. 5) and a

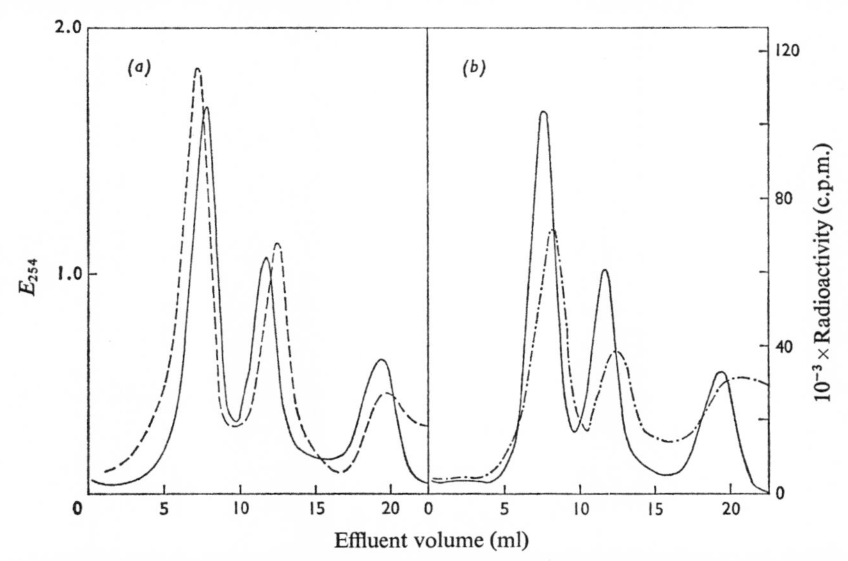


Fig. 5. *Sucrose-density-gradient sedimentation profiles of mixtures of turnip mitochondrial nucleic acid and (a) turnip cytoplasmic [$^{32}$P]RNA, (b) E. coli [$^{32}$P]RNA*

Nucleic acid samples were layered over linear 5–20% sucrose density gradients containing 0.15 M-lithium acetate, pH 6.0, and 0.5% sodium lauryl sulphate. Gradients were centrirfuged fo 18 h at 22000 rev./min. in the Spinco SW25 rotor at a chamber temperature of 5°C. ——, $E_{254}$ (turnip mitochondrial nucleic acid); – – – –, $^{32}$P radioactivity (turnip cytoplasmic [$^{32}$P]RNA); –.–.–, $^{32}$P radioactivity (*E. coli* [$^{32}$P]RNA).

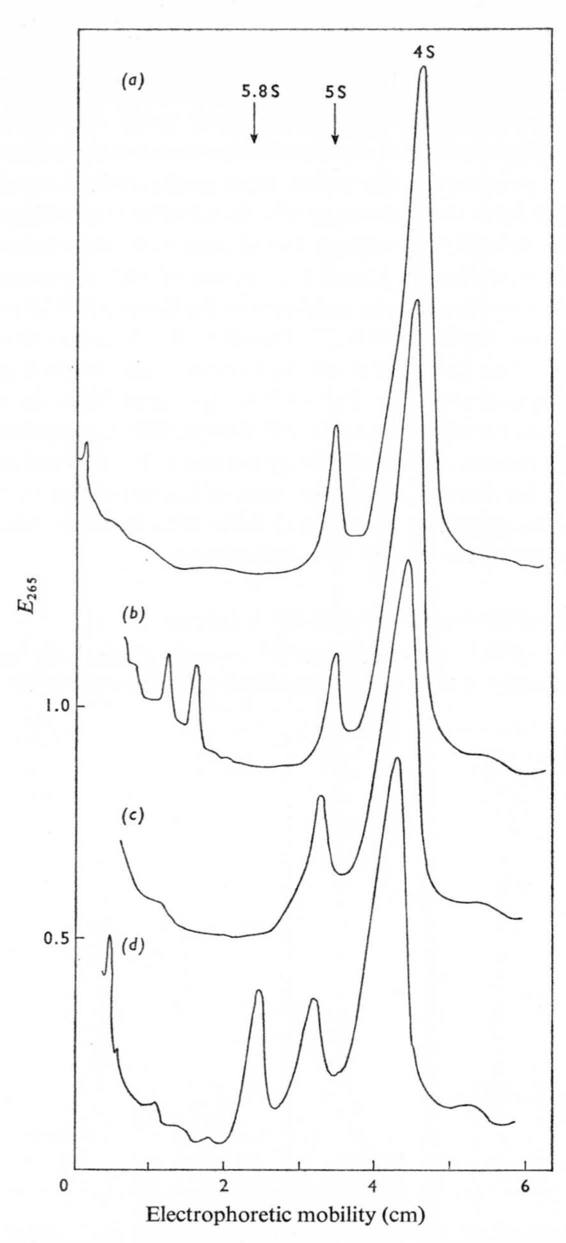


Fig. 6. *Polyacrylamide-gel electrophoresis of the low-molecular-weight RNA components*

Low-molecular-weight-RNA components were purified from (*a*) purified turnip mitochondria, (*b*) as (*a*) but heated at 70°C for 10min and cooled rapidly by immersion in liquid $N_2$ before fractionation, (*c*) total turnip nucleic acid, (*d*) as (*c*) but heated to 70°C before fractionation. The RNA was fractionated on 7.5% (w/v) polyacrylamide gels by electrophoresis for 3h at 9V/cm of gel (Payne & Dyer, 1972).

molecular weight of between 25000 and 40000 (Fig. 2). This RNA accounts for 24–28% of the total mitochondrial nucleic acid as compared with a value of about 10% for the amount of low-molecular-weight RNA in a total cellular nucleic acid preparation. A greater resolution of these low-molecular-weight RNA species was obtained by electrophoresis in 7.5% (w/v) polyacrylamide gels, as described by Loening (1967, 1969). When preparations of either total cytoplasmic nucleic acid (Fig. 6*c*) or pure mitochondrial nucleic acid (Fig. 6*a*) were fractionated two low-molecular-weight components were distinguished. These corresponded to 4S and 5S RNA species and had molecular weights of 25000 and 38750 respectively (4 and 5S RNA species from *E. coli* were used as standards).

When the nucleic acid preparations were heated to 70°C for 10 min, cooled rapidly by immersion in liquid $N_2$ and then fractionated by gel electrophoresis, an additional low-molecular-weight RNA was detected in the total cytoplasmic nucleic acid preparation (Fig. 6*d*), but not in the mitochondrial nucleic acid preparation (Fig. 6*b*). This additional component had a calculated S value of 5.8 and a molecular weight of 50000. These values are in close agreement with those obtained by Payne & Dyer (1972) from a range of higher plants, for the 5.8S rRNA component found hydrogen-bonded to the high-molecular-weight rRNA of the 80S ribosomes of animals and plants. The 70S ribosomes of bacteria, blue–green algae and chloroplasts do not contain this 5.8S RNA species, nor apparently, from our observations, do higher-plant mitochondrial ribosomes. Our failure to demonstrate a 5.8S rRNA component in our preparations would seem to confirm their purity, in that the presence of contaminating cytoplasmic rRNA would have given rise to a 5.8S rRNA on heating.

In addition to the 5S RNA component from total mitochondrial nucleic acid preparations (Fig. 6) we have also recovered a similar 5S rRNA from purified mitochondrial ribosomes (Fig. 7). Calculations from a number of preparations, in which cytoplasmic nucleic acid contamination was negligible as judged by the absence of $1.3 \times 10^6$-dalton cytoplasmic rRNA, indicate that the mitochondrial 5S rRNA is present in approximately equimolar proportions to the mitochondrial rRNA species. We feel that this equimolar relationship is not fortuitous and lends weight to our conclusion that plant mitochondrial ribosomes do contain a 5S rRNA, in contrast with reports that animal and ascomycete mitochondrial ribosomes do not (Lizardi & Luck, 1971).

## Mitochondrial Ribosomes

Purified mitochondria were lysed in Tris–$MgCl_2$–KCl buffer (10mM-Tris–HCl, pH 7.5, 10mM-$MgCl_2$, 50mM-KCl) containing 2.0% (v/v) Triton X-100. The lysate was clarified by centrifugation at 10000*g* for 10min. The resulting supernatant was either layered directly on to a linear sucrose density gradient or the mitochondrial ribosomes were sedimented by centrifugation through a 'cushion' of 1M-sucrose in the Tris–$MgCl_2$–KCl buffer (105000*g* for 4h in the Spinco type 40 rotor), before resuspension and density-gradient analysis. Cytoplasmic ribosomes (80S) and *E. coli* (70S) ribosomes were isolated under essentially similar conditions.

Fig. 7 shows the sedimentation profiles of cytoplasmic and mitochondrial ribosomes from turnip. The mitochondrial ribosomes from turnip, mung bean, cauliflower and french bean sediment more slowly than the homologous cytoplasmic ribosomes, but faster than the 70S ribosomes from *E. coli.* We have tentatively assigned an S value of 77–78 to the mitochondrial ribosomes from these plants. After pulse labelling of isolated, intact mitochondria from mung bean, with radioactive amino acids, the majority of the nascent protein in the mitochondrial lysate was found to be associated with the 77–78 S particles and heavier structures sedimenting up to about 180S (C. J. Leaver & M. A. Harmey, unpublished work).

Isolation and gel electrophoresis of the nucleic acids from the sucrose-density-gradient-purified mitochondrial ribosomes (77–78 S) confirmed the presence of two mit-rRNA species with molecular weights of $1.15 \times 10^6$ and $0.7 \times 10^6$ in turnip (Fig. 7*a*). Gel electrophoresis of the rRNA species from the cytoplasmic

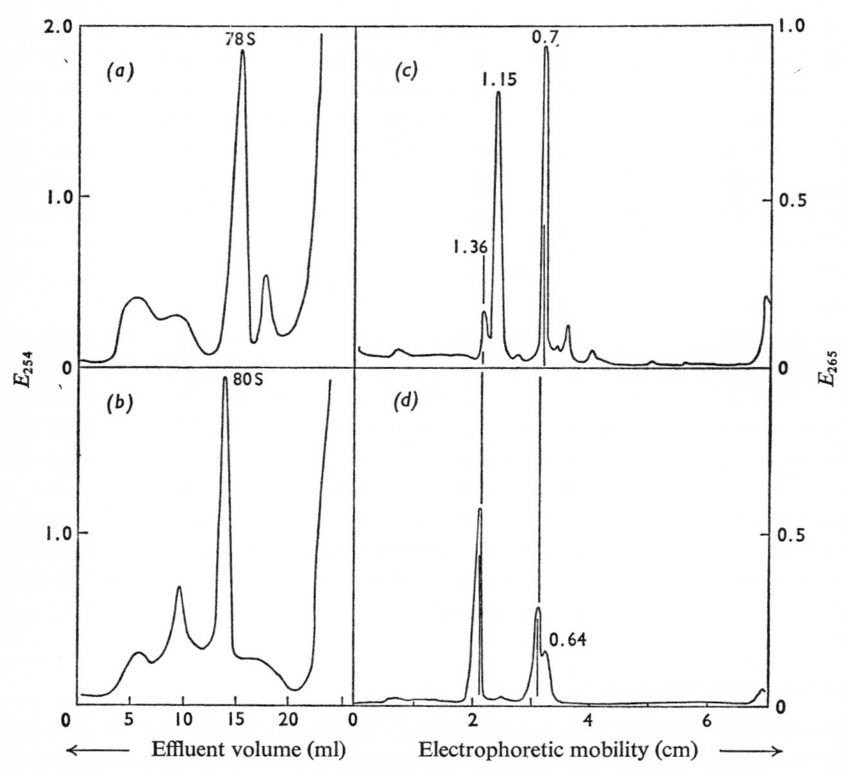


Fig. 7. *Sucrose-density-gradient-sedimentation profiles of* (*a*) *mitochondrial and* (*b*) *cytoplasmic ribosomes from turnip*

The ribosomes (suspended in standard buffer containing 2% Triton X-100) were layered on 10–35% exponential sucrose density gradients containing 10 mM-Tris–HCl buffer, pH 7.5, 50 mM-KCl and 10 mM-$MgCl_2$. Centrifugation was for 5.5h at 24000 rev./min in the Spinco SW25 rotor at a chamber temperature of 0°C. S values were assigned by co-sedimentation of the mitochondrial ribosomes with *E. coli* (70S) and cytoplasmic (80S) ribosomes. RNA was recovered from the monoribosomes and fractionated by gel electrophoresis under standard conditions. (*c*) RNA from the 78 S mitochondrial ribosome, (*d*) RNA from the 80 S cytoplasmic ribosome.

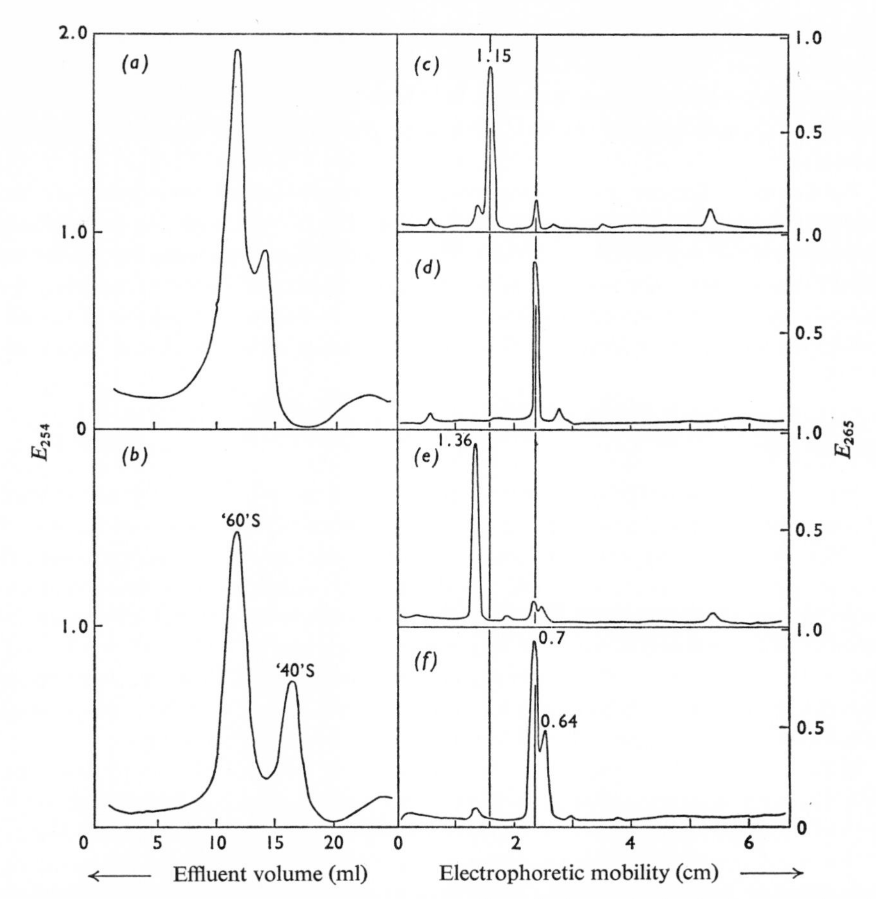


Fig. 8. *Sucrose-density-gradient-sedimentation profiles of ribosomal subunits from turnip mitochondrial ribosomes* (*a*) *and cytoplasmic ribosomes* (*b*)

The purified ribosomes were layered over linear 10–35% sucrose density gradients containing 0.3M-KCl, 3mM-$MgCl_2$, 10mM-Tris–HCl buffer, pH 7.5, and centrifuged in the Spinco SW25 rotor for 17h at 20000rev./min at a chamber temperature of 0°C. RNA was recovered from the subunits and fractionated by gel electrophoresis under standard conditions. (*c*) RNA from the large mitochondrial subunit, (*d*) RNA from the small mitochondrial subunit, (*e*) RNA from the large cytoplasmic subunit, (*f*) RNA from the small cytoplasmic subunit.

ribosomes revealed a further difference from those of the mitochondrial ribosome, this being that whereas the $1.36\times10^6$-dalton heavy rRNA is recovered intact, the $0.7\times10^6$-dalton light rRNA is characteristically cleaved to give a lower-molecular-weight product ($0.64\times10^6$ in turnip) with molecular weights ranging from $0.58\times10^6$ to $0.64\times10^6$ depending on the species.

We have attempted to dissociate mitochondrial ribosomes into their component subunits by a variety of techniques. To date the most successful has involved layering ribosomes on sucrose density gradients containing 0.3–0.5M-KCl and centrifuging in the Spinco SW 25 rotor for 17h at 20000rev./min. Turnip and mung-bean mitochondrial ribosomes dissociate into large and small subunits which can be shown to contain the heavy and light mit-rRNA species. Cosedimentation of cytoplasmic and mitochondrial subunits (Fig. 8) reveals that the large subunits sediment together, whereas the small mitochondrial subunit sediments approximately midway between the large and small cytoplasmic subunits.

To compare further the cytoplasmic and mitochondrial ribosomes, we extracted the proteins present in the monoribosomes of each and compared them by polacrylamide-gel electrophoresis. Plates 1 and 2 show that there are numerous differences between the two sets of ribosomal proteins as has been reported for the cytoplasmic and mitochondrial ribosomal proteins of *Neurospora* (Küntzel, 1969), *Saccharomyces* (Schmitt, 1971) and mung bean (Vasconcelos & Bogorad, 1971).

## Conclusions

We have demonstrated in a range of higher-plant mitochondria, the occurrence of mitochondrial ribosomes with an estimated sedimentation coefficient of 77–78 S, and probably a unique pattern of ribosomal proteins. These ribosomes can be dissociated into large and small subunits which contain discrete high-molecular-weight ribosomal RNA species with molecular weights, estimated by gel electrophoresis and sucrose-density-gradient analysis, of $1.12 \times 10^6$–$1.18 \times 10^6$ and $0.69 \times 10^6$–$0.78 \times 10^6$, depending on the plant species. Further, plant mitochondria contain 4S RNA and from our preliminary data a 5S rRNA component, which has not been demonstrated in mitochondria from other sources.

Mitochondrial ribosomes, in common with the chloroplast and prokaryote, 70S ribosome, do not contain a 5.8S rRNA molecule, which is found hydrogen-bonded to the 25S–28S rRNA in the 80S ribosome of animal and plant cytoplasm.

Our results confirm the idea that mitochondrial ribosomes are unique particles, and that not all ribosomes can be characterized as either bacterial ('70S') or eukaryotic ('80S') in type. The mitochondrial protein-synthesizing system has, however, several functional similarities with the bacterial (and chloroplast) ribosomal system, including the method of initiation, the probable interchangeability of some of the factors, and the effects of inhibitors, which imply a similarity in the ribosomal binding sites of the proteins of the two groups.

It has been calculated that the amount of genetic information present within animal mitochondria does not meet all the needs for a completely autonomous existence of mitochondria in the cell and a conservative hypothesis of the function of mitochondrial DNA would restrict its role to the elaboration of a specialized protein-synthesizing machinery. All the information specifying the sequences of product proteins would be derived from the nucleus.

Having identified some of the major components of the plant mitochondrial protein-synthesizing machinery, one of our major aims must be the elucidation of why plant mitochondria contain so much more genetic information to perform apparently the same function as animal mitochondria. Since the genetic information content of mitochondria is limited, the contribution of the mitochondria to their own biogenesis has to be limited too. It is therefore of importance to find out the relative contributions of nuclear and mitochondrial DNA to the bio-

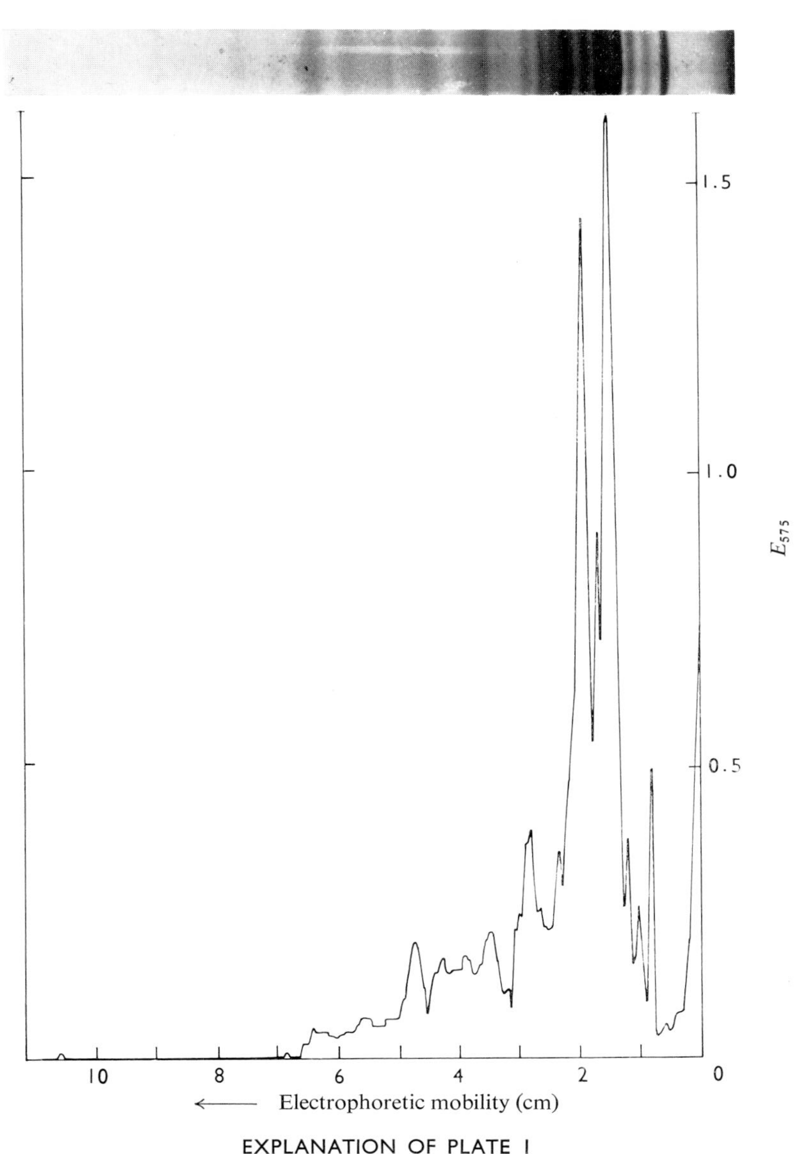


EXPLANATION OF PLATE I

*Polyacrylamide-gel electrophoresis patterns of turnip mitochondrial ribosomal proteins*

Ribosomal proteins were prepared by standard techniques and fractionated on 15% (w/v) polyacrylamide gels containing 6 M-urea and 5% (v/v) acetic acid. Electrophoresis was for 4.5 h at 16 V/cm. The proteins were stained with 0.2% Coomassie Blue and destained in 5% (v/v) acetic acid. They were scanned at 575 nm in a Joyce–Loebl Chromoscan.

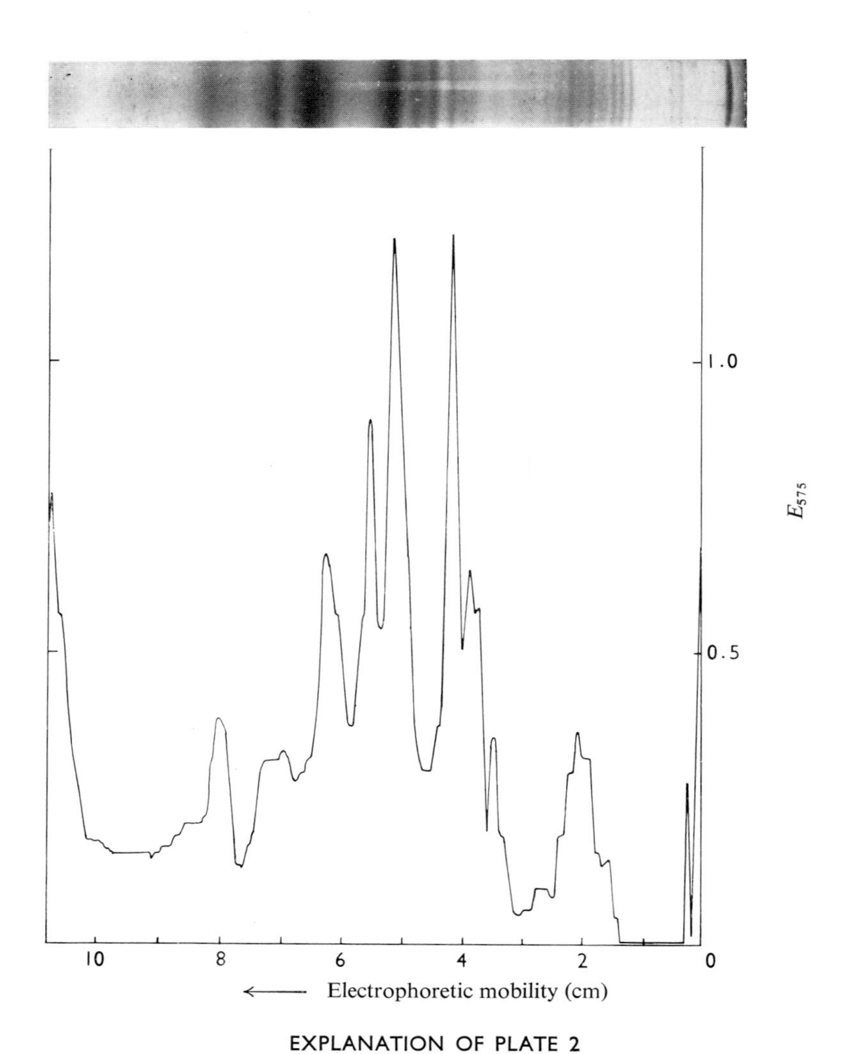


EXPLANATION OF PLATE 2

*Polyacrylamide-gel electrophoresis patterns of turnip cytoplasmic ribosomal proteins*
Ribosomal proteins were prepared and fractionated as in Plate 1.

genesis of the organelle and also how such contributions are co-ordinated during the developmental changes in mitochondrial activity and structure which are known to occur in plants.

M.A.H. is grateful to the Royal Society for the award of an exchange fellowship. We thank Dr. A. Trewavas for assistance and advice with the protein gel electrophoresis, J. M. Sinclair for performing the Spinco model E buoyant-density centrifugations and J. A. Dyer for his excellent technical assistance.

## References

Ashwell, M. & Work, T. S. (1970) *Annu. Rev. Biochem.* **39**, 251–290
Attardi, G., Aloni, Y., Attardi, B., Ojala, D., Pica-Mattoccia, L., Robberson, D. L. & Storrie, B. (1970) *Cold Spring Harbor Symp. Quant. Biol.* **35**, 599–619
Baxter, R. & Bishop, D. H. L. (1968) *Biochem. J.* **109**, 13P–14P
Beattie, D. S. (1971) *Sub-Cell. Biochem.* **1**, 1–23
Bishop, D. H. L., Claybrook, J. R. & Spiegelman, S. (1967) *J. Mol. Biol.* **26**, 373–387
Bonner, W. D. (1967) *Methods Enzymol.* **10**, 126–133
Borst, P. (1970) in *Control of Organelle Development* (Miller, P. L., ed.), pp. 201–225, Cambridge University Press, Cambridge
Borst, P. & Grivell, L. A. (1971) *FEBS Lett.* **13**, 73–88
Boulter, D., Ellis, R. J. & Yarwood, A. (1972) *Biol. Rev. Cambridge Phil. Soc.* **47**, 113–175
Chi, J. C. H. & Suyama, Y. (1970) *J. Mol. Biol.* **53**, 531–556
Dawid, I. B. (1970) in *Control of Organelle Development* (Miller, P. L., ed.), pp. 227–246, Cambridge University Press, Cambridge
Edelman, M., Verma, I. M. & Littauer, U. Z. (1970) *J. Mol. Biol.* **49**, 67–83
Edelman, M., Verma, I. M., Herzog, R., Galun, E. & Littauer, U. Z. (1971) *Eur. J. Biochem.* **19**, 372–378
Epler, J. L. & Barnett, W. E. (1967) *Biochem. Biophys. Res. Commun.* **28**, 328–333
Epler, J. L., Sugart, L. R. & Barnett, W. E. (1970) *Biochemistry* **9**, 3575–3579
Grivell, L. A., Reijnders, L. & Borst, P. (1971) *Eur. J. Biochem.* **19**, 64–72
Groot, P. H. E., Aaij, C. & Borst, P. (1970) *Biochem. Biophys. Res. Commun.* **41**, 1321–1326
Hollenberg, C. P., Borst, P. & Van Bruggen, E. F. J. (1970) *Biochim. Biophys. Acta* **209**, 1–15
Kislev, N., Swift, H. & Bogorad, L. (1965) *J. Cell Biol.* **25**, 327–344
Kolodner, R. & Tewari, K. K. (1972) *Proc. Nat. Acad. Sci.* U.S. **69**, 1830–1834
Kroon, A. M. (1971) *Chimia* **25**, 114–121
Küntzel, H. (1969) *Nature (London)* **222**, 142–146
Küntzel, H. & Noll, H. (1967) *Nature (London)* **215**, 1340–1345
Leaver, C. J. & Ingle, J. (1971) *Biochem. J.* **123**, 235–243
Lizardi, P. M. & Luck, D. J. (1971) *Nature (London) New Biol.* **229**, 140–142
Loening, U. E. (1967) *Biochem. J.* **102**, 251–257
Loening, U. E. (1968) *J. Mol. Biol.* **38**, 355–365
Loening, U. E. (1969) *Biochem. J.* **113**, 131–138
Luck, D. J. L. & Reich, E. (1964) *Proc. Nat. Acad. Sci. U.S.* **52**, 931–938
O'Brien, T. W. & Kalf, G. F. (1967) *J. Biol. Chem.* **242**, 2172–2179
Parish, J. H. & Kirby, K. S. (1966) *Biochim. Biophys. Acta* **129**, 554–562
Payne, P. I. & Dyer, T. A. (1972) *Nature (London) New Biol.* **235**, 145–147
Schmitt, H. (1971) *FEBS Lett.* **15**, 186–190
Smillie, R. M. & Scott, N. S. (1970) *Progr. Mol. Subcell. Bot.* **1**, 136–202
Stegeman, N. J., Cooper, C. S. & Avers, C. J. (1970) *Biochem. Biophys. Res. Commun.* **39**, 69–76
Suyama, Y. & Bonner, W. D. (1966) *Plant Physiol.* **41**, 383–388
Suyama, Y. & Miura, K. (1968) *Proc. Nat. Acad. Sci. U.S.* **60**, 235–242
Tewari, K. K., Vötsch, W., Mahler, H. R. & Mackler, B. (1966) *J. Mol. Biol.* **20**, 453–481
Vasconcelos, A. C. L. & Bogorad, L. (1971) *Biochim. Biophys. Acta* **228**, 492–502
Wells, R. & Birnstiel, M. L. (1969) *Biochem. J.* **112**, 777–786
Wells, R. & Ingle, J. (1970) *Plant Physiol.* **46**, 178–179
Wilson, R. H., Hanson, J. B. & Mollenhauer, H. H. (1968) *Plant Physiol.* **43**, 1874–1877
Wolstenholme, D. R. & Gross, N. J. (1968) *Proc. Nat. Acad. Sci. U.S.* **61**, 245–252

*Biochem. Soc. Symp.* (1973) **38**, 195–215
*Printed in Great Britain*

# A Survey of Cytokinins and Cytokinin Antagonists with Reference to Nucleic Acid and Protein Metabolism

By FOLKE SKOOG

*Institute of Plant Development, Birge Hall, University of Wisconsin, Madison, Wis.* 53706, *U.S.A.*

## Synopsis

Kinds and natural occurrence of cytokinins, their structural properties in relation to growth promoting activity in plants, their origin, their distribution within the plant and their localization in specific tRNA species are reviewed. Possible regulatory functions of cytokinins in nitrogen metabolism are considered with reference to transport and accumulation of nutrients and assimilation products, nucleic acid biosynthesis and function at the transcription and translation levels of gene-controlled polypeptide biosynthesis, and biosynthesis or activation of specific enzymes. In this context the recent development of highly active cytokinin antagonists is briefly reviewed.

## Introduction

In the context of this symposium I shall present a brief review of the cytokinins and cytokinin antagonists with special reference to evidence bearing on their possible functions in nucleic acid and protein metabolism. I regret I cannot report a causal relationship between the latter processes and the role of cytokinins in the regulation of plant growth and development, but I hope I may generate some interest in this topic and perhaps provide some clues to meaningful further investigation.

## Brief Survey of Naturally Occurring and Some Related Cytokinins

### *Structure and biological activity*

Cytokinins were encountered in a search for a factor required for cytokinesis and continued growth of excised tobacco pith parenchyma (Jablonski & Skoog, 1954). They have been arbitrarily defined (Skoog *et al.*, 1965) in terms of their capacity to promote cell division and growth *in vitro* of callus tissue under specified conditions in the same manner as does kinetin, the first substance to be chemically identified (Miller *et al.*, 1956).

Cytokinins like auxins and gibberellins exert regulatory functions in all phases of plant growth and development. Effects on callus growth and organ formation in tobacco tissue cultures are shown in Plate 1. [For general review of effects see Skoog & Schmitz (1972).]

Although by definition the cytokinins are a chemically heterogeneous group, in practice the most potent compounds found so far are the $N^6$-substituted adenine derivatives, and the naturally occurring most active members of this

Table 1. *Naturally occurring and some related cytokinins*

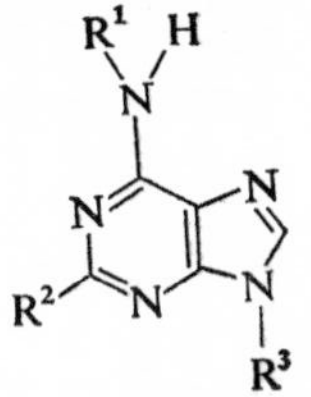

| Substance Chemical name (synonym and/or abbreviation) | Structure $R^1$ | $R^2$ | $R^3$ | Source: Bacteria | Fungi | Higher plants | Animals | No. in Fig. 1 |
|---|---|---|---|---|---|---|---|---|
| 6-(3-Methyl-2-butenylamino)purine ($N^6$-$\Delta^2$-isopentenyladenine; 2iP, $i^6$Ade) | 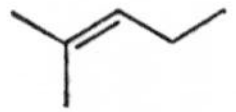 | H | H | + | + | + | ? | 1 |
| 6-(3-Methyl-2-butenylamino)-9-β-D-ribofuranosylpurine [$N^6$-($\Delta^2$-isopentenyl)adenosine; 2iPA, $i^6$A] | 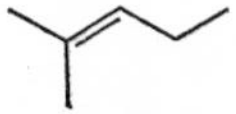 | H | Rib | + | + | + | + | 6 |
| 6-(3-Methyl-2-butenylamino)-2-methylthiopurine (ms2iP, $ms^2i^6$Ade) | 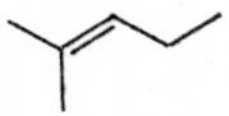 | $H_3CS$ | H | ? | | ? | | 4 |
| 6-(3-Methyl-2-butenylamino)-2-methylthio-9-β-D-ribopuranosylpurine (ms2iPA, $ms^2i^6$A) | 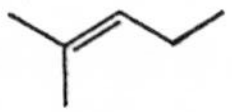 | $H_3CS$ | Rib | + | | + | | 8 |
| 6-(4-Hydroxy-3-methyl-*trans*-2-butenylamino)purine (zeatin, *t*-$io^6$Ade) | 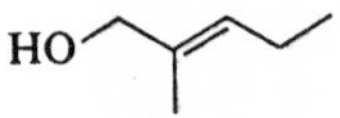 HO | H | H | | + | + | | 1 |
| 6-(4-Hydroxy-3-methyl-*trans*-2-butenylamino)-9-β-D-ribofuranosylpurine (ribosylzeatin, *t*-$io^6$A) | 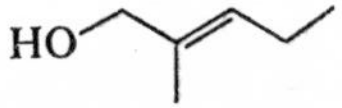 HO | H | Rib | | + | + | | 5 |
| 6-4-Hydroxy-3-methyl-*trans*-2-butenylamino)-2-methylthiopurine (ms-zeatin, $ms^2io^6$Ade) | 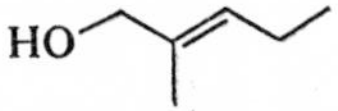 HO | $H_3CS$ | H | | | ? | | 3 |
| 6-(4-Hydroxy-3-methyl-*trans*-2-butenylamino)-2-methylthio-9-β-D-ribofuranosylpurine (ms-ribosylzeatin, *t*-$ms^2io^6$A) | 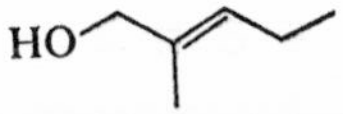 HO | $H_3CS$ | Rib | | | + | | 7 |

| | | | | | | |
|---|---|---|---|---|---|---|
| 6-(4-Hydroxy-3-methyl-*cis*-2-butenylamino)purine (*cis*-zeatin, *c*-io$^6$Ade) | 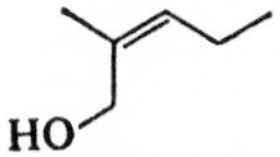 HO | H | H | | ? | III† |
| 6-(4-Hydroxy-3-methyl-*cis*-2-butenylamino)-2-β-D-ribofuranosylpurine (ribosyl-*cis*-zeatin, *c*-io$^6$A) | 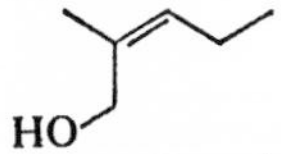 HO | H | Rib | + ? | + ? | XII† |
| 6-(4-Hydroxy-3-methyl-*cis*-butenylamino)-2-methythiopurine (ms-*cis*-zeatin, *c*-ms$^2$io$^6$A) | 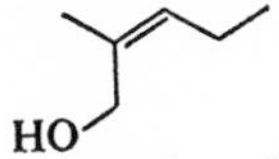 HO | $H_3CS$ | H | | ? | |
| 6-(4-Hydroxy-3-methyl-*cis*-2-butenylamino)-2-methylthio-9-β-D-ribofuranosylpurine (ms-ribosyl-*cis*-zeatin, *c*-ms$^2$io$^6$A) | 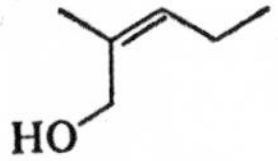 HO | $H_3CS$ | Rib | + ? | + | |
| 6-(3-Methylbutylamino)purine ($N^6$-isoamyladenine; $H_2$iP, hi$^6$Ade) | 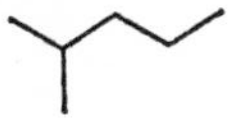 | H | H | | + ? | 10 |
| 6-(3-Methylbutylamino)-9-β-D-ribofuranosylpurine ($N^6$-isoamyladenosine; $H_2$iPA, hi$^6$A) | 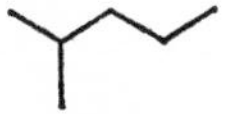 | H | Rib | | | 14 |
| 6-(3-Methylbutylamino)-2-methylthiopurine (ms$H_2$iP, ms$^2$i$^6$Ade) | 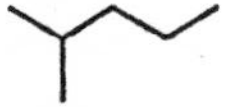 | $H_3CS$ | H | | | 12 |
| 6-(3-Methylbutylamino)-2-methylthio-9-β-D-ribofuranosylpurine (ms-$H_2$iPA, ms$^2$i$^6$A) | 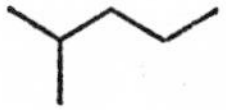 | $H_3CS$ | Rib | | | 16 |
| 6-(4-Hydroxy-3-methylbutylamino)purine (dihydrozeatin, hio$^6$Ade) | 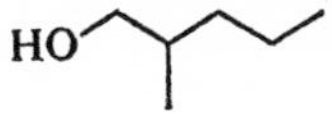 HO | H | H | | + | 9 |
| 6-(4-Hydroxy-3-methylbutylamino)-9-β-D-ribofuranosylpurine (ribosyldihydrozeatin, hio$^6$A) | 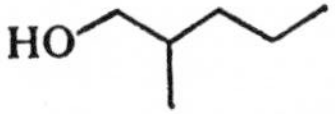 HO | H | Rib | | + | 13 |
| 6-(4-Hydroxy-3-methylbutylamino)-2-methylthiopurine (ms-dihydrozeatin, ms$^2$hio$^6$Ade) | 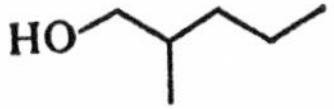 HO | $H_2CS$ | H | | | 11 |

Table 1.—*continued*

| Substance Chemical name (synonym and/or abbreviation) | Structure 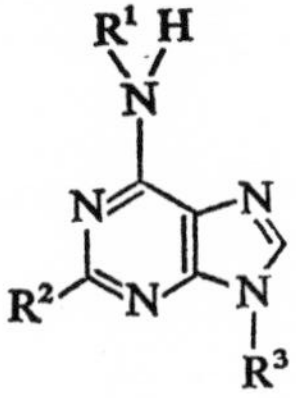 $R^1$ | $R^2$ | $R^3$ | Source: Bacteria | Fungi | Higher plants | Animals | No. in Fig. 1 |
|---|---|---|---|---|---|---|---|---|
| 6-(4-Hydroxy-3-methylbutylamino)-2-methylthio-9-β-D-ribofuranosylpurine (ms-ribosyldihydrozeatin, $ms^2hio^6A$) | 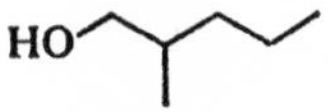 | | | | | | | 15 |
| 6-(3-Hydroxy-3-methylbutylamino)purine (3OHiP)* | 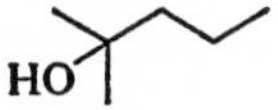 | H | H | ? | ? | ? | ? | |
| 6-(3-Hydroxy-3-methylbutylamino)-9-β-D-ribofuranosylpurine (3OHiPA) | 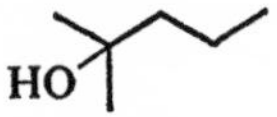 | H | Rib | ? | ? | ? | ? | |
| 6-Furfurylaminopurine (kinetin)* | 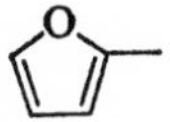 | H | H | ? | ? | ? | ? | |

* Not generally accepted as naturally occurring. † Number of compounds in Fig. 2. Rib=ribosyl.

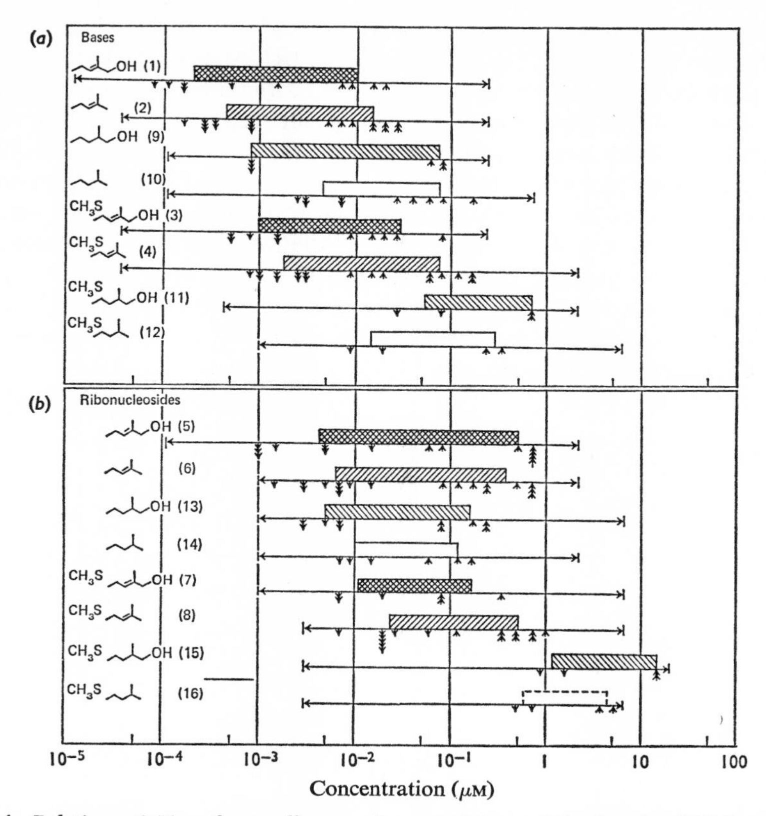


Fig. 1. *Relative activities of naturally occurring cytokinins and closely related N⁶-isopent(en)yl-adenine (a) and adenosine (b) derivatives in the tobacco bioassay*

The compounds are numbered as in Table 1. For easy reference the substituents on the N⁶-position and the methylthio group on the 2-position are indicated on the left-hand side. The base lines represent the tested concentration ranges, the bars represent mean values of the concentration ranges over which growth increases as a nearly linear function of the logarithm of concentration of the cytokinins, and the arrows under the base lines represent the start and end points of these ranges in individual experiments. A broken-line bar indicates that the compound became toxic at concentrations lower than needed for maximum growth. Side-chain structure is indicated by shading of the bars as follows: ▨, isopentenyl group; ▩, 4-OH-substituted isopentenyl group; □, isopentyl group; ▧, 4-OH-substituted isopentyl group. [From Schmitz *et al.* (1972*a*).]

category are 6-isopentenylaminopurine derivatives, of which 6-(3-methyl-2-butenylamino)purine, [$N^6$-($\Delta^2$-isopentenyl)adenine, 2iP], is the most common one. It has been rigorously identified, as the ribonucleoside (2iPA), in tRNA hydrolysates from a broad spectrum of organisms, including bacteria, fungi, higher plants and animals, and it is indicated to be a component of tRNA from insects and human tissues as well. It may be considered as the archetype of all the naturally occurring cytokinins listed in Table 1, and it is the only one so far rigorously characterized from animal sources. In micro-organisms ms2iPA occurs as well as 2iPA and at least in some of the few organisms so far examined, it seems to be the more abundant. In tRNA from higher plants ribosylzeatin appears to be a major component (Hall *et al.*, 1967), perhaps in part because of its high acid stability, but 2iPA, ms-ribosylzeatin and ms2iPA are also present (Burrows *et al.*, 1970, 1971; Vreman *et al.*, 1972). The ribosylzeatin in tRNA generally has been considered to be the *cis*-isomer (Hall *et al.*, 1967), but with the stereospecific synthesis of the two isomers and their separation by chromatography (Leonard *et al.*, 1971) both forms have been demonstrated in light-grown pea shoots, whereas in tRNA preparations from pea roots (Babcock & Morris 1970) as well as from wheat germ and tobacco callus (Playtis & Leonard, 1971) only the *cis*-isomer was obtained. In contrast with this, zeatin found as the free base and ribonucleoside in corn endosperm (Miller, 1961, 1965; Letham *et al.*, 1964) and other plant sources, including a fungus (Miller, 1967), has been the *trans*-isomer. Dihydrozeatin has been isolated from lupin seeds (Koshimizu *et al.*, 1967; Matsubara *et al.*, 1968) and pinto beans (Krasnuk *et al.*, 1971).

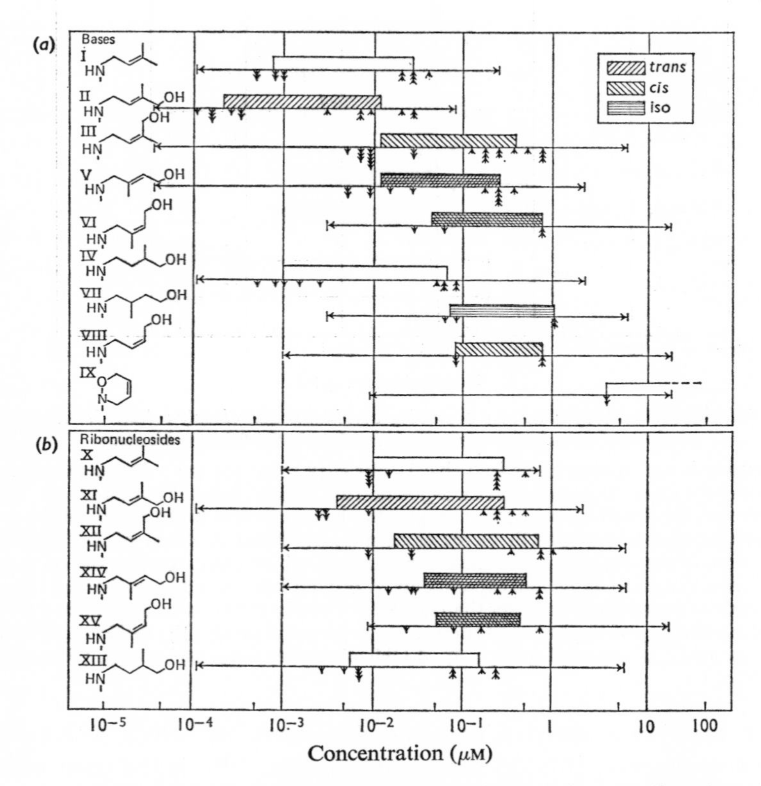


Fig. 2. *Cytokinin activities of geometric and position isomers of zeatin*

(*a*) Adenine bases; (*b*) ribonucleosides. Presentation of data is as in Fig. 1. The compounds are: I, 2iP; II, zeatin; III, *cis*-zeatin; IV, dihydrozeatin; V, *trans*-isozeatin [6-(4-hydroxy-2-methyl-*trans*-butenylamino)purine]; VI, *cis*-isozeatin [6-(4-hydroxy-2-methyl-*cis*-2-butenylamino)-purine]; VII, dihydroisozeatin [6-(4-hydroxy-2-methylbutylamino)purine]; VIII, *cis*-norzeatin [6-(4-hydroxy-*cis*-2-butenylamino)purine]; IX, cyclic-norzeatin [6-(3,6-dihydro-1,2-oxazin-2-yl)purine]; X, 2iPA; XI, ribosylzeatin; XII, ribosyl-*cis*-zeatin; XIII, ribosyldihydrozeatin; XIV, ribosylisozeatin [6-(4-hydroxy-2-methyl-*trans*-2-butenylamino)-9-β-D-ribofuranosylpurine]; XV, ribosyl-*cis*-isozeatin [6-(4-hydroxy-2-methyl-*cis*-2-butenylamino)-9-β-D-ribofuranosylpurine]. An incomplete bar indicates the compound was not tested in high enough concentration for maximum growth. [From Schmitz *et al.* (1972*b*).]

The biological activities of synthetic preparations of naturally occurring cytokinins and likely candidates are compared in Fig. 1. This shows that zeatin (the *trans*-isomer) is the most active compound of this entire group. Saturation of the double bond (Skoog *et al.*, 1967; Leonard *et al.*, 1968), removal or transfer of the hydroxyl group (Schmitz *et al.*, 1972*a*) or any changes detracting from planarity of the side chain (Hecht *et al.*, 1970*b*) tend to lower biological activity. Substitution in the 2-position of the ring may be more general than is now apparent. Its influence varies with the substituent; the $CH_3$- or $CH_3S$-substituted derivatives, as in Fig. 1, are only slightly less active than corresponding unsubstituted compounds, whereas 2-Cl derivatives may be slightly more active and 2-OH derivatives are more than 100 times less active than the parent compounds (Hecht *et al.*, 1970*a*). The 9-ribosyl derivatives are often about ten times less active than the corresponding free bases, but the apparently narrower range in which the ribonucleosides reach maximum activity is due mainly to the exponential scale. The 9-ribosyl moiety in the 9-position is not a requirement for biological activity (Hecht *et al.*, 1971*a*).

Comparisons of *cis*- and *trans*-isomers as well as of iso- and nor-zeatins are shown in Fig. 2(*a*). It is clear that all these modified zeatins have drastically lower activity in the tobacco bioassay. Cyclization of the norzeatin side chain leads to only marginal activity. The ribonucleosides of some of these compounds are compared in Fig. 2(*b*). They are active in a somewhat higher concentration range, but show about the same pattern of change in activity as described for the modified free bases.

*Localization of cytokinins*

Cytokinins occur partly in the form of apparently free bases, ribonucleosides, ribonucleotides and glycosides of undetermined nature in plants, and partly as constituents of tRNA. As far as I know, all unfractionated preparations of RNA and all tRNA preparations which have been thoroughly tested, with only two exceptions, have contained cytokinin activity (see Skoog & Armstrong, 1970). The exceptions are a virus RNA and a tRNA preparation from one of three tested strains of mycoplasma. When the RNA has been further characterized, the activity has been associated with specific tRNA species which are retained longer on columns of benzoylated DEAE-cellulose and constitute a relatively small part of the total tRNA. When these fractions have been further separated and purified the biological activity has been found in the hydrolysates of tRNA species responding to codons starting with uridine and not in those responding to any other codons (Armstrong *et al.*, 1969*b*; Bartz *et al.*, 1970; see also Skoog & Armstrong, 1970).

In the above case of *Escherichia coli* cytokinin-containing tRNA species have been isolated corresponding to all six amino acids (phenylalanine, leucine, serine, tyrosine, cysteine and tryptophan) for which there are codons with the initial letter U (see Fig. 3). The same apparently holds for *Lactobacillus acidophilus* (Peterkofsky & Jesensky, 1969). In tissues of higher organisms studied so far, cytokinin-containing tRNA species apparently respond to codons for fewer amino acids (D. J. Armstrong, unpublished work).

Further, it appears from the work on *E. coli* that species of tRNA without

| 1st letter | 2nd letter U | C | A | G | 3rd letter |
|---|---|---|---|---|---|
| U | Phe | Ser | Tyr | Cys | U |
| | Phe | Ser | Tyr | Cys | C |
| | Leu | Ser | C.T. | C.T. | A |
| | Leu | Ser | C.T. | Try | G |
| C | Leu | Pro | His | Arg | U |
| | Leu | Pro | His | Arg | C |
| | Leu | Pro | Gln | Arg | A |
| | Leu | Pro | Gln | Arg | G |
| A | Ile | Thr | Asn | Ser | U |
| | Ile | Thr | Asn | Ser | C |
| | Ile | Thr | Lys | Arg | A |
| | Met (C.I.) | Thr | Lys | Arg | G |
| G | Val | Ala | Asp | Gly | U |
| | Val | Ala | Asp | Gly | C |
| | Val | Ala | Glu | Gly | A |
| | Val (C.I.) | Ala | Glu | Gly | G |

Fig. 3. *Location of cytokinin-containing E. coli tRNA species in the genetic code*

Striped codons correspond to cytokinin-active tRNA species; boxed-in codons correspond to 6-threoninecarbamoylpurine-containing tRNA species. C.I., Chain-initiating codon; C.T., chain-terminating codon. [Modified from Skoog & Armstrong (1970).]

cytokinin content exist which respond to codons starting with uridine, and there is no evidence that any codon is utilized exclusively by cytokinin-containing tRNA species. In yeast, species of tRNA coding for leucine, serine, tyrosine and cysteine have been found to contain cytokinin-active bases (Armstrong *et al.*, 1969*a*), but phenylalanine tRNA, and apparently also tryptophan tRNA, instead contain base 'Y', a more highly modified adenine derivative (Nakanishi *et al.*, 1971) which becomes active in the tobacco bioassay only after it has been degraded (Hecht *et al.*, 1970*c*).

In cases where it has been established (Zachau *et al.*, 1966*a,b*; Madison & Kung, 1967; see also Skoog & Armstrong, 1970) the location of the cytokinin-active base in the tRNA is adjacent to the 3′ end of the anticodon, as is base Y.

An interesting analogy exists in that Nishimura and co-workers have reported that 6-threoninecarbamoylpurine is located next to the 3′ end of the anticodon of tRNA species responding to most if not all the codons with the initial letter A (see Fig. 3) (Ishikura *et al.*, 1969). Especially if considered in the form of a lactone, the threoninecarbamoylpurine resembles phenylurea derivatives which are active in promoting cell division and growth in several cytokinin bioassays (Bruce & Zwar, 1966), but not in the soya-bean test (Dyson *et al.*, 1972). The resemblance of threoninecarbamoylpurine to cytokinin-active phenylureidopurines (McDonald *et al.*, 1971) is even more striking.

Information on the specific or random association of a given cytokinin with specific tRNA species or specific codons is inconclusive. The distribution of 2iPA and ms2iPA in relatively pure tRNA species from *E. coli* is shown in Table 2. It appears that each cytokinin can be associated with the same codon and with several codons and with the same and several amino acid assignments. The proportions of the two cytokinins differ, however, with the amino acid or codon assignment, and they also change with age of the culture. For example, a marked increase in phenylalanine tRNA in late stages of growth accounts for a relative increase in the 2iPA content at that time.

No reason has been advanced for the presence of several cytokinins in tRNA of a given organism. The possibility is being examined that different cytokinins may be associated with tRNA in different organelles. A model for such behaviour is the distribution of base Y. Barnett and co-workers (Fairfield & Barnett, 1971) have shown that it is present in cytoplasmic phenylalanine tRNA from both *Neurospora* and *Euglena*, but is absent in phenylalanine tRNA from mitochondria of the former and from the chloroplasts of the latter organism.

*Cytokinin biosynthesis*

The origin of cytokinins is not known. It has been proposed that the 'free' cytokinins (bases, ribonucleosides and ribonucleotides) are released by degradation of tRNA (Hall, 1970). Evidence has been presented for enzyme-catalysed formation of cytokinins in tRNA from mevalonate and isopentenyl pyrophosphate as substrates in systems of micro-organisms and higher plants both *in vivo* and *in vitro* (Peterkofsky, 1968; Fittler *et al.*, 1968; Vickers & Logan, 1970; Hall, 1970). However, the report that mevalonate may be substituted for exogenous cytokinin in cytokinin-dependent tobacco cultures (Chen & Hall, 1969) was not confirmed (McChesney, 1970). Possibly mevalonate is utilized by cells or tissues that have become independent of exogenous cytokinin.

A striking demonstration of the difference in biosynthetic capacity between cytokinin-independent and -dependent tissue is the utilization of adenine as a substrate in an experiment by J. W. Einset (personal communication), the results of which are summarized in Fig. 4. At 5h after radioactive adenine was added to the medium three distinct peaks of cytokinin activity were obtained by Sephadex LH-20 chromatography of an alcohol tissue extract from a cytokinin-independent strain of tobacco callus, whereas only a trace, if any, cytokinin activity was obtained from the cytokinin-dependent strain. The active regions correspond to zeatin, 2iP and ms-zeatin respectively, and although they have not yet been identified there is some additional evidence that the second peak associated with the tall peak of radioactivity in this region, and which is not found in the dependent strain, is in fact 2iP. The total biological activity in the extracts suggests a minimum tissue concentration of about 0.1 nM in the cytokinin-independent strain. In another cytokinin-independent callus strain studied by

Table 2. *Distribution of 2iPA and ms2iPA in purified species of E. coli tRNA*
The data are from Bartz *et al.* (1970).

| tRNA sample | Coding properties | Specific amino acid acceptor ($pmol/E_{260}$ unit) | Estimated cytokinin ribonucleosides ($\mu g$/mg of tRNA) 2iPA | ms2iPA |
|---|---|---|---|---|
| Phe | UUY | 1050 | 1.6 | 0.2 |
| Leu | UUG | 1550 | 0.03 | 2.0 |
| $Ser_4$ | UCR | 1400 | — | 2.0 |
| Tyr | UAY | 1650 | — | 2.2 |
| Cys | UGY | 860 | 0.2 | 9.5 |
| $Trp_2$ | UGG | 1400 | 0.01 | 2.4 |

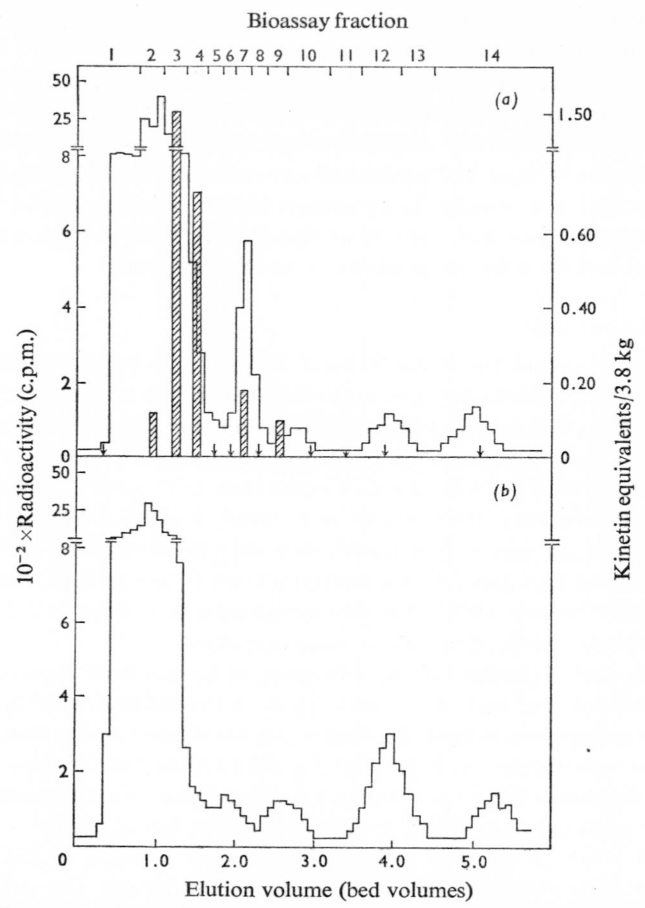


Fig. 4. *Distribution of radioactivity in Sephadex LH-20 column fractions of ethyl acetate extract from (a) cytokinin-independent and (b) cytokinin-dependent tobacco callus tissue*

The tissue was incubated for 5h in 50ml of medium with 0.01 $\mu$mol of [$^{14}$C]adenine (5 $\mu$Ci). The striped bars represent cytokinin activity determined for cytokinin-independent tissue in the indicated eluate fractions in a separate experiment (J. W. Einset, unpublished work).

Dyson & Hall (1972) the tissue concentration of 2iPA, the main active component, was estimated to be about 10$\mu$g/kg or about 30nM whereas the control tissue cultured on the same medium with 2iP had barely the detectable limit (0.085$\mu$g/kg).

In intact plants it is clear that the capacity for cytokinin synthesis varies with the tissue. Relatively rapid rates characteristically are limited to meristems, but slow rates may occur and at least the potential capacity to produce free cytokinins may be present in all living tissues.

## Cytokinin Functions in Nitrogen Metabolism

### *Nutrient translocation and utilization*

A primary function of cytokinins in the intact plant is in correlation phenomena. In some manner the cytokinin is a key factor in the mechanism that determines which parts of the plant will stay alive and grow and which parts will wither and die. It is logical that this regulation is exerted mainly via influences on nucleic acid and protein metabolism, but one important aspect of it is the control of translocation and accumulation of nutrients (Mothes, 1964; Pozsár & Kiraly, 1964). Influences of endogenous cytokinins in determining distribution patterns of metabolic activity, especially of nitrate reduction and amino acid synthesis, and changes occurring in these patterns with age of the plant as well as influences of treatments with exogenous cytokinins have been considered by Kursanov and others (Kursanov, 1963).

In the case of release of apical dominance it has been demonstrated in *Pisum* (Davidson, 1971) that the induction of growth in the inhibited lateral bud treated with cytokinin is preceded by increased production of auxin in it and is associated with decreased production of auxin in the terminal bud. This quantitative shift in biosynthesis and growth results in branching, or, when complete, in the development of a new main axis. Although it is becoming clear that cytokinins play a role in directing the flow of nutrients and assimilated compounds and their accumulation in growing points, the mechanism whereby cytokinin applied to the stem of a decapitated plant may replace the growing point with respect to these processes remains a mystery. Changes in physical properties of cell membranes as well as indirect influences on biosynthesis of enzymes and various metabolites may be involved.

## Cytokinins in Nucleic Acid Metabolism

### *Effects on DNA and RNA synthesis*

One of the earliest observed effects of cytokinins was the induction of DNA synthesis in excised tobacco pith tissue. The role of kinetin in conjunction with auxin (indol-3-ylacetic acid, IAA) in this process and with respect to mitosis and cytokinesis was studied by cytological methods (Patau *et al.*, 1957). The results indicated that both factors were required for continuous growth and cell division and that either one or both might become limiting. However, a 20% increase in DNA content, not accompanied by cell division, was obtained with kinetin treatments in the absence of added auxin.

There is an extensive literature (see Key, 1969; Skoog & Armstrong, 1970) on cytokinin stimulation of RNA as well as DNA biosynthesis in various plant systems. It is generally agreed that cytokinins stimulate and may serve to regulate these processes. In most cases so far only overall increases in total RNA rather than specific kinds and species have been studied. Perhaps most interesting, in terms of cytokinin function as a constituent of tRNA, is the claim by Cherry and co-workers (Anderson & Cherry, 1969) that cytokinin treatment of bean cotyledons preferentially promotes the synthesis of two leucine tRNA subspecies and perhaps decreases that of four others. From comparisons of chromatographic properties with those of leucine tRNA species in micro-organisms one might guess that the former two subspecies respond to UUR codons, but proof of this is lacking.

*Functions of cytokinins in tRNA*

The location of cytokinin-active bases adjacent to the 3′ end of the anticodon in certain tRNA species suggests that they must have some specific function in protein biosynthesis. There is general agreement that cytokinins, like all but one known ribonucleotide in the anticodon loop, do not significantly affect the amino acid acceptor activity of the tRNA molecule.

Three types of experiments, on the other hand, show that the cytokinin component does promote binding of the tRNA to the ribosome–mRNA complex. The first demonstration (Fittler & Hall, 1966) was that $KMnO_4$ treatment of yeast serine tRNA, which apparently specifically removes the isopentenyl chain from adenosine, decreased the efficiency of the ribosome binding. Gefter & Russell (1969) showed that a bacteriophage-induced tyrosine suppressor mutant strain of *E. coli* contains three species of tyrosine tRNA responding to the UAG codon which differ in that one contains unsubstituted adenosine, the second 2iPA and the third ms2iPA in the ribonucleotide adjacent to the 3′ end of the anticodon. The relative ribosome-binding efficiencies of the three tRNA species were found to be in the proportions 7 : 27 : 50; values which were in good agreement with their effectiveness in tests of suppression *in vitro*. It was shown, furthermore, that the tRNA species that lacked a modified base was unable to distinguish between the UAG and CAG codons. Thus both the ribosomal-binding efficiency and the specificity of codon recognition were enhanced by the presence of the cytokinin-active bases. Furuichi *et al.* (1970) have treated yeast tyrosine tRNA with bisulphite to obtain preparations with both the uridine at the 5′ end and the adenine at the 3′ end of the anticodon modified (sample I) and then with weak alkali to leave only the adenine modified (sample III) in Table 3. They were thus able to show that transformation specifically of 2iPA does not affect aminoacyl-acceptor activity, but results in a 65% loss in efficiency of binding the tRNA to the ribosome–mRNA complex.

These results are evidence that the cytokinin-active bases in tRNA species of the UNN codon group function in protein biosynthesis. It seems logical that the 6-threoninecarbamoylpurines might function in an analogous manner in tRNA species responding to ANN codons.

*The question of incorporation of exogenous cytokinins into tRNA*

If exogenous cytokinins exert growth-regulatory functions as constituents of specific tRNA species the question of their incorporation into tRNA becomes a central problem. The frequently quoted evidence for (Fox, 1966; Fox & Chen,

Table 3. *Binding of tRNA species to ribosome–mRNA complex*
The data are from Furuichi *et al.* (1970).

| tRNA | [$^{14}$C]Tyrosyl-tRNA bound to ribosomes (c.p.m.) | | | |
|---|---|---|---|---|
| | +Poly($U_5A$) | −Poly($U_5A$) | Δ | % |
| Untreated | 780 | 264 | 516 | 100 |
| Sample I ($HSO_3$) | 310 | 169 | 141 | 27 |
| Sample II ($HSO_3$+OH) | 517 | 241 | 276 | 53 |
| Sample III ($HSO_3$+OH +DEAE-Sephadex) | 527 | 348 | 179 | 35 |

1967) and against (Kende & Tavares, 1968; also see Kende, 1971) such incorporation is far from conclusive. It is based on experiments, with $^{3}H$- or $^{14}C$-labelled 6-benzylaminopurine and 9-methyl-6-benzylaminopurine supplied to tobacco or soya-bean callus cultures, which in my opinion were inadequate for the purpose both with respect to experimental design and quantities of tissue employed. Joint work on this same problem between Professor Leonard's and my laboratories over the past 4 years has given variable and inconsistent results. The current status of the problem is summarized as follows.

Cytokinin-dependent tobacco callus tissue cultured on media supplied with 6-benzylaminopurine produced tRNA which on hydrolysis yielded the ribonucleosides of zeatin, 2iP and ms-zeatin and probably also of ms2iP. In addition 6-benzylamino-9-β-D-ribofuranosylpurine was obtained from this preparation in low but sufficient quantity for mass-spectrometric identification (Burrows *et al.*, 1971).

In experiments with 6-benzylaminopurine doubly labelled with $^{3}H$ and $^{14}C$ labelled 6-benzylamino-9-β-D-ribofuranosylpurine was recovered from the tRNA hydrolysate, and the $^{3}H/^{14}C$ ratio of the 6-benzyl-9-β-D-ribofuranosylpurine fraction was the same as that of the administered sample of 6-benzylaminopurine. It is concluded, therefore, that the extracted 6-benzylamino-9-β-D-ribofuranosylpurine did not arise from side-chain transfer of the benzyl group from 6-benzylaminopurine to adenine in tRNA. Further, the quantity of 6-benzylamino-9-β-D-ribofuranosylpurine was minute ($<2$ mol/$10^{6}$ mol of ribonucleotide, or about 1 molecule of 6-benzylamino-9-β-D-ribofuranosylpurine/5000 molecules of tRNA). The 6-benzylamino-9-β-D-ribofuranosylpurine found in the tRNA hydrolysate, therefore, most likely was not a constituent of tRNA in the same sense as are cytokinin-active ribonucleosides normally obtained from tRNA preparations.

Of special interest in this connexion is the report (Dyson *et al.*, 1972) of a 6-benzylaminopurine complex, possibly the mononucleotide, capable of releasing 6-benzylamino-9-β-D-ribofuranosylpurine, which accumulates in soya-bean callus tissue supplied with 6-benzylaminopurine and persists as a cytokinin-active entity throughout or even beyond the growth period. Such a complex might be the source of 6-benzylamino-9-β-D-ribofuranosylpurine also in the above experiments with tobacco tissue.

Unfortunately, the above results do not settle the question of whether or not cytokinins need be incorporated into tRNA to function in the regulation of growth. However, since the 6-benzylaminopurine moiety in 6-benzylamino-9-β-D-ribofuranosylpurine is maintained intact the possibility of transient 6-benzylaminopurine incorporation into tRNA and subsequent release of 6-benzylamino-9-β-D-ribofuranosylpurine or a more stable 6-benzylamino-9-β-D-ribofuranosylpurine complex seems remote if not excluded.

Alternative mechanisms of a regulatory function of cytokinins in tRNA metabolism have been reported. It has been suggested (Berridge *et al.*, 1970) that free cytokinins may interfere with the binding of cytokinin-containing tRNA species to the ribosome–mRNA complex. Evidence of higher affinities of biologically highly active as compared with less active cytokinins in ribosome-binding tests are presented to support this view. However, the differences reported seem relatively small, and the critical experiment of competitively excluding tRNA species from the ribosome complex was not performed. Birmingham & Maclachlan (1972) have suggested a growth-regulatory function of auxins and cytokinins in terms of their promotion and inhibition respectively of microsomal ribonuclease activity.

None of the above mechanisms provides the features of specificity and organization at the molecular level that can be envisaged for the functioning of the cytokinins as constituents of tRNA of the UNN-codon group. These unique features are disregarded in concepts interpreting the cytokinin–tRNA relationship merely as one in which the tRNA serves as the source of free cytokinins and according to which the different cytokinins represent merely successive stages of modification of adenine in tRNA. The finding that 2iP serves as a precursor of zeatin (Miura & Miller, 1969) would seem contrary to this concept unless 2iP must first be incorporated into tRNA, in which case the argument becomes circular.

## Cytokinin Effects on Protein Metabolism

Cytokinins are known to stimulate protein biosynthesis in intact plants, in excised tissues and in organelles *in vitro* (see Skoog & Armstrong, 1970; Kulaeva & Romanko, 1967). In tissue cultures amino acid incorporation used as a measure of total protein synthesis is generally promoted by cytokinin treatments in proportion to the effects on growth (see Fig. 5). Recent studies of excised tobacco pith parenchyma (M. Kaminek, S. Swaminathan, R. M. Bock & F. Skoog, unpublished work) show more rapid increases in amino acid label in soluble protein excreted into the medium than in the total tissue protein.

In experiments with matched pairs of amino acids, one labelled with $^3$H and the

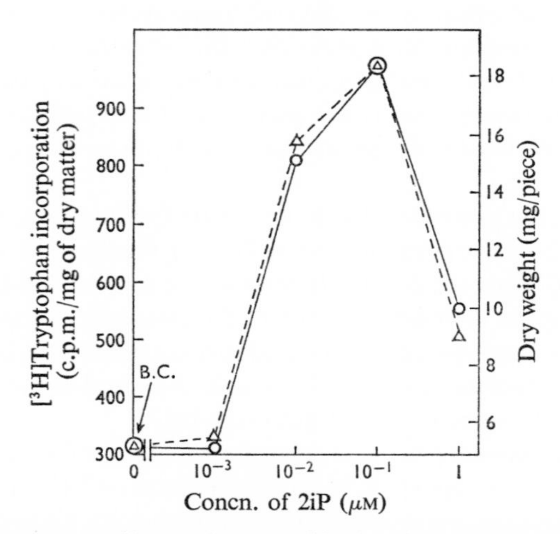


Fig. 5. *Effect of concentration of 2iP on dry-wt. yield* (△) *and incorporation of* [$^3$H]*tryptophan into soluble protein* (○) *of excised tobacco pith culture*

The growth period was 15 days (M. Kaminek, S. Swaminathan, R. M. Bock & F. Skoog, unpublished work).

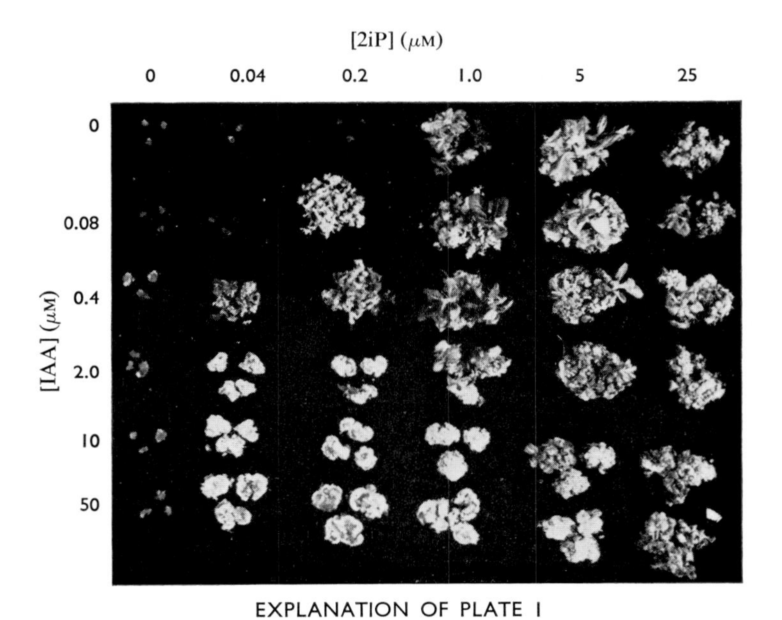


EXPLANATION OF PLATE I

*Effects of serial combinations of indol-3-ylacetic acid (IAA) and 2 iP on growth and organ formation of tobacco tissue cultures*
[From Skoog *et al.* (1967).]

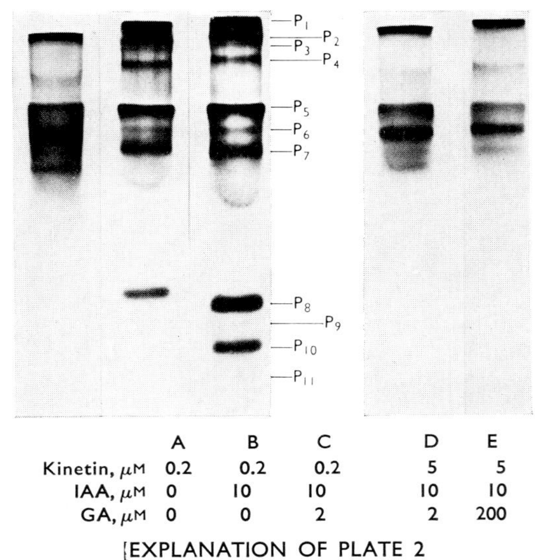


[EXPLANATION OF PLATE 2

*Effect of combined treatments with indolylacetic acid, gibberellic acid and kinetin on peroxidase isoenzymes in tobacco callus cultures*

Tobacco callus tissues were cultured on media containing various amounts of kinetin, gibberellic acid (GA) and indol-3-ylacetic acid (IAA). Peroxidase isoenzymes were extracted and then subjected to gel electrophoresis. [T. T. Lee (unpublished work).]

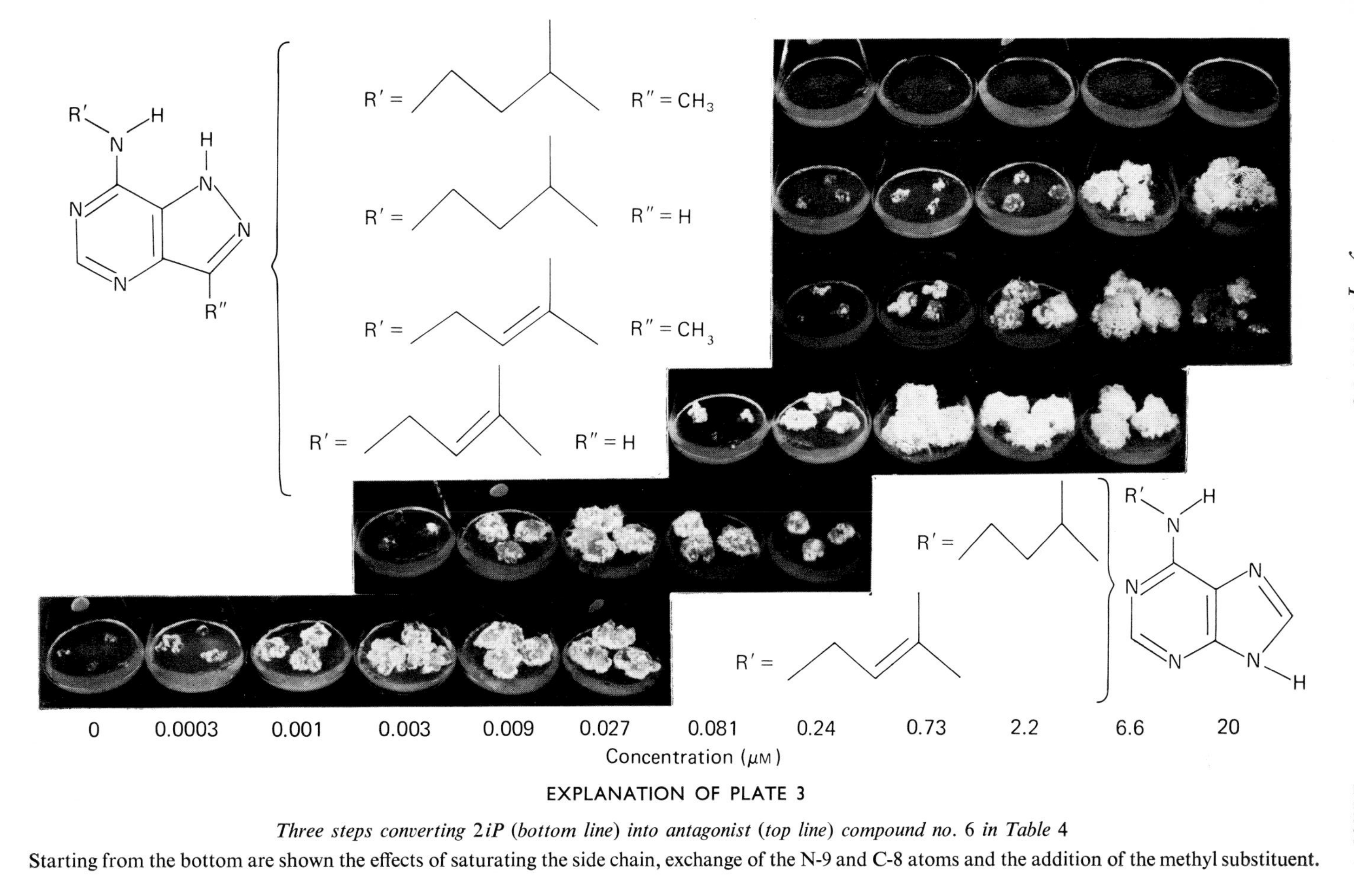


EXPLANATION OF PLATE 3

*Three steps converting* 2*iP* (*bottom line*) *into antagonist* (*top line*) *compound no.* 6 *in Table* 4

Starting from the bottom are shown the effects of saturating the side chain, exchange of the N-9 and C-8 atoms and the addition of the methyl substituent.

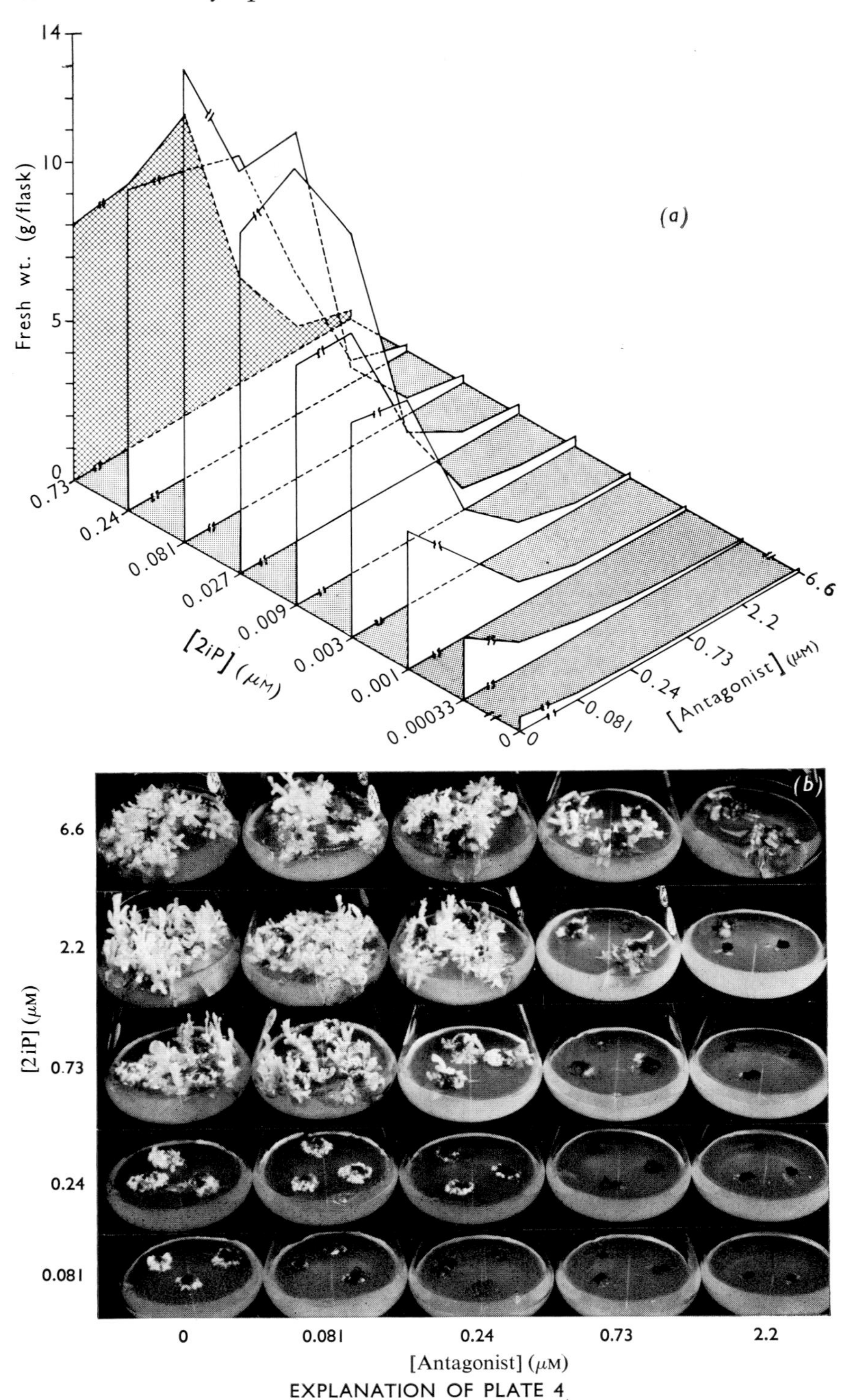


EXPLANATION OF PLATE 4

*Yields of tobacco callus cultured on media with serial combinations of 2iP and antagonist* 3-*methyl*-7-(3-*methylbutylamino*)*pyrazolo*[4,3-*d*]*pyrimidine*, *compound no.* 6 *in Table* 4

The auxin (indol-3-ylacetic acid) concentration was 11.4 μM in (*a*) and 0.1 μM in (*b*). [From Skoog *et al.* (1973).]

F. SKOOG

other with $^{14}C$ and responding to codons starting with different letters, there may have been slightly greater incorporation of those responding to UNN codons into the cytokinin-treated tissue. This was not found invariably and there was no detectable change in amino acid composition of the synthesized protein. In contrast, gel-electrophoretic studies of specific enzymes and isoenzymes which provide for better resolution have revealed marked effects of cytokinin treatments on the presence or absence or on the relative amounts of specific enzymes or isoenzymes both as functions of the cytokinin concentration and of the proportion of cytokinin to other growth factors. Data kindly provided by Dr. T. T. Lee (personal communication) (Plate 2) show effects of cytokinins, in combinations with indol-3-ylacetic acid and gibberellic acid, on peroxidase isoenzymes in tobacco callus cultures. It may be seen that the relative quantities of several bands are quite different for tissues cultured on media with 0.2$\mu$M- and 5$\mu$M-kinetin, some bands present in the former case are absent or practically so in the latter and vice versa. The most detailed study of cytokinin influences on enzyme formation is perhaps the work by Steinhart *et al.* (Steinhart *et al.*, 1964; Mann *et al.*, 1963) on the formation of tyramine methylpherase in germination and early seedling growth of barley. The rapid rise and delayed fall in the activity of this enzyme in response to cytokinin treatment is most striking, especially as most other enzymes do not seem to be similarly affected. However, concomitant effects on some other enzymes were observed. The authors conclude that the cytokinin influence probably is a more general one on protein biosynthesis, and that the stage of development when the treatment is started may determine which particular enzyme is most affected.

As to the mode of action of cytokinins in enzyme activation, evidence from inhibitor studies is much the same as for the work on gibberellin stimulation of $\alpha$-amylase biosynthesis (Varner & Chandra, 1964; Varner & Johri, 1969), i.e. the influence seems to be via promotion of RNA metabolism, but there are indications that biosynthesis of certain specific proteins is a prerequisite to the enhanced RNA metabolism, which in turn activates the bulk protein biosynthesis.

Several recent reports also deal with growth regulation in terms of cytokinin-induced biosynthesis of enzymes as well as in terms of inhibitory effects of cytokinin on the activity of key enzyme systems. Cytokinesin I, but not cytokinins, as exemplified by ribosylzeatin, is claimed to strongly inhibit cyclic AMP phosphodiesterases (Wood *et al.*, 1972). An inhibitory effect of cytokinins on ribonuclease activity in conjunction with and in opposition to a stimulating effect of indolylacetic acid has been proposed as a mechanism at the molecular level for regulating rates of protein biosynthesis and thus for regulating plant growth (Birmingham & Maclachlan, 1972).

Senescence has been investigated perhaps more than any other phase of plant development with reference to cytokinin regulation of protein metabolism. Although some early work suggested that protein degradation might be prevented by cytokinins indirectly via an influence on amino acid accumulation, which in turn through mass action would delay breakdown (Mothes, 1964), much evidence now exists that synthesis of protein *de novo* induced by cytokinins is an essential step in the prevention of cytolysis. Thimann and co-workers (Martin & Thimann, 1972) have stressed the key role of certain hydrolases which they suggest contain

Table 4. *Relation of the structure of 7-substituents to cytokinin and antagonist activities of* 3-*methylpyrazolo*[4,3-*d*]*pyrimidine derivatives*

The results are taken from Skoog *et al.* (1973); testing was done between April 1970 and December 1971. N.A. = Not active. N.R. = Not reached. All values/averages are of two tests except as indicated by numbers in parentheses.

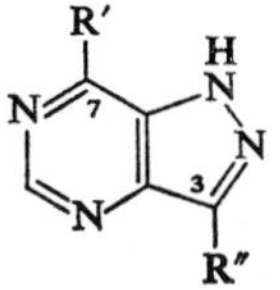

| Compound no. | R″ | R′ | No. of C atoms in R′ | Range tested (μM) | Cytokinin activity Min. concn. (μM) for | | Antagonist activity against 0.003 μM-2iP Min. concn. (μM) for | |
|---|---|---|---|---|---|---|---|---|
| | | | | | Detection | Maximum growth | Detection | Lethal dosage |
| 14 | H | 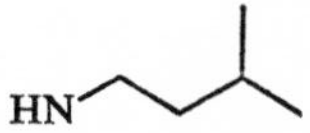 HN | 5 | 0.009–20 | 1.0 | 20 | N.A. | — |
| 15 | H | 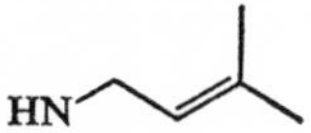 HN | 5 | 0.001–20(3) | 0.08 | 1.0 | N.A. | — |
| 16 | $CH_3$ | HO | 0 | 0.08–20 | N.A. | — | N.A. | — |
| 1 | $CH_3$ | 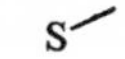 S | 1 | 0.24–20 | N.A. | — | N.A. | — |
| 13 | $CH_3$ | 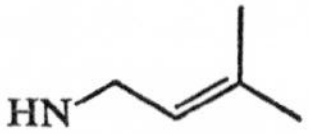 HN | 5 | 0.08–20(4) | 0.24 | 6.6 | N.A. | — |
| 11 | $CH_3$ |  N | 2×4 | 0.24–20 | N.A. | — | N.A. | — |
| 9 | $CH_3$ | 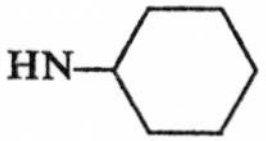 HN | 6 | 0.73–20 | N.A. | — | N.A. | — |
| 7 | $CH_3$ | 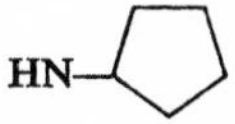 HN | 5 | 0.73–20 | N.A. | — | 6.6 | N.R. |

| | | | | | | | | |
|---|---|---|---|---|---|---|---|---|
| 4 | $CH_3$ | 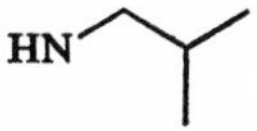 HN | 4 | 0.24–20 | 2.2 | 7* | 6.6 | N.R. |
| 2 | $CH_3$ | 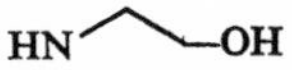 HN OH | 2 | 0.73–20 | N.A. | — | 2.2 | N.R. |
| 12 | $CH_3$ |  HN | 10 | 0.24–20 | N.A. | — | 3.0 | N.R. |
| 10 | $CH_3$ |  HN | 7 | 0.03–6.6 | N.A. | — | 0.2 | 2.2 |
| 3 | $CH_3$ |  HN | 4 | 0.03–6.6 | N.A. | — | 0.1 | 0.73 |
| 6 | $CH_3$ | 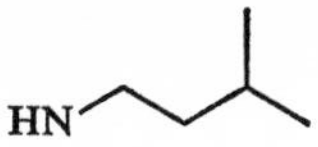 HN | 5 | 0.009–20[4] | N.A. | — | 0.1 | 0.73 |
| 8 | $CH_3$ |  HN | 6 | 0.009–20 | N.A. | — | 0.03 | 0.5 |
| 5 | $CH_3$ |  HN | 5 | 0.009–20[3] | N.A. | — | 0.03 | 0.2 |

* Only slight growth stimulation.

serine, or similarly acting, but less potent, amino acids (cysteine etc.), in their 'active centres' and are inhibited by cytokinins. The enhancement of senescence by these amino acids and its prevention by cytokinins can thus be accounted for. Direct evidence at the molecular level for this interpretation is still lacking; as is also true for any of the relationships between cytokinins and protein metabolism described above.

Recent success in interpreting hormone action in terms of cyclic AMP and membrane functions, and especially the progress in isolating protein receptors for steroid hormones has stimulated work on cytokinin–protein complexes. Of special interest is a protein–kinetin complex which functions specifically at the transcription level of gene-controlled protein biosynthesis (Matthysse, 1969; Matthysse & Abrams, 1970).

## Development of Cytokinin Antagonists

As an approach to locating specific sites of cytokinin action and thus to elucidate their mode of action a systematic search has been made for potent cytokinin antagonists. (This has been a collaborative project in which Dr. S. M. Hecht in Professor Bock's laboratory did the organic syntheses and Dr. R. Y. Schmitz was mainly responsible for biological testing.)

Many substances, especially auxins, gibberellins and abscisic acid are known to 'interact' indirectly with cytokinins, as shown by influences on growth and morphogenesis (see van Overbeek, 1966; Khan, 1971). But until recently we had found no substance closely related structurally to cytokinin-active purine derivatives which might serve as an antimetabolite interfering specifically and reversibly in cytokinin metabolism. One exception to this is 6-methylaminopurine, a weak cytokinin, which in high concentrations also lowers the growth response of callus tissue to 'optimum' concentrations of more active cytokinins (Klämbt *et al.*, 1966) and thus acts both as an antagonist and as a cytokinin.

From studies based on known structure/activity relationships of cytokinins, particularly properties of the side chain conducive to high activity and the relatively low activities of pyrazolo[4,3-*d*]pyrimidine derivatives (Hecht *et al.*, 1971*a*), a series of potent cytokinin antagonists have been developed (Hecht *et al.*, 1971*b*; Skoog *et al.*, 1973). The effects of three structural modifications of 2iP (removal of the $\Delta^2$-double bond, exchange of the C-8 and N-9 atoms and insertion of a methyl substituent on the exchanged C atom) leading to 3-methyl-7-(3-methylbutylamino)pyrazolo[4,3-*d*]pyrimidine, a highly potent antagonist, are shown in Plate 3. Each of these three modifications is essential for antagonist activity, and no two of them would be enough to remove completely cytokinin activity. Further study of side-chain modifications affecting antagonist activity has revealed remarkable specificity in the structural requirements (Table 4). The following features are noteworthy. There is a similarity with cytokinins in that antagonists with side chains 4–6 carbon atoms long are most active, but whereas the isopentenyl chain is optimum for cytokinin activity, the *n*-pentyl chain is the best for antagonist activity; also in the latter case the *n*-hexyl group is superior to the isopentenyl group. It should be noted also that, although the monobutyl substituent in the in the 7-position allows for relatively high antagonist activity, a second butyl group in the 7-position renders the compound inactive. Cyclization or drastic shortening of the chain also tends to eliminate antagonist activity, which is therefore closely dependent on the detailed structure of the chain as well as on the overall configuration of the molecule.

Further testing has been done only with the 3-methyl-7-isopentylpyrazolo[4,3-*d*]pyrimidine derivative which will be referred to below as the antagonist. The effects of serial combinations of 2iP and antagonist on callus growth are shown in Plate 4(*a*). The antagonist counteracts 2iP action not only in the low, growth-promoting range of 2iP but also in the higher, inhibitory range, where the antagonist, therefore, tends to promote growth. The effect of the antagonist on tissue grown on a medium with a 2iP–indolylacetic acid balance conducive to bud formation is shown in Plate 4(*b*).

Concentrations of antagonist higher than about 1 $\mu$M are not effectively counteracted by any concentration of 2iP, or other cytokinin, but the effective range can be more than doubled by adding adenine at concentrations that do not affect the response in treatments with low concentrations or without antagonist. The antagonist, therefore, appears to exert a highly specific anti-cytokinin action when given in low dosages, but in high dosages it probably interferes also in other phases of adenine metabolism.

Preliminary surveys of antagonist effects in various applications to plants have been done only with the isopentylpyrazolo[4,3-*d*]pyrimidine. In relatively low concentrations it effectively prevents germination of seeds, stops growth and kills seedlings, but there is no evidence of significant reversal of these effects by treatments with 2iP or other cytokinins. In fact, in many tests the cytokinins themselves were about as inhibitory as the antagonist. In young plants sprayed with antagonist and exposed to stress conditions the leaves seemed to wilt and turn yellow more rapidly than in controls, but quantitative results are lacking. In the cytokinin bioassay by Fletcher and co-workers (Fletcher & McCullagh, 1971) the antagonist effectively prevents both the greening and expansion of *Cucumis* cotyledons which are promoted by cytokinins, and it strikingly hastens senescence. In this system the cytokinins and antagonist both counteract and re-enforce one another's action in different concomitant processes in visibly clear-cut, but functionally no doubt complex, and as yet unknown, ways. It appears, therefore, that the cytokinin antagonists may be useful in investigations of cytokinin functions in growth and developmental processes, even if these substances prove to be impractical for the control of vegetation on a large scale.

## Conclusion

At the physiological level cytokinins have a regulatory role in many phases of nitrogen metabolism, including the distribution and assimilation of nutrients, the biosynthesis of nucleic acids and of proteins and the development of specific enzyme activities. At the molecular level cytokinins, as constituents of tRNA, are known to function in the regulation of protein biosynthesis by enhancing the binding of tRNA to the ribosome–mRNA complex and by increasing the specificity of codon–anticodon recognition. Nevertheless, the evidence now available does not dictate that endogenous cytokinins in the form of free bases and ribonucleosides, or cytokinins from exogenous sources must be incorporated into tRNA to exert their growth-promoting or morphogenic action.

Current interest centres on reports that low-molecular-weight cytokinin complexes, other than with tRNA, function in the regulation of growth and on reports of specific cytokinin–protein complexes which may function at the transcription level of gene-controlled protein biosynthesis.

It is a pleasure to acknowledge the collaboration of colleagues and students on many phases of work reported here. I am especially indebted to Professor N. J. Leonard and his co-workers at the University of Illinois, and to Professor S. M. Hecht, Massachusetts Institute of Technology who did the chemistry, to Professor R. M. Bock, University of Wisconsin for advice on nucleic acid biochemistry and to Dr. J. D. Armstrong and Dr. R. Y. Schmitz for major contributions on biological aspects. Work done at the University of Wisconsin was supported in part by a Research Grant (GB-25812) from the National Science Foundation.

## References

Anderson, M. B. & Cherry, J. H. (1969) *Proc. Nat. Acad. Sci. U.S.* **62**, 202–209

Armstrong, D. J., Skoog, F., Kirkegard, L. H., Hampel, A., Bock, R. M., Gillam, I. & Tener, G. M. (1969*a*) *Proc. Nat. Acad. Sci. U.S.* **63**, 504–511

Armstrong, D. J., Burrows, W. J., Skoog, F., Roy, K. L. & Söll, D. (1969*b*) *Proc. Nat. Acad. Sci. U.S.* **63**, 834–841

Babcock, D. F. & Morris, R. W. (1970) *Biochemistry* **9**, 3701–3705

Bartz, J., Söll, D., Burrows, W. J. & Skoog, F. (1970) *Proc. Nat. Acad. Sci. U.S.* **67**, 1448–1453

Berridge, M. V., Ralph, R. K. & Letham, D. S. (1970) *Biochem. J.* **119**, 75–84

Birmingham, C. & Maclachlan, G. A. (1972) *Plant Physiol.* **49**, 371–375

Bruce, M. I. & Zwar, J. A. (1966) *Proc. Roy. Soc. Ser. B.* **165**, 245–265

Burrows, W. J., Armstrong, D. J., Kaminek, M., Skoog, F., Bock, R. M., Hecht, S. M., Dammann, L. G., Leonard, N. J. & Occolowitz, J. (1970) *Biochemistry* **9**, 1867–1872

Burrows, W. J., Skoog, F. & Leonard, N. J. (1971) *Biochemistry* **10**, 2189–2194

Chen, C.-M. & Hall, R. H. (1969) *Phytochemistry* **8**, 1687–1695

Davidson, D. R. (1971) Ph.D. Thesis, University of Wisconsin

Dyson, W. H. & Hall, R. H. (1972) *Plant Physiol.* **50**, 616–621

Dyson, W. H., Fox, J. E. & McChesney, J. D. (1972) *Plant Physiol.* **49**, 506–513

Fairfield, S. A. & Barnett, W. E. (1971) *Proc. Nat. Acad. Sci. U.S.* **68**, 2972–2976

Fittler, F. & Hall, R. H. (1966) *Biochem. Biophys. Res. Commun.* **25**, 441–446

Fittler, F., Kline, L. K. & Hall, R. H. (1968) *Biochemistry* **7**, 940–944

Fletcher, R. A. & McCullagh, D. (1971) *Planta* **101**, 88–90

Fox, J. E. (1966) *Plant Physiol.* **41**, 75–82

Fox, J. E. & Chen C.-M. (1967) *J. Biol. Chem.* **242**, 4490–4494

Furuichi, Y., Wataya, Y., Hayatsu, H. & Ukita, T. (1970) *Biochem. Biophys. Res. Commun.* **41**, 1185–1191

Gefter, M. L. & Russell, R. L. (1969) *J. Mol. Biol.* **39**, 145–157

Hall, R. H. (1970) *Progr. Nucl. Acid Res. Mol. Biol.* **10**, 57–86

Hall, R. H., Csonka, L., David, H. & McLennan, B. (1967) *Science* **156**, 69–71

Hecht, S. M., Leonard, N. J., Schmitz, R. Y. & Skoog, F. (1970*a*) *Phytochemistry* **9**, 1173–1180

Hecht, S. M., Leonard, N. J. Schmitz, R. Y. & Skoog, F. (1970*b*) *Phytochemistry* **9**, 1907–1913

Hecht, S. M., Bock, R. M., Leonard, N. J., Schmitz, R. Y. & Skoog, F. (1970*c*) *Biochem. Biophys. Res. Commun.* **41**, 435–440

Hecht, S. M., Bock, R. M., Schmitz, R. Y., Skoog, F., Leonard, N. J. & Occolowitz, J. L. (1971*a*) *Biochemistry* **10**, 4224–4228

Hecht, S. M., Bock, R. M., Schmitz, R. Y., Skoog, F. & Leonard, N. J. (1971*b*) *Proc. Nat. Acad. Sci. U.S.* **68**, 2608–2610

Ishikura, H., Yamada, Y., Murao, K., Saneyoshi, M., & Nishimura, S. (1969) *Biochem. Biophys. Res. Commun.* **37**, 990–996

Jablonski, J. R. & Skoog, F. (1954) *Physiol. Plant.* **7**, 16–24

Kende, H. (1971) *Int. Rev. Cytol.* **31**, 301–338

Kende, H. & Tavares, J. E. (1968) *Plant Physiol.* **43**, 1244–1248

Key, J. L. (1969) *Annu. Rev. Plant Physiol.* **20**, 449–474

Khan, A. A. (1971) *Science* **171**, 853–859
Klämbt, D., Thies, G. & Skoog, F. (1966) *Proc. Nat. Acad. Sci. U.S.* **56**, 52–59
Koshimizu, K., Kusaki, T., Mitsui, T. & Matsubara, S. (1967) *Tetrahedron Lett.* **14**, 1317–1320
Krasnuk, M., Witham, F. H. & Tegley, J. R. (1971) *Plant Physiol.* **48**, 320–324
Kulaeva, O. N. & Romanko, E. G. (1967) *Dokl. Akad. Nauk SSSR (Bot. Sci. Sect.)* **177**, 464–467
Kursanov, A. L. (1963) *Advan. Bot. Res.* **1**, 209–274
Leonard, N. J., Hecht, S. M., Skoog, F. & Schmitz, R. Y. (1968) *Proc. Nat. Acad. Sci. U.S.* **59**, 15–21
Leonard, N. J., Playtis, A. J., Skoog, F. & Schmitz, R. Y. (1971) *J. Amer. Chem. Soc.* **93**, 3056–3058
Letham, D. S., Shannon, J. S. & McDonald, T. R. (1964) *Proc. Chem. Soc.* 230–231
Madison, J. T. & Kung, H.-K. (1967) *J. Biol. Chem.* **242**, 1324–1330
Mann, J. D., Steinhart, C. E. & Mudd, S. H. (1963) *J. Biol. Chem.* **238**, 676–681
Martin, C. & Thimann, K. V. (1972) *Plant Physiol.* **50**, 432–437
Matsubara, S., Koshimizu, K. & Nakahira, R. (1968) *Sci. Rep. Kyoto Prefect. Univ.* **19a**, 19–24
Matthysse, A. G. (1969) *Proc. Int. Bot. Congr. 11th, Seattle*, 143
Matthysse, A. G. & Abrams, C. (1970) *Biochim. Biophys. Acta* **199**, 511–518
McChesney, J. D. (1970) *Can. J. Bot.* **48**, 2357–2359
McDonald, J. J., Leonard, N. J., Schmitz, R. Y. & Skoog, F. (1971) *Phytochemistry* **10**, 1429–1439
Miller, C. O. (1961) *Proc. Nat. Acad. Sci. U.S.* **49**, 170–174
Miller, C. O. (1965) *Proc. Nat. Acad. Sci. U.S.* **54**, 1052–1058
Miller, C. O. (1967) *Science* **157**, 1055–1057
Miller, C. O., Skoog, F., Okumura, F. S. & Strong, F. M. (1956) *J. Amer. Chem. Soc.* **78**, 1375–1380
Miura, G. A. & Miller, C. O. (1969) *Plant Physiol.* **44**, 372–376
Mothes, K. (1964) in *Regulateurs Naturels de la Croissance Végétale* (Nitsch, J., organizer), pp. 131–140, CNRS, Paris
Nakanishi, K., Blobstein, S., Funamizu, M., Furutashi, N., VanLear, G., Grunberger, D., Lanks, K. W. & Weinstein, I. B. (1971) *Nature (London)* **234**, 107–109
Patau, K., Das, N. K. & Skoog, F. (1957) *Physiol. Plant.* **10**, 949–966
Peterkofsky, A. (1968) *Biochemistry* **7**, 472–482
Peterkofsky, A. & Jesensky, C. (1969) *Biochemistry* **8**, 3789–3809
Playtis, A. J. & Leonard, N. J. (1971) *Biochem. Biophys. Res. Commun.* **45**, 1–5
Pozsàr, B. I. & Kiraly, Z. (1964) in *Symp. Host Parasite Relations in Plant Pathology* (Kiraly, Z. & Ubnizoy, G., eds.), pp. 199–210, Hungarian Academy of Sciences, Budapest
Schmitz, R. Y., Skoog, F., Hecht, S. M., Bock, R. M. & Leonard, N. J. (1972*a*) *Phytochemistry* **11**, 1603–1610
Schmitz, R. Y., Skoog, F., Playtis, A. J. & Leonard, N. J. (1972*b*) *Plant Physiol.* **50**, 702–705
Skoog, F. (1971) *Colloq. Int. Cent. Nat. Rech. Sci.* **193**, 115–135
Skoog, F. & Armstrong, D. J. (1970) *Annu. Rev. Plant Physiol.* **21**, 359–384
Skoog, F. & Schmitz, R. Y. (1972) in *Plant Physiology, A Treatise* (Steward, F. C., ed.), vol. 6B, chapter 9, pp. 181–213, Academic Press, New York
Skoog, F., Strong, F. M. & Miller, C. O. (1965) *Science* **148**, 532–533
Skoog, F., Hamzi, A. Q., Szweykowska, A. M., Leonard, N. J., Canaway, K. L., Fujii, T., Helgeson, J. P. & Loeppky, R. N. (1967) *Phytochemistry* **6**, 1169–1192
Skoog, F., Schmitz, R. Y., Bock, R. M. & Hecht, S. M. (1973) *Phytochemistry* **12**, 25–37
Steinhart, C. E., Mann, J. D. & Mudd, S. H. (1964) *Plant Physiol.* **39**, 1030–1038
van Overbeek, J. (1966) *Science* **152**, 721–731
Varner, J. E. & Chandra, G. R. (1964) *Proc. Nat. Acad. Sci. U.S.* **52**, 100–106
Varner, J. E. & Johri, M. M. (1969) in *Biochemistry and Physiology of Plant Growth Substances* (Wightman, F. & Setterfield, G., eds.), pp. 798–814, Runge Press, Ottawa
Vickers, J. D. & Logan, D. M. (1970) *Biochem. Biophys. Res. Commun.* **41**, 741–747
Vreman, H. J., Skoog, F., Frihart, C. R. & Leonard, N. J. (1972) *Plant Physiol.* **49**, 848–851
Wood, H. N., Lin, M. C. & Braun, A. C. (1972) *Proc. Nat. Acad. Sci. U.S.* **69**, 403–406
Zachau, H. G., Dütting, D. & Feldmann, H. (1966*a*) *Angew. Chem.* **78**, 392
Zachau, H. G., Dütting, D. & Feldmann, H. (1966*b*) *Hoppe-Seyler's Z. Physiol. Chem.* **347**, 212–235

*Biochem. Soc. Symp.* (1973) **38**, 217–234
*Printed in Great Britain*

# Regulation of Protein Synthesis in Cotton Seed Embryogenesis and Germination

By LEON S. DURE, III

*Department of Biochemistry, University of Georgia, Athens, Ga.* 30601, *U.S.A.*

## Synopsis

The mRNA for several enzymes that arise during germination in cotton cotyledons is transcribed during embryogenesis. This body of mRNA apparently is not translated during embryogenesis because of the presence of the plant growth regulator abscisic acid. How abscisic acid prevents the translation of this mRNA is not understood, but it appears not to involve changes in the tRNA population of the cotyledon cells. The population of tRNA iso-acceptors in the cotyledon cytosol does not change during embryo development nor during germination, and appears to be identical with that of cotton roots. The only change in tRNA population observed during these developmental stages in cotyledon ontogeny is a 7–8-fold increase in chloroplast tRNA species that occurs during germination and is not light-dependent. Ribonucleoprotein particles containing non-ribosomal RNA that may represent the storage form of the mRNA for the germination enzymes have been isolated.

## Introduction

We have been working for the past several years on the developmental biochemistry of cotton seed embryogenesis and germination, concentrating on the cotton cotyledon as the developing tissue. The cotyledons of seeds such as those of the cotton plant afford an interesting tool for developmental studies in that their embryogenesis is characterized by a massive synthesis of storage protein and lipid, whereas in their subsequent germination a metabolic reversal takes place, and these stored nutrients are rapidly degraded to their basic components to support the growth of the root and shoot axis. During the first several days of the germination process the proplastids of the cotyledon undergo extensive development themselves and readily green after about 3 days. After about 2 weeks germination the cotyledons undergo senescence and are abscised from the growing plantlet.

The embryogenic process in cotton takes about 50 days from anthesis to boll opening. During the first 20 days the embryo develops very slowly while the endosperm and nucella tissue are attaining their maximum size. Between day 20 and day 32 embryo growth is extensive; the embryo developing from a structure with a wet weight of about 5 mg to one with a wet weight of about 90 mg. The bulk of this weight is contributed by the cotyledons which by this stage have encircled the small axis several times. The cotyledons continue to increase in weight but at a much slower rate until about the day 50 after anthesis at which time the embryo averages 125 mg wet weight. At this point the remaining layers of nucella and epidermal tissue sclerify, the boll opens and the embryo quickly dries to a dry seed weight of about 65 mg.

## Concentrations of Nucleic Acids in Cotyledon Development

Our initial studies involved a temporal mapping of the amounts of DNA, rRNA and tRNA in the cotyledons during embryogenesis and germination and a plotting of these values on a per cotyledon pair basis. Two noteworthy observations resulted from these measurements. (1) The amount of DNA/cotyledon does not increase after the embryos have reached 85 mg wet weight. Not only does the DNA content remain the same for the remainder of embryogenesis, but it does not increase during the first 5 days of germination. This observation suggests that the final number of cells in the cotyledons is attained at this point in embryogenesis. We were particularly excited by this finding, since it implied that the biochemical events that characterize the change from biosynthesis to degradation that takes place when germination succeeds embryogenesis could be studied without being complicated by the massive synthesis of all constituents that must occur in dividing cells.

This cessation of DNA synthesis was further verified by our failure to detect any incorporation of $[^{32}P]P_i$ or $[^{14}C]$orotic acid into cotyledon DNA after this point in embryogenesis. (This levelling of DNA content/cotyledon does not necessarily signify that organelle DNA does not continue to increase, since the contribution of organelle DNA to the total DNA may be quite small.)

(2) rRNA and tRNA increase in cotyledons during germination even though the cell number remains the same. However, this increase is the result of a seven- to eight-fold increase in plastid RNA. This fact is easily shown for rRNA by sodium dodecyl sulphate–polyacrylamide-gel electrophoresis (Loening, 1967) of cotyledon RNA extracted at different times during germination, and has been shown for tRNA by our studies on chloroplast tRNA (Merrick & Dure, 1971, 1972) which will be discussed. This observation implies that the metabolic reversal that marks the change from embryonic growth to germinative growth does not require an augmentation in the population of cytosol ribosomes or tRNA.

## Stored mRNA in Mature Cotyledons

Concomitant with this work we undertook the study of the nucleic acid and protein synthesis that is necessary for the dry seed to germinate. We found initially that protein synthesis in the cotyledons was not measurably affected for the first three days of germination when RNA synthesis was prevented by actinomycin D (Dure & Waters, 1965; Waters & Dure, 1966). The subsequent greening of the cotyledons was totally prevented by actinomycin D as was the growth of the hypocotyl and root, which was expected since much of their growth is the result of the production of new cells. This actinomycin D insensitivity of cotyledon protein synthesis suggested that much of the mRNA directing protein synthesis in this tissue in early germination exists in the dry seed and represents stored mRNA that is synthesized at some earlier point in the ontogeny of this tissue. This stored mRNA conceivably could be mRNA for constitutive enzymes that are continuously synthesized in this tissue throughout its development and maturation, and thus is utilized in the latter stages of embryogenesis and carried over into germination intact. In this case the stored mRNA of the cotyledons may not be indicative of a developmental regulatory system. On the other hand, the stored mRNA could be largely composed of mRNA that codes for proteins that are uniquely required for the germination process and not present in the tissue at any earlier developmental stage. Such a temporal separation of transcription from translation would certainly constitute a developmental regulatory mechanism operative in equipping embryonic tissue for germination.

To explore further the function of the stored mRNA in cotyledon tissue during germination, we studied the appearance of isocitrate lyase activity and a carboxypeptidase activity, neither of which is present in the dry seed nor in the tissue during embryogenesis. Isocitrate lyase was chosen because of the requirement in germination that the cotyledon tissue convert much of its stored lipid nutritional reserves into carbohydrate to support the growth of the growing plantlet axis. (The mature cotton embryo has little or no starch.) The proteolytic enzyme was chosen because it could participate in the degradation of the stored protein granules in the cotyledon cells during germination, but conceivably would be deleterious to the process of protein-granule deposition that occurs during embryogenesis. We found that the appearance of both enzyme activities during germination was not influenced by actinomycin D, but was completely inhibited by cycloheximide (Ihle & Dure, 1969, 1970, 1972*b*).

We subsequently purified the carboxypeptidase to homogeneity and demonstrated that its appearance during germination is the result of its synthesis *de novo* from free amino acids (Ihle & Dure, 1972*a*). Further, we were able to demonstrate the point in embryogenesis at which the existence of the mRNA for these two enzymes first becomes demonstrable in cotyledon tissue (Ihle & Dure, 1970, 1972*b*). This was possible because of the ability of the immature cotton embryo to germinate precociously when excised from its surrounding ovule tissue and placed on agar gel or wet filter paper. Visually this precocious germination appears identical with the normal germination of mature seeds. The shoot and root axis elongates, the cotyledons unfurl, expand and green when exposed to light, and the activities of isocitrate lyase and carboxypeptidase develop in the cotyledons. By precociously germinating successively younger embryos in the presence of actinomycin D, we found that the appearance of the two enzyme activities (and precocious germination in general) becomes sensitive to the inhibitor at a point when the embryo weighs about 85 mg (about 65% of its final size). Since about 20 additional days are required for the cotton embryo to reach maturity inside the ovule on the mother plant, the translation of the mRNA for the two enzymes is obviously repressed in some manner during this period.

Since two enzyme activities, both of which appear to be unique to germination, appear to be regulated in an identical manner, we proposed that the synthesis of a number of enzymes that bring about germination are identically regulated (Ihle & Dure, 1972*b*). We consider that isocitrate lyase and carboxypeptidase are representative of this group of germination enzymes.

At this point the following questions could be asked of this system.

1. What prevents the translation of the mRNA for these enzymes during the last 20 days of embryogenesis on the mother plant?
2. How is this delay in translation overcome in germination, both normal and precocious?
3. What induces the transcription of the mRNA for these germination enzymes and the cessation of DNA synthesis and cell division at the 85 mg stage of embryogenesis?
4. What are the physical and chemical properties of the storage form of this body of mRNA?

We explored all these parameters and have been able to expand on several of them. The key to making further observations on how the translation of the mRNA is prevented during the last 20 days of embryogenesis was the finding that the rate of synthesis of the two enzymes during precocious germination was actually stimulated by actinomycin D and that it could be stimulated to the same extent by vigorously washing the embryos in water before placing them in germination dishes. This stimulation of the translation of the mRNA for these two enzymes suggested (*a*) that a translation inhibitor existed in the seed tissues during the last 20 days of embryogenesis, (*b*) that its action required continued RNA synthesis to explain the translation stimulation by actinomycin D, (*c*) that the majority of the inhibitor at any one time existed on the embryo surface rather than inside the embryo, since it could presumably be removed by washing the embryos, and (*d*) that the embryo has the ability to degrade the inhibitor since precocious germination and germination enzyme synthesis take place after the embryos are removed from the surrounding seed tissues.

## Involvement of Abscisic Acid in Inhibition of Translation

These suggestions could be tested and in these experiments we were able to demonstrate that (1) the material washed from the surface of the embryo contained abscisic acid, (2) an aqueous extract of the tissues surrounding the embryo contained abscisic acid, (3) pure abscisic acid (0.1 $\mu$M) completely inhibits precocious germination and the translation of the mRNA for the two enzymes and (4) this inhibition by abscisic acid is completely overcome by actinomycin D (Ihle & Dure, 1970, 1972*b*). We further found that 5-fluorouracil could not replace actinomycin D in overcoming the translation inhibition caused by abscisic acid, suggesting that non-ribosomal RNA synthesis was necessary for the abscisic acid effect (Ihle & Dure, 1972*b*). In addition, neither gibberellic acid ($GA_3$) nor indol-3-ylacetic acid had any effect on the abscisic acid inhibition of enzyme synthesis and precocious germination (Ihle & Dure, 1972*b*). Curiously, we were unable to demonstrate the presence of abscisic acid in extracts of the ovule before the 85 mg stage of development nor in the material washed from embryos that had yet to reach this stage. Embryos smaller than 85 mg will germinate precociously and develop the enzyme activities, but, as has been stressed, all of this is sensitive to actinomycin D. However, these processes in these younger embryos were not stimulated by washing. Therefore, it seems apparent that the production of abscisic acid by the ovule tissues begins at the same time that the transcription of the mRNA for the germination enzymes begins, which also happens to be the same point in embryogenesis that cell division ceases in the embryo.

Since embryos younger than the 85 mg stage will germinate precociously, the transcription of the mRNA for the germination enzymes must be somehow inducible prematurely. By adding actinomycin D to the germination medium of young embryos at successively later time-points during precocious germination we were able to show that at least some of the mRNA for the germination enzymes is produced within 24 h after excision of the embryo from the boll. Abscisic acid does not inhibit the premature induction of the transcription of the mRNA for the germination enzymes that is brought about by precocious germination, but does prevent its translation. Subsequently, we learned that the embryos do not have to be excised from the ovule for this transcription to be induced. Merely removing the boll from the mother plant will induce the mRNA within 24 h (Ihle & Dure, 1972*b*).

Cell division is stopped in the cotyledons of embryos weighing less than 85 mg when they are precociously germinated as shown by the fact that the final cotyledon size attained is progressively smaller with successively younger embryos in precocious germination.

Thus a number of developmental events are seen to occur at about the 85 mg stage of development (30 days after anthesis), some of which can be caused to occur prematurely by excising younger embryos and germinating them precociously. The factor common to both the normal induction of the transcription of the mRNA for the two enzymes and the cessation of cell division that occur at the 85 mg stage and to the premature induction of these events appears to be the rupturing of the vascular connexion between the incipient seed and the mother plant.

We have observed that this rupture takes place normally when embryos reach the 85 mg stage and is caused by the degeneration of the funiculus that attaches the incipient seed to the placenta of the cotton boll. This fact suggests that the developmental changes that occur normally at the 85 mg stage and prematurely in dissected embryos result in part from the loss of regulatory molecules that are components of the vascular flow from the mother plant. The cotton boll itself is apparently not the source of these components, since the mRNA for the two enzymes is induced in cotyledons of embryos smaller than 85 mg when the boll is removed from the mother plant but the embryos allowed to remain undisturbed in the boll.

Based on these data a temporal map of developmental events in the ontogeny of cotton cotyledons can be constructed (Scheme 1). In this scheme the putative maternal factors are considered responsible for maintaining embryo growth by the production of new cells. The loss of these factors resulting from funiculus degeneration is shown to result in a cessation of embryonic growth and a redirection of effort towards a preparation for germination as demonstrated by the transcription of the germination mRNA. However, concomitant with this change in cotyledon development, abscisic acid synthesis is elicited in ovule tissue and the absorption of this compound by the cotyledons prevents the precocious germination of embryos inside the cotton boll during the final stages of embryo-

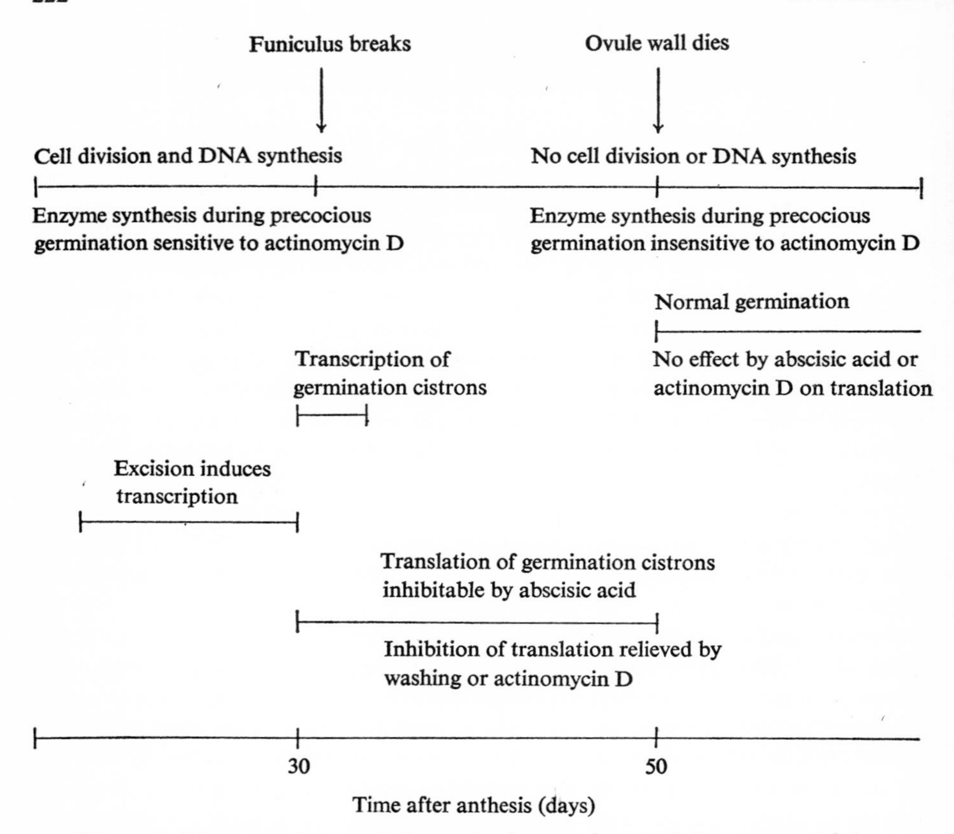


Scheme 1. *Diagrammatic resumé of some developmental events during cotton cotyledon embryogenesis*

genesis. (The synthesis of storage protein and lipid reserves continues during this period unaffected by abscisic acid, and this must be true of other maturation processes as well.) Thus, abscisic acid is considered here as preventing vivipary, which in the case of cotton and other similar plants would seem to be a lethal process. With the death of the remaining ovule tissue to form the sclerified seed coat, the source of the abscisic acid is lost and normal germination takes place when seeds that have reached this stage of development are planted. Vivipary at this point is prevented by the rapid desiccation of the embryo which is no longer protected by living ovule tissue. This scheme of developmental events, although consistent with our results, undoubtedly represents a somewhat over-simplified and naive concept of these developmental processes. Yet it serves as a point of departure for further experimentation.

Although we have implicated the severing of the vascular connexion between the mother plant and the ovule with the cessation of cell division and the induction of the transcription of the germination mRNA, and have implicated the synthesis of abscisic acid in the ovule tissues with the inhibition of the translation of the germination mRNA, we have no further observations on the mode of action of the putative vascular factors nor of abscisic acid. Apparently synthesis of non-ribosomal RNA must continue for abscisic acid to bring about the inhibition of the germination mRNA, but this in itself does not necessarily imply that abscisic acid acts to de-repress specific cistrons.

**Possible Involvement of tRNA in the Regulation of Translation**

Once we had established the sequence of events for the regulation of the appearance of the germination enzymes, we undertook to study the possible involvement of the tRNA-pool composition in the inhibition of germination mRNA translation. At that time the idea of 'codon restriction' (Strehler *et al.*, 1967) as a means of regulating the synthesis of specific groups of proteins enjoyed a great deal of popularity. To study the involvement of tRNA in separating embryogenesis from germination, we examined the tRNA-pool composition of cotyledons of embryos that were less than 85 mg wet weight, of dry seed embryos and of germinated seedlings (both greened and etiolated). To compare the tRNA pools of cotyledons from these developmental stages with a distinctly different tissue the tRNA-pool composition of roots was also studied. Further, to identify tRNA species that originated from the chloroplasts the tRNA-pool composition of partly purified cotton chloroplasts was also studied.

The composition of the tRNA pool was examined in two ways. First the amount of total tRNA that accept each of the 20 amino acids was determined for each of the tRNA preparations, and, second the number and relative amounts of iso-accepting tRNA species were determined for several of the amino acids. By means of these measurements we compared the amounts of tRNA specific for the individual amino acids between the several tRNA preparations. The cotyledon tRNA preparation from embryos smaller than 85 mg represented the tRNA pool before the time of synthesis of germination mRNA, the preparation from dry seed cotyledons represented the pool during the period when the germination mRNA is in existence but not being translated, and the preparation from germinating seedling cotyledons represented the pool at the time of extensive translation of the germination mRNA.

By separating tRNA acylated with a specific amino acid into its iso-accepting species we hoped to visualize the appearance, disappearance or change in relative amounts of individual iso-accepting species that was coincident with the developmental changes at the 85 mg stage or at the commencement of germination. In these types of experiments a large number of control parameters must be established, ranging from proof to total extraction of the tissue tRNA pool to proof of the intactness of the primary structure of the tRNA molecules. Only with a great deal of effort were we able to detect all the artifacts inherent in these types of experiments and satisfy all the requisite parameters (Merrick & Dure, 1971, 1972).

Table 1 shows the composition of the tRNA pools of the cotyledons at the various developmental stages and of the tRNA pools of roots and partly purified chloroplasts with respect to the relative amounts of tRNA for the 20 amino acids. It should be noted that over 90% of the tRNA in these preparations could be aminoacylated. The salient feature of these results is that essentially no difference in the composition of the tRNA pool was found between young embryo cotyledon tRNA and dry seed cotyledon tRNA, and, further, both these pools were essentially identical with that of root tRNA. (There are two exceptions to this; the low amount of asparagine tRNA in young embryo cotyledons and of serine tRNA in roots.

Table 1. *Percentage of tRNA charged with each amino acid*

tRNA was prepared from the tissues indicated and charged to completion with each amino acid as described by Merrick & Dure (1972).

| Amino acid | Source of tRNA | | | | | |
|---|---|---|---|---|---|---|
| | Young embryo cotyledons | Dry seed cotyledons | Roots | Green cotyledons | Etiolated cotyledons | Chloroplasts |
| Ala | 5.1 | 5.1 | 5.0 | 4.8 | 4.7 | 4.7 |
| Arg | 9.0 | 8.7 | 8.7 | 9.3 | 9.3 | 9.5 |
| Asn | *1.4* | 2.5 | 2.5 | 1.3 | 1.4 | 2.1 |
| Asp | 6.8 | 6.5 | 6.6 | 6.1 | 6.0 | 5.0 |
| Cys | 0.8 | 0.7 | 0.8 | 0.7 | 0.8 | 0.9 |
| Gln | 0.2 | 0.2 | 0.2 | 0.2 | 0.2 | 0.2 |
| Glu | 2.0 | 2.1 | 2.3 | 2.2 | 2.3 | 3.0 |
| Gly | 10.0 | 10.1 | 10.3 | 9.7 | 9.6 | 8.6 |
| His | 3.4 | 3.3 | 3.3 | 3.7 | 3.7 | 3.6 |
| Ile | 3.2 | 3.2 | 3.3 | 3.3 | 3.4 | 4.4 |
| Leu | 10.0 | 9.8 | 10.2 | 11.2 | 11.0 | 11.1 |
| Lys | 5.2 | 5.3 | 5.5 | 3.6 | 3.7 | 4.0 |
| Met | 3.5 | 3.3 | 3.5 | 4.7 | 4.7 | 5.8 |
| Phe | 4.6 | 4.6 | 4.7 | 5.0 | 5.1 | 6.7 |
| Pro | 4.6 | 4.6 | 4.4 | 4.1 | 4.0 | 3.2 |
| Ser | 3.4 | 3.2 | *1.3* | 3.8 | 3.7 | 4.1 |
| Thr | 5.6 | 5.6 | 5.7 | 5.4 | 5.5 | 4.7 |
| Trp | 2.1 | 1.9 | 1.9 | 2.3 | 2.3 | 1.9 |
| Tyr* | 2.7 | 2.7 | 2.7 | 2.7 | 2.7 | 3.0 |
| Val | 9.0 | 8.9 | 9.0 | 8.8 | 8.6 | 6.6 |
| Total | 92.6 | 92.3 | 91.9 | 92.9 | 92.7 | 93.1 |

* Estimated from charging with *Escherichia coli* synthetases as explained by Merrick & Dure (1972).

At present, we consider both of these exceptions to be artifactual.) The composition of the tRNA pool appears to change somewhat during cotyledon germination, but we have found that these changes result entirely from the 7- to 8-fold increase in the chloroplast tRNA pool that occurs during germination. Table 1 shows that the chloroplast tRNA pool has a composition distinct from the others.

This constancy in the cytosol tRNA pool of cotyledons during development is shown more dramatically in Figs. 1 and 2 in which the number and relative amounts of the iso-accepting tRNA species for several amino acids have been visualized. Figs. 1 and 2 show the elution profiles from columns of DEAE- and CM-cellulose of [$^{14}$C]aminoacyl-oligonucleotides resulting from the digestion of [$^{14}$C]-aminoacyl-tRNA with ribonuclease $T_1$. We encountered a number of artifacts by using the conventional reversed-phase column separation of acylated iso-accepting tRNA species, such as poor recovery of radioactivity from the column and the generation of false species by nuclease nicking of tRNA during acylation. Consequently we utilized the substituted cellulose column separation of ribonuclease $T_1$-digested aminoacyl-tRNA to visualize iso-accepting species.

Ribonuclease $T_1$ specifically cleaves polynucleotides at the site of guanine residues producing guanosyl 3′-phosphate at the 3′ termini. Thus, the digestion of [$^{14}$C]aminoacyl-tRNA with this enzyme produces only one radioactive oligonucleotide fragment from each aminoacyl-tRNA molecule, i.e. the fragment

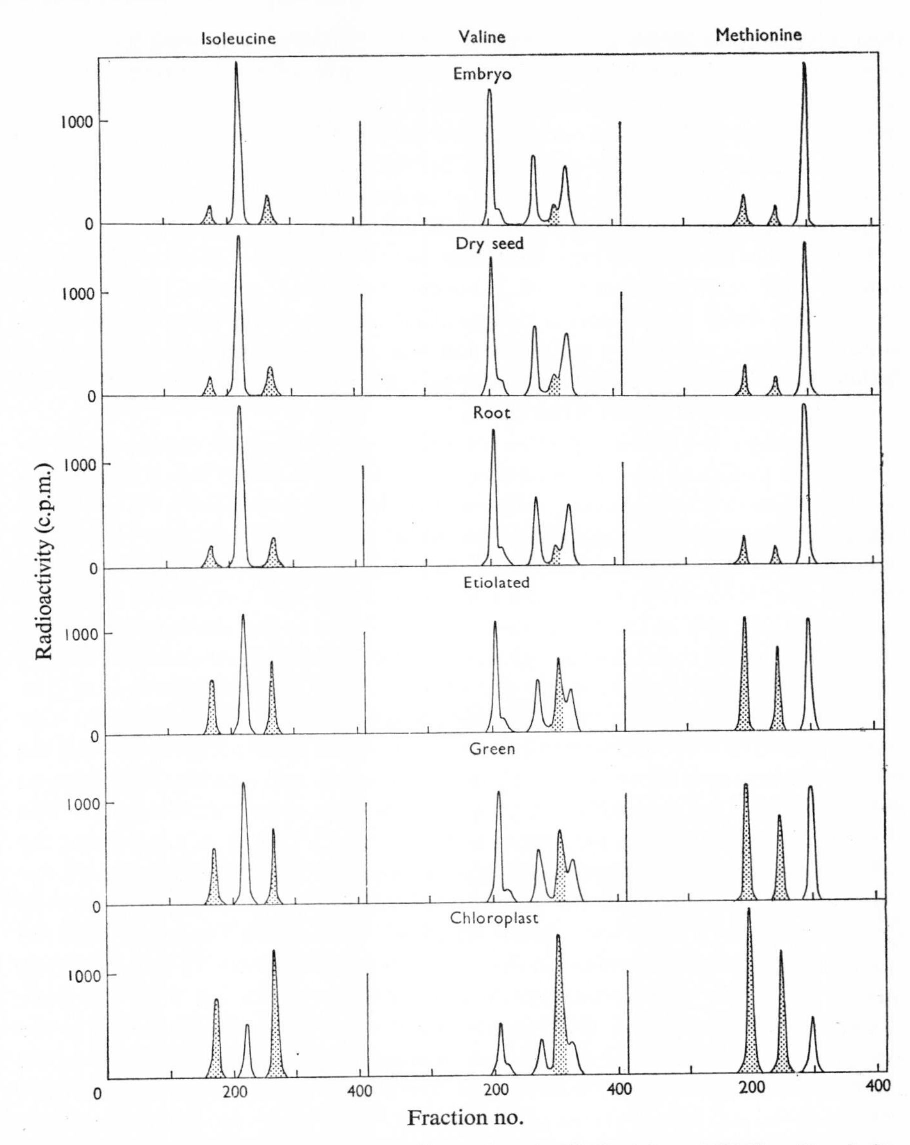


Fig. 1. *Radioactivity elution profiles of [$^{14}$C]isoleucyl-, [$^{14}$C]valyl- and [$^{14}$C]methionyl-oligonucleotides from DEAE-cellulose columns*

tRNA from the cotyledons of the tissues indicated and from roots and chloroplasts was charged with the indicated $^{14}$C-labelled amino acid by a cotton cotyledon enzyme preparation and [$^{14}$C]-aminoacyl-oligonucleotides were prepared from the charged tRNA by digestion with ribonuclease $T_1$ and chromatographed as described by Merrick & Dure (1972). Shaded elution peaks represent aminoacyl-oligonucleotides from chloroplast aminoacyl-tRNA.

containing the amino acid. The nucleotide length and composition of this fragment depends on the position of the guanine residue nearest the CpCpA-amino acid terminus of the molecule. Iso-accepting tRNA species that produce different aminoacyl-oligonucleotides in the digestion can be distinguished and their relative concentrations measured if the [$^{14}$C]aminoacyl-oligonucleotides are chromatographed on DEAE- or CM-cellulose at pH 4.5 without urea. This pH is used to produce about half a net positive charge on cytosine residues and a slight net positive charge on adenine residues. Urea is omitted in this procedure to allow maximum interaction between the nucleoside residues and the column matrix, which promotes the separation of aminoacyl-oligonucleotides that differ in nucleotide composition but not in nucleotide number.

There is obviously a serious limitation to this technique in ascertaining the number and relative amounts of iso-accepting tRNA species. Iso-accepting species that differ in nucleotide composition only in other parts of the polynucleotide chain, including the anticodon region, will generate the same aminoacyl-oligonucleotide fragment on digestion with ribonuclease $T_1$ and thus will not be distinguished by this technique.

Fig. 1 shows the elution profiles of isoleucyl-, valyl- and methionyl-oligonucleotides produced by the ribonuclease $T_1$ digestion. With every tRNA preparation three different isoleucyl-oligonucleotides were produced from isoleucyl-tRNA. These isoleucyl-oligonucleotides indicate that at least three isoleucine tRNA species exist in these tissues and that their relative concentrations are the same in the tRNA from the young embryo and dry-seed cotyledons and from roots. Further, two of these species increase relative to the third during germination, in etiolated cotyledons as well as in green cotyledons, to the extent that they comprise about 50% of the isoleucine tRNA, after 5 days of germination. The same two species are also the predominant species in chloroplast tRNA. Our interpretation of these changes in amounts of isoleucine iso-acceptors is that the species that increase in cotyledons during germination and which predominate in the tRNA from partly purified chloroplasts are chloroplast tRNA species. The alleged chloroplast species are shaded in Figs. 1 and 2. It may be argued that the isoleucyl-oligonucleotides that were found to increase in relative amount are not the result of an increase in the amounts of existing species, but represent the appearance *de novo* of two new isoleucine tRNA species that happen to yield the same isoleucyl-oligonucleotide on digestion with ribonuclease $T_1$ as do existing species. However, the chromatography of isoleucyl-tRNA on reversed-phase columns (Merrick & Dure, 1972) shows that the two minor species found in the dry-seed cotyledons are the species that increase in amount during germination and that are concentrated in chloroplasts. The elution profiles of isoleucyl-tRNA from this reversed-phase column also show that two cytoplasmic isoleucine tRNA species rather than one exist in these tRNA preparations. These two cytoplasmic species apparently produce identical isoleucyl-oligonucleotides on digestion with ribonuclease $T_1$, and thus appear as one species on the DEAE-cellulose column-elution profile.

The information from the two types of chromatography taken together shows that two chloroplast and two cytoplasmic isoleucine tRNA species exist in all the tissues examined (including the young embryo cotyledons and roots) and that the chloroplast species increase markedly during germination. We have found that the amount of tRNA/cell increases approx. 80% in cotyledons during germination, which supports our contention that the amounts of chloroplast species increase per cell during this period, as distinct from there being a decrease in the amount

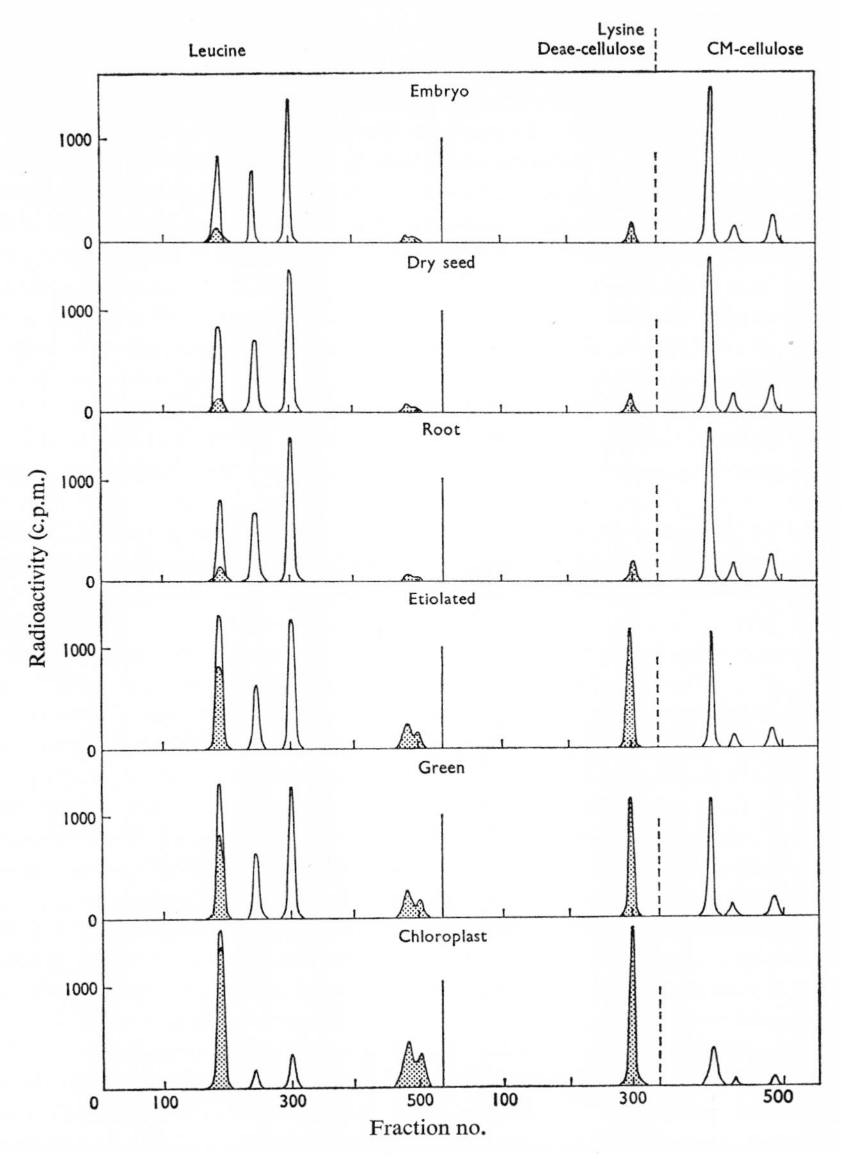


Fig. 2. *Radioactivity elution profiles of* [$^{14}$C]*leucyl- and* [$^{14}$C]*lysyl-oligonucleotides from DEAE- and CM-cellulose columns*

Experimental details were as in the legend to Fig. 1 and in Merrick & Dure (1972).

of cytoplasmic species per cell. However, there appears to be no difference in the amounts of the two cytoplasmic species relative to each other nor in the quantitative relationships between the two chloroplast species in any of the tRNA preparations. (Further, the increase in the chloroplast species is shown not to be light-dependent.) The quantitative identity of the DEAE-cellulose column profiles obtained with young embryo and dry-seed cotyledon tRNA and root tRNA shows that not only do these tissues have the same amounts of total isoleucine tRNA relative to the total tRNA (Table 1), but that they have the same amounts of iso-accepting isoleucine tRNA species.

The same pattern observed for isoleucine tRNA species is reiterated in the case of valine tRNA and methionine tRNA in Fig. 1. There are five valyl-oligonucleotides, one of which appears to be derived from chloroplast tRNA by the criteria given above. Four code words exist for valine and, if chloroplast and cytoplasmic protein synthesis are mutually exclusive in their use of tRNA, and if chloroplast protein synthesis uses the entire genetic code, there should exist at least one other chloroplast valine tRNA. Unfortunately, the reversed-phase column did not resolve valyl-tRNA well enough to determine if the chloroplast valyl-oligonucleotide seen in the DEAE-cellulose column-elution profile was produced by two chloroplast valine tRNA species. Here again the amounts of the cytoplasmic valine tRNA species were found not to change relative to one another during cotyledon development nor to differ from that found in root tRNA.

DEAE-cellulose column chromatography of methionyl-oligonucleotides showed two chloroplast species and one cytoplasmic species. We have found that the methionyl-tRNA that produces the middle methionyl-oligonucleotide elution profile can be formylated to form *N*-formylmethionyl-tRNA by an endogenous cotton enzyme and by *Escherichia coli* transformylase (Merrick & Dure, 1971). This fact further substantiates our assumption that species that increase during germination and that are concentrated in the chloroplast tRNA preparation are chloroplast species, since formylmethionyl-tRNA has been shown to initiate chloroplast but not cytoplasmic protein synthesis in plants. There should be at least two cytoplasmic methionine tRNA species also; one for polypeptide-chain initiation and another for the internal positioning of methionine in translation, but we were not able to demonstrate more than one by reversed-phase chromatography.

Fig. 2 shows the elution profiles of leucyl-oligonucleotides from the DEAE-cellulose column and lysyl-oligonucleotides from the DEAE- and CM-cellulose columns. Five leucyl-oligonucleotides were resolved indicating the existence of at least five leucine tRNA species. However, when these preparations of tRNA were charged with an *E. coli* synthetase preparation and then digested with ribonuclease $T_1$, only a portion of the leucyl-oligonucleotide eluting first from the DEAE-cellulose column was formed. This portion was larger in tRNA from germinated cotyledons than from younger cotyledons and roots and was almost equivalent to that formed with the cotton synthetase preparation in the chloroplast tRNA. This we interpret as indicating that the leucyl-oligonucleotide eluted first is produced from two leucine tRNA species, one cytoplasmic and one chloroplast, and that the *E. coli* synthetase preparation charges only the chloroplast species. Thus the amount that the chloroplast leucine tRNA species contributes to the leucyl-oligonucleotide eluted first (shaded portion of the first peak in Fig. 2) was obtained with the *E. coli* synthetase preparation. From this there appear to be six leucine tRNA species in cotton, three cytoplasmic and three chloroplast. Assuming that these species contain a 'wobble' capability in their anticodons, three species are sufficient to recognize the six leucine code words.

Also, although the chloroplast species increase in amount per cell during germination, there is no change in the amounts of the cytoplasmic species relative to each other nor in the amounts of the chloroplast species relative to each other during cotyledon development nor between young cotyledon and root tRNA. Only one of the lysyl-oligonucleotides was retained by the DEAE-cellulose column, whereas three others were retained by the CM-cellulose column, indicating that they bear a net positive charge at pH 4.5.

Similar results were obtained for several other amino acids (Merrick & Dure, 1972).

Thus these column profiles reinforce the results in Table 1 which predict that there is no pattern of change in the tRNA population as the cotyledon tissue develops and matures, except for the increase in the amount of chloroplast tRNA in the tissue that takes place during germination. The amount of tRNA that accepts each amino acid does not appear to change in cotyledons in their development before germination, and the changes that occur during germination (Table 1) can be attributed to the increased contribution of chloroplast species. There is no cell division in cotyledons during this period and the increase in tRNA/cotyledon that occurs during germination reflects the increase in chloroplast tRNA/cell. This increase in chloroplast tRNA is in step with the increase in chloroplast rRNA (and presumably in chloroplast ribosomes) that occurs during germination. There is about a 7-fold increase in chloroplast rRNA during this period (W. C. Merrick, J. N. Ihle & L. S. Dure, unpublished work), and a commensurate increase in the amount of chloroplast tRNA species is shown in Figs. 1 and 2. We have found that the number of tRNA molecules/ribosome (between 14 and 15) remains constant in both the cytoplasm and chloroplasts during cotyledon development (W. C. Merrick, J. N. Ihle & L. S. Dure, unpublished work).

However, it must be emphasized that our procedure for distinguishing isoaccepting species by the chromatography of ribonuclease $T_1$ digests does not allow modifications of tRNA species by methylation, alkylation or other processes to be manifested, since these modifications are not found on the CpCpA-containing stem. It is conceivable that certain tRNA species become more or less modified during cotyledon development, without influencing the DEAE- or CM-cellulose column-elution profiles of the aminoacyl-oligonucleotides. However, these modifications are not thought to influence directly the capacity of the cell for translating specific code words.

In fact, the idea that a change in tRNA population may play a developmental regulatory role directly by altering the rate of translation of specific code words, thereby altering the amounts of specific proteins, appears to be a poor one. Implicit in this idea is the assumption that synonym code words do not occur randomly in mRNA, but are used with a frequency that is 'in phase' with the change in tRNA population so that the proper change in translation rates of certain mRNA species takes place. Such a restriction imposed on code-word usage would appear incompatible with the idea that the genetic code is highly evolved towards the prevention of lethal mutations. That is, a single base change in DNA could conceivably alter greatly the amounts of specific proteins, if their amounts were regulated by code-word frequency and tRNA population. Perhaps a more likely regulatory role for tRNA molecules would be in the allosteric regulation of steady-state processes that serve to control the internal concentration of cell constituents. In this context, the modification of individual tRNA species by methylases and other enzymes could be considered to regulate the allosteric effectiveness of the molecule.

Although we were not able to relate 'codon restriction' to the prohibition of the translation of the germination mRNA imposed by abscisic acid, we were able to make some notable observations about chloroplast tRNA during cotton cotyledon development. The chromatography of iso-accepting species shows that chloroplast species exist in very young embryo cotyledons and in roots in about the same amounts. It is not surprising that chloroplast tRNA, and presumably the entire chloroplast protein-synthesizing system, exists in the young embryo. Conceptually, it would seem that this system must perpetuate itself during the reproductive cycle, otherwise its re-establishment would have to be carried out in part by the cytoplasmic system. We have no evidence that the tRNA species that we consider to be in the chloroplasts are transcribed from chloroplast DNA. However, it has been found that 35% of the leucyl-tRNA of bean leaves hybridizes specifically with bean leaf chloroplast DNA (Williams & Williams, 1970), which is approximately the amount of chloroplast leucine tRNA that we have found in germinated cotton cotyledons (Fig. 2). The observation that chloroplast tRNA exists in roots strengthens the long-held contention that root amyloplasts are derived from the same proplastid progenitor as are chloroplasts. Finally the results show that the large increase in chloroplast tRNA that occurs during germination does not require an induction by light. In this respect cotton cotyledons differ from chloroplast-bearing unicellular organisms such as *Euglena* in which the induction of chloroplast tRNA synthesis is in response to a light stimulus (Barnett *et al.*, 1969).

**Characterization of the Stored mRNA**

Since tRNA is apparently not involved in the regulation of the translation of the germination mRNA, we next turned our attention to the physical and chemical state of the germination mRNA during the last part of embryogenesis when its translation is prohibited. On the basis of the vast literature about the state of mRNA as it passes from nucleus to cytoplasm, we assumed that the storage form of this body of mRNA would be as ribonucleoprotein particles that had sedimentation properties somewhere between 25S and 60S. Since we can induce the premature transcription of the germination mRNA by precociously germinating embryos smaller than 85mg, we felt we could radioactively label and isolate such putative particles on sucrose density gradients. Since the mRNA is not translated in the presence of abscisic acid we believed that we could stabilize these specific radioactive particles containing the germination mRNA by adding abscisic acid to the germination medium. On removal of the abscisic acid or on addition of actinomycin D the particles should disappear and their radioactive RNA become associated with polyribosomes. We further hoped that by isolating and characterizing ribonucleoprotein particles with these properties we might gain some idea of what aspect of translation initiation is blocked by the presence of abscisic acid. To date we have only made a start in this direction, but some of the results are curious and perhaps encouraging.

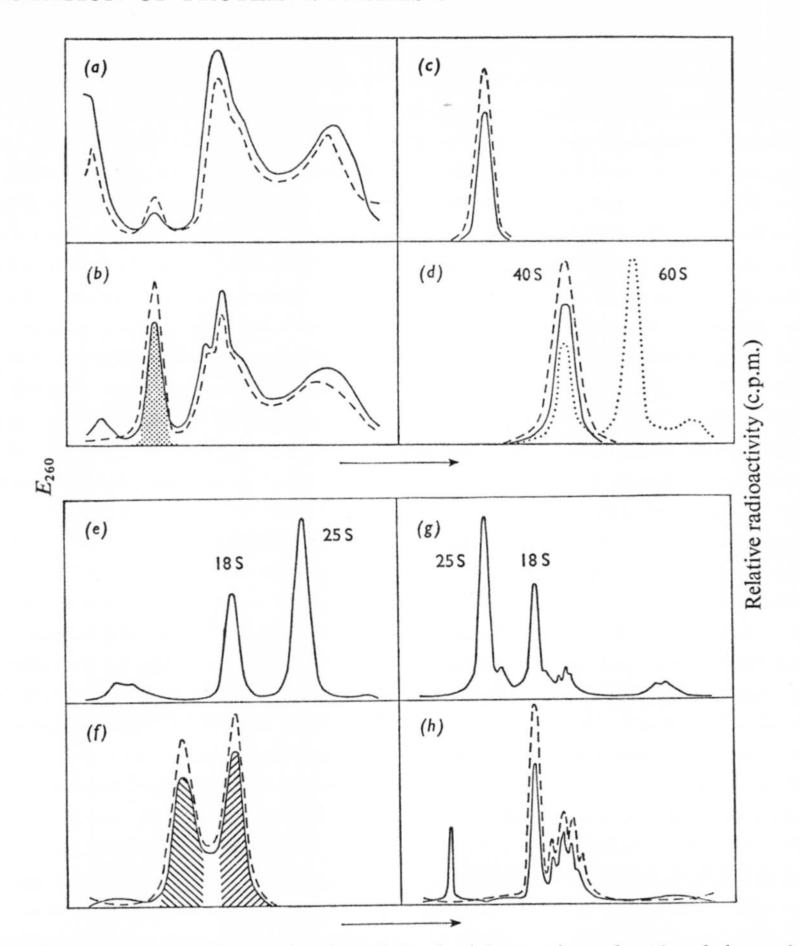


Fig. 3. *Sucrose-density-gradient and sodium dodecyl sulphate–polyacrylamide-gel electrophoresis extinction and radioactivity profiles of various preparations from the cotyledons of* 70 *mg cotton embryos precociously germinated for* 4 *days*

(*a*) Sucrose-density-gradient separation of crude cotyledon supernatant. (*b*) Sucrose-density-gradient separation of a resuspended high-speed centrifugation pellet from the crude cotyledon supernatant. Shaded area denotes fractions pooled. (*c*) Sucrose-density-gradient separation of pooled fraction from (*b*). (*d*) Sucrose-density-gradient separation of pooled fractions from (*b*) treated with EDTA. . . . ., Extinction profile of ribosomal subunits produced by treatment of cotton ribosomes with EDTA. The commonly ascribed sedimentation values of ribosomal subunits are indicated. (*e*) Sucrose-density-gradient separation of RNA from cotton ribosomes. The commonly ascribed sedimentation values of rRNA species are indicated. (*f*) Sucrose-density-gradient separation of RNA from pooled fractions in (*b*). The shaded areas indicate the fractions pooled for the determination of molar base ratios. (*g*) Sodium dodecyl–polyacrylamide-gel electrophoretogram of RNA from cotton ribosomes. The commonly ascribed sedimentation values of rRNA are indicated. (*h*) Sodium dodecyl sulphate–polacrylamide-gel electrophoretogram of RNA from pooled fractions from (*b*). ——, $E_{260}$;– – – –, radioactivity.

To test these possibilities we precociously germinated 70 mg embryos in [$^{32}$P]P$_i$ or [$^{14}$C]orotic acid for 2 days followed by a 2-day 'chase' period during which the precocious germination medium contained abscisic acid. When the cotyledons from these embryos were homogenized, the homogenate was centrifuged briefly at 5000*g* and samples of the resultant supernatant were sedimented through a sucrose density gradient, the extinction and radioactivity profiles shown in Fig. 3(*a*) were found. This sucrose-density separation was designed to sediment the monomeric 80 S ribosomes half-way through the gradient so as to expose more of the material that sediments between 25 S and 60 S. The radioactivity profile shown is the radioactivity that binds to a Millipore filter so as to restrict it to radioactivity contained in particles. Fig. 3(*a*) shows a small peak of extinction that appears in the 40–50S region of the gradient and has a somewhat higher specific radioactivity than the material found in the ribosome–polyribosome region of the gradient. Since this small peak seemed to have some of the characteristics we assumed we should be looking for, we undertook to purify it further and characterize its RNA. We were able to achieve an enrichment of this material simply by centrifuging the crude supernatant for 5 h at 100000 *g*, and by collecting the sediment. This sediment, of course, contains the ribosome–polyribosome complement of the tissue. The ribosomes and polyribosomes from cotton cotyledons will not go back into suspension in their entirety after being sedimented into a compact pellet, whereas the material we were interested in readily goes back into suspension in a solution containing 0.2 M-KCl and 2% Triton X-100. Thus when the resuspended pellet was subjected to the same sucrose-density-gradient centrifugation as used in Fig. 3(*a*), most of the ribosomes and polyribosomes sedimented to the bottom of the tube and the 40–50S material was found to be a dominant peak in both the extinction and radioactivity profiles (Fig. 3*b*). The material in this region of the gradient (shaded in Fig. 3*b*) was pooled from several gradients and used for further characterization.

When the material was resedimented through the same type of sucrose-density gradient as used in Figs. 3(*a*) and 3(*b*), it did not give rise to material of greater or lesser sedimentation rates and appeared to be a relatively homogeneous collection of particles at this stage (Fig. 3*c*). To test the possibility that the particle dissociates when removed from solutions containing 0.01 M-$Mg^{2+}$ in which it had been handled up to this point, a sample of the material was mixed with EDTA (final concentration 0.02 M). After being stirred for 10 min, the mixture was centrifuged through a sucrose density gradient designed to separate the 40 S and 60 S ribosomal subunits. A sample of 80 S monomeric cotton ribosomes was treated in an identical manner and the resulting profiles from both centrifuge tubes are shown in Fig. 3(*d*). The dotted line gives the extinction profile of the ribosomal subunits obtained from the EDTA treatment of the monomeric ribosome. The extinction and radioactivity profiles of the particle preparation show that no further dissociation takes place in the presence of EDTA and the particle appears to sediment in a manner identical with that of the 40 S ribosomal subunit. In fact it would seem at this point that the particle itself is simply the small ribosomal subunit. However, two attributes of the particle were puzzling: (1) its specific radioactivity was greater than that of the ribosomes from the precociously germinating tissue, and (2) there was no evidence in the profiles (Figs. 3*a* and 3*b*) of a commensurate amount of the 60 S ribosomal subunit.

To compare the RNA of the particle with ribosomal RNA, the RNA of the particle preparation and of monomeric 80 S ribosomes was purified by conventional sodium dodecyl sulphate–phenol procedures and subjected to sucrose-density-gradient centrifugation and sodium dodecyl sulphate–polyacrylamide-gel electrophoresis. Fig. 3(*e*) shows the extinction profile of rRNA and Fig. 3(*f*) shows the extinction and radioactivity profiles of the particle RNA from identical sucrose-density-gradient centrifugations. Fig. 3(*e*) shows the expected 18S and 25S rRNA species from the ribosomes. Two types of RNA were found in the putative particle (Fig. 3*f*), they appear to be present in roughly equal amounts, to have roughly the same specific radioactivities, and to have sedimentation values of about 12S and 18S.

At this point it seemed that the particle isolated from the gradient (Fig. 3*b*) was actually a mixture of two types of particle both of which have sedimentation values of about 40S (Fig. 3*d*), but each having different molecular-weight RNA species. The particle with the 18S RNA would seem to be the 40S ribosomal subunit, gel electrophoresis of the RNA also indicated this. Fig. 3(*g*) shows the extincton profile of the RNA from the cotton ribosomes and Fig. 3(*b*) shows the extinction and radioactivity profiles of the particle RNA. Some DNA is often present in this fraction (Fig. 3*h*) (it is not radioactive since precocious germination stops cell division in the cotyledons as has been pointed out). Fig. 3(*h*) also shows that the RNA from the particle fraction contains 18S RNA and RNA with sedimentation values of between 10S and 14S. Thus the 12S RNA seen on the sucrose density gradient (Fig. 3*f*) is shown by gel electrophoresis to be composed of a range of molecular weights. Again the amount of 18S RNA is found to be roughly equivalent to the 10–14S RNA and to have the same specific radioactivity.

Thus, we believed that we had isolated two types of particle from the gradient fractions shown in Fig. 3(*b*), the 40S ribosomal subunit and another group of particles that each appeared to contain a molecule of RNA with a sedimentation value of between 10S and 14S.

Since we had previously determined the molar base ratios of the 18S and 25S RNA of the cotton ribosome, we determined these ratios for the 18S RNA and the 10–14S RNA of the particle fraction, confident that the particle fraction 18S RNA would prove to have the same base composition as does 18S rRNA. We obtained the RNA for these determinations from the fractions of the sucrose density gradient shown in Fig. 3(*f*) indicated by the shaded areas. The molar base ratios for the particle RNA species are given in Table 2 and compared with those of the rRNA species. To our surprise the molar base ratio of the particle 18S RNA proved to be quite distinct from that of 18S rRNA. This observation presented many possible interpretations. One of these is that we are indeed dealing with one type of particle in the fraction obtained from the gradient which contains one molecule of 18S RNA and one molecule of the heterogeneous 10–14S RNA. The apparent equivalence between the amount of each type of

Table 2. *Molar base ratios of cotton rRNA species and particle fraction RNA species*

| | AMP | GMP | CMP | UMP |
|---|---|---|---|---|
| rRNA | | | | |
| 18S | 22.9 | 30.1 | 21.8 | 25.1 |
| 25S | 21.2 | 32.4 | 23.4 | 23.0 |
| Particle fraction RNA | | | | |
| 10–14S | 24.0 | 27.8 | 20.7 | 27.4 |
| 18S | 24.1 | 26.6 | 22.1 | 27.2 |

RNA suggests this as does the apparent equality of specific radioactivity between the two RNA types. In this concept the 18 S RNA could be considered to have a transport or storage role and the 10–14 S RNA could be the body of mRNA that we are pursuing.

As yet this line of investigation has not been taken further. Several other observations have been made which, if anything, confuse the true nature of the particle fractions even further. For instance, 5-fluorouracil which has no effect on the induction of the germination enzymes during precocious germination, lowers the incorporation of radioactive isotopes into both the RNA species of the particle fraction. Further, when these 70 mg embryos are precociously germinated in the presence of [*Me*-$^{14}$C]methionine, radioactivity is incorporated equally into the 18 S and 10–14 S RNA species, indicating that both types of RNA are methylated *in vivo*.

We hope in time to establish the identity and function of the substituents of the particle fraction, and should they prove to be the storage form of the germination mRNA, hopefully they will provide a means for coming to grips with the nature of the abscisic acid-mediated inhibition of the translation of the germination mRNA.

## References


Barnett, W. E., Pennington, C. J. & Fairfield, S. A. (1969) *Proc. Nat. Acad. Sci. U.S.* **63**, 1261–1268

Dure, L. S., III & Waters, L. C. (1965) *Science* **147**, 410–412

Ihle, J. N. & Dure, L. S., III (1969) *Biochem. Biophys. Res. Commun.* **36**, 705–710

Ihle, J. N. & Dure, L. S., III (1970) *Biochem. Biophys. Res. Commun.* **38**, 995–1001

Ihle, J. N. & Dure, L. S., III (1972*a*) *J. Biol. Chem.* **247**, 5034–5040

Ihle, J. N. & Dure, L. S., III (1972*b*) *J. Biol. Chem.* **247**, 5048–5055

Loening, U. (1967) *Biochem. J.* **102**, 251–257

Merrick, W. C. & Dure, L. S., III (1971) *Proc. Nat. Acad. Sci. U.S.* **68**, 641–644

Merrick, W. C. & Dure, L. S., III (1972) *J. Biol. Chem.* **247**, 7988–7999

Strehler, B. L., Hendley, D. D. & Hirsh, G. P. (1967) *Proc. Nat. Acad. Sci.* U.S. **57**, 1751–1758

Waters, L. C. & Dure, L. S., III (1966) *J. Mol. Biol.* **19**, 1–27

Williams, G. R. & Williams, A. S. (1970) *Biochem. Biophys. Res. Commun.* **39**, 858–863

*Biochem. Soc. Symp.* (1973) **38**, 235–246
*Printed in Great Britain*

# Cell Extension Growth: Some Recent Advances

By DAVID L. RAYLE* and MEINHART H. ZENK
*Lehrstuhl für Pflanzenphysiologie, Ruhr University*, 463 *Bochum, German Federal Republic*

## Synopsis

The mechanism by which auxin initiates cell extension growth is not fully known; however, progress has been made. Current evidence indicates that auxin acts at the cell surface and initiates the secretion of protons into the cell-wall region. In a manner yet to be established the auxin-induced increase in the acidity of the cell-wall fluid facilitates cell-wall loosening and extension growth. This mechanism is inconsistent with the previously favoured mechanism involving the synthesis of new macromolecules in response to auxin.

Over the last 40 years there has been a considerable amount of research directed towards understanding the mechanism of action of the plant hormone, indol-3-ylacetic acid, yet we still know relatively little about its mode of action in molecular terms. Perhaps one of the major reasons contributing to our rather low rate of progress is that indol-3-ylacetic acid seems to influence so many physiological parameters: protoplasmic streaming (Thimann & Sweeney, 1937), cell elongation (see reviews by Galston & Purves, 1960; Ray, 1969), cell division (Snow, 1935; Söding, 1934, 1936; Wareing *et al.*, 1964), root formation (Thimann & Behnke, 1950), bud inhibition and apical dominance (Thimann & Skoog, 1933, 1934; Sachs & Thimann, 1967) and abscission (La Rue, 1935; Gardner & Cooper, 1943). As these various effects of auxin appear to be physiologically (and presumably biochemically) so diverse, it becomes rather important to select a particular response or set of responses for careful consideration. It is our opinion, that one of the most logical divisions is to separate the responses on the basis of the rapidity of their appearance; some physiological responses to auxin can be detected within minutes whereas others require hours or even days to become apparent. In the past this distinction has not always been made clear and has undoubtedly contributed to some of the confusion concerning the 'mechanism' of auxin action. Obviously, biochemical modifications of the cell which occur some hours after the application of auxin need not necessarily be related to physiological events which take place minutes after application, but rather may be more closely related to the 'long'-term events mediated by the hormone or simply secondary events of no physiological consequence.

Perhaps the best illustration of this problem concerns the effect of auxin on cell elongation and the role of RNA and protein synthesis. It is our prime purpose in this paper to review the evidence linking auxin-induced cell elongation to the synthesis of macromolecules and to show that such biochemical events are not a

* On leave from Department of Botany, California State University at San Diego, San Diego, Calif. 92115, U.S.A.

prerequisite for auxin-induced growth. Secondly, we shall review some of the recent evidence that would seem to place the site of auxin action at the cell surface. Thirdly, we shall consider the relationship between auxin-induced growth and a model system that utilizes $H^+$ ions as the inducing agent. It is possible that this latter system could provide us with important clues about the actual physical nature of the cell-extension response.

We shall begin by briefly summarizing the three principal lines of evidence that lead to the notion that auxin-induced cell elongation was mediated via an activation of gene transcription and specific protein synthesis. First it is known that long-term auxin-induced growth is inhibited by various antibiotics that inhibit RNA and protein synthesis. This phenomenon was clearly shown by Nooden & Thimann (1963), and since then has been repeated and extended by many investigators (see review by Key, 1969). The second line of evidence is derived from data showing that growth-inducing concentrations of auxin can promote the incorporation of labelled precursors into RNA and protein (Key & Ingle, 1964; Ingle *et al.*, 1965; Lin *et al.*, 1966). This effect of auxins has been most extensively studied by using isolated plant sections (see above references; also Key & Shannon, 1964; Masuda *et al.*, 1967; Tester & Dure, 1967). However, there has been some work with isolated nuclei (Roychoudhury *et al.*, 1965; Matthysse & Phillips, 1969) and chromatin (Mathysse, 1968; O'Brien *et al.*, 1968; Holm *et al.*, 1970). In the latter studies it was shown that chromatin isolated from tissues treated with hormone had an increased capacity to synthesize RNA. It has been suggested that auxins might enhance the production of chromatin-associated RNA polymerase (O'Brien *et al.*, 1968). The speed of the auxin effect on RNA metabolism has generally been reported to have a latent time of about 2h (Trewavas, 1968; Davies *et al.*, 1968), however, latent times as short as 10min have been reported from at least one source (Masuda & Kamisaka, 1969). This report has been criticized, however, on the grounds of experimental variation (see discussion in Warner & Leopold, 1971). Thirdly, it is known that new species of protein are made after treatment of various tissues with auxin (Zenk, 1962; Südi, 1964; Venis & Stoessl, 1969; Fan & MacLachlan, 1967). These three lines of evidence, taken together, constituted the prime basis for the popular belief that auxin-initiated cell elongation by the de-repression of genes and induction of mRNA synthesis which in turn resulted in the synthesis of new enzymes and ultimately in the modification of the cell wall thus allowing for cell expansion. Obviously, such a mechanism has a certain intrinsic lore and glamour about it, and possibly this contributed to the rather extended period of favour that this mechanism enjoyed; a period spanning roughly seven years. Eventually, however, this mechanism was seriously challenged. Evans & Ray (1969) performed an elegant set of kinetic studies in which they argued that the characteristics of the latent period (time between the addition of auxin and the onset of elongation) were not consistent with a mechanism involving the synthesis of informational RNA or of enzymic protein. This initial report challenging the existing dogma was followed by similar but more detailed kinetic studies. The strategy behind such kinetic studies is rather simple; it basically entails a careful examination of the initial kinetics of auxin-induced growth with very sensitive instrumentation over wide-ranging conditions such as concentration and temperature. The data obtained can be compared with the known parameters for enzyme induction in animals, plants and bacteria.

We can perhaps best summarize a rather long story by considering the results in Fig. 1, which shows that under certain conditions auxin can initiate an extension response in *Avena* coleoptiles within a matter of seconds [see Nissl & Zenk (1969) and Durand & Zenk (1972) for further details]. Similar results have been obtained by using coleoptiles from rye and corn (Durand & Zenk, 1972). Other authors using different methods have also obtained very short latent times for auxin-induced growth (Rayle *et al.*, 1970*a*; Morath & Hertel, 1973). Since these growth responses are so rapid, one is forced to conclude that auxin does not initiate cell extension via an effect at the gene level, since the induction of protein synthesis by a chemical effector via transcription and translation of DNA is a time-consuming process and is preceded by a characteristic lag phase. This lag phase may last 3–4 min in the case of $\beta$-galactosidase in *Escherichia coli* (Branscomb & Stewart, 1968) to 2h, the shortest lag observed for a substrate-induced enzyme in higher plants (Afridi & Hewitt, 1964). Auxin induces elongation without such a lag. The validity of this conclusion is supported by detailed kinetic data on the effect of auxin on the physical properties of the cell wall. It is generally recognized that the final site of auxin action is at the level of the cell wall and the rate of cell elongation is regulated by the extensibility of the wall matrix [see Cleland (1971*a*) and references therein]. Therefore, since auxin-induced growth can be detected within minutes after its application, it should be possible, with suitable instrumentation, to detect a similar rapid change in the cell-wall properties. Such a

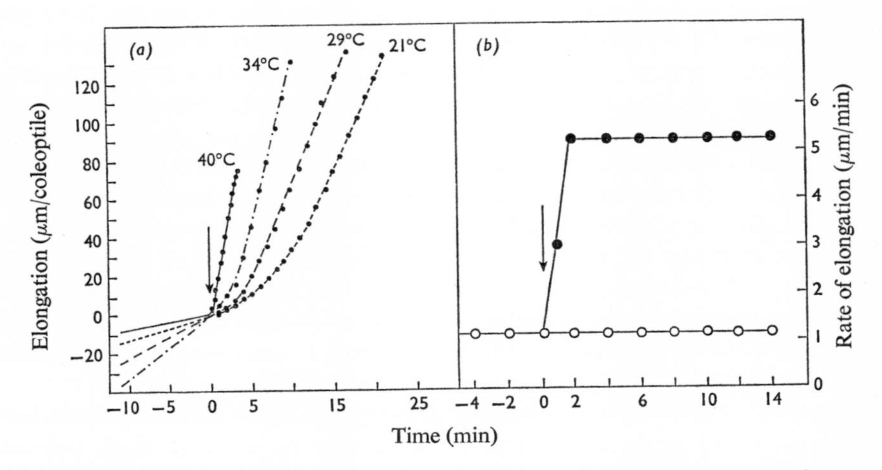


Fig. 1. *Time-course of the auxin response* (*a*) *at different temperatures* (*b*) *plotted as the rate of elongation*

(*a*) The coleoptile cylinders were transferred at the arrow from 0.01 M-$KH_2PO_4$ buffer (pH 4.7) at the temperature indicated, to indol-3-ylacetic acid (5 mM) at the same temperature (from Nissl & Zenk, 1969). (*b*) The cylinders were transferred at the arrow from 0.01 M-$KH_2PO_4$ buffer (pH 4.7) to either 5 mM-indol-3-ylacetic acid (●), or 5 mM-benzoic acid (BA) or 5 mM-3,5-dichlorophenoxyacetic acid (○). The latter two treatments served as a control since these compounds are structurally related to indol-3-ylacetic acid but physiologically inactive. The temperature was maintained at 40°C. (Data from H. Durand and M. H. Zenk, unpublished work).

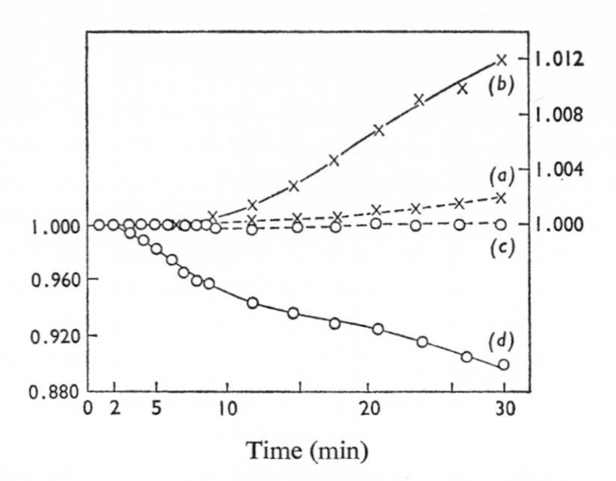


Fig. 2. *Effect of indol-3-ylacetic acid on the length and modulus of Pisum internodes*

At time-zero indol-3-ylacetic was added to *Pisum* internodes at a concentration of 0.1 mM. Values are relative, initially 1.000. (*a*) growth of control (× --- ×); (*b*) growth with auxin (× —— ×); (*c*) modulus of control (○ --- ○); (*d*) modulus with auxin (○ —— ○). Readings were taken every minute up to 9 min, then every third minute (from Burström, 1971).

change in Young's modulus, a parameter related to the elastic properties of the wall (Fig. 2), has been noted [see Burström (1971) and references therein].

A third line of evidence supports the idea that auxin acts on a preformed system. This evidence accrues from studies of the auxin growth response in the presence of cycloheximide (Cleland, 1971*b*; Pope & Black, 1972). In coleoptiles, virtually complete inhibition of protein synthesis is achieved within 5–10 min with cycloheximide at a concentration of 10 p.p.m. Auxin-induced growth, however, can still occur for a limited amount of time in the presence of this inhibitor, i.e. one can inhibit protein synthesis, add auxin, and still induce a growth response which persists for 30–120 min. These results are thought to indicate that auxin somehow activates a preformed protein that is necessary for extension growth. In the absence of further synthesis, this protein is either directly used up in the growth process (a structural protein perhaps), or indirectly exhausted owing to its intrinsic instability. There is little reason to favour one explanation over the other at this time. However, it should be noted that the half-life of the presumed protein (15–80 min) is one of the shortest recorded for plants. This of course seems logical in retrospect since it provides for a finely tuned internal control and nicely explains the earlier reports about the dependence of auxin-induced growth on protein synthesis.

Since some might still argue that the 'growth-limiting proteins' were preferentially synthesized in the presence of cycloheximide by utilizing newly induced mRNA, it is worthwhile to consider a recent investigation with competitive DNA–RNA hybridization (Thomson & Cleland, 1971). In these experiments, reference $^{32}$P-labelled RNA was collected from treated tissue, and its ability to hybridize with DNA in the presence of competing unlabelled RNA obtained from auxin-treated or -untreated tissues was compared. Fig. 3 shows that no difference was observed. Of course, it must be made clear that only certain species of RNA react in hybridization assays (Church & McCarthy, 1968), therefore this particular piece of evidence cannot stand alone as proof that growth-promoting

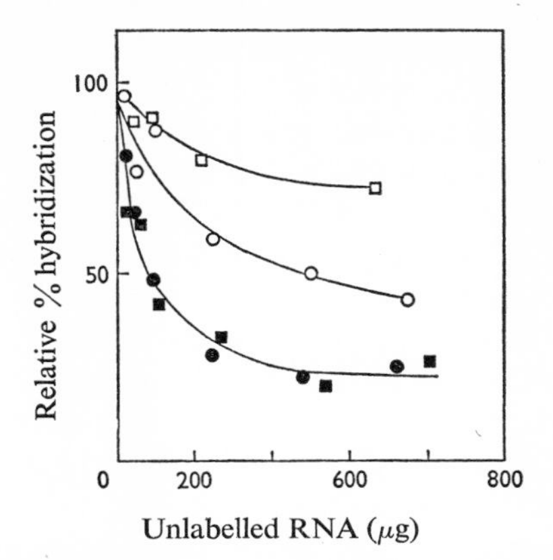


Fig. 3. *Ability of various RNA preparations to compete with the hybridization between* [$^{32}P$]*RNA from auxin-treated pea stem sections and DNA*

Reference RNA (5 μg of [$^{32}$P]RNA) from sections incubated for 6h with 5μM-indol-3-ylacetic acid; competitor RNA species, from pea stem sections incubated for 6h with (■) or without (●) 5 μM-indol-3-ylacetic acid, pea leaves (○) and *Avena* seedlings (□); DNA, 10 μg. The conditions of hybridization were 16h at 67°C in 2 × SSC (15mM-NaCl–1.5mM-sodium citrate buffer, pH 7) (from Thompson & Cleland, 1971).

concentrations of auxin produce no changes in mRNA. Nevertheless, these results coupled with the cycloheximide data, and the fact that auxin can induce growth and cell-wall alternations almost instantaneously, strongly suggests that the primary mode of action of auxin with regard to extension growth is not at the level of gene transcription and translation.

If the primary mode of action of auxin in the case of cell extension is not at the level of the genome, where does it act? Current interest now seems to be centered on the cell membrane as a possible site of auxin attachment and action, and at the present time there seems to be at least some evidence emerging which would not be inconsistent with this view.

Van der Woude *et al.* (1972) have shown that an auxin (2,4-dichlorophenoxy-acetic acid) can increase the activity *in vivo* and *in vitro* of a membrane-bound glucan synthetase in onion. Although the effect is small, (Table 1), these results provide the first direct evidence of an interaction between auxins and membrane-

Table 1. *Stimulation of plasma membrane-associated glucan synthetase by* 2,4-*dichlorophenoxy-acetic acid in vitro*

Plasma membrane fractions were isolated from stem explants. The units of activity of the isolated fraction are nmol of glucose incorporated into total water-insoluble polysaccharide/h per mg of protein. 2,4-Dichlorophenoxyacetic acid (2,4-D) was added to the enzyme assay to test for stimulation of the membrane-associated synthetase *in vitro* (from Van der Woude *et al.*, 1972).

| | Auxin added | | |
|---|---|---|---|
| Expt. no. | None | 5 μM-2,4-D | % increase |
| I | 321 | 344 | 7 |
| II | 384 | 443 | 15 |
| III | 324 | 373 | 15 |
| IV | 342 | 366 | 7 |
| V | 377 | 414 | 10 |
| Average | 350 | 388 | 11 |

bound enzymes at the cell surface. It should be noted that the above is not the first report of an interaction between auxin and a glucan synthetase. A 50% enhancement of glucan synthetase activity in auxin-pretreated pea epicotyl segments was reported by Abdul-Baki & Ray (1971). In this case, however, the activity of the enzyme was located within the Golgi apparatus (Ray *et al.*, 1969). At the present time it is unknown why different intracellular locations should exist for the enzyme in onion and pea. In trying to assess the meaning of these results, it is important to realize that the hypothetical enzymes that regulate the extensibility of the wall may be unrelated to the biosynthetic machinery referred to above. Nevertheless, a response (however small) has been obtained by adding auxin to a system *in vitro* and thus may represent a general response of the enzyme systems within the membrane to auxin.

Obviously, if auxin can affect the activity of enzymes at the cell surface, one should be able to detect direct binding of auxin to plasma membrane fractions, and indeed such binding has been reported. Lembi *et al.* (1971) have found that plasma membrane-rich fractions isolated from maize coleoptiles can bind *N*-1-napthylphthalamic acid, a weak auxin and inhibitor of auxin transport. These experiments have been extended by Hertel *et al.* (1972), who found that plasma membrane fractions can reversibly bind indol-3-ylacetic acid or naphthalene acetic acid, and from saturation kinetics of binding calculate an apparent $K_m$ between $10^{-6}$ and $5 \times 10^{-5}$ M, and a concentration of specific sites of $10^{-6}$ M per tissue volume. Perhaps at this point a word of caution should be inserted, lest one be carried away by the current tendency to ascribe all unknown physiological responses to the mystical properties of the cell membrane. Obviously, binding, even reversible and specific binding, does not necessarily mean that it is physiologically important. This is especially true if we are unsure about similar binding by other organelles or cell surfaces, or the exact chemical nature of the binding and the binding site. Thus, although the prospects in the field look bright, we must be critical of the empirical findings until all the evidence is collated.

At the present time, there is also a growing interest in the early time-course of electrical events which can be induced by auxin (Newmann, 1963; Woodcock & Wilkins, 1969; Morath & Hertel, 1973). Although it is not known precisely how auxin induces potential changes, it seems likely that ion fluxes through the plasma membrane are involved and thus such changes may reflect a fundamental action of auxin on the cell surface. A current report (Morath & Hertel, 1973) suggests that there is an immediate electrical response to auxin which can be either positive or negative depending on the concentration of auxin that is administered. This is followed by a second delayed response which is always positive and much larger in magnitude. Although this report is noteworthy it should be pointed out that the physiological implications are unclear and that there is some controversy about the meaning of such effects [compare Durand & Zenk (1972) with Rayle *et al.* (1970*a*); also see Hild & Hertel (1972)]. In view of the rather suggestive results mentioned above, and considering the current trends in hormone research (auxinology always follows the trends!), it seems safe to say that we can expect to see many more reports about auxins and the plasma membrane in the near future. It seems equally clear that only then will we be in a position clearly to assess their significance.

The phenomenon of rapid growth responses to auxin has indirectly given rise to another expanding area of investigation, which approaches the problem of extension growth in quite a different way. This line of investigation makes use of a 'model' system for cell elongation with $H^+$ as the inducing agent. The strategy behind these investigations is to determine the chemical nature of the load-bearing bonds within the cell wall. Hopefully, answering this question will then lead to a greater understanding of the cell-elongation process in general as well as possibly auxin-induced responses.

The ability of $H^+$ to induce cell elongation (Fig. 4) has been known since the early 1930's (Strugger, 1932; Bonner, 1934). However, only recently has the response been systematically examined (Rayle & Cleland, 1970; Ganot & Reinhold, 1970; Evans *et al.*, 1971; Hager *et al.*, 1971). The response is rapid (a latent time of less than 1 min) and of relatively short duration (1–3 h). The pH optimum for the response is about 3.0 in *Avena* coleoptiles and slightly higher in other species. Of note is the fact that in the *Avena* coleoptile system, the growth rate produced by optimum amounts of $H^+$ is similar to, if not identical with, the rate induced by optimum auxin concentrations (Rayle & Cleland, 1970). As far as tested, the $H^+$ response occurs in all tissues that can be induced to extend with auxin (D. L. Rayle, unpublished work; M. H. Zenk & H. Durand, unpublished work). From a practical as well as a scientific standpoint, the most important aspect of this response may well be that it can be demonstrated in a cell-wall matrix (Rayle *et al.*, 1970*b*; Hager *et al.*, 1971; Rayle & Cleland, 1972). If coleoptiles are frozen and thawed several times and then placed under tension (10–20 g), they will extend at a very low rate if the incubating medium is above pH 5. However, if the medium is then changed to a more acidic solution (e.g. pH 3.6), an extension is initiated (Fig. 5). This technique has certain intrinsic advantages over more conventional methods for the study of cell extension growth. For example, since turgor is replaced by an external load, conditions can be tested that would otherwise be impossible (i.e. high concentrations of osmoticum, membrane-disrupting substances, etc.). Another advantage is that sections that have been frozen and thawed have lost many of their biochemical activities including the

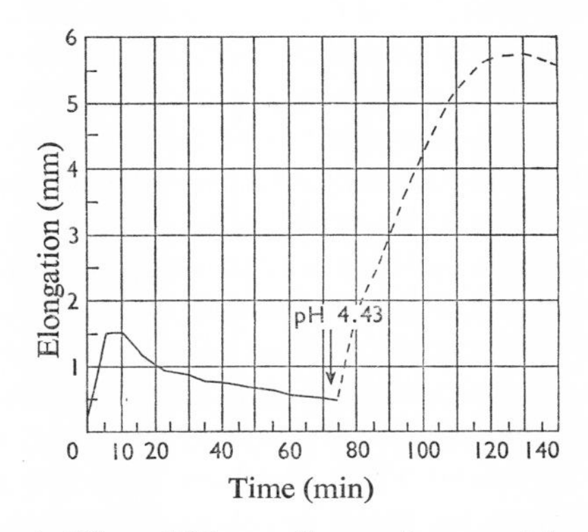


Fig. 4. *Effect of $H^+$ on cell-extension growth in vivo*

——, Response of sunflower hypocotyls in water at pH 7.45; – – – –, response after transfer to buffer at pH 4.43 (from Strugger, 1932).

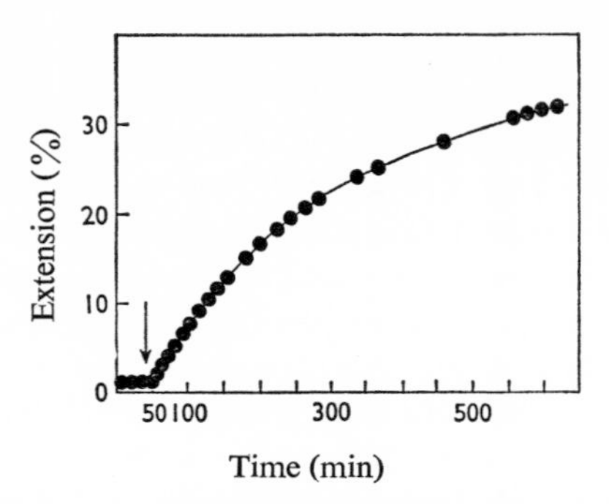


Fig. 5. *Time-course of the $H^+$ response in vitro*

At the arrow, frozen-and-thawed coleoptiles (20g tension) were transferred from buffer at pH 7.0 to buffer at pH 3.6. Elongation was plotted as a percentage of the length when measurements were started, i.e. at the end of the period of viscoelastic creep (from Rayle *et al.*, 1970*b*).

ability to synthesize new wall material (Rayle *et al.*, 1970*b*). This latter fact suggests that the system might be quite useful as a model for studying the biochemistry of cell-wall loosening. Obviously, a cell-extension system such as the one described above would be of rather limited interest if it did not accurately simulate normal cell extension *in vivo*. This does not appear to be a drawback. Indeed, the frozen-and-thawed system closely mimics normal (*in vivo*) $H^+$ or even auxin-induced growth responses, and it is likely that the growth responses *in vitro* and *in vivo* proceed via similar mechanisms (see Rayle & Cleland, 1972). Current research with this system is primarily directed towards answering two questions: how do $H^+$ ions act to cause cell-extension growth and is the auxin response directly related to the acid response? Let us first concentrate on the former question and then proceed to evidence bearing on the second.

In theory acidic solutions either could act to activate some wall- or membrane-bound enzyme which could then act to cleave cell wall bonds; or $H^+$ could participate directly in the cleavage of acid-labile, base-stable bonds or alternatively influence hydrogen bonding within the wall. Unfortunately, at the present time the available evidence is mixed, and therefore one mechanism cannot be singled out as the correct choice to the complete exclusion of the others. Perhaps some specific data should be cited to emphasize our current dilemma. If the response proceeded via an enzymic mechanism, the protein would have to be remarkably stable. For example, it would have to have a pH optimum in the range of 2–3.5 or lower and it would have to be stable at pH values of 12 or higher since frozen-and-thawed sections have such an optimum and can respond in a normal manner after incubation in basic solutions (Rayle & Cleland, 1972; D. L. Rayle, unpublished work). Secondly this hypothetical enzyme would have to be resistant to pre-treatment with urea (6 M), pronase and sodium lauryl sulphate. It should be noted, however, in some cases, the response after such pretreatments is abnormal and of short duration (Rayle & Cleland, 1972). But does this simply reflect a denaturation of the wall, thus making the load bearing relationships abnormal, or is it truly a direct, although incomplete, action on the extension machinery? We cannot say at this time. Hager *et al.* (1971) have claimed that $Cu^{2+}$ (5 mM) can almost completely inhibit the $H^+$ response in a system with sunflower hypocotyls *in vitro*.

By using coleoptiles, we have been able to demonstrate only a slight decrease in the rate with the same concentration of $Cu^{2+}$ (D. L. Rayle & M. H. Zenk, unpublished work). Once again, this effect may only reflect an indirect change in the physical properties of the wall.

On the other hand, one may wonder if it is possible to visualize an acid-labile, base-stable cell-wall bond that could be cleaved at pH values of about 3.5. There are some definite candidates (Lamport, 1970; Cleland, 1971*a*), although it is difficult to single out any particular bond presently as being the most likely. This is because bond breakage only occurs when the wall is under tension (Rayle & Cleland, 1972; Cleland & Rayle, 1972), and therefore the bond stresses might contribute to rather unusual kinetics and render bonds, which would normally be quite stable, sensitive to mildly acidic solutions. It is known that the hydrolysis of collagen by collagenase and elastin by elastase are both markedly speeded up when the polymers are under tension (see discussion in Cleland & Rayle, 1972), thus there is reason to consider this possibility rather seriously.

Obviously, at this time there is ample room and need for further experimentation in this area. It seems to us that this question should be rigorously pursued since the answer could have some bearing on the model discussed below, as well as suggest possible approaches for the direct chemical identification of the bond that is broken.

The second question which bears consideration is the relationship between the auxin-induced response and the $H^+$ response. The first piece of evidence suggesting an interaction stems from the observation that both agents seem to produce approximately the same growth rate at optimum concentrations in several plant species (Rayle & Cleland, 1970; H. Durand & M. H. Zenk, unpublished work). Secondly, both agents cause the plastic extensibility of the wall to increase to the same extent, reaching a maximum value at 2 h after incubation. Thirdly, the two responses show similar profiles when the temperature of the incubation medium is varied (Rayle & Cleland, 1972; H. Durand & M. H. Zenk, unpublished work). Further, both responses require the wall to be under tension for wall loosening to take place (Rayle & Cleland, 1972; Cleland & Rayle, 1972). The one dramatic difference between the responses is that auxin-induced growth requires normal metabolic activities (Bonner, 1936, 1950; Hackett & Thimann, 1953), whereas the $H^+$ response does not. These observations, among others, have led to the suggestion that $H^+$ and auxin may ultimately affect the same cell-wall bonds, but the mechanism leading up to bond breakage might be quite different (Hager *et al.*, 1971; Rayle & Cleland, 1972). Clearly it would be possible to devise several different models to incorporate these ideas, however, the point can be adequately made if we consider a recent model put forth by Hager *et al.* (1971). This model (Scheme 1) incorporates ideas obtained from a comparison of the $H^+$ and auxin responses as well as current ideas regarding possible interactions of auxin with the cell membrane. It visualizes auxin entering the cell and interacting with a high-energy compound such as GTP. This complex then co-operatively interacts with an adenosine triphosphatase bound to the cell surface. It is proposed that this auxin–enzyme interaction then potentiates the pumping of $H^+$ into the region of the cell wall. The $H^+$ would then presumably initiate wall loosening directly or indirectly via enzyme activation. One might actually consider this model as a sort of

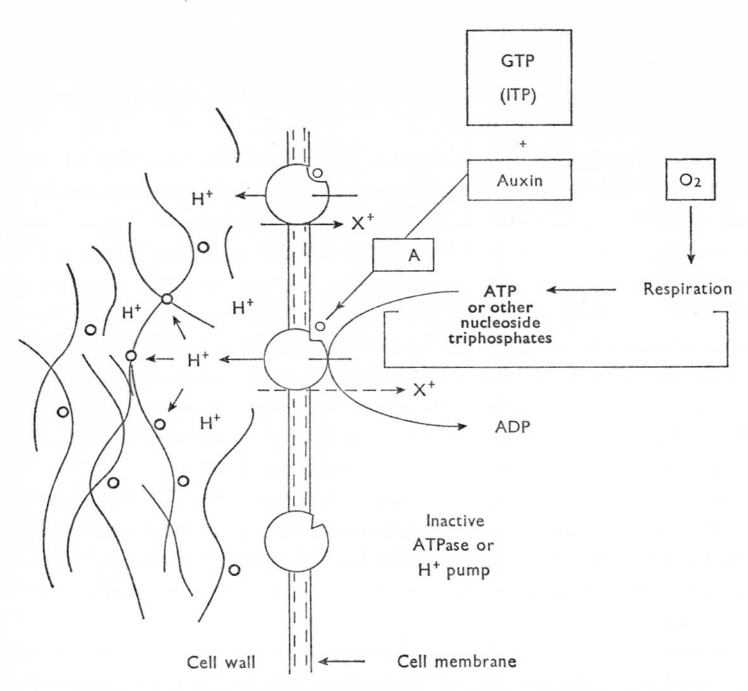


Scheme 1. *Model of indol-3-ylacetic acid-enhanced cell elongation*

In combination with GTP (ITP) auxin acts as an effector of a membrane-bound anisotropic adenosine triphosphatase (ATPase) or proton pump. This pump increases (with the aid of respiratory energy) the proton concentration in a compartment, in or close to the cell wall. This leads to cell-wall loosening and therefore cell extension (after Hager *et al.*, 1971).

reverse Mitchell effect (Mitchell, 1961; Greville, 1969). Clearly, a model such as this is far from proven, but it does seem to integrate some of the current thinking regarding auxin action and certainly suggests ample areas for investigation. Can one measure an auxin-induced pumping of $H^+$? Could this give rise to the bioelectric potentials induced by auxin? Do hydrogen ions activate an enzyme system within the wall or membrane or do they act directly on the wall? Can one detect an interaction of auxin with an adenosine triphosphatase at the cell surface? Does auxin even have to enter the interior of the cell to induce an extension response? And perhaps the most crucial question, what cell-wall bonds are broken during $H^+$ and auxin-induced responses?

Hopefully some of these questions will be answered before this review is actually published, but others will undoubtedly require considerably more effort. Nevertheless, it would appear that we are at last making some progress towards understanding the mechanism by which auxin initiates rapid cell-extension growth. In closing, we wish to stress that the mechanisms which we have emphasized here refer only to the initial phases of auxin-induced growth. For the so-called long-term responses, quite different mechanisms involving induction of protein synthesis are possible. There may not be one master reaction in the mechanism of auxin action, but rather there is the possibility that auxin triggers off physiological responses via several completely different and unrelated events.

This work was supported in part by a grant from Deutsche Forschungsgemeinschaft, Bonn-Bad Godesberg, and the Landesamt für Forschung des Landes Nordrhein-Westfalen, Düsseldorf.

## References

Abdul-Baki, A. A. & Ray, P. M. (1971) *Plant Physiol.* **47**, 537–545
Afridi, M. M. R. K. & Hewitt, E. J. (1964) *J. Exp. Bot.* **15**, 251–271
Bonner, J. (1934) *Protoplasma* **21**, 406–423
Bonner, J. (1936) *J. Gen. Physiol.* **20**, 1–11
Bonner, J. (1950) *Plant Physiol.* **25**, 181–184
Branscomb, E. W. & Stuart, R. N. (1968) *Biochem. Biophys. Res. Commun.* **32**, 731–738
Burström, H. G. (1971) *Endeavour* **30**, 87–90
Church, R. B. & McCarthy, B. J. (1968) *Biochem. Genet.* **2**, 55–73
Cleland, R. (1971*a*) *Annu. Rev. Plant Physiol.* **22**, 197–222
Cleland, R. (1971*b*) *Planta* **99**, 1–11
Cleland, R. & Rayle, D. L. (1972) *Planta* **106**, 61–71
Davies, D. D., Patterson, B. D. & Trewavas, A. J. (1968) *SCI Monog.* **31**, 208–223
Durand, H. & Zenk, M. H. (1972) in *Plant Growth Substances* (Carr, D. J., ed.), pp. 62–67, Springer-Verlag, Berlin
Evans, M. L. & Ray, P. M. (1969) *J. Gen. Physiol.* **53**, 1–20
Evans. M. L., Reinhold, L. & Ray, P. M. (1971) *Plant Physiol.* **41**, 335–341
Fan, D. F. & MacLachlan, G. A. (1967) *Plant Physiol.* **42**, 1114–1122
Galston, A. W. & Purves, W. K. (1960) *Annu. Rev. Plant Physiol.* **11**, 239–276
Ganot, D. & Reinhold, L. (1970) *Planta* **95**, 62–68
Gardner, F. E. & Cooper, W. C. (1943) *Bot. Gaz.* (*Chicago*) **105**, 80–89
Greville, G. D. (1969) *Curr. Top. Bioenerg.* **3**, 1–78
Hackett, D. P. & Thimann, K. V. (1953) *Amer. J. Bot.* **40**, 183–188
Hager, A., Henzel, H. & Krauss, A. (1971) *Planta* **100**, 47–75
Hertel, R., St.-Thomson, K. & Russo, V. E. A. (1972) *Planta* **107**, 325–340
Hild, V. & Hertel, R. (1972) *Planta* **108**, 245–258
Holm, R. E., O'Brien, T. J. & Cherry, J. L. (1970) *Plant Physiol.* **45**, 41–45
Ingle, J., Key, J. L. & Holm, R. E. (1965) *J. Mol. Biol.* **11**, 730–746
Key, J. L. (1969) *Annu. Rev. Plant Physiol.* **20**, 449–474
Key, J. L. & Ingle, J. (1964) *Proc. Nat. Acad. Sci. U.S.* **52**, 1382–1388
Key, J. L. & Shannon, J. C. (1964) *Plant Physiol.* **39**, 360–364
Lamport, D. T. A. (1970) *Annu. Rev. Plant Physiol.* **21**, 235–270
La Rue, C. D. (1935) *Amer. J. Bot.* **22**, 908–909
Lembi, C. A., Morré, D. J., St.-Thompson, K. & Hertel, R. (1971) *Planta* **99**, 37–42
Lin, C. Y., Ley, J. L. & Bracker, C. E. (1966) *Plant Physiol.* **41**, 946–982
Masuda, Y. & Kamisaka, S. (1969) *Plant Cell Physiol.* **10**, 79–86
Masuda, Y., Tanimoto, E. & Wada, S. (1967) *Plant Physiol.* **20**, 713–719
Matthysse, A. G. (1968) *Plant Physiol. Suppl.* **43**, S-42
Matthysse, A. G. & Phillips, C. (1969) *Proc. Nat. Acad. Sci. U.S.* **63**, 897–903
Mitchell, P. (1961) *Nature* (*London*) **191**, 144–148
Morath, M. & Hertel, R. (1973) *Planta* in the press
Newmann, I. (1963) *Aust. J. Biol. Sci.* **16**, 629–646
Nissl, D. & Zenk, M. H. (1969) *Planta* **89**, 323–341
Nooden, L. D. & Thimann, K. V. (1963) *Proc. Nat. Acad. Sci. U.S.* **50**, 194–200
O'Brien, I. J., Jaruis, B. C., Cherry, J. H. & Hanson, J. B. (1968) *Biochim. Biophys. Acta* **169**, 35–43
Pope, D. & Black, M. (1972) *Planta* **102**, 26–36
Ray, P. M. (1969) *Develop. Biol. Suppl.* **3**, 172–205
Ray, P. M., Shininger, T. L. & Ray, M. M. (1969) *Proc. Nat. Acad. Sci. U.S.* **64**, 605–609
Rayle, D. L. & Cleland, R. (1970) *Plant Physiol.* **46**, 250–253
Rayle, D. L. & Cleland, R. (1972) *Planta* **104**, 282–290
Rayle, D. L., Evans, M. L. & Hertel, R. (1970*a*) *Proc. Nat. Acad. Sci. U.S.* **65**, 184–191
Rayle, D. L., Haughton, P. M. & Cleland, R. (1970*b*) *Proc. Nat. Acad. Sci. U.S.* **67**, 1814–1817
Roychoudhury, R., Datla, A. & Sen, S. P. (1965) *Biochim. Biophys. Acta* **107**, 346–351
Sachs, T. & Thimann, K. V. (1967) *Amer. J. Bot.* **51**, 136–144
Snow, R. (1935) *New Phytol.* **34**, 347–360
Söding, H. (1934) *Jahrb. Wiss. Bot.* **79**, 231–255

Söding, H. (1936) *Ber. Deut. Bot. Ges.* **54**, 291–304
Strugger, S. (1932) *Ber. Deut. Bot. Ges.* **50**, 77–93
Südi, J. (1964) *Nature* (*London*) **201**, 1009–1010
Tester, C. F. & Dure, L. S. (1967) *Biochemistry* **6**, 2532–2557
Thimann, K. V. & Behnke, J. (1950) *The Use of Auxins in the Rooting of Woody Cuttings*, Cabot Found. Publ. no. 1, Petersham, Mass.
Thimann, K. V. & Skoog, F. (1933) *Proc. Nat. Acad. Sci. U.S.* **19**, 714–716
Thimann, K. V. & Skoog, F. (1934) *Proc. Roy. Soc. Ser. B* **114**, 317–339
Thimann, K. V. & Sweeney, B. M. (1937) *J. Gen. Physiol.* **21**, 123–135
Thompson, W. F. & Cleland, R. (1971) *Plant Physiol.* **48**, 663–670
Trewavas, A. (1968) *Progr. Phytochem.* **1**, 113–160
Van der Woude, W. J., Lemi, C. A. & Morré, D. J. (1972) *Biochem. Biophys. Res. Commun.* **46**, 245–253
Venis, M. A. & Stoessl, A. (1969) *Biochem. Biophys. Res. Commun.* **36**, 54–56
Wareing, P. F., Hanney, C. E. A. & Digby, J. (1964) in *Formation of Wood in Forest Trees* (Zimmermann, M., ed.), pp. 323–344, Academic Press, New York
Warner, H. L. & Leopold, A. C. (1971) *Biochem. Biophys. Res. Commun.* **44**, 989–994
Woodcock, A. E. R. & Wilkins, M. B. (1969) *J. Exp. Bot.* **20**, 687–697
Zenk, M. H. (1962) *Planta* **58**, 75–94

*Biochem. Soc. Symp.* (1973) **38**, 247–275
*Printed in Great Britain*

# Biosynthesis of Auxins in Tomato Shoots

By F. WIGHTMAN

*Department of Biology, Carleton University, Ottawa, Ont. K1S 5B6, Canada*

## Synopsis

Radioactive metabolism experiments with shoot cell-free extracts and with excised whole shoots have shown that both [3-$^{14}$C]tryptophan and [2-$^{14}$C]tryptamine can be converted into [$^{14}$C]indol-3-ylacetic acid by enzyme systems present in tomato shoots. The amounts of radioactivity found in [$^{14}$C]indol-3-ylacetic acid and in the expected intermediates obtained when each radioactive precursor was supplied separately indicate that the primary pathway for indol-3-ylacetic acid synthesis in tomato is via indol-3-ylpyruvic acid, rather than via tryptamine. The site(s) of auxin synthesis in tomato has been investigated and contrary to the widely accepted view, it was found that indol-3-ylacetic acid synthesis is not confined to the tissues of the shoot apex but readily occurs in all the expanding and mature leaves of the shoot system. The metabolism of L-[3-$^{14}$C]-phenylalanine has been examined in excised tomato shoots *in vivo* and the formation of the growth-promoting substance, phenylacetic acid, has been demonstrated. Since this substance has also been shown to occur naturally in tomato shoots, the possible importance of phenylacetic acid as a new plant auxin is considered.

## Introduction

Research work in my laboratory has been directed for several years to the elucidation of the biochemical pathway(s) involved in the formation of indol-3-ylacetic acid in higher plants. Because of its close chemical similarity to indol-3-ylacetic acid and universal occurrence in plants, the amino acid L-tryptophan has generally been considered to be the primary precursor of this important growth hormone. Thimann (1935) was the first to implicate tryptophan in the biosynthesis of indol-3-ylacetic acid when he showed that cultures of *Rhizopus suinus* readily formed auxin when supplied with the indole amino acid. He further observed that the formation of auxin was an aerobic process and suggested that the first reaction in the metabolism of tryptophan to indol-3-ylacetic acid was the oxidative deamination of the amino acid to produce indol-3-ylpyruvic acid via the unstable intermediate, indol-3-yliminopyruvic acid.

The next naturally occurring substance to be implicated in the biosynthesis of indol-3-ylacetic acid was the indole amine, tryptamine. This compound was shown by Skoog (1937), and later by other workers, to give rise to an auxin-like response when supplied to *Avena* coleoptiles and a range of other young plant tissues, and it was suggested that the growth-promoting activity of the amine was due to its conversion into indol-3-ylacetic acid. Not all the tissues examined, however, appeared able to convert tryptamine into indol-3-ylacetic acid, and since it is now known that this amine does not occur universally throughout the plant kingdom (Schneider *et al.*, 1972), it would seem that the conversion of tryptophan into tryptamine, presumably via a decarboxylation reaction, is not an essential first step in the main pathway leading to the formation of indol-3-ylacetic acid in all plants.

The remaining intermediary steps in the early proposed pathway(s) for the biosynthesis of indol-3-ylacetic acid became more clear when Larsen (1944) showed the presence of a neutral indole in the shoots of etiolated pea seedlings, which he identified as indol-3-ylacetaldehyde. This substance was later found in extracts of pineapple leaves and was shown to be enzymically produced from tryptophan by homogenates of this leaf tissue (Gordon & Sanchez Nieva, 1949), and by pea epicotyl tissue (Larsen, 1951). Indol-3-ylacetaldehyde was proposed as the immediate precursor of indol-3-ylacetic acid when it was found that homogenates of pineapple leaves (Gordon & Sanchez Nieva, 1949), *Avena* coleoptiles (Larsen, 1951) and mung-bean seedlings (Gordon, 1956) could all rapidly convert preparations of indol-3-ylacetaldehyde into an acidic product which showed growth-promoting properties identical with those of indol-3-ylacetic acid. It was suggested that indol-3-ylacetaldehyde probably arose in plants from the decarboxylation of indol-3-ylpyruvic acid, rather than from the oxidative deamination of tryptamine, since Gordon (1956) showed that in pea stem tissue the presence of amine oxidase inhibitors prevented the conversion of tryptamine into indol-3-ylacetic acid, whereas its formation from tryptophan was unimpaired. However, in those plants that are able to form tryptamine from tryptophan and also possess an amine oxidase system, it is clear from the work of Clarke & Mann (1957) that the product of tryptamine oxidase activity is indol-3-ylacetaldehyde, so this route for the biosynthesis of indol-3-ylacetic acid (Pathway II) cannot be ignored for those plants that are known to contain the indole amine (Schneider *et al.*, 1972).

Evidence for the possible occurrence of another pathway for indol-3-ylacetic acid biosynthesis in higher plants came with the demonstration by Jones *et al.* (1952) of the occurrence of indol-3-ylacetonitrile in immature cabbage plants. The same workers later showed the presence of indol-3-ylacetonitrile in several other Cruciferae plants (Henbest *et al.*, 1953) and suggested that the acetonitrile might be formed from tryptophan via a pathway involving the intermediary formation of indol-3-yliminopropionic acid (Jones *et al.*, 1952). The potential importance of indol-3-ylacetonitrile as an immediate precursor of indol-3-ylacetic acid was clearly suggested by the high growth-promoting activity shown by the acetonitrile in *Avena* coleoptile tests (Bentley & Housley, 1952), and this conversion was soon demonstrated in *Avena* (Thimann, 1953; Stowe & Thimann, 1954) and *Triticum* coleoptile tissues (Seeley *et al.*, 1956). Although indol-3-ylacetonitrile has now been found as a natural constituent in a range of *Brassica* species, evidence for its occurrence in non-cruciferous plants is either negative or not convincing, and it would appear that the formation of this indole nitrile from tryptophan and its role as a potential precursor of indol-3-ylacetic acid is confined to Cruciferae plants.

Since these early studies and the resultant proposal of three potential pathways for the conversion of tryptophan into indol-3-ylacetic acid (Reinert, 1954;

Gordon, 1954, 1956; Stowe, 1959; Mahadevan, 1963), several metabolism experiments *in vivo* and *in vitro* carried out mainly during the 1960's have clearly demonstrated that radioactivity from [$^{14}$C]tryptophan can be incorporated into indol-3-ylacetic acid in fruit tissue (Dannenburg & Liverman, 1957; Khalifa, 1967), in *Lens* root and cucumber hypocotyl segments (Pilet, 1964; Sherwin & Purves, 1969), in excised cabbage, tomato and tobacco shoots (Wightman, 1962, 1964; Kutácek & Kefeli, 1968; von Kindl, 1968; Phelps & Sequeira, 1967), in sterile corn and oat coleoptiles (Libbert *et al.*, 1968; Libbert & Silhengst, 1970; Black & Hamilton, 1971) and in cell-free systems from mung-bean and pea plants (Wightman & Cohen, 1968; Moore & Shaner, 1967, 1968; Moore, 1969). These tracer investigations have also provided good, though not conclusive evidence in all cases, for the occurrence of the following four pathways (I–IV) in higher plants for the biosynthesis of this growth hormone.

I Tryptophan → indol-3-ylpyruvic acid → indol-3-ylacetaldehyde → indol-3-ylacetic acid

II Tryptophan → tryptamine → indol-3-ylacetaldehyde → indol-3-ylacetic acid

III Tryptophan → glucobrassicin → indol-3-ylacetonitrile → indol-3-ylacetic acid

IV Tryptophan → indol-3-ylacetaldoxime → indol-3-ylacetonitrile → indol-3-ylacetic acid

The main objectives of the work I shall describe in this paper were, first, to determine which of these pathways is operating in the biosynthesis *in vivo* of indol-3-ylacetic acid in the shoots of a typical dicotyledonous plant, tomato and secondly, to determine the location of the main site(s) of auxin formation in a growing, vegetative shoot system.

Four lines of evidence will be presented which throw light on these problems.

1. Identification of the radioactive indole compounds obtained in metabolism experiments *in vitro* with [3-$^{14}$C]tryptophan and [2-$^{14}$C]tryptamine when these were provided as substrates for the enzymes present in cell-free extracts prepared from tomato shoot-tip tissues.

2. Demonstration of the presence of several of the enzymes that are probably involved in the biosynthesis of indol-3-ylacetic acid in cell-free preparations of tomato shoots, and also characterization of the aminotransferase and decarboxylase enzymes catalysing the initial reactions in Pathways I and II for tryptophan metabolism.

3. Identification of the radioactive indole compounds obtained in metabolism experiments with tomato plants *in vivo* in which [3-$^{14}$C]tryptophan and [2-$^{14}$C]tryptamine were fed separately to excised, 6-week-old vegetative shoots.

4. Demonstration of the capacity of tomato shoot tips and leaves of different ages to form radioactive indol-3-ylacetic acid when supplied with [3-$^{14}$C]-tryptophan.

As a further development from these studies on the conversion of tryptophan into indol-3-ylacetic acid in tomato shoots, in which we found that the aminotransferase catalysing the first reaction step in Pathway I is not specific for

L-tryptophan but exhibits multispecificity for all three aromatic amino acids, it was decided to examine the possibility that tomato shoots might be able to convert L-phenylalanine into phenylacetic acid by a sequence of reactions similar to those involved in the formation of indol-3-ylacetic acid via Pathway I or Pathway II. If such a conversion of L-phenylalanine into phenylacetic acid occurs in shoot tissues, then it may have important physiological implications, since phenylacetic acid is known to exhibit growth-promoting activity in a range of vegetative and fruit tissues (Zimmerman & Wilcoxon, 1935; Zimmerman *et al.*, 1936; Muir & Hansch, 1953; Pybus *et al.*, 1959; Chamberlain & Wain, 1971). The final part of this paper will deal with our recent study of [3-$^{14}$C]phenylalanine metabolism in tomato plants in which we have shown (Rauthan & Wightman, 1972) that [$^{14}$C]phenylacetic acid is indeed formed in appreciable amounts by vegetative shoot tissues. The implication of this finding with respect to the hormonal regulation of plant growth and development will be briefly considered.

## Materials and Methods

### *Plant material*

The plant material used in these experiments was the shoots of 6-week-old tomato plants (*Lycopersicon esculentum* var. Big Boy) grown in an air-conditioned greenhouse at 24°C under natural daylight supplemented to 18h with incandescent lighting. Plants were approx. 38cm high and at the eight-leaf stage when they were harvested for metabolism experiments *in vivo* or *in vitro*.

### *Enzyme assays*

The presence or absence of the enzymes thought to be involved in the conversion of L-tryptophan into indol-3-ylacetic acid, via Pathway I or II, was determined in cell-free systems prepared from the same shoot material as that used for the metabolism experiments *in vivo*. Only shoot tips were used, comprising the apical stem tissues and the first two expanding leaves, in the preparation of the cell-free extracts. Enzyme activities were assayed in 35000*g* supernatant fractions prepared as follows: excised tomato shoot-tips were first chilled in a cold-room at 3°C for 30min, and were then homogenized for 1 min in a Waring Blendor with 2 vol. of grinding medium containing 0.1M-potassium phosphate buffer, 0.01M-EDTA, 0.01M-2-mercaptoethanol and 0.1mM-pyridoxal phosphate, adjusted to a final pH of 8.0. The homogenate was filtered through cheesecloth and centrifuged at 35000*g* for two successive 20min periods. The combined supernatant fractions from the second centrifugation were freed from soluble low-molecular-weight compounds by passing through columns (3.5cm × 30cm) of Sephadex G-25 (fine) equilibrated with elution buffer. This buffer solution contained 0.05M-potassium phosphate, 0.01M-2-mercaptoethanol and 0.1mM-pyridoxal phosphate at a final pH of 8.0. Immediately before addition to the reaction flasks, the supernatant fraction was sterilized by passage through a Millipore filter with a 0.45$\mu$m pore size.

Tryptophan aminotransferase (L-tryptophan–$\alpha$-oxoglutaric acid aminotransferase) aldehyde dehydrogenase (indol-3-ylacetaldehyde–NAD oxidoreductase) and alcohol dehydrogenase (tryptophol–NAD oxidoreductase) were assayed by the methods of Wightman & Cohen (1968). Tryptophan decarboxylase (L-tryptophan carboxy-lyase) and indol-3-ylpyruvic acid decarboxylase were assayed by the method of Gibson *et al.* (1972*b*). In all these enzyme assays, the substrates were incubated separately with the sterile supernatant fraction, in the presence of pyridoxal phosphate (100 $\mu$g/ml) for 6 h at pH 8.0. The final concentrations of the different substrates in the reaction mixtures for each enzyme assay were: 0.04 M-L-tryptophan, 0.01 M-$\alpha$-oxoglutaric acid, 4 mM-indol-3-ylpyruvic acid and 4 mM-indol-3-ylacetaldehyde. The reaction products were extracted into ether, isolated by paper chromatography and the amounts were determined quantitatively by the colorimetric methods of Schneider *et al.* (1972). The product of tryptophan decarboxylase activity, tryptamine, was extracted into ether at pH 11.0, and the indol-3-ylacetic acid formed by indol-3-ylacetaldehyde dehydrogenase activity was extracted into ether at pH 3.0. Owing to the instability of indol-3-ylacetaldehyde, which is the expected decarboxylation product of indol-3-ylpyruvic acid, the decarboxylase activity of the tomato supernatant fraction with this substrate was assayed by measuring the formation of tryptophol, by using the method of Wightman & Cohen (1968). Since tryptophol was readily formed from indol-3-ylacetaldehyde by the supernatant fraction, the amount of tryptophol found in the final reaction mixture was considered a reliable, indirect measure of the amount of indol-3-ylacetaldehyde formed from indol-3-ylpyruvic acid by decarboxylase activity.

The protein content of each supernatant fraction was determined by the method of Lowry *et al.* (1951), after dialysis of the fraction for 24 h to remove phosphate. Bovine serum albumin was used as a standard.

*$(NH_4)_2SO_4$ precipitation of tryptophan decarboxylase and tryptophan aminotransferase*

Finely ground $(NH_4)_2SO_4$ (enzyme grade) was added to a supernatant preparation, with constant stirring, to give a rise of 10% saturation in 30 min. The solution was stirred for a further 15 min and then centrifuged at 20000 *g* for 15 min. The precipitate obtained with each 10% rise in saturation was dissolved in elution buffer and passed through a column (2.5 cm × 30 cm) of Sephadex G-25 (fine grade) before being assayed for tryptophan decarboxylase and tryptophan transaminase activity.

*Partial purification of tryptophan decarboxylase*

A 25–35%-satd.-$(NH_4)_2SO_4$ precipitate was dissolved in a minimal quantity of elution buffer at pH 8.0. This solution was centrifuged for 30 min at 35000 *g* and the supernatant passed through a column (2.5 cm × 30 cm) of Sephadex G-25 (fine grade). A 15 ml sample of the protein eluate was then placed on a column (2.5 cm × 100 cm) of Sephadex G-200, and again eluted with elution buffer at pH 8.0. The eluate was collected in 5 ml fractions and assayed for tryptophan decarboxylase activity. The peak of activity, which appeared between fractions 30 and 40, was concentrated to 17 ml with dry Sephadex G-25. A 2 ml sample of this solution was used for enzyme assay, and the remainder of the fraction was further purified by rechromatography on Sephadex G-200, as described above. All purification procedures were carried out at 2–4°C.

*Partial purification of tryptophan aminotransferase*

A 45–70%-satd.-$(NH_4)_2SO_4$ precipitate was dissolved in a minimal quantity of standard elution buffer at pH 8.0 and dialysed against the buffer overnight at 2°C. The dialysed fraction was concentrated to about 20 ml by dehydration with Aquicide II, and was then applied to a column (85 cm × 2.5 cm) of Sephadex G-100. Protein was eluted with elution buffer at a flow rate of 20 ml/h. The effluent from the column was monitored at 280 nm with an ISCO UA-2 analyser and 5 ml fractions were collected. All fractions were assayed for aminotransferase activity by using L-tryptophan, L-phenylalanine and L-tyrosine as substrates and α-oxoglutarate as the amino group acceptor. All the fractions showing activity were pooled and concentrated to about 25 ml by dehydration with Aquicide II. This fraction was applied to a column (45 cm × 2.5 cm) of DEAE-Sephadex and after it had been adsorbed, the column was washed overnight with 150–200 ml of standard elution buffer at pH 8.0. Protein was then eluted by using a buffer gradient produced by having 250 ml of 0.05 M-potassium phosphate in the cylindrical mixing vessel and 250 ml of 0.05 M-potassium phosphate plus 0.5 M-KCl in the reservoir of the gradient-making apparatus. The flow rate of the column was about 10 ml/h and 5 ml fractions were collected. All fractions were tested for aminotransferase activity, as before, and the active fractions were pooled and the volume decreased to about 10 ml with Aquicide II. The concentrated fraction was dialysed for about 12 h against elution buffer additionally containing 0.1 M-KCl, and was then applied to a column (85 cm × 1.5 cm) of Sephadex G-200. Protein was eluted with standard elution buffer at pH 8.0 and the flow rate of the column was 10–15 ml/h. All fractions (5 ml) collected were tested for aminotransferase activity as before. At this stage of purification, tryptophan aminotransferase activity was found to occur together with phenylalanine aminotransferase and tyrosine aminotransferase activities in several 5 ml fractions containing a single, small protein peak running just ahead of the main band of protein. All procedures used in this enzyme purification were carried out in a cold-room at 2°C.

*Procedure for feeding radioactive compounds to excised shoots*

Tomato shoots were excised at ground level and after removal of a further 5 cm of stem, under water, to exclude the possibility of air locks in the vascular system, the shoots were placed in 250 ml pharmaceutical-type graduated cylinders containing the feeding solutions. DL-[3-$^{14}$C]Tryptophan, [2-$^{14}$C]tryptamine bisuccinate and DL-[3-$^{14}$C]phenylalanine were fed at the rate of 0.25 mg of compound per g fresh wt. of shoot, and the specific radioactivity of the solutions was adjusted to 0.5 μCi/mg by dilution with unlabelled compound. The plants took up the radioactive solution within 2 h, and were then supplied with water and maintained in a growth chamber for 24 h under constant light at 24°C.

To examine for the possible presence of impurities, particularly in the radiochemicals, the following controls were incorporated into all experiments.

1. A portion of the $^{14}$C-labelled feeding solution, equivalent to that fed to

100 g fresh wt. of plant tissue, was extracted in the same way as the plant material examined for possible contaminants by chromatography.

2. A similar portion of the $^{14}C$-labelled feeding solution was added to 100 g of freshly excised shoots after maceration of the tissue in methanol. This extract was then examined for possible radioactive artifacts incurred during extraction and chromatography.

No significant radioactive contaminants were, in fact, found in any of these control treatments. As an additional control, untreated plants supplied with water only were also included in each experiment and these were examined to determine whether any changes occurred in the amounts of native indole compounds during the 24 h feeding period.

*Extraction procedure and identification of metabolites in extracts*

The methods used for extraction, fractionation, chromatography and quantitative determination of indole compounds have been described in detail by Schneider *et al.* (1972). Briefly, the plants were macerated and extracted with methanol, the extract was concentrated in a rotary evaporator and the precipitated chlorophyll was removed from the remaining aqueous solution by filtration through Celite. The aqueous extract was then partitioned against successive volumes of freshly distilled ether, first at pH 7.0 to give the neutral ether fraction, then at pH 3.0 to give the acid ether fraction, and finally at pH 11.0 to give the basic ether fraction. The indole compounds present in these ether fractions and in the residual aqueous fraction were then separated by paper chromatography and t.l.c., by using the following solvent systems.

Paper chromatography: I, propan-2-ol–$NH_3$–water (8:1:1, by vol.); II, butan-1-ol–acetic acid–water (12:3:2, by vol.); III, propan-2-ol–benzene–water (55:30:11, by vol.), with paper buffered at pH 6.5 and dried before use; IV, benzene–acetic acid–water (2:2:1, by vol., upper phase); V, 8% (w/v) NaCl.

T.l.c.: VI, benzene–acetic acid–ethyl acetate (18:1:1, by vol.); VII, ethyl acetate–propanol-2-ol–$NH_3$ (9:7:4, by vol.); VIII, benzene; IX, chloroform–ethyl acetate–formic acid (7:11:2, by vol.).

Most of the indole compounds present in the different fractions were determined quantitatively by densitometry of the chromatograms, after treatment with either dimethylaminocinnamaldehyde reagent for paper chromatograms, or with Ehrlich reagent for thin-layer chromatograms, as described by Schneider *et al.* (1972).

*Preparation of 2,4-dinitrophenylhydrazones of radioactive indole metabolites formed in metabolism experiments in vivo*

Acid ether fractions, equivalent to 400 g fresh wt. of tissue, were obtained from shoots fed with both [3-$^{14}C$]tryptophan and [2-$^{14}C$]tryptamine and 100 $\mu$g of unlabelled indol-3-ylpyruvic acid was added to each fraction. The extracts were evaporated to dryness, and the residues dissolved in 5 ml of phosphate buffer at pH 8.0. A saturated solution of Dnp-hydrazine in 2M-HCl was then added to each solution until precipitates ceased to form. The precipitated hydrazones were collected by centrifugation, dissolved in ethyl acetate, and applied to preparative (0.3 mm) silica-gel thin-layer plates. Marker spots of authentic indol-3-

ylpyruvic acid Dnp-hydrazone were added, and the plates were run in the solvent system VI. The two suspected radioactive indol-3-ylpyruvic acid Dnp-hydrazone bands (corresponding to the *cis–trans* isomers) were eluted and re-chromatographed on a second plate in the same solvent system to concentrate the sample and remove other contaminating hydrazones.

Radioactive Dnp-hydrazones were prepared in a similar way from the neutral ether fractions, after addition of carrier indol-3-ylacetaldehyde (100$\mu$g) and indol-3-ylaldehyde (50$\mu$g) to each fraction. The precipitate of neutral Dnp-hydrazones was very fine and difficult to collect by centrifugation, and better yields were obtained if the hydrazones were extracted from the phosphate buffer with ethyl acetate. The ethyl acetate solution was applied to a preparative thin-layer plate and chromatographed in benzene. The yellow bands of suspected radioactive indol-3-ylacetaldehyde Dnp-hydrazone and indol-3-ylaldehyde Dnp-hydrazone were then eluted and re-chromatographed in solvent systems VI and VIII solvents, together with authentic marker compounds.

*Determination of radioactivity*

The radioactive compounds present on paper chromatograms were located by means of a Nuclear–Chicago Actigraph II scanner and by using a Panax Radiochromatogram Scanner System with thin-layer chromatograms. For quantitative measurements, radioactive bands were either cut out of paper chromatograms, or scraped off thin-layer chromatograms and placed in scintillation vials containing Omnifluor solution (New England Nuclear Corp.). Total radioactivity was then determined by using a Packard Tri-Carb liquid-scintillation spectrometer, with a counting efficiency of 89%.

*Metabolism experiments with excised shoot tips and leaves of different ages in vivo*

To determine the site(s) of auxin synthesis in vegetative tomato shoots, the metabolism of DL-[3-$^{14}$C]tryptophan was examined (1) in shoot-tip tissues and individual leaves of excised, entire tomato shoots, and (2) in excised shoot tips and leaves of different ages. The procedure for feeding radioactive tryptophan solutions, the length and conditions of the metabolism period, and the methods used for extracting and characterizing the radioactive indole metabolites formed in the different organs were as described above. [3-$^{14}$C]Tryptophan was supplied to each batch of excised shoots, groups of excised leaves or shoot tips at the rate of 0.25mg/g fresh wt. of tissue, and the specific ratioactivity of all feeding solutions was adjusted to 0.5$\mu$Ci/mg by dilution with unlabelled L-tryptophan.

## Results and Discussion

*Metabolism experiments in vitro with cell-free extracts of tomato shoots*

Since in our earlier studies of [3-$^{14}$C]tryptophan metabolism in cell-free systems from mung-bean seedlings (Wightman & Cohen, 1968) it had been possible to demonstrate not only the formation of appreciable amounts of [$^{14}$C]indol-3-ylacetic acid, but also of the two radioactive intermediates expected in Pathway I, namely, [$^{14}$C]indol-3-ylpyruvic acid and [$^{14}$C]indol-3-ylacetaldehyde, it was decided to examine first the biosynthesis of indol-3-ylacetic acid in cell-free preparations of tomato shoot tissues. Sterile 35000*g* supernatant fractions were prepared from homogenates of shoot-tip tissues and 50ml quantities were mixed with equal volumes of sterile buffer solution containing either 0.08M-[3-$^{14}$C]-tryptophan (total radioactivity, 50$\mu$Ci) or 0.08M-[2-$^{14}$C]tryptamine (total radioactivity, 50$\mu$Ci). Small amounts of the cofactors pyridoxal 5-phosphate, thiamin pyrophosphate and NAD$^+$ were included in each substrate solution, and 0.04M-$\alpha$-oxoglutarate was also included in the solution containing radioactive tryptophan to serve as the amino group acceptor during transamination of the amino acid. Reaction mixtures were adjusted to pH8.0 and metabolism of the radioactive substrates was allowed to proceed for 6h in the dark at 35°C. Each final reaction mixture was then divided into two equal portions; the radioactive metabolites in one portion were isolated by sequential ether extraction, and the metabolites with reactive carbonyl groups were isolated from the second portion by the formation of stable Dnp-hydrazones, which were then extracted into ethyl acetate. Samples of the neutral, acid and basic ether fractions, and of the ethyl acetate–hydrazone fraction were analysed by paper chromatography and t.l.c. and the location of each radioactive metabolite was revealed by scanning for $^{14}$C and colorimetric tests for indole compounds.

The results presented in Figs. 1 and 2 show the distribution of radioactive compounds on paper chromatograms of the acid ether fractions from the [$^{14}$C]-tryptophan and [$^{14}$C]tryptamine reaction systems, after development of the

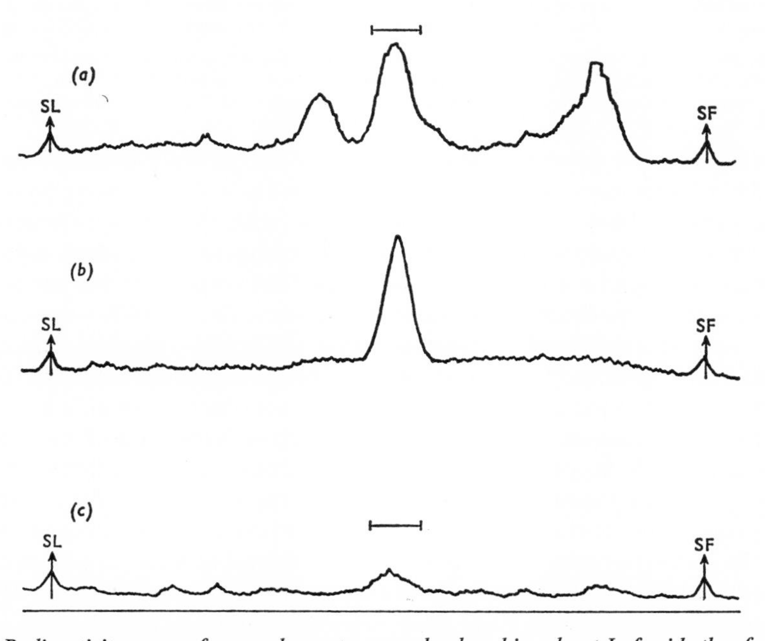


Fig. 1. *Radioactivity scans of paper chromatograms developed in solvent I of acid ether fractions from tomato cell-free extracts supplied with* (*a*) [3-$^{14}$*C*]*tryptophan or* (*c*) [2-$^{14}$*C*]*tryptamine*

Scan (*b*) shows a $^{14}$C radioactivity scan of a similar chromatogram containing a marker sample of authentic [2-$^{14}$C]indol-3-ylacetic acid. Chromatograms were scanned at the 1000c.p.m. range. SL, Starting line; SF, solvent front.

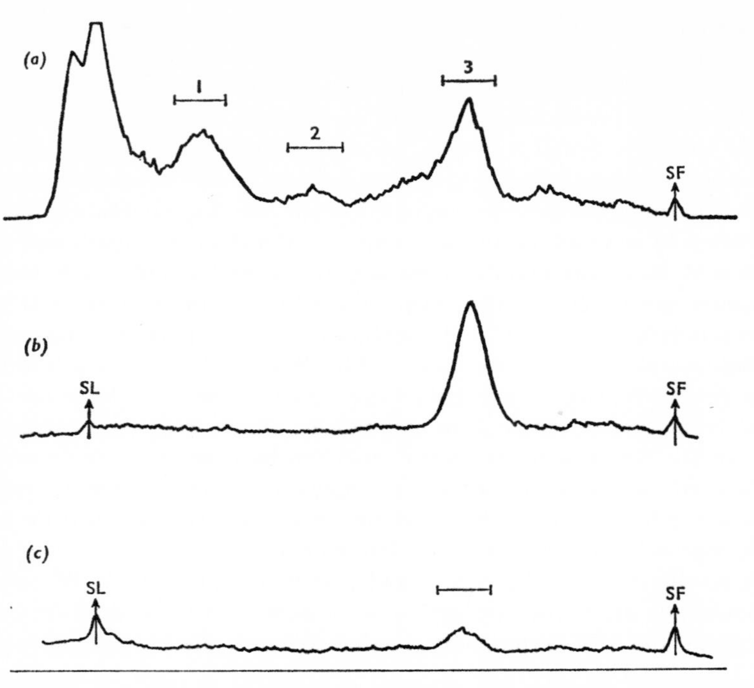


Fig. 2. *Radioactivity scans of paper chromatograms developed in solvent IV of acid ether fractions from tomato cell-free extracts supplied with either* (*a*) [3-$^{14}$*C*]*tryptophan or* (*c*) [2-$^{14}$*C*]*tryptamine* Scan (*b*) shows a $^{14}$C radioactivity scan of similar chromatogram containing a marker sample of authentic [2-$^{14}$C]indol-3-ylacetic acid. Chromatograms were scanned at the 1000c.p.m. range. Peak 1, indol-3-yl-lactic acid; peak 2, indol-3-ylpyruvic acid; peak 3, indol-3-ylacetic acid.

chromatograms in solvent systems I or IV. Clearly, radioactivity from both $^{14}$C-labelled substrates was incorporated into an acidic metabolite which had an $R_F$ value in both solvent systems identical with that shown by authentic [2-$^{14}$C]indol-3-ylacetic acid. This provisional identification was fully confirmed by elution and rechromatography of the metabolite in other solvent systems when it was found to co-chromatograph with authentic [2-$^{14}$C]indol-3-ylacetic acid in all cases. It was therefore concluded that [$^{14}$C]indol-3-ylacetic acid can be formed from both [3-$^{14}$C]tryptophan and [2-$^{14}$C]tryptamine by the soluble enzymes present in tomato shoot extracts. However, the relative difference in the amount of radioactivity incorporated from the two substrates (Figs. 1 and 2) indicates that [$^{14}$C]tryptophan was a better precursor of auxin than [$^{14}$C]tryptamine. This observation suggests that in such tomato extracts, the metabolism of tryptophan via Pathway I, rather than via Pathway II, is the predominant route for indol-3-ylacetic acid formation. On the other hand, a proportion of [$^{14}$C]tryptophan metabolism must have been proceeding via Pathway II since chromatographic analysis of the basic ether fraction from this reaction system revealed the presence of [$^{14}$C]tryptamine (Fig. 3). The identification of this metabolite was confirmed by co-chromatography with authentic tryptamine and by comparing u.v. and fluoresence emission spectra of the suspected and authentic compounds, when each set of spectra were found to be identical.

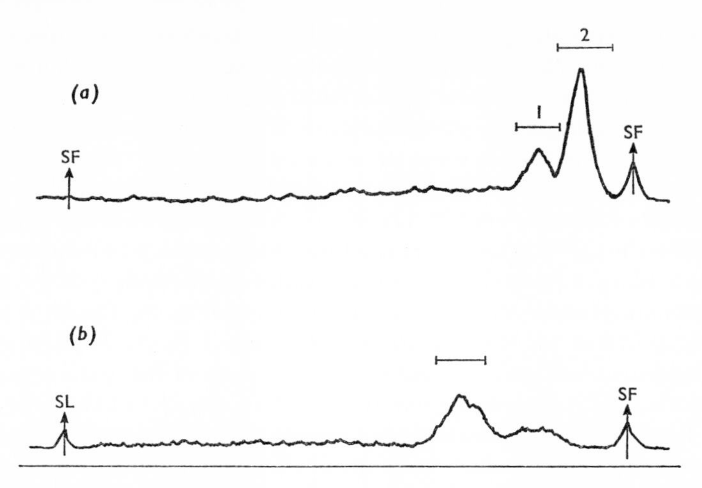


Fig. 3. *Radioactivity scans of paper chromatograms of* (*a*) *the neutral ether fraction and* (*b*) *the basic ether fraction from tomato cell-free extracts supplied with* [3-$^{14}$*C*]*tryptophan*

The chromatograms were developed in (*a*) solvent I, and (*b*) solvent II. Chromatograms were scanned at the 1000c.p.m. range. Peak 1, indol-3-ylacetaldehyde; peak 2, tryptophol and indol-3-ylaldehyde; peak 3, tryptamine.

Additional evidence that the metabolism of [3-$^{14}$C]tryptophan was indeed proceeding via Pathway I in these tomato shoot extracts was obtained from three sources. First, from the observation that a radioactive band corresponding in $R_F$ to that of authentic indol-3-ylpyruvic acid was found on the chromatogram of the acid ether fraction from the [3-$^{14}$C]tryptophan reaction system after

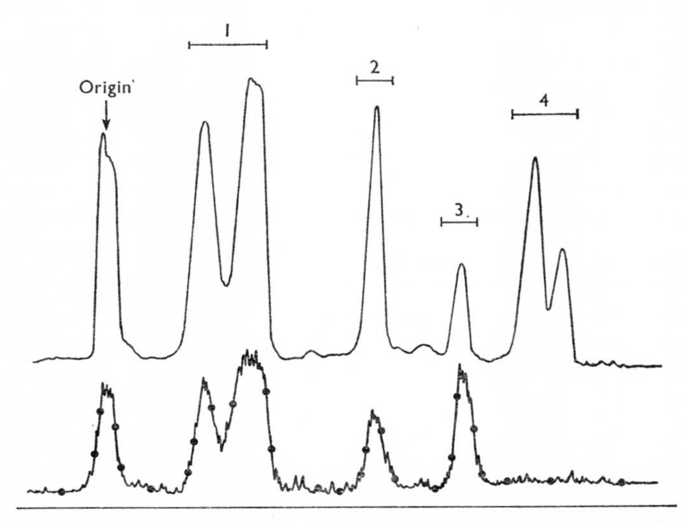


Fig. 4. *Densitometer and radioactivity scans of a thin-layer chromatogram of ethyl acetate extract of Dnp-hydrazones formed in a tomato cell-free extract supplied with* [3-$^{14}$*C*]*tryptophan*

The chromatogram was developed in solvent VI. ——, Densitometer trace; ●, radioactivity scan. Peak 1, *cis–trans* isomers of indol-3-ylpyruvic acid Dnp-hydrazone; peak 2, indol-3-ylaldehyde Dnp-hydrazone; peak 3, indol-3-ylacetaldehyde Dnp-hydrazone; peak 4, Dnp-hydrazine.

development in solvent system IV (Fig. 1). This band was not present on the corresponding chromatogram developed in solvent system I since indol-3-ylpyruvic acid is unstable under ammoniacal chromatographic conditions, nor was the band present on any chromatogram of the acid ether fraction from the [2-$^{14}$C]tryptamine reaction system, since the metabolism of this substrate would not be expected to give rise to radioactive indol-3-ylpyruvic acid. The second piece of evidence was provided by chromatographic analysis of the neutral ether fraction from the [3-$^{14}$C]tryptophan reaction system, when a radioactive metabolite corresponding in $R_F$ to that of authentic indol-3-ylacetaldehyde was observed on the chromatogram developed in solvent system I (Fig. 3). The third and most convincing evidence for the operation of Pathway I in the [3-$^{14}$C]tryptophan reaction system came from chromatographic analysis of the hydrazones formed on the addition of a Dnp-hydrazine solution to a portion of the final reaction mixture. T.l.c. of an ethyl acetate extract revealed the presence of four radioactive hydrazone bands (Fig. 4) which were identical in colour and $R_F$ value with those shown by authentic samples of the two isomers of indol-3-ylpyruvic acid Dnp-hydrazone, indol-3-ylacetaldehyde Dnp-hydrazone and indol-3-ylaldehyde Dnp-hydrazone.

Final proof of the identity of the suspected [$^{14}$C]indol-3-ylpyruvic acid Dnp-hydrazone and [$^{14}$C]indol-3-ylacetaldehyde Dnp-hydrazone was obtained by

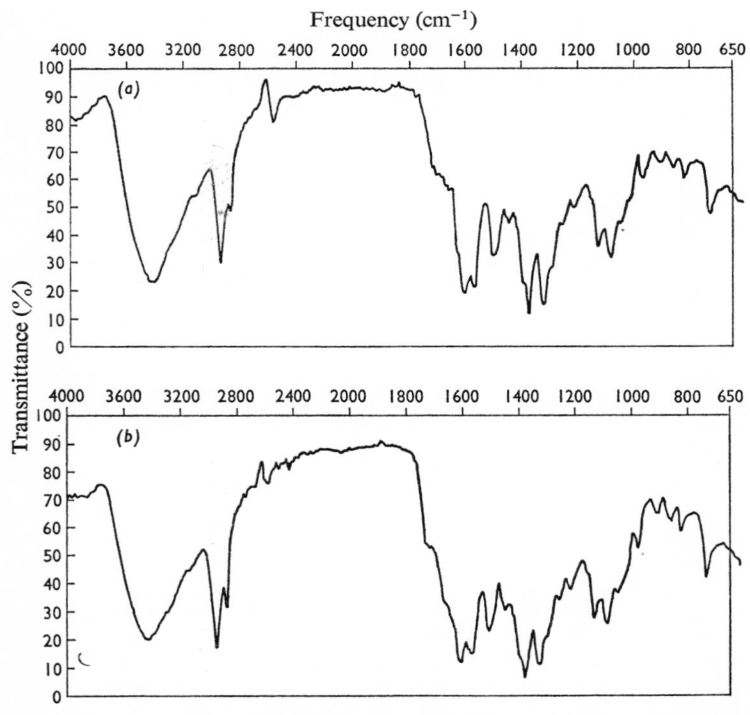


Fig. 5. *I.r. spectrum of* (*a*) *suspected indol-3-ylpyruvic acid Dnp-hydrazone isolated from tomato cell-free extract supplied with* [3-$^{14}$*C*]*tryptophan*

(*b*) I.r. spectrum of authentic indol-3-ylpyruvic acid Dnp-hydrazone.

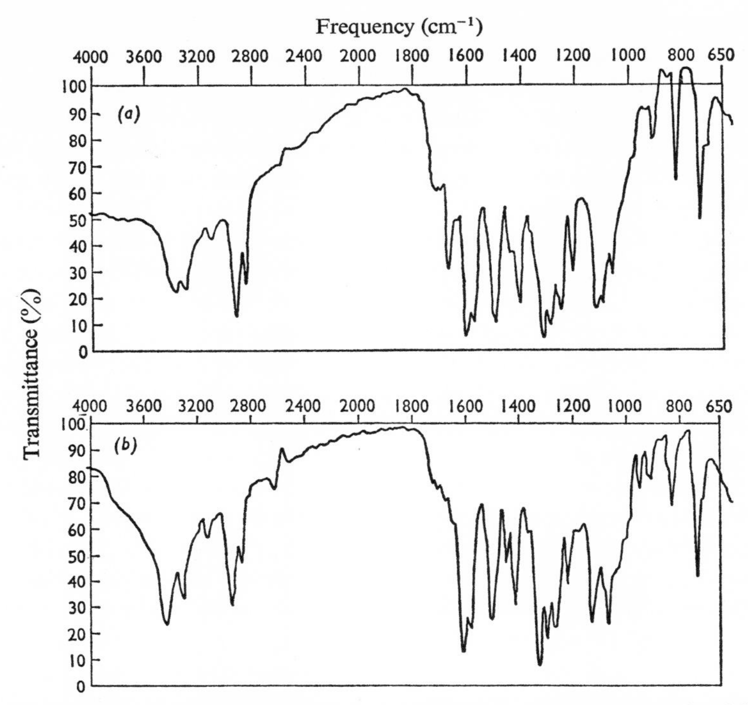


Fig. 6. *I.r. spectrum of* (*a*) *suspected indol-3-ylacetaldehyde Dnp-hydrazone isolated from tomato cell-free extract supplied with* [3-$^{14}$*C*]*tryptophan*

(*b*) I.r. spectrum of authentic indol-3-ylacetaldehyde Dnp-hydrazone.

i.r. spectrophotometric analysis of the eluted compounds when their absorption spectra were found to be identical with the spectra obtained from the corresponding authentic hydrazones (Figs. 5 and 6). This unequivocal evidence for the formation of radioactive indol-3-ylpyruvic acid in tomato cell-free extracts during the metabolism of [3-$^{14}$C]tryptophan is in complete agreement with the findings from earlier investigations with similar reaction systems prepared from mung-bean seedlings (Wightman & Cohen, 1968) and pea seedlings (Moore & Shaner, 1968).

T.l.c. analysis of the Dnp-hydrazone fraction prepared from the [2-$^{14}$C]tryptamine reaction system revealed, as expected, the presence of [$^{14}$C]indol-3-ylacetaldehyde Dnp-hydrazone and [$^{14}$C]indol-3-ylaldehyde Dnp-hydrazone. However, the amount of the acetaldehyde derivative was less than that found on equivalent chromatograms of the hydrazone fraction prepared from the [3-$^{14}$C]-tryptophan reaction system, and this finding indicates that the formation of [$^{14}$C]indol-3-ylacetaldehyde in the cell-free systems supplied with [3-$^{14}$C]tryptophan was primarily due to the metabolism of the amino acid via Pathway I.

*Demonstration of several of the enzymes for indol-3-ylacetic acid biosynthesis in tomato cell-free extracts*

Evidence for the presence in tomato shoots of all the enzymes required for the biosynthesis of indol-3-ylacetic acid via Pathways I and II was sought by assaying the activity of each enzyme in sterile, 35000*g* supernatant preparations of shoot-tip tissues. Table 1 shows that all the enzymes necessary for the metabolism of tryptophan via either of these pathways were present in the cell-free preparations, with the exception of an amine oxidase system presumably required for catalysing the conversion of tryptamine into indol-3-ylacetaldehyde (Clarke & Mann, 1957). The nature of the enzyme catalysing this reaction in tomato shoots is currently under investigation.

Tryptophan aminotransferase was the most active enzyme in these tomato preparations, its rate of activity was comparable with that previously found for the same enzyme in mung-bean cell-free extracts (Wightman & Cohen, 1968). The main indole product formed in the transamination reaction between L-tryptophan and α-oxoglutarate was identified as indol-3-ylpyruvic acid, which agrees with our earlier studies of tryptophan transamination in mung-bean extracts (Wightman & Cohen, 1968) and with the findings by Moore & Shaner (1968) using cell-free extracts of pea seedlings. Tryptophan decarboxylase activity was also present in the tomato supernatant preparation and, like the tryptophan aminotransferase activity, was found to be dependent on the presence of pyridoxal phosphate for optimum activity. The product of the decarboxylase activity was shown to be tryptamine (Gibson *et al.*, 1970, 1972*a*) which agrees with the results of Sherwin (1970), but not with those of Reed (1968), who examined the activity of tryptophan decarboxylase preparations from cucumber hypocotyl and pea epicotyl tissues respectively. Table 1 shows that the rate of activity of tryptophan decarboxylase in these tomato preparations was about 20-fold less than that of the corresponding aminotransferase.

High enzyme activities were also detected in the extracts for indol-3-ylacetaldehyde dehydrogenase and tryptophol dehydrogenase (Table 1). The aldehyde dehydrogenase was assayed by measuring the formation of indol-3-ylacetic acid from indol-3-ylacetaldehyde, when $NAD^+$ was included in the reaction mixture, whereas alcohol dehydrogenase activity was assayed by determining the rate of conversion of indol-3-ylacetaldehyde to tryptophol when NADH was added to the reaction system. The presence of indol-3-ylpyruvic acid decarboxylase activity was demonstrated indirectly by observing the conversion of indol-3-ylpyruvic acid into tryptophol, which presumably was formed by the action of

Table 1. *Activities of some of the enzymes probably involved in indol-3-ylacetic acid biosynthesis in cell-free preparations of tomato shoots*

| Enzyme assayed | Activity (μg of product formed/h per mg of protein) |
|---|---|
| Tryptophan aminotransferase (tryptophan → indol-3-ylpyruvic acid) | 48.0 |
| Tryptophan decarboxylase (tryptophan → tryptamine) | 2.3 |
| Indol-3-ylpyruvic acid decarboxylase (indol-3-ylpyruvic acid → indol-3-ylacetaldehyde) | Present, but activity was measured indirectly |
| Indol-3-ylacetaldehyde dehydrogenase (indol-3-ylacetaldehyde → indol-3-ylacetic acid) | 7.8 |
| Tryptophol dehydrogenase (indol-3-ylacetaldehyde → tryptophol) | 17.4 |

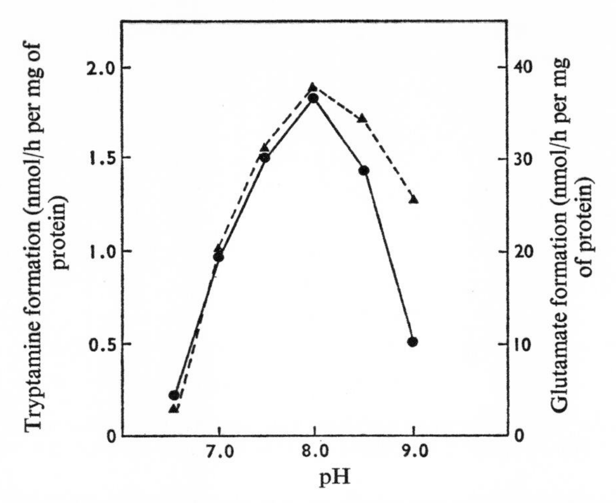


Fig. 7. *Effect of pH on tryptamine formation by tryptophan decarboxylase (●) and glutamic acid formation by tryptophan aminotransferase (▲) in tomato shoot cell-free extracts*

tryptophol dehydrogenase on the indol-3-ylacetaldehyde produced by indol-3-ylpyruvic acid decarboxylase activity. Both pyridoxal phosphate and NADH were included in this assay system, but it was found that the metabolism of indol-3-ylpyruvic acid to tryptophol was not dependent on the presence of pyridoxal phosphate. The overall rate of these two reactions was calculated as about 11$\mu$g of tryptophol formed/h per mg of protein, but since the production of tryptophol from indol-3-ylpyruvic acid also required the presence of tryptophol dehydrogenase, it is possible that the activity of the indol-3-ylpyruvic acid decarboxylase might have been higher than the present results indicate.

In view of the possible role that tryptophan aminotransferase and tryptophan decarboxylase might play in regulating the conversion of tryptophan into indol-3-ylacetic acid in tomato via Pathways I and II, and also because of the similarity in their general properties (e.g. pH optima, substrate concentration optima,

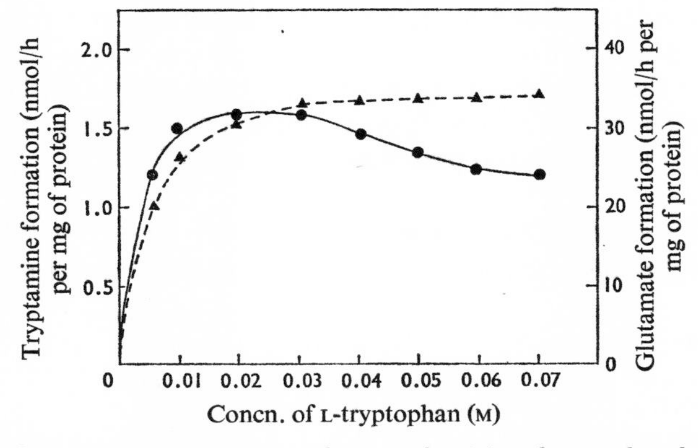


Fig. 8. *Effect of substrate concentration on the rate of activity of tryptophan decarboxylase (●) and tryptophan aminotransferase (▲) in tomato shoot cell-free extracts*

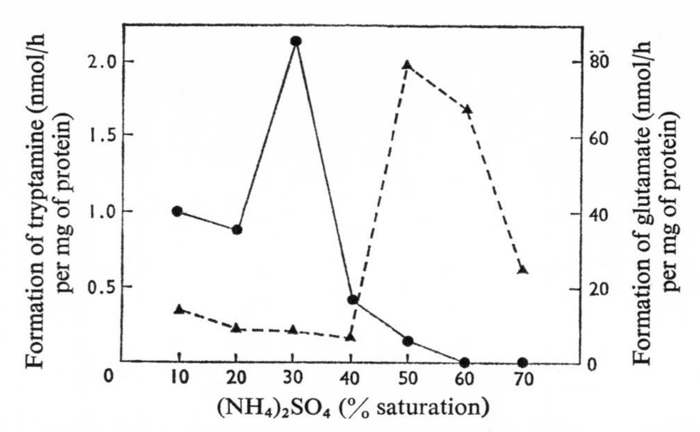


Fig. 9. *Distribution of tryptophan decarboxylase (●) and tryptophan aminotransferase (▲) activities in different $(NH_4)_2SO_4$ fractions of tomato shoot cell-free extracts*

pyridoxal phosphate dependence) as shown by their activity in 35000*g* supernatant preparations (Figs. 7 and 8), it was decided to isolate and purify the two enzymes to characterize their activities more fully. Separation of the enzymes was achieved by $(NH_4)_2SO_4$ fractionation of a supernatant preparation (Fig. 9) and further purification was then carried out by sequential Sephadex chromatography (Gibson *et al.*, 1972*b*). The amount of purification obtained with each enzyme

Table 2. *Purification of* L-*tryptophan–α-oxoglutarate aminotransferase from tomato shoot homogenate*

The enzyme specific activity was calculated as μmol of glutamate formed/h per mg of protein.

| Purification step | Specific activity | Purification (fold) | Recovery (%) |
|---|---|---|---|
| 35000*g* supernatant | 0.30 | 1.0 | 100 |
| 40–70%-satd.-$(NH_4)_2$ $SO_4$ precipitation | 1.35 | 4.5 | 74 |
| Sephadex G-100 | 4.65 | 15.5 | 64 |
| DEAE-Sephadex | 22.20 | 74.0 | 53 |
| Sephadex G-200 | 45.30 | 151.0 | 40 |

at different steps in these procedures are shown in Tables 2 and 3. Tryptophan aminotransferase was found to be a very stable enzyme and a purification of about 150-fold was achieved during a 4–5 day period, with a final recovery of nearly 40% of the original activity. In contrast, tryptophan decarboxylase was a much less stable enzyme and 80% of its original activity was lost within 48h, by which time the enzyme had been purified about 85-fold.

Table 3. *Purification of* L-*tryptophan decarboxylase from tomato shoot homogenate*

The enzyme specific activity was calculated as nmol of tryptamine formed/h per mg of protein.

| Purification step | Specific activity | Purification (fold) | Recovery (%) |
|---|---|---|---|
| 35000*g* supernatant | 1.21 | 1.0 | 100 |
| 20–40%-satd.-$(NH_4)_2$ $SO_4$ precipitation | 3.2 | 2.6 | 91 |
| Sephadex G-200 | 21.0 | 17.6 | 69 |
| Sephadex concentration | 21.0 | 17.6 | 39 |
| Sephadex G-200 | 104.0 | 86.0 | 20 |

Table 4. *Amino acid substrate specificity of partly purified* (150-*fold*) *aromatic aminotransferase system from tomato shoots*

| Amino acid tested (0.04 M) | Aminotransferase activity with 0.01 M-α-oxoglutarate (μmol of glutamate/h per mg of protein) | (%) |
|---|---|---|
| L-Tryptophan | 45.3 | 100 |
| L-Phenylalanine | 34.2 | 75 |
| L-Tyrosine | 26.3 | 58 |
| D-Tryptophan | 0 | 0 |
| D-Phenylalanine | 0 | 0 |
| D-Tyrosine | 0 | 0 |

Examination of the substrate specificity of these purified fractions revealed that the aminotransferase preparation was able to catalyse transamination of the L-isomers of all three aromatic amino acids (Table 4), whereas the decarboxylase preparation showed activity only with L-tryptophan and its closely related derivative, L-5-hydroxytryptophan (Table 5). Although the aminotransferase preparation used in these substrate-specificity determinations was not completely pure, the consistency of the transamination results obtained with the preparation at each step in the purification procedure strongly suggests that the aminotransferase present in the 150-fold purified fraction was, in fact, a multispecific enzyme capable of catalysing the transamination of all three aromatic L-amino acids. This aromatic aminotransferase from tomato shoots was thus very similar in its substrate specificity to the multispecific aminotransferase recently purified from bushbean seedlings by Forest & Wightman (1972). The partly purified decarboxylase preparation, on the other hand, showed high substrate specificity, particularly for the L-isomer of tryptophan. Whereas other aromatic amino acid decarboxylases may occur in tomato, it is evident from the present study that the conversion of tryptophan into tryptamine by Pathway I is catalysed in tomato shoots by a highly specific L-tryptophan decarboxylase.

*Metabolism experiments with excised tomato shoots in vivo*

In these experiments, excised shoots of 6-week-old tomato plants were fed with solutions of either [3-$^{14}$C]tryptophan or [2-$^{14}$C]tryptamine (100 μCi/200 g fresh wt. of shoot tissues). The shoots were allowed to metabolize the indole compound for 24 h, after which they were homogenized in cold methanol and the radioactive metabolites extracted into neutral, acid and basic ether fractions.

Table 5. *Amino acid substrate specificity of a partly purified* (20-*fold*) *tryptophan decarboxylase preparation from tomato shoots*

| Amino acid tested (0.02 M) | Decarboxylase activity (nmol of product formed/h per mg of protein) | (%) |
|---|---|---|
| L-Tryptophan | 21.5 | 100 |
| L-Phenylalanine | 0 | 0 |
| L-Tyrosine | 0 | 0 |
| L-5-Hydroxytryptophan | 11.0 | 52 |
| D-Tryptophan | 0 | 0 |
| D-Phenylalanine | 0 | 0 |
| D-Tyrosine | 0 | 0 |

Table 6. *Distribution of radioactivity in the indole compounds isolated from tomato shoots 24 h after feeding with either* [3-$^{14}$C]*tryptophan or* [2-$^{14}$C]*tryptamine*

| Compound | Unfed Zero time (μg/g wt.) | Unfed 24 h (μg/g wt.) | Feeding soln. (d.p.m./μg) | Shoots fed with [3-$^{14}$C]tryptophan (μg/g fresh wt.) | Shoots fed with [3-$^{14}$C]tryptophan (c.p.m.*/g fresh wt.) | Shoots fed with [3-$^{14}$C]tryptophan (Dilution†) | Shoots fed with [2-$^{14}$C]tryptamine (μg/g fresh wt.) | Shoots fed with [2-$^{14}$C]tryptamine (c.p.m./g fresh wt.) | Shoots fed with [2-$^{14}$C]tryptamine (Dilution) |
|---|---|---|---|---|---|---|---|---|---|
| Tryptophan | 12.0 | 12.0 | 2750 | 313.0 | 497000 | 1.5 | 14.0 | 600 | 52.2 |
| Tryptamine | 1.0 | 1.0 | 2600 | 1.6 | 710 | 5.4 | 160.0 | 28000 | 1.3 |
| Indol-3-yl-lactic acid | 0.005 | 0.005 | — | 0.102 | 127 | 2.0 | 0.005 | — | — |
| Tryptophol | 0.005 | 0.005 | — | 0.014 | 185 | 1.9 | 0.066 | 154 | 1.0 |
| Indol-3-ylacetic acid | 0.05 | 0.05 | — | 0.06 | 53 | 2.8 | 0.056 | 19 | 6.6 |

* Corrected for a counting efficiency of 89%.

† $\text{Dilution} = \dfrac{\text{Sp. radioactivity of feeding soln. (d.p.m./}\mu\text{g)}}{\text{Sp. radioactivity of compound (c.p.m./}\mu\text{g)}}$

Table 6 shows the amounts of radioactivity found in the indole metabolites identified in the various ether fractions obtained from the plants fed with [$^{14}$C]-tryptophan and [$^{14}$C]tryptamine. Radioactivity from both labelled substrates was incorporated into indol-3-ylacetic acid, confirming the results from our earlier enzyme experiments that both [$^{14}$C]tryptophan and [$^{14}$C]tryptamine can serve as precursors of the growth hormone in tomato shoot tissue. The relative differences in the amount of radioactivity incorporated into indol-3-ylacetic acid during metabolism of the two $^{14}$C-labelled substrates indicates that tryptophan was a much better precursor of auxin than was tryptamine. These results thus confirm the earlier findings with cell-free extracts that the predominant route of indol-3-ylacetic acid biosynthesis in tomato occurs via Pathway I.

Evidence for the formation of all the intermediates required for the operation of Pathways I and II for tryptophan metabolism *in vivo* was obtained in these experiments with excised shoots. After extraction of the radioactive metabolites from plants fed on [$^{14}$C]tryptophan, carrier indol-3-ylpyruvic acid was added to the acid ether fraction and carrier indol-3-ylacetaldehyde to the neutral ether fraction, after which the Dnp-hydrazone of each compound was prepared in the ether fraction and then isolated by ethyl acetate extraction followed by t.l.c. Both hydrazones were found to be radioactive (Fig. 10). Therefore, all the compounds proposed as intermediates in the conversion of tryptophan into indol-3-ylacetic acid via the indol-3-ylpyruvic acid pathway (I) were found to be labelled in tomato shoots after the administration of [3-$^{14}$C]tryptophan.

Table 6 shows that radioactivity from [3-$^{14}$C]tryptophan was also found in tryptamine. When [2-$^{14}$C]tryptamine was fed to shoots, radioactivity could be found not only in indol-3-ylacetic acid, but also in indol-3-ylacetaldehyde, when

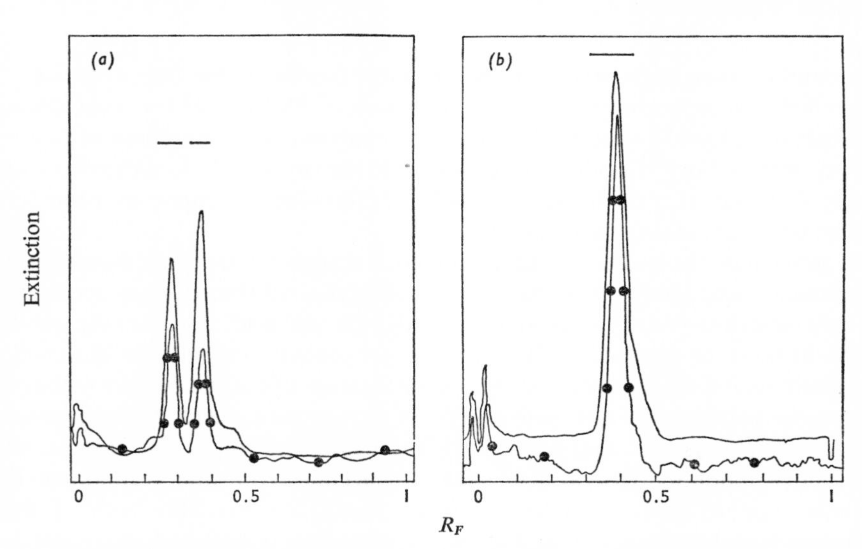


Fig. 10. *Densitometer and radioactivity scans of thin-layer chromatograms of suspected* (*a*) [$^{14}$*C*]*indol-3-ylpyruvic acid Dnp-hydrazone and* (*b*) [$^{14}$*C*]*indol-3-ylacetaldehyde Dnp-hydrazone isolated from the appropriate ether fraction of tomato shoots fed with* [3-$^{14}$*C*]*tryptophan*

——, Densitometer scan at 410nm; ●, radioactivity scan. The solvent systems used were (*a*) solvent VI and (*b*) solvent VIII.

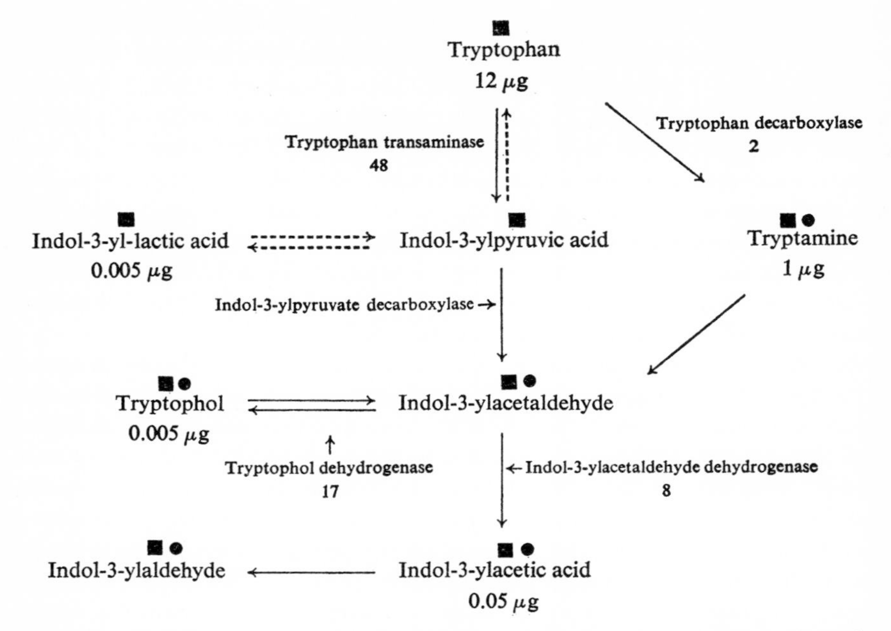


Scheme 1. *Summary of evidence for the pathways of indol-3-ylacetic acid biosynthesis from tryptophan in tomato shoots*

The numbers below the names of indole compounds show the concentration of each compound in normal, unfed shoots (μg/g fresh wt.). The numbers below the names of enzymes show the rate of activity of each enzyme (μg of product formed/h per g fresh wt.). ■, Metabolites labelled with $^{14}C$ after feeding [3-$^{14}C$]tryptophan to shoots; ●, metabolites labelled with $^{14}C$ after feeding [2-$^{14}C$]tryptamine to shoots.

this compound was isolated from the neutral ether fraction as the Dnp-hydrazone. These findings are consistent with the operation of Pathway II for tryptophan metabolism in tomato shoots, although a comparison of the amounts of radioactivity obtained in [$^{14}C$]indol-3-ylacetic acid in the two metabolism experiments (Table 6) indicate that the tryptamine pathway (II) is the less-important route for the formation of indol-3-ylacetic acid.

To summarize the evidence so far presented, it is apparent from our metabolism experiments with shoot-tip preparations *in vitro* that all the enzymes necessary for the conversion of L-tryptophan into indol-3-ylacetic acid, via either the indol-3-ylpyruvate (I) or tryptamine (II) pathways, are present in tomato shoot tissues. The qualitative evidence obtained from identification of the radioactive metabolites produced during the metabolism of [3-$^{14}C$]tryptophan and [2-$^{14}C$]tryptamine by the cell-free systems, and particularly the demonstration of the presence of radioactive indol-3-ylpyruvic acid, indol-3-ylacetaldehyde and tryptamine in systems supplied with [3-$^{14}C$]tryptophan, clearly suggests that both of the proposed biosynthetic pathways for the conversion of tryptophan into indol-3-ylacetic acid may be operating in tomato plants (Scheme I). The results from our metabolism experiments with excised shoots strongly support this view, and the quantitative results showing the amounts of radioactivity present in indol-3-ylacetic acid and in the expected intermediates formed when either

[3-$^{14}$C]tryptophan or [2-$^{14}$C]tryptamine was fed to the shoots, further indicate that the indolylpyruvate pathway is the predominant route in tomato for indol-3-ylacetic acid synthesis *in vivo*.

*Experiments to determine the site(s) of auxin synthesis in vegetative shoots*

Our most recent studies on auxin metabolism have been concerned with attempts to determine the main site(s) of indol-3-ylacetic acid synthesis in vegetative shoots of 6-week-old tomato plants. A series of experiments have been carried out in which solutions of [3-$^{14}$C]tryptophan were fed to excised whole shoots and to groups of shoot tips and leaves of different ages excised from similar shoots before the feeding procedure. The whole shoots and the groups of shoot tips and leaves were allowed to metabolize the absorbed radioactive tryptophan for 24h under the standard conditions described above. At the end of this time, the shoot tip and all the leaves on the batch of whole shoots were excised, collected into groups and together with the remaining stems were subjected separately to the methanol extraction and ether fractionation procedures used in the earlier metabolism experiments *in vivo*. The groups of shoot tips and leaves of different ages excised from shoots before feeding with [3-$^{14}$C]tryptophan were also subjected, after the 24h metabolism period, to the same extraction and ether fractionation procedures to isolate [$^{14}$C]indol-3-ylacetic acid and other radioactive indole products. The acid ether fractions from the different tissues were all initially analysed by paper chromatography in solvent I, and the radioactive band on each chromatogram corresponding to the position of marker [2-$^{14}$C]indol-3-ylacetic acid was then eluted and the metabolite re-chromatographed on buffered paper (pH 7.0) with butanol–ethanol–water (4:4:1, by vol.) as the developing solvent. This second chromatography completely separated [$^{14}$C]indol-3-ylacetic acid from all other possible acidic metabolites (e.g. [$^{14}$C]indol-3-yl-lactic acid) and so allowed an accurate estimation of the relative amounts of [$^{14}$C]indol-3-ylacetic

Table 7. *Determination of the relative amounts of* [$^{14}$*C*]*indol-3-ylacetic acid extracted from shoot tips, stems and leaves of different ages excised from whole shoots after feeding with* [3-$^{14}$*C*]*tryptophan for 24 h*

Values are also given for the determination of [$^{14}$C]indol-3-ylacetic acid extracted from excised shoot organs fed separately with [3-$^{14}$C]tryptophan for 24h

| | Area of [$^{14}$C]indol-3-ylacetic acid peak on chromatograms | |
|---|---|---|
| Whole shoot feeding | (mm$^2$/g fresh wt.) | (mm$^2$/organ) |
| Shoot tip | 6.9 | 44 |
| Leaf 1 | 39.0 | 236 |
| Leaf 2 | 25.5 | 186 |
| Leaf 3 | 20.0 | 152 |
| Leaf 4 | 14.7 | 132 |
| Stem | 4.6 | 46 |
| Separate organ feeding | | |
| Shoot tip | 15.7 | 94 |
| Leaf 1 | 66.6 | 399 |
| Leaf 2 | 33.0 | 264 |
| Leaf 3 | 24.7 | 222 |
| Leaf 4 | 17.0 | 151 |

Table 8. *Determination of the relative amounts of* [$^{14}$*C*]*indol-3-ylacetic acid extracted from cell-free preparations of tomato shoot tips, leaves and stems after metabolism of* [3-$^{14}$*C*] *tryptophan for* 6*h*

| | Area of [$^{14}$C]indol-3-ylacetic acid peak on chromatograms | |
|---|---|---|
| | (cm$^2$/mg of protein) | (cm$^2$/organ) |
| Shoot tip | 0.41 | 16.5 |
| Leaf 1 | 0.46 | 23.5 |
| Leaf 2 | 0.35 | 19.2 |
| Leaf 3 | 0.29 | 16.6 |
| Leaf 4 | 0.22 | 10.9 |
| Stem | 0.39 | 19.6 |

acid formed in shoot-tip tissues and in leaves of different ages in the two experimental situations.

The results (Table 7) show the relative amounts of [$^{14}$C]indol-3-ylacetic acid produced by each of the shoot organs, expressed in terms of the peak area obtained on chromatograms/g fresh wt. of tissue, and per organ. The results clearly show that shoot tips and all the leaves examined in the two feeding situations were able to synthesize [$^{14}$C]indol-3-ylacetic acid from radioactive tryptophan, and that young expanding leaves (leaves 1 and 2) rather than shoot-tip tissues (i.e. stem apex and surrounding very young leaves) were the most active sites of auxin synthesis. These findings have been confirmed in two other similar feeding experiments, and a study of the relative abilities of cell-free supernatant preparations of tomato shoot tips, stem tissue and leaves of different ages to metabolize [3-$^{14}$C]tryptophan also showed (Table 8) that the enzymic systems present in expanding leaves 1 and 2 had the highest capacity to synthesize [$^{14}$C]indol-3-ylacetic acid.

*Demonstration of the biosynthesis and natural occurrence of phenylacetic acid in tomato shoots*

In view of our earlier finding that the enzyme in tomato shoots catalysing the conversion of L-tryptophan into indol-3-ylpyruvic acid is a multispecific aminotransferase capable of catalysing the transamination of all three L-aromatic amino acids, and bearing in mind the observation of earlier workers that L-phenylalanine, phenylpyruvic acid and phenyl-lactic acid are interconvertible in wheat shoots (Gamborg & Neish, 1959) and that L-[$^{14}$C]phenylalanine can be metabolized to radioactive phenylacetic acid when supplied to growing cultures of the fungus, *Schizophyllum commune* (Moore & Towers, 1967), we decided to investigate the possibility that L-phenylalanine might be metabolized in tomato to phenylacetic acid via a sequence of reactions similar to those involved in the conversion of L-tryptophan into indol-3-ylacetic acid by Pathways I and II. If formation of phenylacetic acid occurs in growing vegetative shoots, then it may have important physiological implications since, as stated earlier, phenylacetic acid is known to exhibit growth-promoting activity in a range of vegetative and fruit tissues.

Excised 6-week-old tomato shoots were fed, in the usual manner, with solutions of DL-[3-$^{14}$C]phenylalanine and after 24h, the radioactive metabolites were

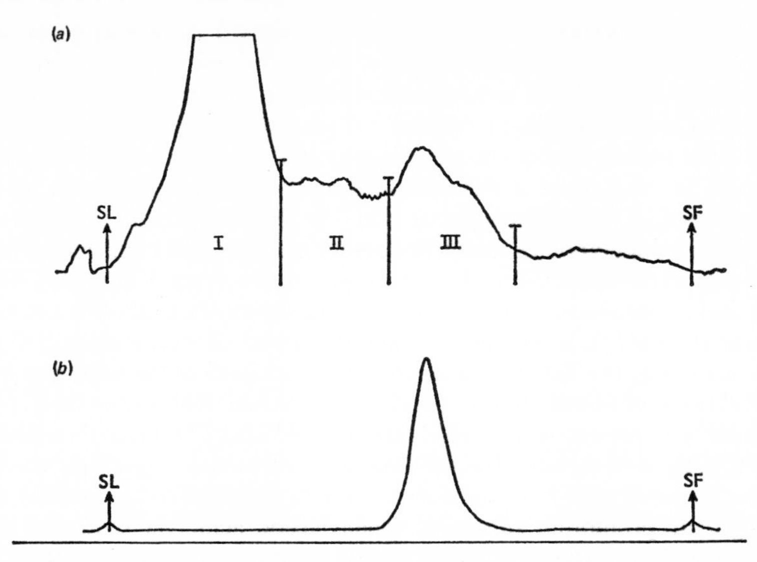


Fig. 11. *Radioactivity scans of chromatograms of* (*a*) *an acid ether fraction from tomato shoots fed with* [3-$^{14}$C]*phenylalanine and* (*b*) *authentic* [2-$^{14}$C]*phenylalanine*

Both chromatograms were developed in solvent I.

extracted into methanol and isolated into acidic, basic and neutral ether fractions. All fractions were analysed by paper chromatography or t.l.c. and the presence of radioactive compounds on the chromatograms was revealed by using radio-chromatogram-scanning equipment.

Fig. 11 shows the radioactivity scan obtained from the paper chromatogram of the acid ether fraction developed in solvent system I. From its $R_F$ value and subsequent other chromatographic properties, the metabolite forming the main radioactive band (I) on this chromatogram was identified as [$^{14}$C]*N*-malonyl-phenylalanine. This finding is consistent with the earlier results of Rosa & Neish

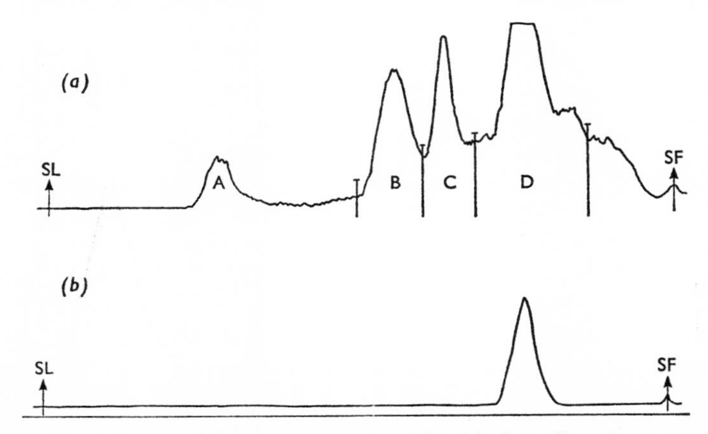


Fig. 12. *Radioactivity scans of chromatograms containing* (*a*) *the radioactive compounds eluted from bands II and III of Fig.* 11(*a*) *and* (*b*) *authentic* [2-$^{14}$C]*phenylalanine*

Both chromatograms were multiple developed (three times) in solvent I.

(1968) on the formation of such malonyl derivatives in several plant species supplied with radioactive phenylalanine. When radioactive bands II and III were eluted, combined and rechromatographed in solvent system I by using the technique of multiple development, the $^{14}C$ scan shown in Fig. 12 was obtained. After further elution and re-chromatography in three other solvent systems, the compound forming band B was identified as [$^{14}C$]phenyl-lactic acid, and that forming band C as [$^{14}C$]phenylpyruvic acid. The $^{14}C$ scan of band D indicated that two radioactive metabolites might be present and this was confirmed when the compounds were eluted and rechromatographed in solvent V (Fig. 13). Further elution and re-chromatography of these two radioactive metabolites established that band (1) was [$^{14}C$]cinnamic acid, and that band (2) was, indeed, [$^{14}C$]phenylacetic acid (Fig. 14). Similar chromatographic analysis of the basic and neutral ether fractions obtained in this experiment showed that radioactive phenylethylamine was present in the basic fraction and that [$^{14}C$]phenylacetaldehyde occurred in the neutral ether fraction. All these observations are fully consistent with our proposal that L-phenylalanine may be metabolized in tomato shoots to the growth-promoting substance, phenylacetic acid, via sequence of biochemical reactions (Scheme 2) which appear identical with those involved in the conversion of L-tryptophan into indol-3-ylacetic acid.

This noteworthy finding led us to investigate the possible natural occurrence of phenylacetic acid in untreated tomato shoots. Since phenylacetic acid, unlike indol-3-ylacetic acid does not react with all the known colour reagents used to visualize aromatic compounds on chromatograms, the presence of phenylacetic

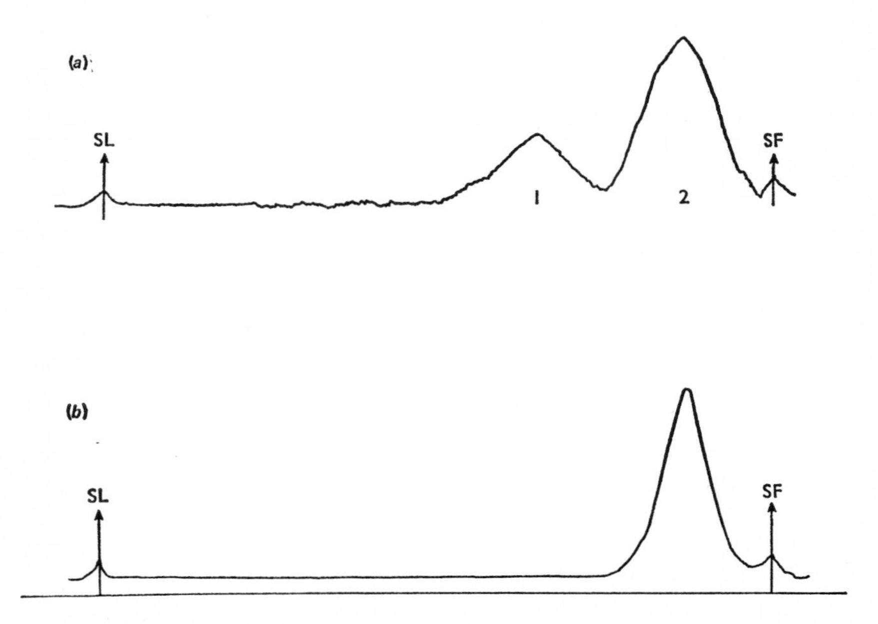


Fig. 13. *Radioactivity scans of chromatograms containing (a) the radioactive compounds eluted from band D from Fig.* 12(*a*) *and* (*b*) *authentic* [2-$^{14}$C]*phenylalanine*

Both chromatograms were developed in 8% (w/v) NaCl.

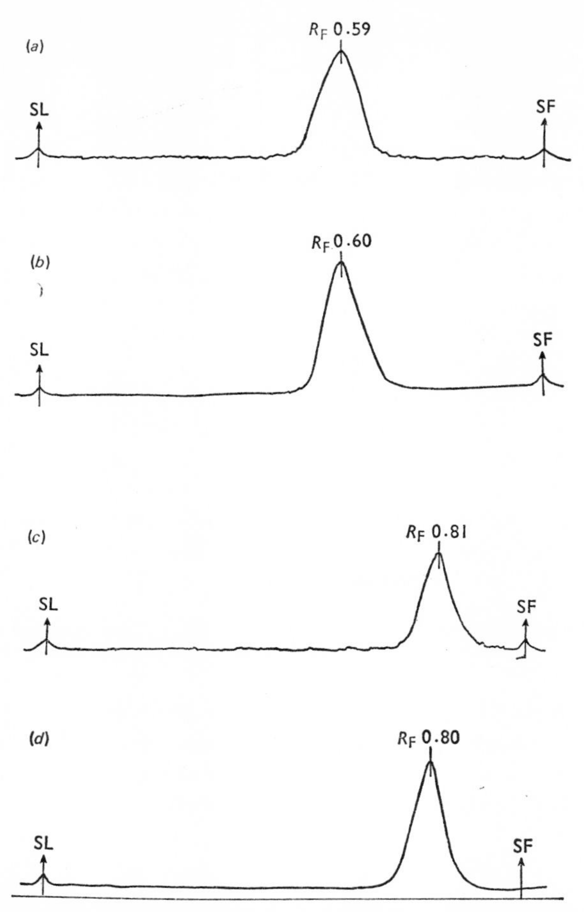


Fig. 14. *Radioactivity scans of chromatograms containing suspected [$^{14}$C]phenylacetic acid eluted from band* (2) *in Fig.* 13 (*a and c*) *and authentic* [2-$^{14}$C]*phenylalanine* (*b and d*)

Chromatograms were developed in solvent system I (*a* and *b*), and in carbon tetrachloride–acetic acid (50 : 1, v/v) (*c* and *d*).

acid together with indol-3-ylacetic acid in acid ether extracts of kg quantities of 6-week-old tomato shoots was demonstrated by using the chromatogram-bioassay procedure described by Fawcett *et al.* (1960). The results obtained when paper chromatograms of such acid ether extracts were bioassayed by using the wheat coleoptile-segment test are shown in Fig. 15. Clearly, two regions showing growth-promoting activity were present on the chromatogram, the first, $R_F$ value 0.45–0.55, coincided with the position of authentic [2-$^{14}$C]indol-3-ylacetic acid and the second $R_F$ 0.60–0.70 coincided with the position of authentic [2-$^{14}$C]phenylacetic acid. When the compounds in these two regions on equivalent chromatograms were eluted, each eluate re-chromatographed separately in solvent system I and the two chromatograms examined with the wheat coleoptile-bioassay procedure, the two compounds again exhibited growth-promoting activity (Fig. 16). Since the $R_F$ values of these two active zones also coincided with the position of marker samples of [2-$^{14}$C]indol-3-ylacetic acid and [2-$^{14}$C]phenylacetic acid,

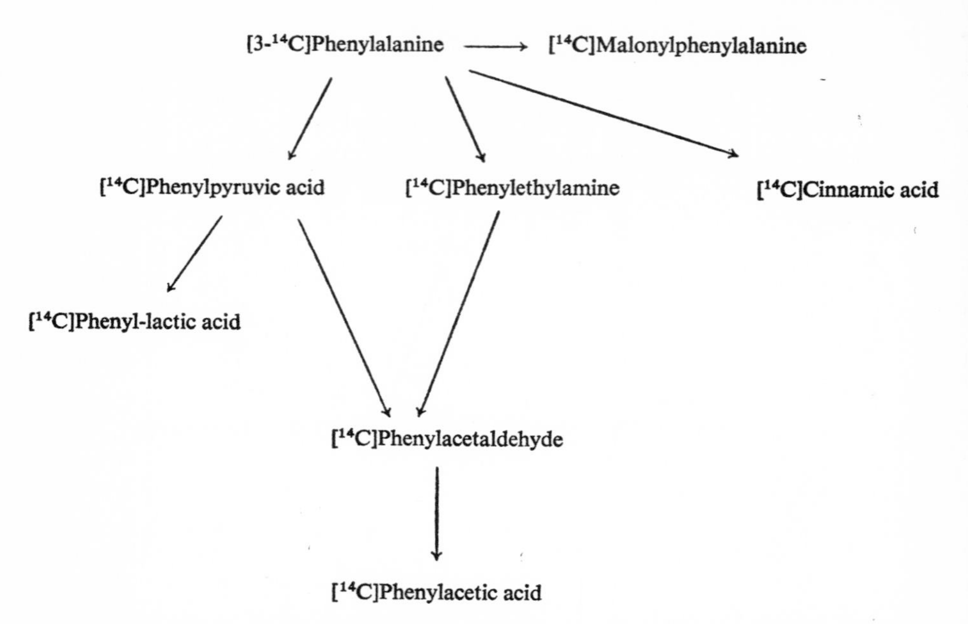


Scheme 2. *Biochemical pathways for the metabolism of* DL-[3-$^{14}$*C*]*phenylalanine to* [$^{14}$*C*]*phenylacetic acid and other phenyl compounds in tomato shoots*

it is concluded from this bioassay evidence that phenylacetic acid together with indol-3-ylacetic acid are natural auxins in tomato shoots. Conclusive proof of this finding should come from g.l.c. analysis of ether extracts of tomato shoots and this work is presently in progress in my laboratory.

## Conclusions

The qualitative evidence obtained from the identification of the radioactive metabolites produced during the metabolism of [3-$^{14}$C]tryptophan and [2-$^{14}$C]-tryptamine by cell-free extracts of tomato shoots, and the enzyme activities demonstrated in such preparations, all combine to indicate that both the indolyl-pyruvate and tryptamine pathways (I and II) for the biosynthesis of indol-3-ylacetic acid can operate in tomato shoots. The qualitative results from metabolism experiments with excised shoots *in vivo* fully support this conclusion, and the

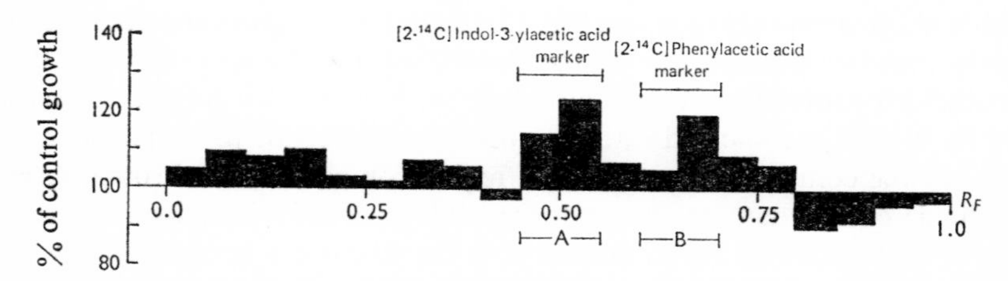


Fig. 15. *Activity recorded in the wheat coleoptile bioassay of a chromatogram of acid ether fraction of untreated tomato shoots*

The chromatogram was developed in solvent system I.

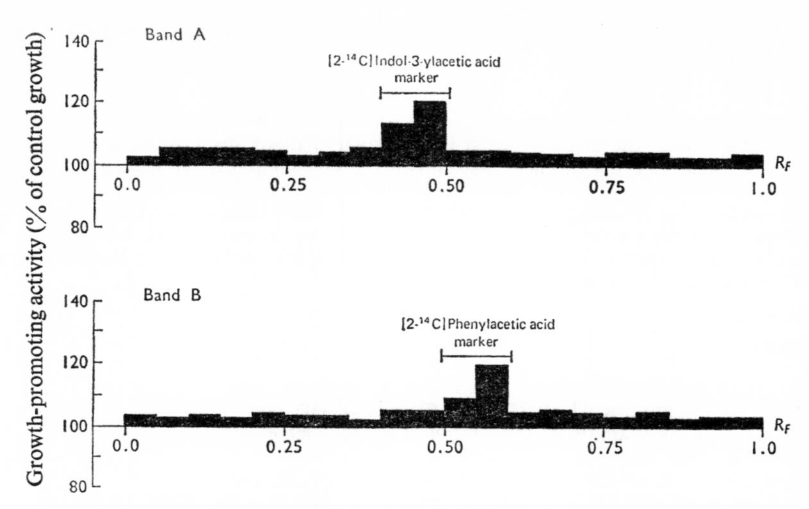


Fig. 16. *Activity recorded in the wheat coleoptile bioassay of chromatograms containing the growth-promoting substances eluted from bands A and B in Fig.* 15

Both chromatograms were developed in solvent system I.

quantitative results showing the amounts of radioactivity present in indol-3-ylacetic acid and in the expected intermediates formed when [3-$^{14}$C]tryptophan or [2-$^{14}$C]tryptamine was fed to shoots, further indicate that the pathway via indol-3-ylpyruvic acid is the predominant route *in vivo* in tomato for the conversion of L-tryptophan into indol-3-ylacetic acid.

The results from our experiments to investigate the site(s) of auxin synthesis in tomato shoots have shown that, contrary to the generally accepted view, the formation of indol-3-ylacetic acid is not confined to tissues at the shoot apex, but readily also occurs in all developing and mature leaves of a vegetative shoot system. The rapidly expanding leaves in the upper part of the shoot appear to have the highest capacity for auxin synthesis, and this finding agrees well with the results of Scott & Briggs (1960) who examined the natural auxin content of leaves and internodes on shoots of 'Alaska' pea seedlings.

The qualitative evidence from metabolism experiments *in vivo* with L-[3-$^{14}$C]-phenylalanine indicate that this amino acid can be converted in tomato shoots into the growth-promoting substance, phenylacetic acid. Our demonstration of the formation of radioactive phenylpyruvic acid and phenyl-lactic acid, phenylethylamine and phenylacetaldehyde in shoots during these experiments strongly suggests that the metabolic pathways involved in the formation of phenylacetic acid from phenylalanine are the same, or very similar to those known to be responsible for the conversion of L-tryptophan into indol-3-ylacetic acid. The results from the bioassay examination of acid ether extracts of normal, unfed shoots indicate that phenylacetic acid, like indol-3-ylacetic acid, is a naturally occurring substance in tomato shoots. These experiments provide the first evidence for the biosynthesis and natural occurrence of phenylacetic acid in higher plants, and in view of the known growth-promoting activity of this compound, future considerations of the hormonal regulation of growth and development in vegetative shoots should take into account the presence of this new natural auxin.

I acknowledge the major contributions made to this work by my research assistants and graduate students over the last five years. In particular, I thank my Postdoctoral Associate, Dr. Elnora A. Schneider, my excellent assistants R. A. Gibson and K. Stanley, and my Ph.D. students, J. C. Forest and B. S. Rauthan. I also thank Mr. Hank Datema, our Greenhouse Supervisor, for his continuous production of excellent plant material, and also Dr. E. F. Schneider of the Canada Department of Agriculture, Ottawa, for his assistance in the determination of i.r. spectra. This work was supported by a Grant-in-aid of Research to the author from the National Research Council of Canada.

## References

Bentley, J. A. & Housley, S. (1952) *J. Exp. Bot.* **3**, 393–405
Black, R. C. & Hamilton, R. H. (1971) *Plant Physiol.* **48**, 603–606
Chamberlain, V. K. & Wain, R. L. (1971) *Ann. Appl. Biol.* **69**, 65–72
Clarke, A. J. & Mann, P. J. G. (1957) *Biochem. J.* **65**, 763–774
Dannenburg, W. N. & Liverman, J. L. (1957) *Plant Physiol.* **32**, 263–269
Fawcett, C. H., Wain, R. L. & Wightman, F. (1960) *Proc. Roy. Soc. Ser. B* **152**, 231–254
Forest, J. C. & Wightman, F. (1972) *Can. J. Biochem.* **50**, 813–829
Gamborg, O. L. & Neish, A. C. (1959) *Can. J. Biochem. Physiol.* **37**, 1277–1285
Gibson, R. A., Barrett, G. & Wightman, F. (1970) *Proc. Can. Soc. Plant Physiol.* **10**, 47
Gibson, R. A., Schneider, E. A. & Wightman, F. (1972*a*) *J. Exp. Bot.* **23**, 381–399
Gibson, R. A., Barrett, G. & Wightman, F. (1972*b*) *J. Exp. Bot.* **23**, 775–786
Gordon, S. A. (1954) *Annu. Rev. Plant Physiol.* **5**, 341–378
Gordon, S. A. (1956) in *The Chemistry and Mode of Action of Plant Growth Substances* (Wain, R. L. & Wightman, F., eds.), pp. 65–75, Butterworths Scientific Publications, London.
Gordon, S. A. & Sanchez Nieva, F. (1949) *Arch. Biochem.* **20**, 367–385
Henbest, H. B., Jones, E. R. H. & Smith, G. F. (1953) *J. Chem. Soc. London* 3796–3801
Jones, E. R. H., Henbest, H. B., Smith, G. F. & Bentley, J. A. (1952) *Nature (London)* **169**, 485–487
Khalifa, R. A. (1967) *Physiol. Plant.* **20**, 355–360
Kutácek, M. & Kefeli, V. I. (1968) in *Biochemistry and Physiology of Plant Growth Substances* (Wightman, F. & Setterfield, G., eds.), pp. 127–152, Runge Press, Ottawa
Larsen, P. (1944) *Dan. Bot. Ark.* **11**, 1–132
Larsen, P. (1951) *Annu. Rev. Plant Physiol.* **2**, 169–198
Libbert, E. & Silhengst, P. (1970) *Physiol. Plant.* **23**, 480–487
Libbert, E., Wichner, S., Duerst, E., Kunert, R., Kaiser, W., Manicki, A., Mateuffel, R., Riecke, E. & Schroder, R. (1968) in *Biochemistry and Physiology of Plant Growth Substances* (Wightman, F. & Setterfield, G., eds.), pp. 213–230, Runge Press, Ottawa
Lowry, O. H., Rosebrough, N. J., Farr, A. L. & Randall, R. J. (1951) *J. Biol. Chem.* **193**, 265–275
Mahadevan, S. (1963) in *Modern Methods of Plant Analysis* (Linskens, H. F. & Tracy, M. V., eds.), vol. 7, pp. 238–259, Springer-Verlag, Berlin
Moore, T. C. (1969) *Phytochemistry* **8**, 1109–1120
Moore, T. C. & Shaner, C. A. (1967) *Plant Physiol.* **42**, 1787–1796
Moore, T. C. & Shaner, C. A. (1968) *Arch. Biochem. Biophys.* **127**, 613–621
Moore, K. & Towers, G. H. N. (1967) *Can. J. Biochem.* **45**, 1659–1665
Muir, R. M. & Hansch, C. (1953) *Plant Physiol.* **28**, 218–232
Phelps, R. H. & Sequeira, L. (1967) *Plant Physiol.* **42**, 1161–1163
Pilet, P. E. (1964) in *Régulateurs Naturels de la Croissance Végétale* (Nitsch, J. P., ed.), vol. 123, pp. 543–558, C.N.R.S., Paris
Pybus, M. B., Wain, R. L. & Wightman, F. (1959) *Ann. Appl. Biol.* **47**, 593–600
Rauthan, B. S. & Wightman, F. (1972) *Proc. Can. Soc. Plant Physiol.* **12**, 45
Reed, D. J. (1968) in *Biochemistry and Physiology of Plant Growth Substances* (Wightman, F. & Setterfield, G., eds.), pp. 243–258, Runge Press, Ottawa
Reinert, J. (1954) *Fortschr. Bot.* **16**, 330–341
Rosa, N. & Neish, A. C. (1968) *Can. J. Biochem.* **46**, 797–806
Schneider, E. A., Gibson, R. A. & Wightman, F. (1972) *J. Exp. Bot.* **23**, 152–170
Scott, T. K. & Briggs, W. R. (1960) *Amer. J. Bot.* **47**, 492–499

Seeley, R. C., Fawcett, C. H., Wain, R. L. & Wightman, F. (1956) in *The Chemistry and Mode of Action of Plant Growth Substances* (Wain, R. L. & Wightman, F., eds.), pp. 234–247, Butterworths Scientific Publications, London
Sherwin, J. E. (1970) *Plant Cell Physiol.* **11**, 865–872
Sherwin, J. E. & Purves, W. K. (1969) *Plant Physiol.* **44**, 1303–1309
Skoog, F. (1937) *J. Gen. Physiol.* **20**, 311–334
Stowe, B. B. (1959) *Fortschr. Chem. Org. Naturst.* **17**, 248–297
Stowe, B. B. & Thimann, K. V. (1954) *Arch. Biochem. Biophys.* **51**, 499–516
Thimann, K. V. (1935) *J. Biol. Chem.* **109**, 279–291
Thimann, K. V. (1953) *Arch. Biochem. Biophys.* **44**, 242–243
von Kindl, H. (1968) *Hoppe-Seyler's Z. Physiol. Chem.* **349**, 519–520
Wightman, F. (1962) *Can. J. Bot.* **40**, 689–718
Wightman, F. (1964) in *Régulateurs Naturels de la Croissance Végétale* (Nitsch, J. P., ed.), vol. 123, pp. 191–212, C.N.R.S., Paris
Wightman, F. & Cohen, D. (1968) in *Biochemistry and Physiology of Plant Growth Substances* (Wightman, F. & Setterfield, G., eds.), pp. 273–288, Runge Press, Ottawa
Zimmerman, P. W. & Wilcoxon, F. (1935) *Contrib. Boyce Thompson Inst.* **7**, 209–229
Zimmermann, P. W., Hitchcock, A. E. & Wilcoxon, F. (1936) *Contrib. Boyce Thompson Inst.* **8**, 105–112

*Biochem. Soc. Symp.* (1973) **38**, 277–302

# Biosynthesis of Cyanogenic Glycosides

By ERIC E. CONN

*Department of Biochemistry and Biophysics, University of California, Davis, Calif.* 95616, *U.S.A.*

## Synopsis

The cyanogenic glycosides are a group of secondary plant products that is again attracting the interest of investigators in the life sciences. In the present review the cyanogenic glycosides of known structure are classified on the basis of their probable biosynthetic origin. The principles on which the methods of detection are based are discussed. Earlier biosynthetic studies on these compounds are reviewed, and some emphasis is placed on the fact that the established intermediates in the biosynthetic pathway (aldoximes, nitriles and α-hydroxynitriles) are not encountered in the metabolism of amino acids in animals and micro-organisms. More recent findings involving stereochemical aspects of biosynthesis are presented here, in outline form, for the first time. In addition, some initial findings in the genetic control of cyanogenic glucoside biosynthesis are described. Finally, the possible implication of two of these compounds as a cause of chronic cyanide poisoning in one major plant (cassava) food source is discussed.

## Introduction

The cyanogenic glycosides may be defined as glycosidic derivatives of α-hydroxynitriles. They have a wide distribution among the higher plants, but are also found in ferns and in some moths and millipedes. Cyanogenic glycosides will release hydrocyanic acid on treatment with dilute acids. However, the phenomenon of 'cyanogenesis', the production of HCN from these compounds is due to the action of enzymes usually present in the tissues of cyanophoric plants. The action of the enzymes is initiated by crushing or otherwise destroying the cellular structure of the plant. Although cyanogenic glycosides are undoubtedly the origin of the HCN in most plants, an alternate source of HCN has recently been described, the cyanolipids whose occurrence to date appear restricted to the family *Sapindaceae* (Mikolajczak *et al.*, 1970). The cyanolipids, which are esters of α-hydroxynitriles, will produce HCN on hydrolysis.

The scientific interest in the cyanogenic glycosides arises from at least three different areas. First, the toxicity of many cyanophoric plants can be directly attributed to the high concentration of HCN which they can produce. The leaves of sorghum and cherry laurel or the roots of cassava can contain between 50 and 250 mg of HCN/100 g of tissue. These and other plants have been responsible for many cases of acute cyanide poisoning of animals including man (Kingsbury, 1964; Montgomery, 1969). Secondly, the unusual chemical structure of these compounds has attracted the interest of organic chemists for more than a century. More recently plant biochemists have concentrated on the metabolism of these compounds in the plants in which they are found. A number of reviews on the chemistry and the metabolism of cyanogenic compounds have appeared (Conn, 1969; Conn & Butler, 1969; Eyjolfsson, 1970). Thirdly, considerable interest has recently been centred on the possible role of two of these compounds in the aetiology of tropical ataxic neuropathy and goitre. Put simply, these pathological conditions may be the result of chronic cyanide intoxication. The present review will concentrate primarily on recent developments in the biosynthesis of cyanogenic glycosides in higher plants, but brief reference to other areas of interest will be made.

## Chemical Structure

The most familiar of the cyanogenic glycosides is amygdalin (Fig. 1) a β-glycoside of D(−)-mandelonitrile or benzaldehyde cyanohydrin. This glycoside, which is found in many members of the Rosaceae, on hydrolysis yields 1 mol each of benzaldehyde and HCN and 2 mol of glucose. In this last feature, amygdalin is atypical for only it and two others of the 20 known cyanogenic glycosides contain disaccharide residues, the remainder being simple β-glucosides.

In his recent review of the cyanogenic glycosides, Eyjolfsson (1970) listed 18 compounds whose chemical properties have been described in various degrees of detail. These were classified into four groups on the basis of the chemical structures of the aglycones. Rather than employ Eyjolfsson's (1970) classification, Table 1 shows the 14 cyanogenic glycosides whose aglycones are known to be formed, or may be assumed to be formed, from structurally related protein amino acids. Table 1 shows that the aromatic amino acids, phenylalanine and tyrosine, are, or may be assumed to be, the precursors of the aglycones of all but three of the 14 listed. The variations which give rise to these 11 are several: first, amygdalin, vicianin and lucumin are disaccharide forms of the β-glucoside prunasin. If the biosynthesis of these disaccharides is analogous to that of other disaccharides, prunasin will be an immediate precursor of these compounds. Secondly, the presence of a chiral centre in mandelonitrile provides the opportunity for two β-glucosides, prunasin and sambunigrin, to occur. Both compounds exist, but do not appear to occur in identical species. Thus prunasin occurs in the families Rosaceae, Myrtaceae, Polypodiaceae, Saxifragaceae and Schrophulariaceae and sambunigrin occurs in the Caprifoliaceae, Mimosaceae and Olacaceae.

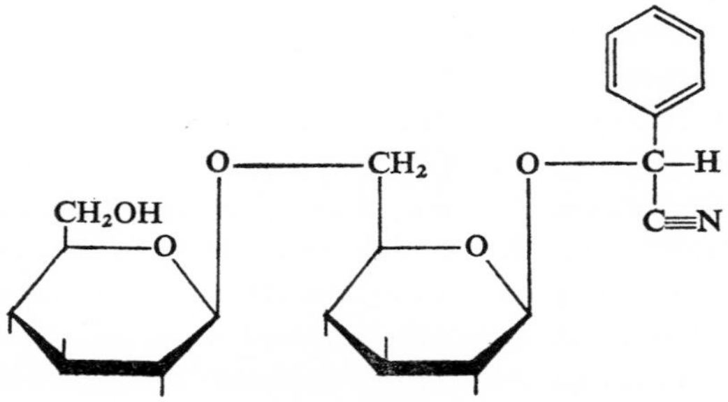

Fig. 1. *Structure of amygdalin*

Table 1. *Cyanogenic glycosides whose aglycones are formed or presumably are formed from protein amino acids*

All the compounds listed are β-glucosides (presumably pyranosides) except amygdalin, vicianin and lucumin. For structures of the last two as well as detailed references consult Eyjolfsson (1970). Where the configuration of the aglycone is established, it is specifically stated.

| Amino acid | Glycoside | Aglycone | Reference |
|---|---|---|---|
| L-Phenylalanine | Prunasin | (*R*)-Mandelonitrile | Herissey (1906); Mentzer & Favre-Bonvin (1961) |
| | Amygdalin | (*R*)-Mandelonitrile | Robiquet & Boutron-Charlard (1830); Abrol (1967) |
| | Vicianin | (*R*)-Mandelonitrile | Bertrand (1906); Tschiersch (1966) |
| | Lucumin | (*R*)-Mandelonitrile | Bachstez *et al.* (1948); Eyjolfsson (1971) |
| | Sambunigrin | (*S*)-Mandelonitrile | Bourquelot & Danjou (1905) |
| | Holocalin | *m*-Hydroxy-(*R*)-Mandelonitrile | Gmelin *et al.* (1973); Nahrstedt (1973) |
| | Zierin | *m*-Hydroxy-(*S*)-mandelonitrile | Finnemore & Cooper (1936); Nahrstedt (1973) |
| L-Tyrosine | Dhurrin | *p*-Hydroxy-(*S*)-mandelonitrile | Towers *et al.* (1964); Conn & Butler (1969) |
| | Taxiphyllin | *p*-Hydroxy-(*R*)-mandelonitrile | Towers *et al.* (1964) |
| | Nandina glucoside* | | Abrol *et al.* (1966) |
| | Proteacin | *p*-Glucosyloxymandelonitrile | Young & Hamilton (1966) |
| L-Valine | Linamarin | 2-Hydroxyisobutyronitrile | Jorissen & Hairs (1891); Butler & Butler (1960) |
| L-Isoleucine | (*R*)-Lotaustralin | (*R*)-2-Hydroxy-2-methylbutyronitrile | Bissett *et al.* (1969); Butler & Butler (1960) |
| L-Leucine | Acacipetalin | Dimethylketen cyanohydrin | Rimington (1935) |

* The sugar is bonded in glycosidic linkage to the *p*-hydroxyl group of *p*-hydroxymandelonitrile.

The chiral centre in *p*-hydroxymandelonitrile gives rise to the two epimeric cyanogenic glycosides that derive their aglycones from tyrosine. These compounds similarly do not occur in closely related families, dhurrin being reported only in the genus *Sorghum* of the Gramineae, whereas taxiphyllin occurs in members of the Taxaceae, Proteaceae and Euphorbiaceae. The aglycone *p*-hydroxymandelonitrile possesses an additional site for glycosylation and this position is thus occupied in proteacin and in the nandina glucoside. The latter compound is unique among the glycosides listed in Table 1 in that its α-hydroxyl group is not bound in glycosidic linkage. As such therefore the nandina glucoside is unstable and dissociates readily into *p*-glucosyloxybenzaldehyde and HCN.

Table 2 lists those glycosides whose aglycones are not derived, at least directly, from a proteinaceous amino acid. The structure given for triglochinin represents a recent revision (Ettlinger & Eyjolfsson, 1972) of an earlier proposal. The revised structure makes it possible to see triglochinin as being derived from a diphenolic 3,4-dihydroxymandelonitrile glucoside by oxidative cleavage between the aromatic hydroxyl groups. It may be predicted that triglochinin will be derived from tyrosine.

Table 2. *Cyanogenic glycosides whose aglycones are not formed directly from protein amino acids*

| Glycoside | Structure | Reference |
| --- | --- | --- |
| Triglochinin | 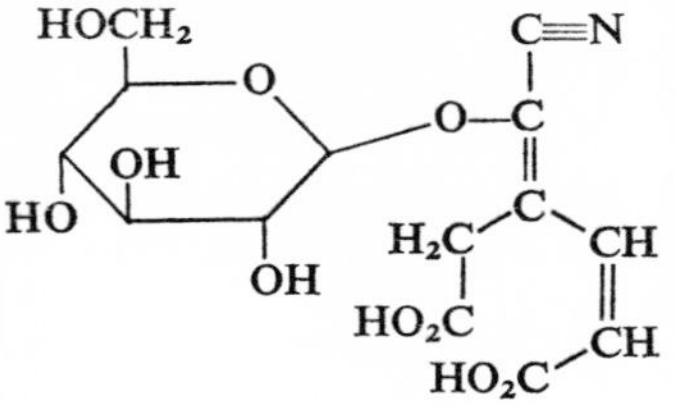 | Ettlinger & Eyjolfsson (1972) |
| Gynocardin | 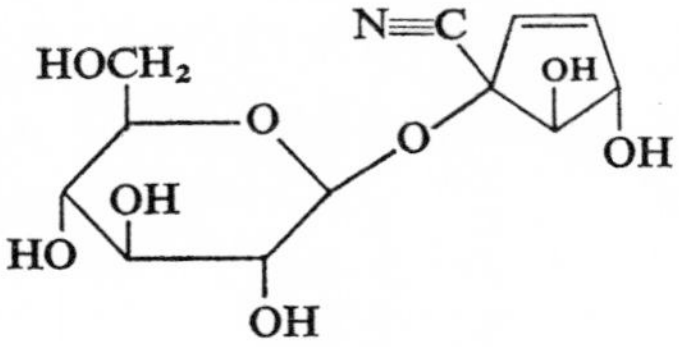 | Kim *et al.* (1970) |
| Deidaclin; tetraphyllin A | 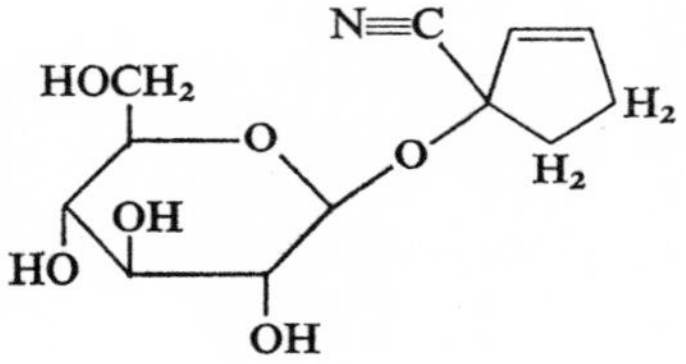 | Clapp *et al.* (1970); Russell & Reay (1971) |
| Barterin; tetraphyllin B | 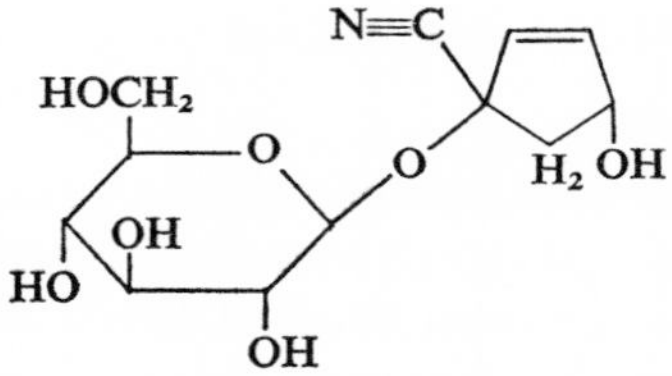 | Paris *et al.* (1969); Russell & Reay (1971) |
| Polydesmus glycoside | 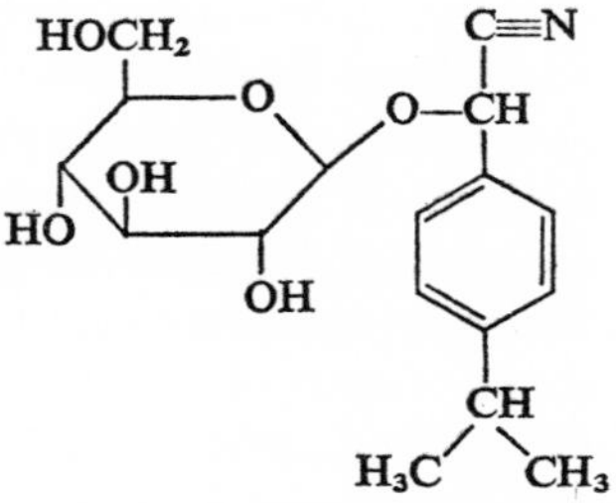 | Pallares (1945) |
| Acalyphin | Structure unknown | Rimington & Roets (1937) |

A group of cyanogenic glucosides (tetraphyllin A and B, deidaclin and barterin), that occur in the closely related families Flacourtiaceae and Passifloraceae and are related to gynocardin known since 1904, have recently been discovered. Although the structures of these compounds are known in considerable detail, the configuration of the carbon atom attached to the glucose moiety needs, with the exception of gynocardin (Kim *et al.*, 1970), to be established. The aglycones of these compounds could be formed from L-2-cyclopentene-1-glycine. This amino acid, however, has not yet been shown to occur in nature. Further work on the structure of acalyphin and the polydemus glycoside has not occurred since these compounds were first reported in the literature.

**Detection**

The phenomenon of cyanogenesis is dependent on (1) the presence of a cyanogenic substance in the organism being examined and in most cases, (2) the co-occurrence of an enzyme system that accomplishes the release of HCN from the cyanogen. Only in those cases where the cyanogen is unstable (e.g. the nandina glucoside) will HCN be produced in the absence of a degradative enzyme. A brief discussion of the enzymic degradation of these compounds is therefore in order; the enzymic degradation of dhurrin can serve as an example.

The initial reaction is the hydrolysis, by a $\beta$-glucosidase, of the glycosidic bond to form the free sugar and the aglycone, an $\alpha$-hydroxynitrile (cyanohydrin). The $\beta$-glucosidases of three cyanophoric plants, flax, sorghum and almond have been purified and their properties reviewed (Conn, 1969). The second reaction is the dissociation of the $\alpha$-hydroxynitrile to HCN and an aldehyde or ketone. Although this dissociation can occur non-enzymically, cyanophoric plants contain hydroxynitrile lyases which catalyse the reversible dissociation of the $\alpha$-hydroxynitrile into its dissociation products.

The hydroxynitrile lyase of sorghum (Seely *et al.*, 1966) and those of several members of the Rosaceae including almonds (Becker & Pfeil, 1966) have been highly purified and characterized. A particularly noteworthy aspect of this work is the observation (Gerstner & Pfeil, 1972) that the lyases purified from the members of the Rosaceae (which contain mandelonitrile glucoside) require FAD as a prosthetic group. The lyase of sorghum on the other hand, which uses as its preferred substrate *p*-hydroxymandelonitrile, is devoid of flavin. Since the dissociation of the hydroxynitrile does not involve an oxidation–reduction reaction, it is not clear why the enzymes from the members of the Rosaceae should be flavoproteins, although a structural role has been proposed for the flavin cofactor. It will be important to purify the hydroxynitrile lyase from a plant which contains sambunigrin to determine whether the sorghum lyase is atypical in this regard. The hydroxynitrile lyases in plants that contain linamarin and lotaustralin should also be examined.

The presence of a cyanogenic glycoside is indicated in a specimen when positive qualitative tests for the presence of HCN are obtained. Alternatively, tests for the presence of the aldehyde or ketone moiety may be performed, e.g. the characteristic 'almond-like' odour of benzaldehyde may be detected when glycosides of mandelonitrile are present. The fact that HCN may be easily detected by simple qualitative tests involving picric acid or nitrobenzaldehyde–dinitrobenzene [see Eyjolfsson (1970) for review] has, however, led to a problem in the literature of cyanogenesis. When the literature is examined, one finds that more than 800 species of plants representing 70 families are listed as cyanophoric. However, the nature of the cyanogenic compound which gave rise to the HCN is known in probably no more than 50 species presumably because the characterization of the aglycone of the glycoside was more difficult and time consuming than the simple detection of HCN. As with other secondary plant products, there is much interest in the use of the cyanogenic compounds as a taxonomic tool (Gibbs, 1963). However, until more precise information is available regarding the nature of the aglycones which give rise to the HCN, it is predictable that progress in this area will be slow.

Other obvious factors need to be taken into account when classifying a plant as cyanogenic. Whereas cyanide has been found in probably all structures of a plant, any one species may contain the cyanogen in only one portion of its structure. It is also well known that the amount of glycoside which a plant possesses at a given time can be subject to environmental factors such as soil, geographical location, climate, as well as the age of the plant.

Another important factor (discussed briefly below), concerns the genetics of cyanogenesis. Although this phenomenon does not seem to be widely distributed, in at least two species, *Lotus corniculatus* and *Trifolium repens*, the production of cyanide from a cyanogen is dependent on two independently inherited genes. Unless the plant possesses the genetic capacity to synthesize a glycosidase that hydrolyses the glucoside which is present in the plant, cyanide will not be released when the plant tissues are crushed even though an adequate amount of cyanogenic glycoside is present in the tissue.

## Distribution

The phenomenon of cyanogenesis occurs widely among the higher plants. The papers of Hegnauer [see Hegnauer (1971) for review] as well as his treatise *Chemotaxonomie der Pflanzen* (Hegnauer, 1962–69) are primary sources for determining whether a given species is cyanophoric. Other reviews (Dilleman, 1958; Kingsbury, 1964) have been helpful to the author.

Cyanogenesis also occurs in bacteria and fungi, but it is doubtful whether the immediate source of the HCN is one of the cyanogenic glycosides listed in Table 1. Although linamarin and lotaustralin were detected in an HCN-producing psychrophilic basidiomycete (Stevens & Strobel, 1968) the conclusion that these compounds are the source of the HCN must be re-examined. Ward *et al.* (1971) have in effect repeated and refined the experiments of Stevens & Strobel (1968) that indicated that valine and isoleucine were the precursors of the HCN produced by the fungus. Their results clearly substantiate the earlier work of Ward & Thorn (1966) and demonstrate that the methylene carbon of glycine is the source of the carbon atom in the HCN produced by this fungus.

Several species in the bacterial family Pseudomonadaceae are cyanogenic (Eyjolfsson, 1970). In *Chromobacterium violaceum*, the source of the cyanide is the methylene carbon of glycine (Michaels *et al.*, 1965). Moreover, in double-labelled experiments patterned after earlier experiments in higher plants (Butler & Conn, 1964) the methylene C–N bond of glycine was shown to remain intact as those atoms of glycine were converted into HCN (Brysk *et al.*, 1969).

No evidence has been reported in bacteria for a cyanogenic glycoside that breaks down to HCN, an aldehyde or ketone and a sugar. But this is what would be expected if the amino acid glycine undergoes conversion into a nitrile according to the biosynthetic sequence proposed for the production of cyanogenic glycosides from closely related amino acids in higher plants (refer to Scheme 1). Glycine would be converted into HCN by a relatively simple reaction sequence involving formaldoxime (and possible *N*-hydroxyglycine). Perhaps the unstable cyanogenic fraction reported by Ward (1964) in the fungus which he has studied will prove to be one of these two compounds.

Eyjolfsson (1970) has reviewed the subject of cyanogenesis in animals. To his remarks can be added the observation that there is evidence that the HCN produced by millipedes can arise both from an intact cyanogenic glycoside (Pallares, 1945) and from a $\alpha$-hydroxynitrile that is not glycosidically bound to a sugar (Eisner *et al.*, 1963). In the latter case the $\alpha$-hydroxynitrile could dissociate non-enzymically to produce HCN. However, evidence has been presented showing that the vestibule of the cyanogenic glandular apparatus of *Apheloria corrugata* contains an enzyme which triggers the cyanogenesis (Eisner *et al.*, 1963). From work on cyanogenesis in plants it would be expected that the enzyme in the vestibule would be an $\alpha$-hydroxynitrile lyase that catalyses the dissociation of the cyanohydrin.

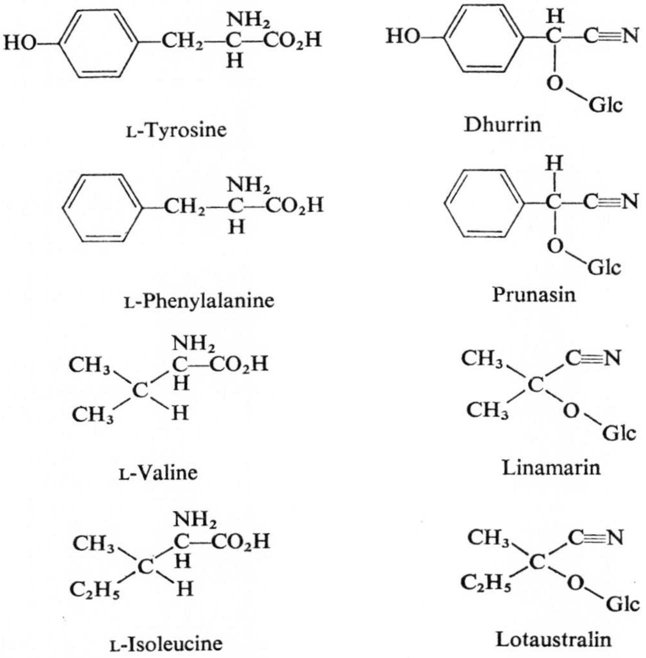

L-Tyrosine — Dhurrin

L-Phenylalanine — Prunasin

L-Valine — Linamarin

L-Isoleucine — Lotaustralin

Fig. 2. *Precursor–product relationship between four amino acids and cyanogenic glycosides*

## Biosynthesis

### *Early studies with labelled compounds*

Earlier reviews (Conn, 1969; Conn & Butler, 1969) have discussed the evidence which demonstrates a precursor–product relationship (Fig. 2) between four proteinaceous amino acids and the aglycones of several of the glycosides listed in Table 1. The evidence consisted of feeding amino acids radioactively labelled in one or more atoms to cyanophoric plants (or parts thereof) and measuring the incorporation of isotope into the aglycone of the glucoside. Table 3 contains a list of glycosides and plants on which experiments of this nature have been performed.

With dhurrin as a specific example (Fig. 3), work in several laboratories has shown that the $\alpha$-carbon of tyrosine becomes the nitrile carbon of the glycoside and the $\beta$-carbon of the amino acid becomes the aglycone carbon that bears the glucosyl group [see Conn & Butler (1969) for review]. The nitrogen atom of the aglycone is derived from the amino nitrogen of the amino acid. In the case of dhurrin (Uribe & Conn, 1966), linamarin (Butler & Conn, 1964) and taxiphyllin (Bleichert *et al.*, 1966) double-labelled experiments involving the use of amino acids labelled with $^{14}C$ in the $\alpha$-carbon and $^{15}N$ in the amino group have shown that, as the amino acid is converted into the glucoside, there is little change in the ratio of specific radioactivities of the two isotopes. This has been taken as evidence that the bond between those two atoms is not severed during the conversion and therefore that all intermediates in the biosynthetic pathway must be nitrogenous in nature. Further, in the case of dhurrin, tyrosine double labelled in the $\alpha$ and $\beta$ carbon atoms was administered and the ratio of specific radioactivities of the corresponding atoms in the aglycone determined (Koukol *et al.*, 1962). As there

Table 3. *Glycosides on which biosynthetic studies have confirmed that protein amino acids are precursors of the aglycones*

| Glycoside | Plant | Amino acid administered | Reference |
|---|---|---|---|
| Amygdalin | Bitter almond | DL-[3-$^{14}$C]Phenylalanine* | Abrol (1967) |
| Prunasin | Cherry laurel | DL-[3-$^{14}$C]Phenylalanine | Mentzer & Favre-Bonvin (1961) |
| | | L-[U-$^{14}$C]Phenylalanine | Tapper & Butler (1971) |
| | Peach | DL-[3-$^{14}$C]Phenylalanine | Ben-Yehoshua & Conn (1964) |
| | | L-[U-$^{14}$C]Phenylalanine | Tapper & Butler (1971) |
| Vicianin | Common vetch | DL-[2-$^{14}$C]Phenylalanine | Tschiersch (1966) |
| Dhurrin | Sorghum | L-[2-$^{14}$C]Tyrosine* | Gander (1958); Conn & Akazawa (1958) |
| Taxiphyllin | Yew | L-[U-$^{14}$C]Tyrosine | Bleichert *et al.* (1966) |
| Nandina glucoside | Nandina | L-[U-$^{14}$C]Tyrosine* | Abrol *et al.* (1966) |
| | Thalictrum | L-[U-$^{14}$C]Tyrosine* | Sharples & Stoker (1969) |
| Linamarin | Linen flax | L-[U-$^{14}$C]Valine | Butler & Butler (1960) |
| | Cassava | L-[U-$^{14}$C]Valine | Nartey (1968) |
| | Lotus | L-[U-$^{14}$C]Valine | Abrol & Conn (1966) |
| | Poppy | L-[U-$^{14}$C]Valine | Abrol (1966) |
| Lotaustralin | Linen flax | L-[U-$^{14}$C]Isoleucine | Butler & Butler (1960) |
| | Cassava | L-[U-$^{14}$C]Isoleucine | Nartey (1968) |
| | Lotus | L-[U-$^{14}$C]Isoleucine | Abrol & Conn (1966) |
| | Poppy | L-[U-$^{14}$C]Isoleucine | Abrol (1966) |
| Proteacin | Thalictrum | L-[U-$^{14}$C]Tyrosine | Sharples & Stoker (1969) |

* Amino acid with $^{14}$C in other positions was also used here.

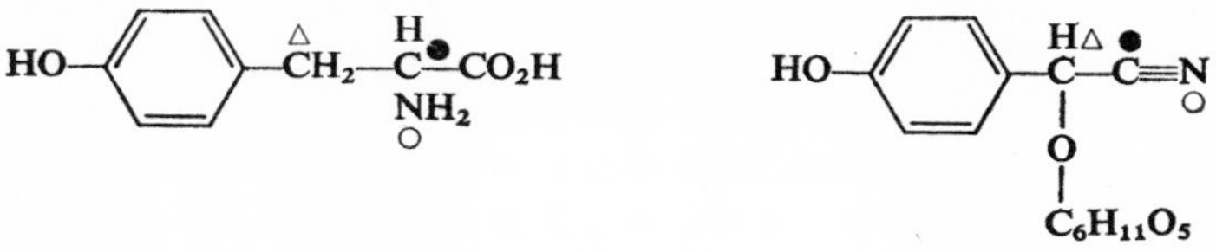

Fig. 3. *Relationship between certain atoms in* L-*tyrosine and the aglycone of dhurrin*

was no significant difference in the isotope ratio for those two atoms, it was concluded that the covalent bond linking the $\alpha$ and $\beta$ carbon atoms was not severed during the biosynthesis.

These observations required a biosynthetic pathway in which the carboxyl carbon of the precursor amino acid is lost while the $\alpha$-carbon is oxidized to the level of a nitrile (a 4-electron oxidation) and the nitrogen atom is retained. In addition, the $\beta$-carbon undergoes a 2-electron oxidation to form the hydroxyl group to which the sugar is attached. These requirements and the known non-enzymic conversion of 2-oximino acids into nitrile compounds by a concerted dehydration and decarboxylation led to the testing of $^{14}C$-labelled oximes and nitriles as possible precursors of the cyanogenic glycosides in flax and sorghum [see Conn & Butler (1969) for review]. When radioactivity from these compounds was effectively incorporated into the cyanogens the pathway presented in Scheme 1 was postulated (Hahlbrock *et al.*, 1968) and efforts directed to testing it.

Unfortunately, the direct demonstration of the proposed aldoximes, nitriles and cyanohydrins as intermediates in extracts of untreated cyanophoric plants has not, with one exception, been successful. This could be due to analytical methods not being sufficiently sensitive to detect very small amounts. An alternative possibility, of course, is that the intermediates remain bound on the surface of the enzyme system(s) that carry out the biosynthesis.

In the exceptional case, Tapper & Butler (1972) have been able to obtain direct evidence for isobutyraldoxime as an intermediate in the pathway leading from

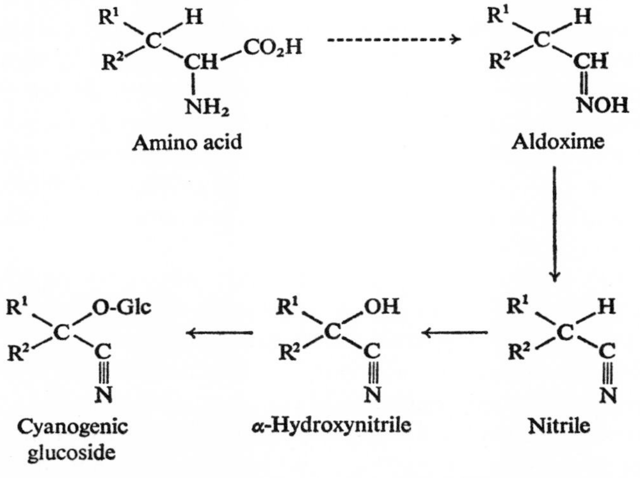

Amino acid → Aldoxime → Nitrile → $\alpha$-Hydroxynitrile → Cyanogenic glucoside

Scheme 1. *A biosynthetic pathway for the formation of cyanogenic glycosides* valine to linamarin in flax. This compound and its *O*-$\beta$-glucoside accumulate when flax seedling shoots are treated with valine and three compounds that inhibit linamarin biosynthesis; DL-*O*-methylthreonine, DL-allo-*O*-methylthreonine and DL-2-methoxypropionaldoxime. Indeed, their observation, made several years ago, provided the major impetus for the extensive amount of work required in the labelling and trapping experiments described in the next section. In one other type of experiment (Tapper *et al.*, 1972), [$^{14}$C]isobutyraldoxime can be detected as an isolatable intermediate when [$^{14}$C]valine is administered to flax shoots simultaneously with unlabelled 2-hydroxyisobutyraldoxime. The latter compound appears to inhibit the normal metabolsim of isobutyraldoxime.

The evidence for the pathway shown in Scheme 1 consists of three kinds. The first consists of experiments in which the appropriate $^{14}$C-labelled aldoxime, nitrile and $\alpha$-hydroxynitrile were administered to cyanophoric plants and the incorporation of isotope from these postulated intermediates into the cyanogen compared with that incorporated from [$^{14}$C]valine. The second type of evidence consists of 'trapping' experiments in which the suspected intermediate (an aldoxime or nitrile) is administered together with the $^{14}$C-labelled precursor amino acid to the appropriate plant. After a period of metabolism the intermediate is then re-isolated from the plant, purified and examined for radioactivity. In those instances where radioactivity is found in the intermediate being examined in the trapping experiment, one can conclude that the intermediate can be formed from the precursor amino acid and therefore participate in the biosynthetic pathway. The third type of evidence consists of the direct demonstration of enzymes involved in the postulated pathway.

We may summarize the evidence acquired to date as follows. Tapper & Butler (1971) have recently published their extensive data, some of which has been cited earlier, showing that $^{14}$C-labelled aldoximes, nitriles and $\alpha$-hydroxynitriles are effectively converted, when compared with the precursor amino acids, into linamarin in flax shoots and into prunasin in cherry-laurel shoots. Note should be made of the extensive incorporation of the administered labelled compounds into the product glycosides. As much as 25% of the labelled valine administered to flax plants was incorporated, whereas the incorporation of the corresponding aldoximes, nitriles and cyanohydrins varied between 9 and 28%. Similarly, in the case of prunasin biosynthesis, the amount of labelled phenylalanine incorporated ranged between 8 and 16% in different experiments whereas 43 and 63% of the phenylacetaldoxime and phenylacetonitrile was incorporated respectively. To these experiments can be added data showing that *p*-hydroxyphenylacetaldoxime and *p*-hydroxyphenylacetonitrile are readily incorporated into dhurrin when compared with tyrosine (Table 4).

In 'trapping' experiments isobutyraldoxime and isobutyronitrile become labelled with $^{14}$C when administered simultaneously to flax with [$^{14}$C]valine (Tapper & Butler, 1972). Unpublished 'trapping' experiments in our laboratory also yielded positive results when *p*-hydroxyphenylacetaldoxime and *p*-hydroxyphenylacetonitrile were fed simultaneously to sorghum shoots with [$^{14}$C]tyrosine. The $\alpha$-hydroxynitriles have not been examined in trapping experiments owing to difficulties in interpreting results that might be obtained. These difficulties arise because of the presence of $\beta$-glucosidases which hydrolyse the cyanogenic

Table 4. *Incorporation of* $^{14}C$*-labelled compounds into dhurrin in shoots of Sorghum vulgare*

| Compound administered | | Dhurrin produced | | |
|---|---|---|---|---|
| | Radioactivity administered | Amount | Specific radioactivity | Precursor converted |
| Expt. | ($\mu$Ci) | ($\mu$mol) | (nCi/$\mu$mol) | (%) |
| 1. L-[U-$^{14}$C]Tyrosine | 0.45 | 22.2 | 2.4 | 11.8 |
| *trans-p*-Hydroxyphenyl[1-$^{14}$C]acetaldoxime | 0.54 | 25.5 | 9.5 | 44.6 |
| 2. L-[U-$^{14}$C]Tyrosine | 0.74 | 19.0 | 4.3 | 12.5 |
| *p*-Hydroxyphenyl[1-$^{14}$C]acetonitrile | 0.75 | 19.4 | 7.3 | 3.3* |

* Average of three samples.

glucosides into glucose and the $\alpha$-hydroxynitriles (see below). Thus if an $\alpha$-hydroxynitrile becomes labelled with $^{14}$C after feeding the $^{14}$C-labelled precursor amino acid to the plant, the radioactivity in the $\alpha$-hydroxynitrile may result from its becoming labelled as a true intermediate. Alternatively, the labelled $\alpha$-hydroxynitrile may have been produced on hydrolysis of the $^{14}$C-labelled cyanogenic glucoside which of course is formed from the precursor $^{14}$C-labelled amino acid.

As the third type of evidence supporting the pathway in Scheme 1, we have previously cited the detection, partial purification and characterization of a glucosyltransferase in flax that catalyses the formation of linamarin and lotaustralin, the final step in the pathway. Further studies on this enzyme will be presented below as well as the detection, partial purification and characterization of a similar enzyme in sorghum that catalyses the formation of dhurrin.

*Hydroxyaldoximes as intermediates*

As pointed out earlier (Conn, 1969; Conn & Butler, 1969) more than one scheme can be proposed for the conversion of an aldoxime into a cyanogenic glycoside. One such scheme, which the highly efficient incorporation of simple nitriles supports, is that presented in Scheme 1 where the aldoxime is dehydrated to form the nitrile which is subsequently oxidized and glucosylated. This and another possible scheme in which the aldoxime is first oxidized to an $\alpha$-hydroxyaldoxime and then dehydrated to the $\alpha$-hydroxynitrile are presented in Scheme 2 for the appropriate compounds leading to linamarin formation. Some consideration has been given this second possibility because the $\alpha$-hydroxyaldoxime could arise, not only by direct hydroxylation but, as suggested by Ettlinger & Kjaer (1968), by rearrangement of the *aci*-nitro compound or oxime *N*-oxide. That is, the oxygen which eventually bears the sugar moiety would be introduced by oxidation initially at the N atom of isobutyraldoxime. A nitro compound 1-nitro-2-phenylethane has recently been shown to be a precursor of the glucosinolate compound in *Tropaeolum majus*. The *aci*-tautomer, which can presumably arise from the nitro compound, would then constitute one further common intermediate in the biosynthesis of cyanogenic glycosides and glucosinolates (see below).

The two possibilities indicated in Scheme 2 were tested by synthesizing 2-hydroxy[1,3-$^{14}$C]isobutyraldoxime and comparing its effectiveness as a precursor of linamarin with [U-$^{14}$C]isobutyraldoxime and L-[U-$^{14}$C]valine (Tapper

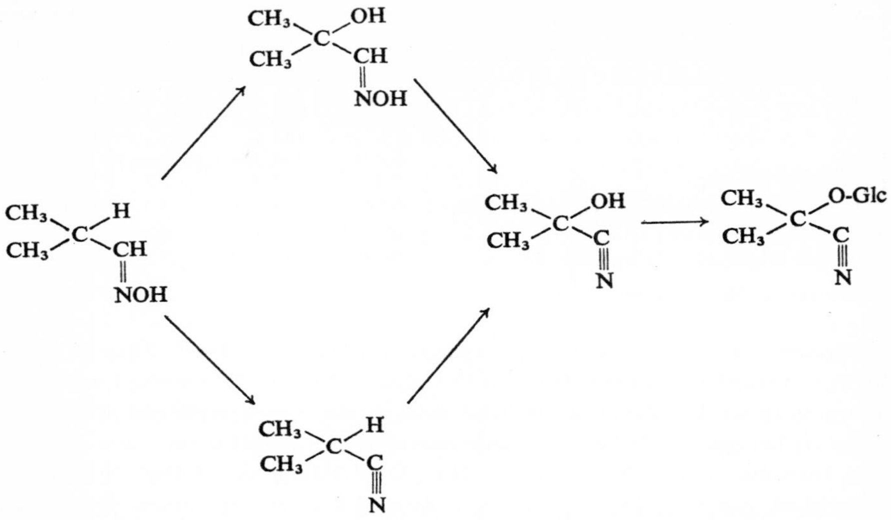

Scheme 2. *Two alternative routes for the conversion of isobutyraldoxime into* 2-*hydroxyisobutyronitrile*

*et al.*, 1972). The results showed (Table 5) that the labelled α-hydroxyisobutyraldoxime is converted into linamarin somewhat less effectively than valine when comparable amounts were administered. However, the incorporation was sufficient to indicate that the α-hydroxyaldoxime might be an alternative intermediate to the isobutyronitrile which was incorporated to about the same extent. To obtain additional information regarding the role of this compound, unlabelled 2-hydroxyisobutyraldoxime and isobutyronitrile were administered to batches of flax shoots in separate trapping experiments together with [$^{14}$C]valine. After 7 h metabolism, the plants were extracted appropriately to obtain the volatile nitrile and α-hydroxyaldoxime and these compounds were separated by g.l.c. and the radioactivity they contained was examined simultaneously. The results, which are described in detail elsewhere (Tapper *et al.*, 1972), showed that radioactivity

Table 5. *Incorporation of* $^{14}$C-*labelled compounds into linamarin in flax seedlings*

| Expt. | Compounds administered | Amount (μmol) | Linamarin: Precursor converted (%) | Linamarin: Dilution of $^{14}$C |
|---|---|---|---|---|
| 1. | L-[U-$^{14}$C]Valine | 1 | 34† | 39† |
| | 2-Hydroxy[1,3-$^{14}$C]isobutyraldoxime | 4.2 | 12 | 25 |
| 2.* | L-[U-$^{14}$C]Valine | 1 | 23† | 55† |
| | [1-$^{14}$C]Isobutyronitrile | 10 | 11 | 10 |
| | 2-Hydroxy[1-$^{14}$C]isobutyronitrile | 2.5 | 28 | 17 |
| 3. | L-[U-$^{14}$C]Valine | 1 | 27† | — |
| | L-[U-$^{14}$C]Valine in the presence of 25 μmol of α-hydroxyisobutyraldoxime | 1 | 22† | — |

* Data from Hahlbrock *et al.* (1968).
† Corrected for loss of carboxyl carbon.

administered to the plant as L-[U-$^{14}$C]valine was readily incorporated into the isobutyronitrile, thereby confirming the results of earlier 'trapping' experiments performed by Tapper & Butler (1972). However, in the 'trapping' experiment with 2-hydroxyisobutyraldoxime, only a very small amount of radioactivity was detected when that compound was isolated from the plant. Moreover, significant amounts of radioactivity were detected in isobutyraldoxime which under these conditions could be isolated from the plants.

These results were further confirmed with competition experiments in which a relatively large amount (25$\mu$mol) of unlabelled 2-hydroxyisobutyraldoxime was fed to flax shoots simultaneously with 1$\mu$mol of L-[U-$^{14}$C]valine. If the hydroxyaldoxime were an intermediate, a greatly decreased incorporation of $^{14}$C into the product should have been observed. Instead the incorporation of $^{14}$C decreased only slightly (Table 5).

Because of the equivocal nature of this result, DL-2-hydroxy[2-$^{3}$H]phenylacetaldoxime was synthesized and its role as an intermediate in the synthesis of prunasin in cherry-laurel shoots examined. Whereas the [$^{3}$H]hydroxyaldoxime was incorporated into the cherry-laurel glucoside, competition experiments with L-[U-$^{14}$C]phenylalanine and unlabelled 2-hydroxyphenylacetaldoxime were negative (Tapper *et al.*, 1972). Thus only conflicting results are available for these series of experiments with the $\alpha$-hydroxyaldoximes. Whereas the administration of labelled hydroxyaldoximes indicate that this compound can be converted into a cyanogenic glycoside, 'trapping' and competition experiments do not support this conclusion.

These results involving competition experiments and the 'trapping' of intermediates of course can only be valid if the administered compounds exchange freely with pools of the corresponding compounds within the plant. Without proof of this being so, these types of experiments are limited in their usefulness for they do not allow a decision on whether $\alpha$-hydroxyaldoximes are physiological intermediates in the biosynthesis of cyanogenic glycosides. It will be necessary eventually to answer such questions by working with enzymes.

### *Early intermediates in the pathway*

Some comment regarding the nature of intermediates between the precursor amino acid and aldoxime is appropriate since this portion of the biosynthetic pathway may well be identical with the pathway leading from amino acids to the structurally related glucosinolate compounds (Underhill *et al.*, 1973). The conversion of an amino acid into the related oxime represents a four-electron oxidative decarboxylation. There is precedence for such reactions being catalysed by a single-enzyme system (Mazelis & Ingraham, 1962), although in the reference cited the product is an amide. The overall conversion can also occur by two 2-electron oxidation steps followed by decarboxylation of the 2-oximino acid thus formed (Scheme 3).

To summarize the evidence regarding these compounds as intermediates, Tapper & Butler (1971) found that the 2-oximino acid could serve as a very efficient precursor of linamarin in flax and of prunasin in cherry-laurel and peach shoots. However, because of the facile, non-enzymic conversion of 2-oximino

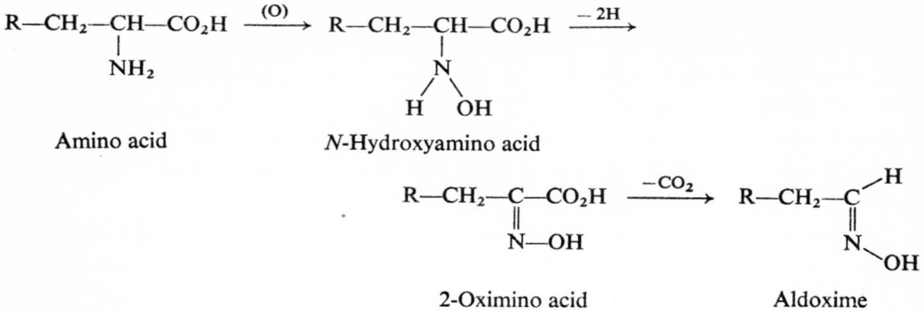

$$\underset{\text{Amino acid}}{R{-}CH_2{-}\underset{\substack{|\\NH_2}}{CH}{-}CO_2H} \xrightarrow{(O)} \underset{N\text{-Hydroxyamino acid}}{R{-}CH_2{-}\underset{\substack{|\\N\\H\quad OH}}{CH}{-}CO_2H} \xrightarrow{-2H}$$

$$\underset{\text{2-Oximino acid}}{R{-}CH_2{-}\underset{\substack{\|\\N{-}OH}}{C}{-}CO_2H} \xrightarrow{-CO_2} \underset{\text{Aldoxime}}{R{-}CH_2{-}\underset{\substack{\|\\N{-}OH}}{C}{-}H}$$

Scheme 3. *Route for the conversion of an amino acid into the related aldoxime*

acids into nitriles (Ahmad & Spenser, 1961), compounds which are also efficient precursors of cyanogenic glucosides, Tapper & Butler (1971) concluded that the role of the oximino acids was uncertain. This is probably appropriate in view of the fact that the oximino acids do not serve as precursors of the structurally related glucosinolates.

With regard to the *N*-hydroxyamino acid, Kindl & Underhill (1968) have provided evidence that *N*-hydroxyphenylalanine can be oxidatively decarboxylated by an enzyme preparation from *Tropaeolum majus* to form phenylacetaldoxime. This reaction is by no means established in the formation of the cyanogenic glycosides. Numerous attempts to demonstrate the enzymic conversion of *N*-hydroxyvaline into isobutyraldoxime by specific enzyme preparations from flax seedlings were negative, although this reaction was weakly catalysed by $Fe^{2+}$ ions and preparations of horseradish peroxidase (B. A. Tapper & E. E. Conn, unpublished work). Other attempts to demonstrate the enzymic formation of *N*-hydroxyvaline and *N*-hydroxytyrosine (from valine and tyrosine respectively) have not been successful (C. L. Tipton & E. E. Conn, unpublished work; P. F. Reay & E. E. Conn, unpublished work).

### *Studies with [$^{18}O$]oxygen*

To learn more about the nature of the reaction in which the glucosidic-linkage oxygen is introduced into the aglycone, experiments on the origin of that atom in linamarin and lotaustralin have been performed (Zilg *et al.*, 1972). For these two compounds, two mechanisms may be considered for the introduction of the 2-hydroxy moiety before transglycosylation. The appropriate precursor isobutyronitrile (or isobutyraldoxime) could be oxidized to the corresponding isobutene derivative which in turn could add water. Alternatively, an oxygenase might directly introduce a hydroxyl group on the designated carbon atom. This question was examined by growing flax seedlings in an atmosphere enriched in $^{18}O_2$ for 12h and simultaneously administering [$^{14}C$]valine. At the conclusion of the experiment the linamarin and lotaustralin were isolated, purified by various techniques, and then converted into their corresponding tetra-*O*-trimethylsilyl derivatives. This derivative of the glucoside was chosen because, on subjecting the derivative to analysis in a mass spectrometer, suitable fragment ions for determination of the $^{18}O$ in the glucosidic bond can be found (Scheme 4). The

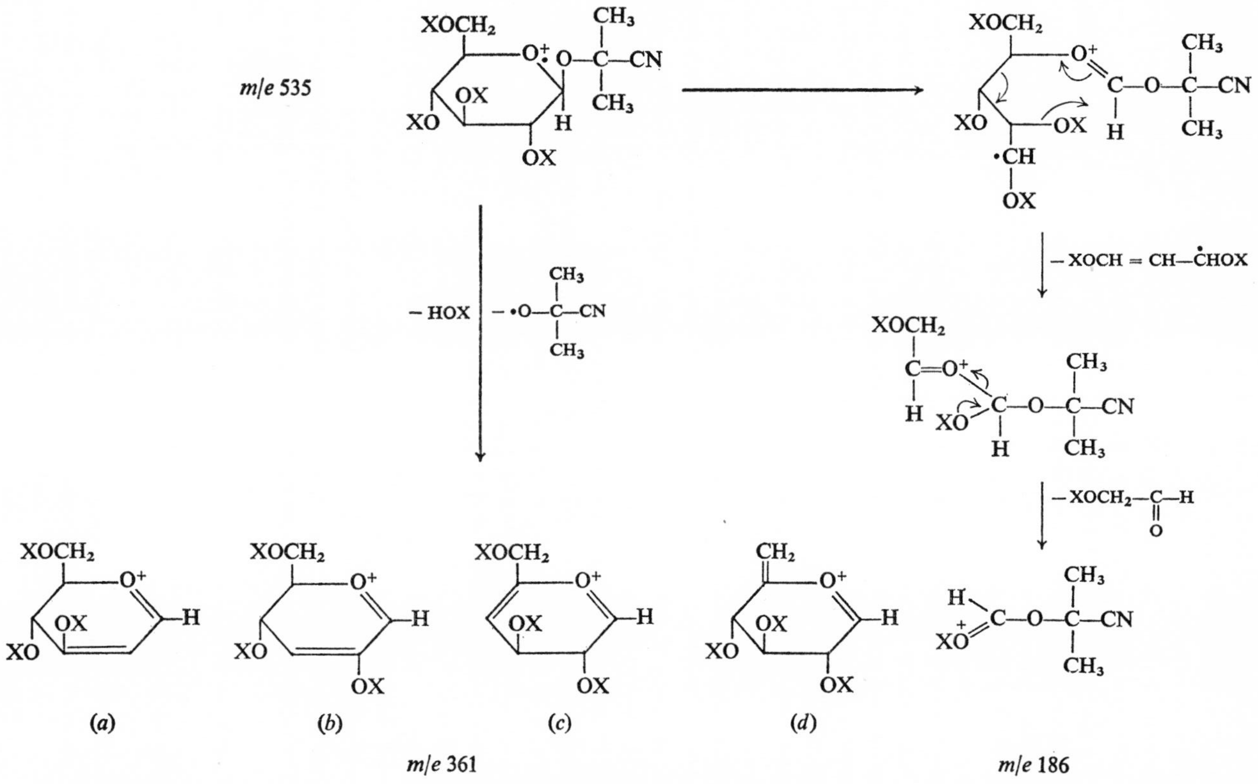

m/e 535

$-$XOCH = CH—ĊHOX

$-$HOX

$-$XOCH$_2$—C(=O)—H

(a) (b) (c) (d)

m/e 361

m/e 186

X, trimethylsilyl

Scheme 4. *Proposed schemes for the formation of ions m/e* 361 *and m/e* 186 *in the mass spectrum of tetra-O-trimethylsilyl-linamarin* fragment ion *m/e* 186 contains the aglycone including the glucosidic oxygen. Four structures *a, b, c* and *d* can contribute to the fragment ion *m/e* 361 which represents the glucosyl moiety after loss of the aglycone with its glucosidic linkage oxygen and a molecule of trimethylsilanol. Since the *m/e* 186 ion also contains one oxygen atom from the glucose moiety, the *m/e* 361 ion was examined to determine any significant incorporation of $^{18}O$ into the glucose moiety. By examination of the $^{18}O$ content of these two ions therefore, one could determine whether $^{18}O$ had been incorporated into the glucosidic-linkage oxygen.

The determination of the *m/e* 186 and *m/e* 361 and the corresponding $m^{2+}$ ions (for $^{18}O$) was carried out on a prototype of the LKB-9000 combination gas chromatograph–mass spectrometer (Waller, 1968) put at our disposal by George R. Waller, Department of Biochemistry, Oklahoma State University, Okla., U.S.A. The appropriate mass ratios obtained for linamarin in the presence of $^{18}O$ in the light and in the dark were compared with that of natural linamarin (Table 6). The results show that the incorporation of $^{18}O$ at any position other than that of glucosidic linkage oxygen was insignificant. This is indicated by the nearly identical values of the mass ratio 363/(361 + 363) obtained with all the trimethylsilyl-linamarin samples. Any one of the non-glucosidic-linkage oxygen atoms appears in at least three of the four contributing isomers of fragment ion *m/e* 361. An abnormally high *m/e* 363 would therefore indicate an incorporation of $^{18}O$ in a position other than the glucosidic-linkage oxygen, which was not found. The enrichment observed with the *m/e* 186, 188 pair, is therefore due to $^{18}O$ being present solely in the oxygen of the glucosidic linkage. Moreover, the dilution of this oxygen is remarkably low in both the light and the dark experiments. The dilution observed rules out the participation of water in the step involving the introduction of the hydroxyl group on to the aglycone as only approx. 1 mmol of $^{18}O$ was present in the experiment initially. If this were incorporated by any mechanism involving exchange with the 80 mmol (approx.) of water present in the plant tissue, a dilution value well in excess of 40-fold would have been expected. As can be seen, the dilution was only about 6-fold. The dilution of the $^{18}O$ may be compared with that of $^{14}C$ by means of the $^{14}C/^{18}O$ dilution ratio. In the dark, both isotopes were diluted nearly equally, the $^{14}C$ being diluted slightly less. In experiments performed in the light where some photosynthesis could occur, dilution within the plant tissue of the $^{18}O$ would be expected. The relative dilution value would be expected to decrease therefore, and

Table 6. *Incorporation of* [$^{18}O$]*oxygen and* L-[*U*-$^{14}C$]*valine into linamarin*

Incorporation of [$^{18}O$]oxygen starting from 68.2 atoms percent excess was determined by mass spectrometry of trimethylsilyl-linamarin.

| Source of linamarin | Fragment ion intensity ratio: 363/(361 + 363) | 188/(186 + 188) | $^{18}O$ in glucosidic linkage: Enrichment (atoms percent excess) | Dilution | $^{14}C/^{18}O$ dilution ratio |
|---|---|---|---|---|---|
| $^{18}O_2$, light experiment | 0.194 | 0.173 | 11.7 | 5.83 | 0.70 |
| $^{18}O_2$, dark experiment | 0.194 | 0.183 | 12.8 | 5.34 | 0.92 |
| Natural linamarin | 0.191 | 0.068 | — | — | — |

this was observed. Essentially similar results were obtained when lotaustralin was isolated, purified, converted into various derivatives and examined in the mass spectrometer, and these data have been published (Zilg *et al.*, 1972).

It is clear, then, that the glucosidic-linkage oxygen atoms of linamarin and lotaustralin are derived from oxygen gas. These results exclude, therefore, a mechanism in which a methacryl intermediate is hydrated. Although epoxidation by molecular oxygen of such a methacryl intermediate could account for the $^{18}O$ enrichment observed, the possibility of an epoxide intermediate is unlikely since Tapper & Butler (1972) failed to demonstrate any intermediate role for methacrylonitrile or methyacrylein oxime in linamarin biosynthesis. It is concluded, therefore, that the 2-hydroxyl moiety is derived from molecular oxygen by the action of an oxygenase on isobutyronitrile (or isobutyraldoxime).

### *Enzymic studies*

Enzymic studies on the biosynthesis of cyanogenic glycosides have been restricted almost exclusively to the last step in the proposed pathway, the step which catalyses the synthesis of the cyanogen from the corresponding α-hydroxynitrile (cyanohydrin). An enzyme in flax that catalyses the glucosylation of a series of aliphatic α-hydroxynitrile compounds in the presence of UDP-glucose as a glucose donor has been partly purified and characterized (Hahlbrock & Conn,

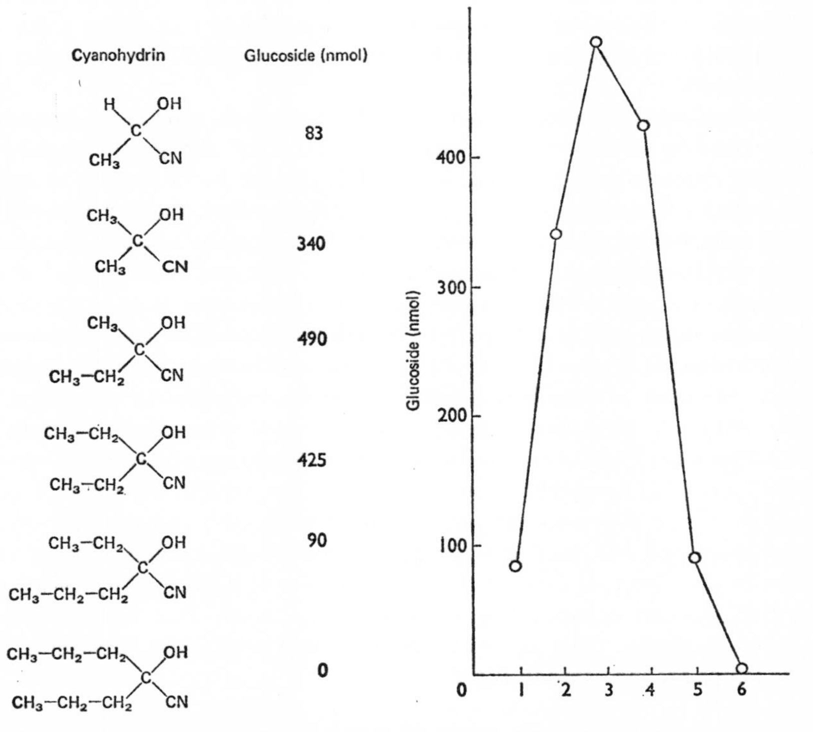


Fig. 4. *Rate of glucosylation of aliphatic cyanohydrins by the UDP-glucose glucosyltransferase of flax seedlings*

1970). This enzyme exhibits maximal rates of reaction with the α-hydroxynitrile compounds that give rise to linamarin and lotaustralin on glucosylation (Fig. 4) but is inactive with aromatic α-hydroxynitrile compounds such as mandelonitrile. The product of the action of the enzyme on α-hydroxyisobutyronitrile was identified as linamarin. At the time this work was performed, the nature of the product formed with the next higher homologue was not identified other than to show it was a β-glucoside capable of being hydrolysed by linamarase. It was assumed to be lotaustralin.

Since 2-hydroxy-2-methylbutyronitrile contains a chiral centre (and the racemic mixture of the two isomers was present in the enzyme studies), it would have been possible for the flax enzyme to glucosylate both of the enantiomers and produce the diastereomeric glucosides. These two glucosides, which have been termed methyl-linamarins, have been synthesized by Bissett *et al.* (1969). The compound having the (*R*) configuration at the glucosidic carbon has been designated as (*R*)-lotaustralin and is the glucoside that occurs naturally in *Trifolium repens*. The epimer, which to date has not been found as a naturally occurring glucoside, was designated as (*S*)-epi-lotaustralin.

The work of Bissett *et al.* (1969) made it possible to identify the nature of the lotaustralin that occurs naturally in flax as well as to examine the products synthesized by the glycosyltransferase of flax. This was done by developing a t.l.c.-separation procedure for the tetra-acetates of (*R*)-lotaustralin and (*S*)-epi-lotaustralin that is based on a similar separation of the tetra-acetates on a silica-gel column. With this procedure, the lotaustralin of flax was shown to have the (*R*)-configuration and therefore is identical with that of *Trifolium repens* (Zilg & Conn, 1974).

It was noteworthy to observe that both (*R*)-lotaustralin and (*S*)-epi-lotaustralin were formed in approximately equal amounts, when the partly purified flax glucosyltransferase was incubated with UDP-glucose and a racemic mixture of the 2-hydroxy-2-methyl[1-$^{14}$C]butyronitriles. Apparently, the flax enzyme does not distinguish between the two enantiomers of 2-hydroxy-2-methylbutyronitrile and glycosylates both at equal rates. This might have been predicted since Hahlbrock & Conn (1970) showed that the flax enzyme readily glucosylates 2-hydroxyisobutyronitrile, 2-hydroxy-2-methylbutyronitrile and 2-hydroxy-2-ethylbutyronitrile *in vitro* (Fig. 4). This was pursued further by administering the racemic mixture of labelled 2-hydroxy-2-methylbutyronitrile to intact flax shoots. When the linamarin sample was examined it consisted of both (*R*)-lotaustralin and (*S*)-epi-lotaustralin in the ratio 60:40, or nearly equal amounts.

Since only (*R*)-lotaustralin occurs in flax, the glucosyltransferase is obviously presented *in vivo* with only the one α-hydroxynitrile as a substrate for glycosylation. Moreover, this fact must be determined by the configuration of the β-carbon of the (3*S*)-$L_s$-isoleucine that serves as a precursor. If (3*R*)-$L_s$-isoleucine with the opposite configuration on that carbon atom were fed the cyanogen produced by the flax plant should be the unnatural or (*S*)-epi-lotaustralin. These relationships are indicated in Fig. 5.

Since (3*R*)-$L_s$-[$^{14}$C]isoleucine was not readily available, advantage was taken of the fact that the sugar moieties of the cyanogenic glucosides are labelled when [$^{14}$C]glucose is administered either separately or simultaneously with their

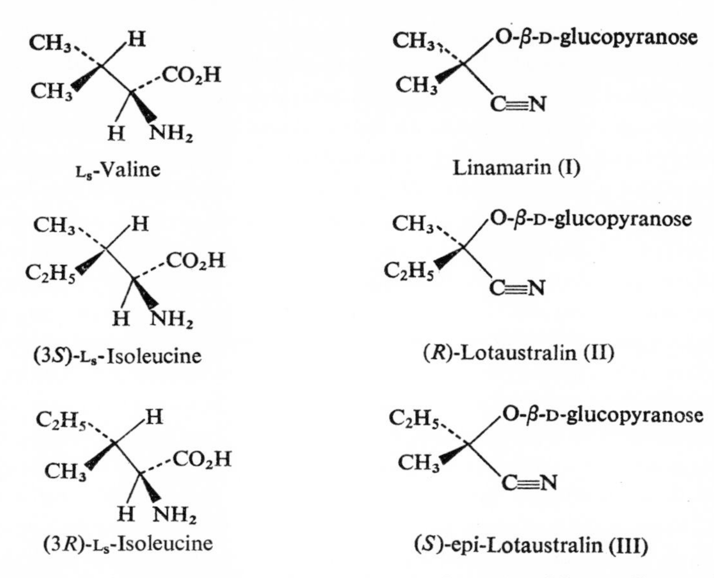


Fig. 5. *Stereochemical relationships between the cyanogenic glycosides of flax seedlings and their precursor amino acids*

amino acid precursor. Thus, when D-[U-$^{14}$C]glucose was administered to flax shoots, both linamarin and lotaustralin became labelled, the latter compound being composed of only (*R*)-lotaustralin (Table 7). When D-[U-$^{14}$C]glucose was administered together with (3*S*)-$L_s$-isoleucine, again both the linamarin and lotaustralin isolated from the plant contained $^{14}$C in the glucose moiety and again the lotaustralin formed was only the (*R*)-isomer. When however, (3*R*)-$L_s$-isoleucine was administered together with D-[U-$^{14}$C]glucose, the lotaustralin produced was a mixture which contained 64% (*S*)-epi-lotaustralin and 36% (*R*)-lotaustralin. The fact that some (*R*)-lotaustralin was formed in the latter experiment can be explained in part by the fact that the sample of (3*R*)-$L_s$-isoleucine employed in this work contained about 20% (3*S*)-$L_s$-isoleucine. In addition, the plant obviously contained an endogenous source of (3*S*)-$L_s$-isoleucine that would give rise to (*R*)-lotaustralin in the same manner that the linamarin becomes labelled from an endogenous source of valine.

Several noteworthy conclusions can be drawn from these experiments. First, the results show that, in the case of flax, the configuration of the (*R*)-lotaustralin is determined by the configuration of the β-carbon in the isoleucine that serves as

Table 7. *Incorporation of* D-[*U*-$^{14}$*C*]*glucose into cyanogenic glucosides in flax seedlings with simultaneous application of amino acids* (% *incorporation*)

| Amino acid administered | Linamarin + lotaustralin | Linamarin | Lotaustralin | |
|---|---|---|---|---|
| | | | (*R*)-Lotaustralin | (*S*) epi-Lotaustralin |
| (3*S*)-$L_s$-Isoleucine | 4.25 | 1.43 | 2.82 | <0.03 |
| (3*R*)-$L_s$-Isoleucine | 4.67 | 1.27 | 1.07 | 2.33 |
| None | 4.61 | 3.08 | 1.53 | <0.03 |

its precursor. Secondly, the configuration (*S*) on that carbon is retained as the L-isoleucine is converted into *R*-lotaustralin. Thirdly, the set of enzymes that convert isoleucine [or (3*R*)-$L_s$-isoleucine] into lotaustralin (or epi-lotaustralin) obviously tolerate an exchange of the ethyl and methyl groups on the β-carbon of the amino acid and all other intermediates in the biosynthetic pathway. Such results argue for a single set of enzymes being responsible for linamarin and lotaustralin biosynthesis in plants, a fact in agreement with the conclusion that only one glycosyltransferase is required for the synthesis of the two aliphatic cyanogens found in flax (Hahlbrock & Conn, 1971). Fourthly, these observations are consistent with the $^{18}O$ isotope studies on the origin of the glucosidic oxygen on these glycosides. The only reaction in the biosynthetic pathway involving the β-carbon atom is the introduction of oxygen on that atom. The results of the work with $^{18}O$ cited earlier suggest that an oxygenase is involved; the hydroxylation reactions catalysed by such enzymes proceed with retention of configuration.

The glucosyltransferase which catalyses the last step in the synthesis of dhurrin in sorghum seedlings has been purified approx. 80-fold and characterized (P. F. Reay & E. E. Conn, unpublished work). As with the flax enzyme, it was necessary to develop a purification procedure which separated the glucosyltransferase from the glucosidase which catalyses the hydrolysis of dhurrin. In addition, the sorghum seedlings contain a second transferase which transfers glucose to the aromatic hydroxyl group of *p*-hydroxybenzaldehyde and forms *p*-glucosyloxybenzaldehyde. This activity was largely removed during purification of the enzyme that makes dhurrin.

The stereospecificity of the sorghum enzyme has been examined and was shown to contrast sharply with that of the flax enzyme. Again, because the substrate *p*-hydroxymandelonitrile contains a chiral carbon (and the aqueous enzyme reaction mixture contains the racemic pair), it was possible that both enantiomers of *p*-hydroxymandelonitrile would serve as substrate. If the (*S*)-enantiomer were glucosylated, dhurrin would be formed; if the (*R*) enantiomer was acted on, taxiphyllin would be the product. Studies on differences in the n.m.r. spectra of these two glycosides (Towers *et al.*, 1964) together with g.l.c. provided a means for answering this question, and Reay has shown that the product of the sorghum glycosyltransferase is exclusively the single cyanogen dhurrin. Because recovery of the dhurrin synthesized by the enzyme reaction was only about 20%, after purification for n.m.r. studies, Reay then took care to show that the apparently exclusive formation of dhurrin was not due to selective degradation of taxiphyllin nor to separation of the two isomers during preliminary purification. This was important in view of the ability of the flax enzyme to glucosylate both enantiomers of 2-hydroxy-2-methylbutyronitrile. It was accomplished by adding taxiphyllin and dhurrin standards to a second incubation mixture which contained only one substrate and the sorghum enzyme and showing by g.l.c. that the two glucosides remained in the same proportion after the isolation and purification procedure had been carried out.

The sorghum enzyme obviously differs from the flax enzyme in that it utilizes aromatic rather than aliphatic cyanohydrins as the preferred substrates. Mandelonitrile will serve as effectively as a substrate as does *p*-hydroxymandelonitrile (Table 8). A mixture of *p*-hydroxybenzaldehyde and HCN is also rapidly glucosy-

Table 8. *Specificity of sorghum glucosyltransferase for cyanohydrins*

The glucoside formed is expressed as a percentage of dhurrin formation from *p*-hydroxymandelonitrile, at substrate concentrations of 4mM or 20mM.

| Substrate | Glucoside formed 4mM | 20mM |
|---|---|---|
| *p*-Hydroxymandelonitrile | — | 100 |
| Mandelonitrile | 96 | 106 |
| *p*-Hydroxybenzaldehyde + HCN | — | 102 |
| *p*-Hydroxybenzaldehyde | 0.7 | 1.2 |
| Acetone cyanohydrin | $<0.3$ | $<0.3$ |
| Acetaldehyde cyanohydrin | $<0.3$ | $<0.3$ |
| *p*-Hydroxybenzoic acid | 3.2 | 4.2 |

lated, showing that the association of these two compounds to form the cyanohydrin in aqueous solution must be extremely rapid. The aromatic hydroxyl groups of *p*-hydroxybenzaldehyde and *p*-hydroxybenzoic acid can be glucosylated, but at much lower rates.

In other respects, the flax and sorghum enzymes are similar. Both form $\beta$-glucosides. Both have an absolute specificity for UDP-glucose as the glucosyl donor. They have similar pH optima and have $K_m$ values for UDP-glucose of less than 1mM. The sorghum enzyme will also utilize quinol (hydroquinone) and *p*-hydroxybenzyl alcohol as substrates; this it does not require a nitrile group in its substrate.

## Genetic Control of Cyanogenic Glycoside Biosynthesis

Of all the plant species that are cyanogenic only a few are polymorphic for the cyanogenetic character [see Jones (1972) for review]. The principle features of cyanogenesis in *Trifolium repens* were worked out by Corkill (1942) and his co-workers and may be summarized as follows: the two cyanogenic glucosides linamarin and (*R*)-lotaustralin occur in white-clover leaves together with the enzyme linamarase which is capable of hydrolysing the $\beta$-glucosides to form glucose and the unstable 2-hydroxyisobutyronitrile and 2-hydroxy-2-methylbutyronitrile. The presence or absence of linamarase is determined by alleles of a single gene (*Li*); similarly the presence or absence of both the glucosides is determined by alleles of another independently inherited gene (*Ac*). Only plants that possess at least one dominant allele of both genes are cyanogenic.

The availability of labelled intermediates for the biosynthesis of linamarin, together with some knowledge of the biosynthetic pathway, provided an opportunity to investigate which step in the pathway is affected by the *Ac* gene (M. A. Hughes & E. E. Conn, unpublished work). Employed in this study were white clover plants which lacked linamarase activity but which contained linamarin and lotaustralin [phenotype cyanoglucoside positive, linamarase negative ($lin^{+ve}$)] and plants which had the phenotype cyanogenic glucoside negative, linamarase negative ($lin^{-ve}$). (The former plants contained dominant alleles of the *Ac* gene whereas the latter contained only recessive alleles of the gene.)

Mutations in a gene that controls the activity of an enzyme have three effects on the organisms containing mutant alleles: first, the enzyme activity itself is missing; secondly, the product of the enzyme is not formed; thirdly, the substrate of the enzyme may accumulate. Since only the enzyme catalysing the last step in the biosynthesis of the cyanogens has been characterized and that done only with flax and not *Trifolium repens*, it has been necessary to investigate the second and third effects of mutant alleles in the $lin^{-ve}$ plants. Initially, therefore, L-[U-$^{14}$C]valine was administered to leaves and shoots of $lin^{+ve}$ and $lin^{-ve}$ plants and after a period of metabolism the plants were extracted and examined for radioactive linamarin as well as radioactive isobutyraldoxime; the other intermediates isobutyronitrile and 2-hydroxyisobutyronitrile were considered too volatile or labile to detect readily. In the $lin^{+ve}$ plants, from 4 to 6% of the valine administered was converted into linamarin as could be expected. The $lin^{-ve}$ plants could not carry out this conversion. In neither of the phenotypes was [$^{14}$C]isobutyraldoxime nor any other intermediates identified. *O*-Methylthreonine was then administered simultaneously with L-[U-$^{14}$C]valine to both phenotypes and the results observed. The synthesis of linamarin in the $lin^{+ve}$ plants was inhibited about 70%. However, neither isobutyraldoxime nor its *O*-β-D-glucoside accumulated in the $lin^{+ve}$ phenotype. It would appear, therefore, that the utilization of isobutyraldoxime is not inhibited by *O*-methylthreonine in the $lin^{+ve}$ plants and the effect of *O*-methylthreonine must be due to inhibition at some other point in the pathway.

To obtain more information on this point, [U-$^{14}$C]isobutyraldoxime was fed

Table 9. *Incorporation of* [$^{14}$C]*isobutyraldoxime and 2-hydroxy*[1,3-$^{14}$C]*isobutyraldoxime into linamarin*

In Expt. 1 0.90 μCi of [U-$^{14}$C]isobutyraldoxime (specific radioactivity 0.74 μCi/μmol) was administered. In Expt. 2 0.23 μCi of 2-hydroxy[1,3-$^{14}$C]isobutyraldoxime (specific radioactivity 0.28 μCi/μmol) was given.

| Plant | Fresh weight (g) | Compound administered | % converted into linamarin |
|---|---|---|---|
| Expt. 1 | | | |
| $lin^{+ve}$ | | | |
| S100/1 | 0.7 | [U-$^{14}$C]Isobutyraldoxime | 1.7 |
| S100/10 | 0.7 | [U-$^{14}$C]Isobutyraldoxime | 1.6 |
| S100/10 | 0.9 | [U-$^{14}$C]Isobutyraldoxime | 2.0 |
| S100/1 | 1.0 | [U-$^{14}$C]Isobutyraldoxime | 2.4 |
| $lin^{-ve}$ | | | |
| DWW/1 | 0.6 | [U-$^{14}$C]Isobutyraldoxime | 0 |
| DWW/3 | 0.7 | [U-$^{14}$C]Isobutyraldoxime | 0 |
| DWW/3 | 0.7 | [U-$^{14}$C]Isobutyraldoxime | 0 |
| DWW/1 | 1.0 | [U-$^{14}$C]Isobutyraldoxime | 0 |
| $lin^{+ve}$ | | | |
| S100/1 | 0.6 | [U-$^{14}$C]Isobutyraldoxime + 20 μmol of *O*-methylthreonine | 1.6 |
| S100/10 | 0.6 | [U-$^{14}$C]Isobutyraldoxime + 20 μmol of *O*-methylthreonine | 2.0 |
| Expt. 2 | | | |
| $lin^{+ve}$ | | | |
| S100/1 | 0.6 | 2-Hydroxy[1,3-$^{14}$C]isobutyraldoxime | 0 (17.2*) |
| S100/10 | 0.6 | 2-Hydroxy[1,3-$^{14}$C]isobutyraldoxime | 0 (19.6*) |
| $lin^{-ve}$ | | | |
| DWW/1 | 0.7 | 2-Hydroxy[1,3-$^{14}$C]isobutyraldoxime | 0 (0*) |
| DWW/3 | 0.5 | 2-Hydroxy[1,3-$^{14}$C]isobutyraldoxime | 0 (0*) |

* Percentage glucosylated.

to both *lin*$^{+ve}$ and *lin*$^{-ve}$ plants. The incorporation of isotope into linamarin in *lin*$^{+ve}$ plants is shown in Table 9, but there was no evidence for the synthesis of linamarin in the *lin*$^{-ve}$ plants. These results indicate that at least one of the steps after the aldoxime is missing in such plants. Table 9 also shows that *O*-methylthreonine does not inhibit the incorporation of [U-$^{14}$C]isobutyraldoxime into linamarin. This observation explains the lack of accumulation of isobutyraldoxime when *O*-methylthreonine was fed in the absence of L-[U-$^{14}$C]valine and suggests that this inhibitor functions at a different point in clover than in flax, possibly at an earlier step in the biosynthetic pathway.

Some evidence that the step missing after the aldoxime in *lin*$^{-ve}$ plants might be the glycosyltransferase was obtained when 2-hydroxy[1,3-$^{14}$C]isobutyraldoxime was administered to both phenotypes. As with flax plants, the hydroxyl group of the hydrozyaldoxime was estensively glucosylated by the *lin*$^{+ve}$ clover plants to form 2-(glucosyloxy)isobutyraldoxime (Table 9, Expt. 2). However, the *lin*$^{-ve}$ plants failed to show any of this compound and instead the 2-hydroxy-[1,3-$^{14}$C]isobutyraldoxime was recovered largely unchanged. Whereas such results imply that the *lin*$^{-ve}$ plants lack the glucosylation enzyme, it is desirable to demonstrate directly the presence of the glucosyltransferase in the *lin*$^{+ve}$ clover plants and then confirm its presence or absence in the *lin*$^{-ve}$ phenotype. Such experiments are planned.

If the only step missing in the *lin*$^{-ve}$ phenotype is the glycosylation reaction, then it should have been possible to demonstrate the formation of isobutyraldoxime from valine in both *lin*$^{+ve}$ and *lin*$^{-ve}$ phenotypes. An attempt was therefore made to detect isobutyraldoxime by performing trapping experiments with L-[U-$^{14}$C]valine and unlabelled isobutyraldoxime in both kinds of plants. Table 10 shows that it was possible to detect significant amounts of radioactivity in the isobutyraldoxime trap in the case of the *lin*$^{+ve}$ plant but none was detected in *lin*$^{-ve}$ plants. These experiments indicate that the *lin*$^{-ve}$ plants are also unable to convert the valine into isobutyraldoxime, presumably because they lack the necessary enzyme(s). When those enzymes have been identified in cyanophoric plants, the direct experiments on the two phenotypes of *Trifolium repens* can be performed.

The conclusion to be drawn from these studies is that the *lin*$^{-ve}$ plants lack the enzymes necessary to carry out the first and last steps of the pathway presented in Scheme 1. In a similar study of the genetic control of *o*-hydroxycinnamic acid in *Melilotus alba*, Haskins & Kosuge (1965) concluded that

Table 10. *Incorporation of* [$^{14}$C]*valine into linamarin in shoots treated with isobutyraldoxime*

| Plant | Fresh weight (g) | Compound administered: L-[U-$^{14}$C]-Valine* ($\mu$Ci) | Compound administered: Isobutyraldoxime ($\mu$mol) | Incorporation from [$^{14}$C]valine: (% converted into linamarin)† | Incorporation from [$^{14}$C]valine: (% trapped as isobutyraldoxime) |
|---|---|---|---|---|---|
| *lin*$^{+ve}$ | | | | | |
| S100/10 | 1.5 | 3.68 | 43.2 | 2.9 | 0.3 |
| S100/1 | 0.6 | 1.84 | 21.6 | 3.5 | 0.7 |
| *lin*$^{-ve}$ | | | | | |
| DWW/1 | 1.3 | 3.68 | 43.2 | 0 | 0 |
| DWW/3 | 0.7 | 1.84 | 21.6 | 0 | 0 |

* Specific radioactivity of 200 $\mu$Ci/$\mu$mol.

† Corrected for an assumed loss of carboxyl atom.

the *Cu* locus controls the *o*-hydroxylation of *trans*-cinnamic acid. In this system, which has many similarities to the cyanogenic polymorphism in white clover, the *Cu* locus appears to control the activity of a single enzyme, and this is consistent with the genetic evidence that the *Cu* locus is a single gene.

This preliminary evidence indicating the loss of two enzyme activities in *acac* white clover suggests that the single-gene nature of the *Ac* locus should be questioned. There are precedents in microbial genetics for mutations in a single gene affecting more than one enzyme activity. Mutations at the operator gene ($O^{\circ}$) in the lactose operon in *Escherichia coli* result in the loss of three enzymes (Jacob *et al.*, 1965) and polarity mutants are known in the histidine biosynthetic operon (Ames & Hartman, 1963) which cause a decrease of all enzyme activities in the pathway. Although the genetic evidence in white clover indicates that the *Ac* locus is a single gene, the poor resolution of higher-plant genetics means that the *ac* allele may represent several mutations in a number of linked genes. In any of these proposals the close linkage of the structural genes for linamarin biosynthesis is implied. An alternative explanation is that this pathway is catalysed by a multi-enzyme complex and the *Ac* locus controls a protein which is necessary for the functional integrity of this complex. This type of protein has been proposed as an explanation of polarity mutants affecting the aromatic synthetic enzymes in *Neurospora crassa* (Giles *et al.*, 1967).

**Toxicology of Cyanogenic Plants**

The subject of cyanide poisoning of human populations whether it results from polluted atmospheres or dietary sources, has attracted interest in recent years. Many examples of acute poisoning by HCN of plant origin have been cited (Montgomery, 1965, 1969). With education, with plant selection and with government regulation of plant importation, the problem has been controlled in many countries. The possibility of chronic cyanide poisoning remains, however, in several areas of the world where cyanogenic plants such as cassava or beans, comprise the major item in the diet (Montgomery, 1969).

The poisonous nature of the cassava root has been known for centuries. Traditional methods of preparation involve the crushing and soaking of the root to remove the cyanogen and HCN before the tissue is dried and pounded into a flour. Alternatively, the root may simply be sliced and dried in the sun, a procedure which surely is less effective in removing the toxic constituents. Some cassava preparations will contain significant quantities of HCN (5mg/100g) which is present primarily as the glucosides linamarin and lotaustralin. In some areas of Nigeria, the consumption of such flour may reach 750g/day, corresponding to 35mg of HCN. That amount is approx. one-half of the lethal dose of HCN when taken at one time.

These observations have led to the suggestion that certain neurological conditions encountered in areas where cassava is consumed may result from chronic cyanide poisoning [see Montgomery (1969) for review]. Osuntokun (1971) has summarized his recent evidence that chronic cyanide intoxication is the major factor in the etiology of Nigerian tropical ataxic neuropathy. The detoxification of cyanide by animals through its conversion into thiocyanate is a mixed blessing because of the well-known goitrogenic nature of this material. The study by Delange & Ermans (1971) of the causative agent of endemic goitre in the Eastern Congo implicates a dietary factor, possibly cassava. The proceedings of a recent workshop on this subject have been published (Nestel & MacIntyre, 1973).

The author's investigations on cyanogenic glycosides have been supported, in part, by Grant GM-05301 from the National Institute of General Medical Sciences, U.S. Public Health Service, and by Grant 30319X from the National Science Foundation.

## References

Abrol, Y. P. (1966) *Indian J. Chem.* **4**, 251–252
Abrol, Y. P. (1967) *Indian J. Chem.* **5**, 54–55
Abrol, Y. P. & Conn, E. E. (1966) *Phytochemistry* **5**, 237–242
Abrol, Y. P., Conn, E. E. & Stoker, J. R. (1966) *Phytochemistry* **5**, 1021–1027
Ahmad, A. & Spenser, I. D. (1961) *Can. J. Chem.* **39**, 1340–1359
Ames, B. N. & Hartmann, P. E. (1963) *Cold Spring Harbor Symp. Quant. Biol.* **28**, 349–356
Bachstez, M., Prieto, E. S. & Gaja, A. M. C. (1948) *Ciencia (Mexico City)* **9**, 200–202
Becker, W. & Pfeil, E. (1966) *Biochem. Z.* **346**, 301–321
Ben-Yehoshua, S. & Conn, E. E. (1964) *Plant Physiol.* **39**, 331–333
Bertrand, G. (1906) *C. R. Acad. Sci.* **143**, 832–834
Bissett, F. H., Clapp, R. H., Coburn, R. A., Ettlinger, M. G. & Long, L., Jr. (1969) *Phytochemistry* **8**, 2235–2247
Bleichert, E. F., Neish, A. C. & Towers, G. H. N. (1966) *Colloq. Biosynthesis of Aromatic Compounds: Proc. FEBS Meet. 2nd* **3**, 119–127
Bourquelot, E. & Danjou, E. (1905) *J. Pharm. Chim.* **22**, 219–222
Brysk, M. M., Lauinger, C. & Ressler, C. (1969) *Biochim. Biophys. Acta* **184**, 583–588
Butler, G. W. & Butler, B. G. (1960) *Nature (London)* **187**, 780–781
Butler, G. W. & Conn, E. E. (1964) *J. Biol. Chem.* **239**, 1674–1679
Clapp, R. C., Ettlinger, M. G. & Long, L., Jr. (1970) *J. Amer. Chem. Soc.* **92**, 6378–6379
Conn, E. E. (1969) *J. Agr. Food Chem.* **17**, 519–526
Conn, E. E. & Akazawa, T. (1958) *Fed. Proc. Fed. Amer. Soc. Exp. Biol.* **17**, 205
Conn, E. E. & Butler, G. W. (1969) in *Perspectives in Phytochemistry* (Harborne, J. B. & Swain, T., eds.), pp. 47–74, Academic Press, London
Corkill, L. (1942) *N. Z. J. Sci. Technol. Sect. B* **23**, 178–193
Delange, F. & Ermans, A. M. (1971) *Amer. J. Clin. Nutr.* **24**, 1354–1360
Dilleman, G. (1958) in *Handbuch der Pflanzenphysiologie* (Ruhland, W., ed.), vol. 8, pp. 1050–1075, Springer, Berlin
Eisner, T., Eisner, H. E., Hurst, J. J., Kafatos, F. C. & Meinwald, J. (1963) *Science* **139**, 1218–1220
Ettlinger, M. G. & Eyjolfsson, R. (1972) *J. Chem. Soc. Chem. Commun.* 572–573
Ettlinger, M. G. & Kjaer, A. (1968) *Recent Advan. Phytochem.* **1**, 59–144
Eyjolfsson, R. (1970) *Fortschs. Chem. Org. Naturst.* **28**, 74–108
Eyjolfsson, R. (1971) *Acta Chem. Scand.* **25**, 1898–1900
Finnemore, H. & Cooper, J. M. (1936) *J. Proc. Roy. Soc. N.S.W.* **70**, 175–182
Gander, J. E. (1958) *Fed. Proc. Fed. Amer. Soc. Exp. Biol.* **17**, 226
Gerstner, E. & Pfeil, E. (1972) *Hoppe-Seyler's Z. Physiol. Chem.* **353**, 271–286
Gibbs, R. D. (1963) in *Chemical Plant Taxonomy* (Swain, T., ed.), pp. 41–88, Academic Press, London
Giles, N. H., Case, M. E., Partridge, C. W. H. & Ahmed, S. I. (1967) *Proc. Nat. Acad. Sci. U.S.* **58**, 1453–1460
Gmelin, R., Schuler, M. & Bordas, E. (1973) *Phytochemistry* **12**, 457–461
Hahlbrock, K. & Conn, E. E. (1970) *J. Biol. Chem.* **245**, 917–922
Hahlbrock, K. & Conn, E. E. (1971) *Phytochemistry* **10**, 1019–1023
Hahlbrock, K., Tapper, B. A., Butler, G. W. & Conn, E. E. (1968) *Arch. Biochem. Biophys.* **125**, 1013–1016
Haskins, F. A. & Kosuge, T. (1965) *Genetics* **52**, 1059–1068
Hegnauer, R. (1962–69) *Chemotaxonomie der Pflanzen*, vols. 1–5, Berkhauser, Basel
Hegnauer, R. (1971) *Pharm. Acta Helv.* **46**, 585–601
Herissey, H. (1906) *J. Pharm. Chim.* **23**, 1–4
Jacob, F., Ullmann, A. & Monod, J. (1965) *J. Mol. Biol.* **13**, 704–719

Jones, D. A. (1972) in *Phytochemical Ecology* (Harborne, J. B., ed.), pp. 103–124, Academic Press, London
Jorissen, A. & Hairs, E. (1891) *Bull. Acad. Roy. Sci. Belg.* **21**, 529–539
Kim, H. S., Jeffrey, G. A., Panke, D., Clapp, R. C., Coburn, R. A. & Long, L., Jr. (1970) *J. Chem. Soc. Chem. Commun.* 381–382
Kindl, H. & Underhill, E. W. (1968) *Phytochemistry* **7**, 745–756
Kingsbury, J. M. (1964) *Poisonous Plants of the United States and Canada*, Prentice Hall, Englewood Cliffs
Koukol, J., Miljanich, P. & Conn, E. E. (1962) *J. Biol. Chem.* **237**, 3223–3228
Mazelis, M. & Ingraham, L. L. (1962) *J. Biol. Chem.* **237**, 109–112
Mentzer, C. & Favre-Bonvin, J. (1961) *C. R. Acad. Sci.* **253**, 1072–1074
Michaels, R., Hankes, L. V. & Corpe, W. A. (1965) *Arch. Biochem. Biophys.* **111**, 121–125
Mikolajczak, K. L., Smith, C. R., Jr. & Tjarks, L. W. (1970) *Lipids* **5**, 812–817
Montgomery, R. D. (1965) *Amer. J. Clin. Nutr.* **17**, 103–113
Montgomery, R. D. (1969) in *Toxic Constituents of Plant Food Stuffs* (Liener, I. E., ed.), pp. 143–157, Academic Press, New York
Nahrstedt, A. (1973) *Phytochemistry* **12**, 2799–2800
Nartey, F. (1968) *Phytochemistry* **7**, 1307–1312
Nestel, B. & MacIntyre, R. (1973) *Chronic Cassava Toxicity* (*Publication IRDC*-010*e*), International Development Research Centre, Ottawa
Osuntokun, B. O. (1971) *Trans. Roy. Soc. Trop. Med. Hyg.* **65**, 454–479
Pallares, E. S. (1945) *Arch. Biochem.* **9**, 105–108
Paris, M., Bouquet, A. & Paris, R. R. (1969) *C. R. Acad. Sci. Ser. D* **268**, 2804–2806
Rimington, C. (1935) *Onderstepoort J. Vet. Sci. Anim. Ind.* **5**, 445–464
Rimington, C. & Roets, G. C. S. (1937) *Onderstepoort, J. Vet. Sci. Anim. Ind.* **9**, 193–201
Robiquet, P. J. & Boutron-Charlard, A. F. (1830) *Ann. Chim. Phys.* **44**, 352–382
Russell, G. B. & Reay, P. F. (1971) *Phytochemistry* **10**, 1373–1377
Seely, M. K., Criddle, R. S. & Conn, E. E. (1966) *J. Biol. Chem.* **241**, 4457–4462
Sharples, D. & Stoker, J. R. (1969) *Phytochemistry* **8**, 597–601
Stevens, D. L. & Strobel, G. A. (1968) *J. Bacteriol.* **95**, 1094–1102
Tapper, B. A. & Butler, G. W. (1971) *Biochem. J.* **124**, 935–941
Tapper, B. A. & Butler, G. W. (1972) *Phytochemistry* **11**, 1041–1046
Tapper, B. A., Zilg, H. & Conn, E. E. (1972) *Phytochemistry* **11**, 1047–1053
Towers, G. H. N., McInnes, A. G. & Neish, A. C. (1964) *Tetrahedron* **20**, 71–77
Tschiersch, B. (1966) *Flora* (*Jena*) *Abt. A* **157**, 43–50
Underhill, E. W., Wetter, L. R. & Chisholm, M. D. (1973) *Biochem. Soc. Symp.* **38**, 303–326
Uribe, E. G. & Conn, E. E. (1966) *J. Biol. Chem.* **241**, 92–94
Waller, G. R. (1968) *Proc. Okla. Acad. Sci.* **47**, 271–292
Ward, E. W. B. (1964) *Can. J. Bot.* **42**, 319–327
Ward, E. W. B. & Thorn, G. D. (1966) *Can. J. Bot.* **44**, 95–104
Ward, E. W. B., Thorn, G. D. & Starratt, A. N. (1971) *Can. J. Microbiol.* **17**, 1061–1066
Young, R. L. & Hamilton, R. A. (1966) *Proc. Annu. Meet. Hawaii Macadamia Prod. Ass. 6th* 27–30
Zilg, H. & Conn, E. E. (1974) *J. Biol. Chem.* in the press
Zilg, H., Tapper, B. A. & Conn, E. E. (1972) *J. Biol. Chem.* **247**, 2384–2386

*Biochem. Soc. Symp.* (1973) **38**, 303–326
*Printed in Great Britain*

# Biosynthesis of Glucosinolates

By E. W. UNDERHILL, L. R. WETTER and M. D. CHISHOLM

*National Research Council of Canada, Prairie Regional Laboratory, Saskatoon, Sask., Canada*

## Synopsis

The aglycone moieties of a number of glucosinolates are directly derived from amino acids whereas others are formed through glucosinolate interconversion. Many of the amino acids are formed by single or multiple chain elongation of 'common' amino acids via the methyl carbon of acetate. A common pathway has been established from the amino acids to glucosinolates involving several nitrogenous and sulphur-containing intermediates; some of the associated enzymes have been isolated.

## Introduction

The glucosinolates or mustard-oil glucosides constitute a unique class of natural plant products which are found primarily in plants of the family Cruciferae. The use of many of these plants as vegetables, condiments and medicines dates back to antiquity (Vaughan & Hemingway, 1959); undoubtedly one of the primary reasons for this culinary and medicinal interest was the pungent, sharp flavour and irritant principles encountered when these plants were crushed. The first concentrates of these volatile substances were derived from mustard plants and thus the name 'mustard oils' was given to these early isothiocyanate concentrates. About 1830 it was reported that these mustard oils did not exist *per se* in the plants, but were derived from parent substances (the mustard-oil glucosides or glucosinolates) when the plant tissue was crushed (Gildemeister & Hofmann, 1927).

The first structural formula for a glucosinolate was proposed by Gadamer (1897) for sinigrin (allylglucosinolate). In spite of the fact that a number of chemical properties peculiar to this class of glucoside could not be accounted for by his formula (Challenger, 1959), it remained until Ettlinger & Lundeen (1956) proposed the correct structural formula for sinigrin (1*a*, allylglucosinolate) and sinalbin (1*b*, *p*-hydroxybenzylglucosinolate). Their formula for sinigrin has been confirmed by X-ray crystallographic studies (Waser & Watson, 1963; Marsh & Waser, 1970) which also established the configuration around the C=N bond, namely, the side-chain R and the sulphate are *anti*. To date no glucosinolates have been isolated whose structures deviate from the general structure (1), in which

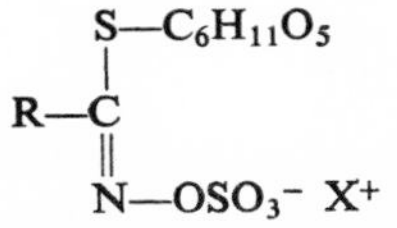

$$\begin{array}{l} \quad S{-}C_6H_{11}O_5 \\ \quad | \\ R{-}C \\ \quad \| \\ \quad N{-}OSO_3^-\ X^+ \end{array}$$

(1)

| | R | X |
|---|---|---|
| 1*a* | $CH_2{=}CH{-}CH_2$ | K |
| 1*b* | $HO{-}C_6H_4{-}CH_2$  | Sinapine |

$C_6H_{11}O_5$ represents a β-D-1-glucopyranosyl moiety. (The structures of compounds are numbered in the order in which they appear in the figures and the table.)

Most of the naturally occurring glucosinolates are difficult to crystallize and fewer than half have been characterized as crystalline glucosides or glucoside derivatives. The structures of the other glucosinolates have been inferred on the basis of the products formed by enzymic and chemical degradation. Before 1948 only nine glucosinolates had been reported; since then the number of naturally derived glucosinolates has increased to over 65 due mainly to the brilliant work of Kjaer and his colleagues in Denmark. A number of reviews which summarize the work and present the structures of the isolated glucosinolates have appeared (Kjaer, 1960, 1963; Underhill & Wetter, 1966; Ettlinger & Kjaer, 1968; Gmelin, 1969*a,b*). The structures of those glucosides isolated since 1968 are listed in Table 1.

The nomenclature employed to designate individual members of this class of thioglucosides before 1961 was based on a system of trivial names. Apart from the first glucosides, sinigrin and sinalbin, the trivial names of these glucosides were derived from the Latin name of the plant from which the glucoside was first isolated, prefixed with 'gluco' and suffixed with 'in', e.g. glucotropaeolin from *Tropaeolum* species. To identify more precisely the ever increasing number of newly discovered thioglucosides, Ettlinger & Dateo (1961) introduced a semi-systematic method of nomenclature in which the anion of the salt (1) (R = H), is designated by the term 'glucosinolate' and the chemical name of the radical 'R' is used as a prefix; designation of the cation is facilitated by using this system, e.g. sinapine *p*-hydroxybenzylglucosinolate (1*b*).

The enzyme thioglucoside glucohydrolase (EC 3.2.3.1), also known by the trivial name thioglucosidase (myrosinase), was first isolated from plants of the mustard family in about 1840. It is now considered to co-exist in those glucosinolate-containing plants of the families Capparidaceae, Cruciferae, Resedaceae, Tovariaceae, Moringaceae, Limnanthaceae, Tropaeolaceae, Caricaceae, Euphorbiaceae, Gyrostemonaceae and Salvadoraceae (Ettlinger & Kjaer, 1968). Its presence has also been reported in fungi (Reese *et al.*, 1958), bacteria (Oginsky *et al.*, 1965) and insects (MacGibbon & Allison, 1971).

The specificity of the thioglucosidase for the aglycone represented by 'R' in (1) appears to be low; many glucosinolates have been tested in the authors' laboratory and all were completely hydrolysed. The specificity towards the glucose or

$$\underset{(1)}{\mathrm{R{-}C(SGlc){=}N{-}OSO_3^-}} \xrightarrow{\beta\text{-Thioglucosidase}} \begin{cases} \xrightarrow{\text{pH 3}} \underset{(3)}{\mathrm{R{-}C{\equiv}N + S}} \\ \xrightarrow{\text{pH 7}} \underset{(2)}{\mathrm{R{-}N{=}C{=}S}} \end{cases} + \mathrm{HSO_4^-} + \text{D-Glucose}$$

sulphate moieties is much higher (Gaines & Goering, 1962). In addition to D-glucose and bisulphate, the enzymic hydrolysis of glucosinolates results in the formation of several different products depending on the composition of the glucosinolate aglycone and the pH of the media. At neutral or slightly acidic pH isothiocyanates (2) are formed, presumably by a Lossen type of re-arrangement after glucose and sulphate cleavage; at acidic pH values organic nitriles (3) and inorganic sulphur result. Isothiocyanates obtained from a number of glucosinolates including *p*-hydroxybenzylglucosinolate and indol-3-ylmethylglucosinolate are unstable in neutral or alkaline media and give rise to thiocyanate ion (Gmelin & Virtanen, 1960; Virtanen, 1962, 1965; Barothy & Neukom, 1965). Substituted oxazolidine-2-thiones or tetrahydro-1,3-oxazine-2-thiones are obtained by spontaneous cyclization of isothiocyanates derived from glucosinolates with an hydroxyl group at $C_2$ or $C_3$ (Kjaer & Boe Jensen, 1958). A number of these hydrolytic products of glucosinolates are known to elicit toxic effects in mammals; specifically, isothiocyanates, thiocyanate ion and oxazolidine-2-thiones possess goitrogenic activity (Van Etten, 1969).

Over the years there has been a controversy whether one or two enzymes are involved in glucosinolate hydrolysis. Early workers suggested there were two, a thioglucosidase and a sulphatase (von Euler & Eriksson, 1926; Sandberg & Holley, 1932; Ishimoto & Yamashina, 1949) and Neuberg & von Schoenebeck (1933) reported their separation. Ettlinger & Lundeen (1956) suggested that the liberation of the sulphate moiety occurred during Lossen re-arrangement after enzymic cleavage of the thioglucoside bond and thus provided support for a single enzyme. This one enzyme concept was supported by Nagashima & Uchiyama (1959) and by Reese *et al.* (1958) for the thioglucosidase isolated from *Aspergillus sydowi.*

Gaines & Goering (1960) separated a purified enzyme preparation by electrophoresis and DEAE-cellulose chromatography (Gaines & Goering, 1962) and reported the presence of two enzymes, one releasing sulphate, the other glucose. About this time the observation was made that ascorbic acid greatly enhanced the enzymic cleavage rate of allylglucosinolate (Nagashima & Uchiyama, 1959). Ettlinger *et al.* (1961) suggested that indeed, there were two enzymes in the complex, one activated by ascorbic acid, the other not. This activation by ascorbic acid was confirmed by Tsuruo & Hata (1967). Recently, Vose (1972) isolated two thioglucosidases from *Sinapis alba* seed by isoelectric focusing. One had an absolute requirement for ascorbic acid, the other had no such requirement.

The $\beta$-thioglucosidase–sulphatase concept reopened by Gaines & Goering (1962) led to a re-investigation of this problem. Calderon *et al.* (1966) were unable to demonstrate the presence of a sulphatase when they repeated the work of Gaines & Goering (1962). Others (Tsuruo *et al.*, 1967) in a detailed investigation of an enzyme obtained from *Brassica juncea* failed to find support for the two-enzyme concept. The same conclusion was reached by Ohtsuru *et al.* (1969), using an enzyme isolated from fungi. An abstract of a recent paper suggests that the enzyme complex isolated from *Sinapis alba* contains a $\beta$-thioglucosidase and a sulphatase (Iversen, 1972). Indeed, it would appear that this controversy is still not completely resolved and awaits further clarification.

Several plants containing glucosinolates are of commercial importance (Van Etten *et al.*, 1969). Some are important because of the glucosinolates they contain such as *Sinapis alba* and *Brassica nigra.* The pungent principles of these condiments as well as the flavour of other plants such as cabbage, turnip and radish are mainly derived from the enzymic-degradation products of glucosinolates. Other plants would have a greater economic importance if they did not contain glucosinolates, specifically the rapeseed species, *Brassica napus* and *Brassica campestris*, sources of edible vegetable oil.

An important by-product of oil extraction is the protein-rich rapeseed meal which is used as a protein supplement in livestock and poultry feeds. Unfortunately the use of this meal for feeding purposes can result in various manifestations of toxicity since hydrolysis of the glucosinolates produces a number of goitrogenic compounds (Kingsbury, 1964; Bowland *et al.*, 1965; Van Etten, 1969). Our group at the Prairie Regional Laboratory undertook the study of the formation of these deleterious glucosides since rapeseed, an important crop in Western Canada, would be of much greater commercial value if it were devoid of glucosinolates.

## Biosynthetic Studies

The similarity between the carbon skeletons of a number of 'common' amino acids and some glucosinolates led to the suggestion that amino acids may be natural progenitors of the aglycone moiety of these glucosides (Kjaer, 1954; Ettlinger & Lundeen, 1956). Exactly 10 years have passed since the first experimental confirmation of this hypothesis (Kutáček *et al.*, 1962; Underhill *et al.*, 1962; Benn, 1962). During this time, the biosynthesis of more than 15 members of this class of natural products has been studied. Without exception all have been shown to be derived from amino acids, several intermediates between the amino acids and the glucosinolates have been demonstrated and some of the enzymes involved have been isolated and studied. Most of the studies have involved the administration of variously labelled compounds ($^{3}H$, $^{14}C$, $^{15}N$ or $^{35}S$) to plants and the assessment of their relative efficiencies as precursors on the basis of the extent of incorporation of isotope into the glucosinolate.

### *Glucosinolates from 'common' amino acids*

In the first published report on the biosynthesis of a glucosinolate Kutáček *et al.* (1962) demonstrated that [3-$^{14}$C]tryptophan (4) was converted into indol-3-ylmethylglucosinolate (5). Both [3-$^{14}$C]tryptophan and indol-3-yl[1-$^{14}$C]acetonitrile were fed to cabbage (*Brassica oleracea*) and paper chromatograms of the

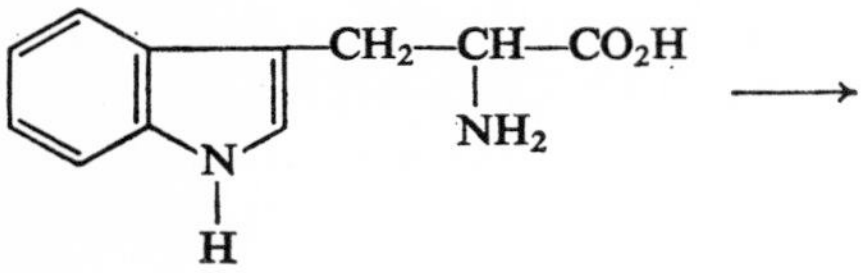

$CH_2—CH(NH_2)—CO_2H$ (indole, N—H) →

(4)

$CH_2—C(SGlc)=N—OSO_3^-$ (indole, N—H) + $CH_2—C(SGlc)=N—OSO_3^-$ (indole, N—$OCH_3$)

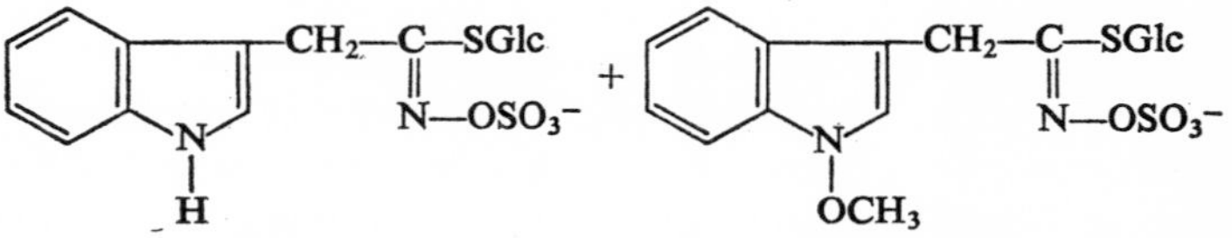

(5) (6)

plant extract showed the presence of the radioactive glucosinolate derived from the amino acid but not from the nitrile. Their work has been corroborated by Schraudolf & Bergmann (1965) who fed DL-[3-$^{14}$C]tryptophan to *Sinapis alba* hypocotyls and demonstrated that the labelled amino acid was converted into both indol-3-ylmethylglucosinolate and its *N*-methoxy analogue, 1-methoxyindol-3-ylmethylglucosinolate (6).

The biosynthesis of benzylglucosinolate (8) in *Tropaeolum majus* was reported independently by Underhill *et al.* (1962) and by Benn (1962). Both groups found phenylalanine was incorporated into the glucosinolate with high efficiency, up to 12% of the administered $^{14}$C was recovered in the aglycone carbons. The incorporation of $^{14}$C from phenylalanine (7) was specific; the activity from phenyl[2-$^{14}$C]alanine was located in the thiohydroximate carbon whereas that from phenyl[3-$^{14}$C]alanine was found in the benzyl carbon. No activity was present in the glucosinolate when phenyl[1-$^{14}$C]alanine was fed.

$C_6H_5$—$\overset{\bullet}{C}H_2$—$\overset{*}{C}H(\overset{\triangledown}{N}H_2)$—$\overset{\circ}{C}O_2H$ ⟶ $C_6H_5$—$\overset{\bullet}{C}H_2$—$\overset{*}{C}(SGlc)$=$\overset{\triangledown}{N}$—$OSO_3^-$

(7) (8)

To learn more about the source of the glucosinolate nitrogen, Underhill & Chisholm (1964) fed doubly labelled L-[$^{14}$C,$^{15}$N]phenylalanine (7) to *Tropaeolum majus* and found that the aglycone moiety of benzylglucosinolate (8) possessed the same $^{14}$C/$^{15}$N ratio as the amino acid. From this it was established that the amino nitrogen and the carbon skeleton of L-phenylalanine, except for the carboxyl carbon, were incorporated intact into benzylglucosinolate, thus making it mandatory that all intermediates between the amino acid and the glucosinolate be nitrogenous. A similar conclusion was reached by Meakin (1965).

Kindl (1964, 1965) studying the biosynthesis of *p*-hydroxybenzylglucosinolate (11) in *Sinapis alba*, found that *p*-[3-$^{14}$C]coumaric acid (9) was converted into the glucosinolate much more efficiently than either [2- or 3-$^{14}$C]tyrosine (10), or [$^{14}$C]phenylalanine. This result appeared to be at variance with the observation by Underhill *et al.* (1962) who found that [3-$^{14}$C]cinnamic acid was only poorly converted into benzylglucosinolate. Initially these results suggested that a basically different pathway may exist in the formation of these two aromatic glucosinolates. However, in a subsequent study Kindl & Schiefer (1969) demonstrated the reversible transformation of *p*-coumaric acid into tyrosine in detached leaves of *Sinapis alba* and by tyrosine ammonia-lyase preparations derived from this plant. This reversible transformation of *p*-coumaric acid into tyrosine accounts for the

HO—$C_6H_4$—$\overset{\triangledown}{C}H$=CH—$CO_2H$ ⇌

(9)

HO—$C_6H_4$—$\overset{\triangledown}{C}H_2$—$\overset{*}{C}H(NH_2)$—$CO_2H$ ⟶ HO—$C_6H_4$—$\overset{\triangledown}{C}H_2$—$\overset{*}{C}(SGlc)$=N—$OSO_3^-$

(10) (11)

unexpectedly high incorporation of tracer from *p*-coumaric acid in *Sinapis alba* and, here too, the glucosinolate is considered to be derived from the structurally related amino acid. L. R. Wetter (unpublished work) has also fed labelled *p*-coumaric acid, L-tyrosine and L-phenylalanine to *Sinapis alba* but he found that tyrosine was converted into *p*-hydroxybenzylglucosinolate considerably more efficiently than the other two labelled compounds. There has been no definitive explanation to account for the differences in relative precursor efficiency of tyrosine and *p*-coumaric acid as found by these two groups, except perhaps for plant variation in tyrosine ammonia-lyase activity.

The biosynthesis of three other glucosinolates from structurally related α-amino acids has been noted. The aglycone moiety of isopropylglucosinolate (13) has been demonstrated to be formed from valine (12) in both *Tropaeolum peregrinum*

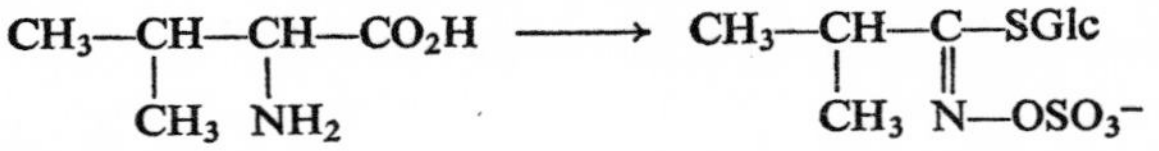

$$CH_3{-}CH(CH_3){-}CH(NH_2){-}CO_2H \longrightarrow CH_3{-}CH(CH_3){-}C({=}N{-}OSO_3^-){-}SGlc$$

(12) (13)

(Benn & Meakin, 1965) and *Cochlearia officinalis* (Tapper & Butler, 1967). Underhill & Kirkland (1972*b*) reported the efficient conversion of $^{14}C$ from L-[$^{14}C$]leucine (14) into 2-hydroxy-2-methylpropylglucosinolate (15) in *Conringia orientalis*. The biosynthesis of 3-methoxycarbonylpropylglucosinolate (18), which occurs in a few species of *Erysimum*, has been studied by M. D. Chisholm

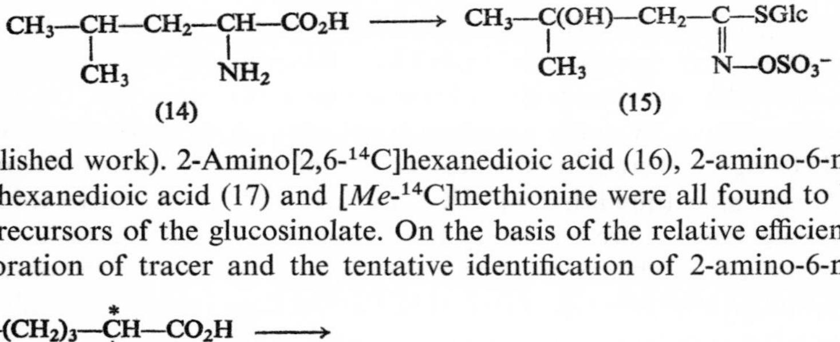

$$CH_3{-}CH(CH_3){-}CH_2{-}CH(NH_2){-}CO_2H \longrightarrow CH_3{-}C(OH)(CH_3){-}CH_2{-}C({=}N{-}OSO_3^-){-}SGlc$$

(14) (15)

(unpublished work). 2-Amino[2,6-$^{14}C$]hexanedioic acid (16), 2-amino-6-methyl-[2-$^{14}C$]hexanedioic acid (17) and [*Me*-$^{14}C$]methionine were all found to be efficient precursors of the glucosinolate. On the basis of the relative efficiencies of incorporation of tracer and the tentative identification of 2-amino-6-methyl-

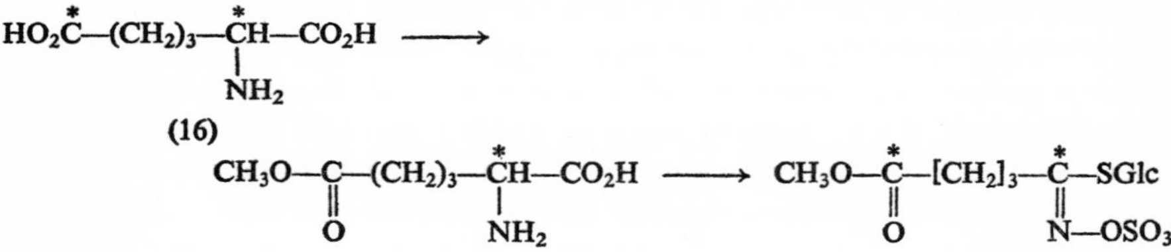

$$HO_2\overset{*}{C}{-}(CH_2)_3{-}\overset{*}{C}H(NH_2){-}CO_2H \longrightarrow$$

(16)

$$CH_3O{-}C({=}O){-}(CH_2)_3{-}\overset{*}{C}H(NH_2){-}CO_2H \longrightarrow CH_3O{-}\overset{*}{C}({=}O){-}[CH_2]_3{-}\overset{*}{C}({=}N{-}OSO_3^-){-}SGlc$$

(17) (18)

hexanedioic acid as a natural amino acid in the plant, it would appear that the methoxy carbon of the glucosinolate is derived at the amino acid stage of biosynthesis.

*Glucosinolates from 'chain-elongated' amino acids*

The fact that the structures of many glucosinolates e.g. allylglucosinolate (26), 2-phenylethylglucosinolate (32) and the aliphatic glucosinolates in rapeseed (38, 39, 40, 42, 43, 44) were not related in an obvious manner to commonly occurring amino acids suggested that a pathway other than that from amino acids might exist for their formation. However, studies on their biosynthesis have demonstrated that they too are derived from amino acids, previously unrecognized as natural plant constituents. These are α-amino acid homologues, derived from 'common' amino acids by chain-extension, whose carbon skeleton and amino nitrogen are specifically incorporated into the glucosinolates.

In their preliminary investigations on the biosynthesis of allylglucosinolate in horseradish leaves, *Armoracea lapathifolia*, it was demonstrated (Underhill *et al.*, 1962) that both [$^{14}$C]aspartic acid and [2-$^{14}$C]acetate were incorporated into the aglycone moiety to a greater extent than D-[$^{14}$C]glucose. The methyl carbon of acetate specifically labelled the thiohydroximate carbon. There was negligible incorporation of activity from [1-$^{14}$C]acetate, [2-$^{14}$C]glutamic acid and [2-$^{14}$C]-glycine. Considering these results and the conversion in *Escherichia coli* of aspartic acid into homoserine, a progenitor of methionine (Sayre & Greenberg, 1956; Stadtman *et al.*, 1961), Chisholm & Wetter (1964) fed a number of labelled compounds including specifically labelled methionine (19) and acetate (21). Portions of both of these compounds were efficiently incorporated into allylglucosinolate (26). It was confirmed that the label in the thiohydroximate carbon was derived from the methyl carbon of acetate and the allyl carbons from $C_2$, $C_3$ and $C_4$ of methionine. Neither the carboxyl of acetate and methionine nor the methyl carbon of methionine were incorporated. An independent study by Matsuo & Yamazaki (1964) using *Brassica juncea* also demonstrated the derivation of the thiohydroximate carbon of allylglucosinolate from [2-$^{14}$C]acetate.

On the basis of these findings and the observation (Underhill *et al.*, 1962)

$$CH_3{-}S{-}\overset{*}{C}H_2{-}\overset{*}{C}H_2{-}\overset{\bullet}{C}H(NH_2){-}\overset{\circ}{C}O_2H \rightleftharpoons CH_3{-}S{-}CH_2{-}CH_2{-}C({=}O){-}CO_2H \xrightarrow[(21)]{\overset{\triangledown}{C}H_3{-}\overset{\circ}{C}O_2H}$$

(19) (20)

$$CH_3{-}S{-}CH_2{-}CH_2{-}C(OH)(CO_2H){-}CH_2{-}CO_2H \longrightarrow$$

(22)

$$CH_3{-}S{-}CH_2{-}CH_2{-}CH(CO_2H){-}CH(OH){-}CO_2H \xrightarrow{-2H-CO_2}$$

(23)

$$CH_3{-}S{-}CH_2{-}CH_2{-}CH_2{-}C({=}O){-}CO_2H \rightleftharpoons$$

(24)

$$CH_3{-}S{-}CH_2{-}CH_2{-}CH_2{-}\overset{\triangledown}{C}H(\overset{\triangledown}{N}H_2){-}CO_2H \longrightarrow \longrightarrow \overset{*}{C}H_2{=}\overset{*}{C}H{-}\overset{\bullet}{C}H_2{-}\overset{\triangledown}{C}({=}\overset{\triangledown}{N}{-}OSO_3^-){-}SGlc$$

(25) (26)

that the thiohydroximate carbon of 2-phenylethylglucosinolate was also derived from C-2 of acetate, Chisholm & Wetter (1964) proposed that allylglucosinolate may arise from an amino acid formed from methionine and acetate by a chain-lengthening pathway (19–25) similar to the formation of leucine from valine and acetate (Strassman & Ceci, 1963).

Confirmation of this hypothesis was provided by Chisholm & Wetter (1966) and by Matsuo & Yamazaki (1966). Both groups reported that [2-$^{14}$C]homomethionine (25) was incorporated with greater efficiency than [2-$^{14}$C]methionine, and the activity from this higher homologue was specifically incorporated into the thiohydroximate carbon. Matsuo & Yamazaki (1966) also fed [2-$^{14}$C,$^{15}$N]homomethionine and concluded that both the carbon and nitrogen atoms were incorporated intact into the glucosinolate. Additional support for the intermediacy of homomethionine in the biosynthesis of allylglucosinolate in horseradish was obtained when Chisholm & Wetter (1966) found this amino acid to be both naturally occurring as well as derived from [2-$^{14}$C]methionine. Both allyl[2-$^{14}$C]-glycine (2-amino-4-pentenoic acid), the amino acid structurally related to allylglucosinolate, and 2-amino-5-hydroxy[2-$^{14}$C]valeric acid were poorly incorporated thus indicating that the cleavage of the methylthio group of homomethionine, resulting in the terminal unsaturation of the glucosinolate, takes place at a later stage in the biosynthetic sequence.

Concurrently with the allylglucosinolate studies, Underhill *et al.* (1962) observed the labelled carbons from [2-$^{14}$C]acetate (21) and phenyl[2,3-$^{14}$C]alanine (7) were incorporated specifically and with high efficiency into the aglycone carbons of 2-phenylethylglucosinolate (32). No radioactivity was present in the aglycone when either [1-$^{14}$C]acetate or phenyl[1-$^{14}$C]alanine were fed. These data, consistent with the chain-elongation hypothesis, led to the feeding of 2-amino-4-phenyl[2,3-$^{14}$C]butyric acid (31), the next higher homologue of phenylalanine

$$C_6H_5-CH_2-CH(NH_2)-CO_2H \;(7) \rightleftharpoons C_6H_5-CH_2-C(=O)-CO_2H \;(27) \xrightarrow[(21)]{CH_3-CO_2H}$$

$$C_6H_5-CH_2-C(OH)(CO_2H)-CH_2-CO_2H \;(28) \longrightarrow C_6H_5-CH_2-CH(CO_2H)-CH(OH)-CO_2H \;(29) \xrightarrow{-2H-CO_2}$$

$$C_6H_5-CH_2-CH_2-C(=O)-CO_2H \;(30) \rightleftharpoons$$

$$C_6H_5-CH_2-CH_2-CH(NH_2)-CO_2H \;(31) \longrightarrow C_6H_5-CH_2-CH_2-C(SGlc)=N-OSO_3^- \;(32)$$

(Underhill, 1965*a*). Approx. 40% of the administered radioactivity was converted into the aglycone carbons of the glucosinolate, providing further experimental data supporting the chain-extension hypothesis.

Both the carbon and nitrogen of L-2-[$^{15}$N]amino-4-phenyl[2-$^{14}$C]butyric acid (31) were incorporated as a unit into the glucosinolate (32). However, when the racemate was fed, the incorporation of $^{14}$C was twice that of the $^{15}$N and Underhill suggested that the D-isomer may have been converted through the oxo acid into the L-isomer with the resulting loss of $^{15}$N, before incorporation. He also examined the amino acid fraction of *Nasturtium officinale* and found the chain-elongated 2-amino-4-phenylbutyric acid to be a natural constituent of this plant, although in trace amounts (Underhill, 1968). Evidence was also provided for the existence of one of the intermediates in the chain-elongation pathway, 3-benzylmalic acid. As might be predicted the efficiency of incorporation of 3-benzyl[1,2-$^{14}C_2$]-malic acid (29) into 2-phenylethylglucosinolate was midway between that of phenyl[2-$^{14}$C]alanine and 2-amino-4-phenyl[2-$^{14}$C]butyric acid.

A similar pattern of incorporation into (*S*)2-hydroxy-2-phenylethylglucosinolate was found (Underhill, 1965*b*) when acetate, phenylalanine and 2-amino-4-phenylbutyric acid were fed to *Reseda luteola*. When racemic 2-[$^{15}$N]amino-4-phenyl[2-$^{14}$C]butyric acid was fed, dissimilar dilution values were observed for the two isotopes and a conclusion regarding the source of the glucosinolate nitrogen was not possible.

Another series of experiments was designed (Underhill & Kirkland, 1972*a*), to learn more about the source of the glucosinolate nitrogen and to determine whether the glucosinolate hydroxyl group was introduced at the amino acid or some later stage of the biosynthetic sequence. L-2-[$^{15}$N]Amino-4-phenyl[2-$^{14}$C]-butyric acid, as well as (2*S*,4*S*) and (2*S*,4*R*) 2-[$^{15}$N]amino-4-hydroxy-4-phenyl-[2,3-$^{14}C_2$]butyric acid were fed and in each instance the $^{14}$C and the $^{15}$N were incorporated as a unit into the glucosinolate. However, repeated feeding experiments consistently demonstrated greater incorporation of tracer from 2-amino-4-phenylbutyric acid than from either of the isomers of the hydroxylated amino acid structurally related to the glucosinolate. On the basis of this greater incorporation and the data obtained from feeding [$^{14}$C]2-phenylethylglucosinolate, they suggested the naturally functioning pathway in *Reseda luteola* is phenylalanine $\rightarrow\rightarrow$ 2-amino-4-phenylbutyric acid $\rightarrow\rightarrow$ 2-phenylethylglucosinolate $\rightarrow$ 2-hydroxy-2-phenylethylglucosinolate. Several possibilities can be considered to account for the intact incorporation of carbon and nitrogen from the isomers of the hydroxylated amino acid namely: (1) the isomers may have been converted into the isomeric (*S*) and (*R*)2-hydroxy-2-phenylethylglucosinolate as a result of low enzyme specificity; (2) the isomers may have first been reduced and incorporated via 2-amino-4-phenylbutyric acid; (3) two separate pathways for the formation of the glucosinolate may exist, one from 2-amino-4-phenylbutyric acid and the other from its hydroxylated analogue.

An unusual characteristic exhibited by this class of plant glucosides is the large number of homologous glucosinolates each differing by a single methylene carbon. More than 25% of the known glucosinolates can be subdivided into these four series (33, 34, 35 and 36).

Biosynthetic studies have been reported on representatives of three of the four

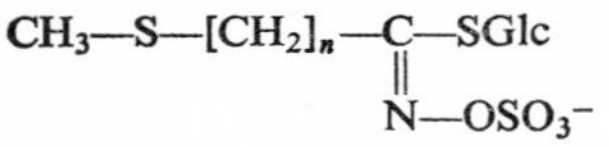

$$CH_3—S—[CH_2]_n—C(=N—OSO_3^-)—SGlc$$
(33)
$n = 3–8$

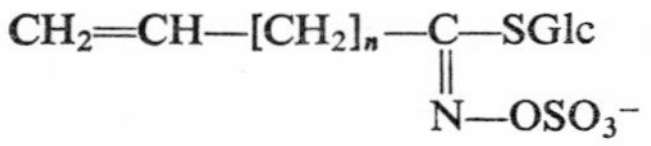

$$CH_2{=}CH—[CH_2]_n—C(=N—OSO_3^-)—SGlc$$
(34)
$n = 1, 2, 3$

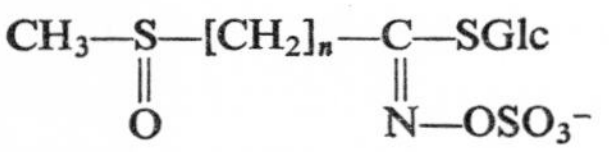

$$CH_3—S(=O)—[CH_2]_n—C(=N—OSO_3^-)—SGlc$$
(35)
$n = 3–11$

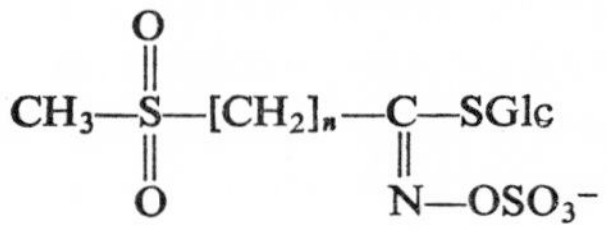

$$CH_3—S(=O)_2—[CH_2]_n—C(=N—OSO_3^-)—SGlc$$
(36)
$n = 3, 4$

classes of homologues and in each instance data have been found consistent with the theory that the glucosinolate precursors are amino acids derived by either single or multiple chain-extension of methionine by the methyl carbon of acetate. In a study on the biosynthesis of 3-butenylglucosinolate (39), 2-hydroxy-3-butenylglucosinolate (40) and 4-pentenylglucosinolate (43) in *Brassica campestris*, Chisholm & Wetter (1967) found that the radioactivity from [2-$^{14}$C]methionine, [3,4-$^{14}C_2$]methionine, [2-$^{14}$C]acetate and [2-$^{14}$C]homomethionine was converted into each of these three major rapeseed glucosinolates. Here too, there was little or no incorporation of radioactivity from [1-$^{14}$C]acetate, [1-$^{14}$C]methionine, allyl-[2-$^{14}$C]glycine or from 2-amino-5-hydroxy[2-$^{14}$C]valeric acid. From the differences in the extent of incorporation of $^{14}$C into the thioglucosides, they suggested

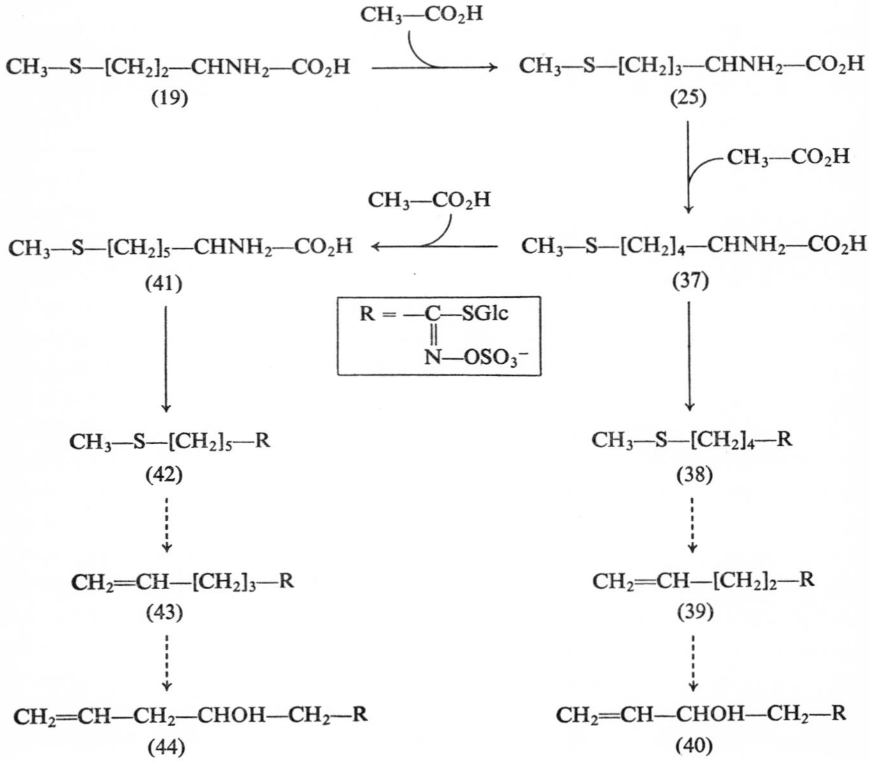

$$CH_3—S—[CH_2]_2—CHNH_2—CO_2H \;(19) \xrightarrow{CH_3—CO_2H} CH_3—S—[CH_2]_3—CHNH_2—CO_2H \;(25)$$

$$CH_3—S—[CH_2]_3—CHNH_2—CO_2H \;(25) \xrightarrow{CH_3—CO_2H} CH_3—S—[CH_2]_4—CHNH_2—CO_2H \;(37)$$

$$CH_3—S—[CH_2]_4—CHNH_2—CO_2H \;(37) \xrightarrow{CH_3—CO_2H} CH_3—S—[CH_2]_5—CHNH_2—CO_2H \;(41)$$

$$R = —C(=N—OSO_3^-)—SGlc$$

$$(41) \rightarrow CH_3—S—[CH_2]_5—R \;(42) \dashrightarrow CH_2{=}CH—[CH_2]_3—R \;(43) \dashrightarrow CH_2{=}CH—CH_2—CHOH—CH_2—R \;(44)$$

$$(37) \rightarrow CH_3—S—[CH_2]_4—R \;(38) \dashrightarrow CH_2{=}CH—[CH_2]_2—R \;(39) \dashrightarrow CH_2{=}CH—CHOH—CH_2—R \;(40)$$

that 3-butenylglucosinolate (39) and its hydroxylated analogue (40) may be derived from 2-amino-6-methylthiohexanoic acid (37) and 4-pentenylglucosinolate (43) from 2-amino-7-methylthioheptanoic acid (41); it was suggested that each of these amino acids was derived from methionine (19) by multiple chain-extensions via the methyl carbon of acetate.

Studies on the formation of 2-hydroxy-3-butenylglucosinolate (40), in rutabaga (*Brassica napobrassica*) leaves, have provided experimental data confirming its formation by the above pathway. A partial degradation of the aglycone derived from [2-$^{14}$C]acetate (Serif & Schmotzer, 1968) afforded the thiohydroximate carbon which contained approx. 38% of the total aglycone radioactivity, an amount which is close to that predicted on the basis of a methionine chain-extension by means of two condensations with acetate. In subsequent publications Lee & Serif (1968, 1970) reported the specific incorporation of $^{14}$C from 2-[$^{15}$N]-amino-6-methylthio[2-$^{14}$C]hexanoic acid (37) with retention of the amino nitrogen into the aglycone moiety of 2-hydroxy-3-butenylglucosinolate. By using *Brassica napus*, Josefsson (1971) found that 2-amino-6-methylthio[2-$^{14}$C]hexanoic acid was a precursor of 3-butenylglucosinolate, 2-hydroxy-3-butenylglucosinolate and 4-pentenylglucosinolate.

Noteworthy is the observation (M. D. Chisholm, unpublished work) that *S*-methylcysteine was not chain-elongated to form methionine in horseradish leaves. It was incorporated with equal, although low, efficiency into both the aglycone and glucosyl moieties of allylglucosinolate suggesting its prior conversion into pyruvate and then into glucose.

The principal plant families containing glucosinolates derived by single and multiple chain-extension of methionine and phenylalanine are the Cruciferae and Resedaceae (Ettlinger & Kjaer, 1968). On the basis of the reasonable assumption that higher glucosinolate homologues are also derived from amino acids, one may anticipate that several new amino acids of related carbon structure will be found in these plant families.

Although not related directly to glucosinolate biosynthesis another example of homologization of carbon chains of amino acids by means of acetate comes from studies on the biosynthesis of 2-amino-4-methylhex-4-enoic acid in *Aesculus californica* (Fowden & Mazelis, 1971; Boyle & Fowden, 1971). Data from their feeding experiments indicate that [$^{14}$C]isoleucine is first degraded to tiglyl-CoA which, on condensation with acetate followed by transamination, yields 2-amino-4-methylhex-4-enoic acid.

*'Intermediates' between amino acids and glucosinolates*

A number of nitrogenous compounds have been found to be efficient precursors of glucosinolates and they are considered to form part of the biosynthetic pathway between amino acids and glucosinolates; it is in this sense we refer to them as 'intermediates'. Conclusions derived from precursor feeding experiments alone can only be taken as tentative since it is generally recognized that (1) the precursor may be in enzymic or chemical equilibrium with the true intermediate and cannot be distinguished from it by such experiments, and (2) unnatural compounds can serve as precursors of natural products as a result of enzyme induction or lack of enzyme specificity. To establish that such a compound is in fact an intermediate, it is necessary to demonstrate that it is both formed and utilized *in vivo* and to isolate the enzymes involved in these transformations (Tapper & Butler, 1972). A number of compounds mentioned here have met some of these requirements; none have, as yet, met all.

The intact incorporation of labelled carbon and amino nitrogen into glucosinolates, demanding that intermediates between amino acids and glucosinolates be nitrogenous, led to the testing of a number of $^{14}$C-labelled nitrogen-containing compounds as precursors of glucosinolates. However, there was little or no incorporation of $^{14}$C into benzylglucosinolate with 2-phenylethylamine, phenylacetamide, benzyl cyanide and phenylacetohydroxamic acid (Underhill & Chisholm, 1964) or into allylglucosinolate with 3-methylthiopropionamide and 4-methylthiobutyramide (Matsuo & Yamazaki, 1966). The first demonstration of a nitrogenous intermediate was reported by Tapper & Butler (1967) and Underhill (1967); each found phenyl[1-$^{14}$C]acetaldoxime (45, $R=C_6H_5CH_2$) a more efficient precursor of benzylglucosinolate than [$^{14}$C]phenylalanine in *Lepidium sativum* and *Tropaeolum majus* respectively. Phenylacetaldoxime was also found to be a naturally occurring aldoxime in *Tropaeolum majus* formed from phenyl-[2-$^{14}$C]alanine. The extent of incorporation of $^{14}$C from [$^{14}$C]isobutyraldoxime [45, $R=[CH_3]_2CH$] and 3-phenyl[1-$^{14}$C]propionaldoxime (45, $R = C_6H_5C_2H_4$) into isopropylglucosinolate (Tapper & Butler, 1967) and 2-phenylethylglucosinolate (Underhill, 1967) respectively was equal to or greater than the incorporation from their corresponding amino acids. Since these initial findings both Matsuo

$$R-\overset{*}{C}H(NH_2)-CO_2H \longrightarrow R-\overset{*}{C}(=NOH)-H \longrightarrow R-\overset{*}{C}(=N-OSO_3^-)-SGlc$$

**(45)**

(1968*b*) and Lee & Serif (1971) have noted the intact conversion of carbon and nitrogen of the oxime moiety of 4-methylthio[1-$^{14}$C,$^{15}$N]butyraldoxime (45, $R=CH_3S[CH_2]_3$) into allylglucosinolate and 5-methylthio[1-$^{14}$C,$^{15}$N]valeraldoxime (45, $R=CH_3S[CH_2]_4$) into 2-hydroxy-3-butenylglucosinolate. Kindl (1968) has demonstrated the natural derivation of indol-3-ylacetaldoxime from $^{14}CO_2$ in *Brassica oleracea*, *Reseda luteola* and *Sinapis alba* and has also reported its formation both *in vivo* and *in vitro* from tryptophan. The conversion of [$^3$H]indol-3-ylacetaldoxime into indol-3-ylmethylglucosinolate and 1-sulpho-indol-3-ylmethylglucosinolate in *Isatis tinctoria* has been confirmed by Mahadevan & Stowe (1972). Again, Kindl & Schiefer (1969) found *p*-hydroxyphenyl-[1-$^{14}$C]acetaldoxime (45, $R=HOC_6H_4CH_2$) was converted into *p*-hydroxybenzylglucosinolate ten times more efficiently than was L-[3-$^{14}$C]tyrosine. Clearly, aldoximes are naturally occurring compounds derived from amino acids and precursors of glucosinolates.

Not only are aldoximes involved in the formation of glucosinolates but also in the biosynthesis of nitrile compounds in higher plants including indolylacetonitrile (Mahadevan, 1963; Shukla & Mahadevan, 1968), the cyanogenic glucosides (Tapper *et al.*, 1967; Tapper & Butler, 1971) and of substituted benzonitriles in bacteria (Milborrow, 1963). The aromatic alcohols, tyrosol and *p*-methoxyphenylethyl alcohol, have also been shown to be derived from aldoximes (Kindl & Schiefer, 1971).

2-Oximino acids, which in theory could be progenitors of aldoximes by decarboxylation, have been considered as possible precursors of glucosinolates and cyanogenic glucosides. Both 2-oximino-3-phenyl[2-$^{14}$C]propionic acid (Underhill & Chisholm, 1964; Meakin, 1965; Kindl & Underhill, 1968) and 2-oximino[$^{14}$C]-isovaleric acid (Tapper & Butler, 1967) were found to be considerably less effective as precursors of benzylglucosinolate and isopropylglucosinolate than the corresponding amino acids and neither are considered as intermediates of these glucosinolates. The conversion of oximino acids into cyanogenic glucosides is regarded as inconclusive (Tapper & Butler, 1971) since they may have first been converted, non-enzymically, into the next lower nitriles, which are known precursors of this class of glucosides. The formation of nitriles in aqueous solution is a general reaction of α-oximino acids (Ahmad & Spenser, 1961).

Studies on the biosynthesis of the hydroxamic acids, hadacidin (Stevens & Emery, 1966) and ferrichrome (Emery, 1966) have focused attention on *N*-

$C_6H_5$—$CH_2$—CH($NH_2$)—$CO_2H$ (7) ⟶ $C_6H_5$—$CH_2$—C*H(NHOH)—$CO_2H$ (46) ⟶

$C_6H_5$—$CH_2$—C*(=NOH)—H (47) ⟶ $C_6H_5$—$CH_2$—C*(=N—$OSO_3^-$)—SGlc (8)

hydroxyamino acids as intermediates in plant metabolism. In their attempts to find intermediates between phenylalanine and phenylacetaldoxime, Kindl & Underhill (1968) fed *N*-hydroxyphenyl[2-$^{14}$C]alanine (46) to *Tropaeolum majus* and noted that the efficiency of conversion of $^{14}$C into the aglycone moiety of benzylglucosinolate was higher than that from phenyl[2-$^{14}$C]alanine and was comparable with that observed previously from phenyl[1-$^{14}$C]acetaldoxime (47). Although they were unable to establish if the labelled *N*-hydroxyphenylalanine had been incorporated into the glucosinolate by prior reduction to phenylalanine, (*N*-hydroxyamino acids are known to afford both α-oximino and α-amino acids by disproportionation; Spenser & Ahmad, 1961) Kindl & Underhill (1968) concluded on the basis of the somewhat greater incorporation of $^{14}$C, that *N*-hydroxyphenylalanine may be an intermediate between the amino acid and phenylacetaldoxime. They were also able to demonstrate the formation of phenylacetaldoxime from *N*-hydroxyphenylalanine by using cell-free preparations from *Sinapis alba*, *Tropaeolum majus* and *Nasturtium officinale*. This transformation of *N*-hydroxyphenylalanine involving both dehydrogenation and decarboxylation may have some parallel with the oxidative decarboxylation of amino acids to the next lower acid amide (Mazelis & Ingraham, 1962; Kosuge *et al.*, 1966). Efforts to demonstrate either the formation of *N*-hydroxyphenylalanine in cell-free systems or its occurrence in *Tropaeolum majus* have been without success (E. W. Underhill, unpublished work).

The incorporation of sulphur is an integral part of the biosynthetic sequence leading from aldoximes to glucosinolates. All of the known glucosinolates contain at least two sulphur atoms and many contain three. The incorporation of $^{35}S$ into allylglucosinolate in *Armoracea lapathifolia* has been studied independently by Matsuo (1963) and by Wetter (1964). Matsuo (1963) found sodium [$^{35}S$]sulphate incorporated into both sulphur atoms, 89% of the activity appearing in the sulphate sulphur and the remainder in the thiohydroximate sulphur. Wetter (1964) observed a similar distribution of activity when he fed $^{35}S$-labelled sulphate, thiosulphate and sulphide, suggesting that these inorganic sources were incorporated by a common sulphur pathway. By using *Sinapis alba*, similar distributions of [$^{35}S$]sulphate have been obtained in *p*-hydroxybenzylglucosinolate (Kindl, 1964, 1965), and in indol-3-ylmethylglucosinolate and 1-methoxyindol-3-ylmethylglucosinolate (Schraudolf & Bergmann, 1965). From the relative specific radioactivities obtained after feeding $^{35}SO_4{}^{2-}$, Elliott & Stowe (1971*b*) suggested that indol-3-ylmethylglucosinolate is the precursor of both the 1-methoxy and 1-sulpho analogues. $^{35}SO_2$ has also been incorporated into both sulphur atoms of glucosinolates found in *Brassica oleracea* (Kutáček *et al.*, 1966; Spálený & Kutáček, 1969).

The distribution of radioactivity in the glucosinolate sulphur atoms derived from $^{35}S$-labelled amino acids, was in sharp contrast with the distribution observed when inorganic sulphur compounds were fed. Approx. 80% of the sulphur incorporated into allylglucosinolate from both [$^{35}S$]methionine and [$^{35}S$]cysteine was located in the thiohydroximate sulphur (Wetter, 1964; Matsuo, 1968*a*). The most effective source of the thiohydroximate sulphur of glucosinolates has been found to be cysteine (Kindl, 1965; Wetter & Chisholm, 1968; Matsuo, 1968*a*). It is noteworthy that inorganic sulphide labelled predominantly the bisulphate moiety of glucosinolates whereas cysteine which can be formed from *O*-acetylserine and sulphide (Giovanelli & Mudd, 1967) labelled the thiohydroximate sulphur.

Two pathways were proposed for the formation of the thioglucosyl unit of glucosinolates (Meakin, 1965, 1967) namely, (1) the formation of a thiohydroximic acid and its subsequent glucosylation and (2) the formation of 1-thioglucose and its incorporation as a unit into some unknown intermediate. No experimental support for the latter possibility could be found by Matsuo (1968*a*) who fed $^{35}S$-labelled methionine, cysteine, $SO_4{}^{2-}$ and 1-thioglucose to horseradish and found the incorporation of sulphur from 1-thioglucose was negligible by comparison. Similar results were obtained by Wetter & Chisholm (1968) who reported that the incorporation of $^{14}C$ from [U-$^{14}C$]glucose into allylglucosinolate was eight times more efficient than the incorporation from 1-thio[U-$^{14}C$]glucose, clearly an indication that the thioglucosyl moiety of this glucosinolate is not derived directly from 1-thioglucose. They also showed that the incorporation of $^{35}S$ from 1-[$^{35}S$]thioglucose into various glucosinolates was low, thus verifying the above observations.

Data confirming the precursor role of thiohydroximates in the biosynthesis of glucosinolates were reported by Underhill & Wetter (1969) who found high conversions of tracer into benzylglucosinolate not only from phenyl[2-$^{14}C$]aceto-[$^{35}S$]thiohydroximate (48) but also from desulphobenzylglucosinolate [*S*-(*β*-

D-glucopyranosyl)phenylacetothiohydroximic acid] (49). The intact conversion of carbon and sulphur from phenylacetothiohydroximate was indicated since both $^{14}C$ and $^{35}S$ were incorporated to the same extent (35%) into the glucosinolate. These results have been confirmed and extended by the demonstration of

$$\longrightarrow C_6H_5\text{—}CH_2\text{—}\overset{*}{C}(=NOH)\text{—}H \;(47) \rightarrow\rightarrow C_6H_5\text{—}\overset{\bullet}{C}H_2\text{—}\overset{*}{C}(=NOH)\text{—}\overset{\circ}{S}^- \;(48) \xrightarrow{\text{UDP-}\overset{\triangledown}{\text{glucose}} \;\to\; \text{UDP}}$$

$$C_6H_5\text{—}CH_2\text{—}\overset{*}{C}(=NOH)\text{—}S\overset{\triangledown}{G}lc \;(49) \longrightarrow C_6H_5\text{—}\overset{\bullet}{C}H_2\text{—}\overset{*}{C}(=N\text{—}OSO_3^-)\text{—}\overset{\circ}{S}\overset{\triangledown}{G}lc \;(8)$$

phenylacetothiohydroximate as a natural product in *Tropaeolum majus* (Matsuo & Underhill, 1969). The formation of phenylacetothiohydroximate from phenylacetaldoxime was also demonstrated when radioactive phenylacetothiohydroximate (48) was isolated in a trapping experiment with *Tropaeolum majus* shoots which had been simultaneously given phenyl[1-$^{14}C$]acetaldoxime (47) and inactive thiohydroximate. Matsuo & Underhill (1969, 1971) have isolated an enzyme from leaves of *Tropaeolum majus*, *Sinapis alba*, *Nasturtium officinale* and *Armoracea lapathifolia* which catalysed the formation of desulphobenzylglucosinolate by glucosyl transfer from UDP-glucose to phenylacetothiohydroximate. The enzyme (UDP-glucose–thiohydroximate glucosyltransferase) was shown to possess a high degree of specificity for the thiohydroximate functional group, the nucleotide donor (UDP) and the sugar transferred (glucose). Little specificity was exhibited for the side chain of the thiohydroximate since, with the exception of acetothiohydroximate, all of the thiohydroximates tested served as substrates.

Biosynthetic studies of other glucosinolates have provided further evidence that →→ thiohydroximate → desulphoglucosinolate → glucosinolate is a generally occurring biosynthetic pathway. Kindl & Schiefer (1969) have demonstrated the natural occurrence of desulpho-*p*-hydroxybenzylglucosinolate in shoots of *Sinapis alba* and found that approximately three times more radioactivity accumulated in it than in the corresponding glucosinolate isolated after feeding *p*-hydroxyphenyl[1-$^{14}C$]acetaldoxime. High conversion of radioactivity from both thiohydroximates and desulphoglucosinolates has been recorded by Chisholm & Matsuo (1972); more than 50% of the $^{14}C$ from 4-methylthio[1-$^{14}C$]butyrothiohydroximate and desulpho-3-methylthiopropyl[$^{14}C$=N]glucosinolate was converted into allylglucosinolate in horseradish leaves during a 24h metabolic period. In their studies on the biosynthesis of indole glucosinolates in *Isatis tinctoria*, Mahadevan & Stowe (1972) fed [$^3H$]indol-3-ylacetaldoxime and isolated a radioactive neutral sulphur-containing compound 'X' which they have tentatively identified as desulphoindol-3-ylmethylglucosinolate. Compound 'X' accumulated when selenate, an inhibitor of biological sulphation reactions, was fed to leaves of *Isatis tinctoria*. When partially purified $^3H$-labelled compound 'X' was fed to the leaves over 60% of the administered radioactivity was converted into indol-3-ylmethylglucosinolate and 1-sulphoindol-3-ylmethylglucosinolate. From the accumulated data it would appear that the sequence aldoxime → thiohydroximate → desulphoglucosinolate → glucosinolate may be considered a general pathway in the biosynthesis of glucosinolates.

Only recently have investigations been made on the formation of thiohydroximates from aldoximes and the mechanism of incorporation of the reduced or thiohydroximate sulphur. Noting that thiohydroximic acids are formed by base-catalysed reactions of thiols and primary nitro compounds (Copenhaver, 1957), Ettlinger & Kjaer (1968) suggested that *S*-substituted thiohydroximic acids may be derived biologically by the addition of a mercaptide to an *aci*-nitro tautomer by the mechanism (50) → (51) → (52) → (53) → (54). Only a few nitro compounds have been detected in plants but studies on their biosynthesis have, in general, shown amino acids as their precursors (Birch *et al.*, 1960; Burrows *et al.*, 1965; Shaw, 1967; Larsen, 1968; Venulet & Van Etten, 1970). Although considerable evidence indicates the amino nitrogen as the source of the nitro group (Gatenbeck & Forsgren, 1964; Lancini *et al.*, 1966; Shaw & McCloskey, 1967), intermediates with nitrogen at a higher degree of oxidation than the amino nitrogen have not been reported.

Matsuo *et al.* (1972) obtained data supporting the suggestion of Ettlinger & Kjaer (1968) that primary nitro compounds may be involved in the formation of glucosinolates. Although the extent of incorporation of tracer from 1-nitro-2-[1,2-$^{14}C_2$]phenylethane into benzylglucosinolate was equal to or slightly less than that from phenyl[3-$^{14}C$]alanine the incorporation of $^{14}C$ was specific. 1-Nitro-2-phenylethane was demonstrated to be a natural product of *Tropaeolum majus* and by means of a trapping experiment it was shown to be derived from labelled phenylacetaldoxime. On the basis of the formation of 1-nitro-2-phenylethane and phenylacetothiohydroximate from phenylacetaldoxime, and assuming cysteine [52, $R{=}CH_2CH(NH_2)CO_2H$] as the source of the thiohydroximate sulphur, the following pathway has been suggested (Ettlinger & Kjaer, 1968; Matsuo *et al.*, 1972) for the biosynthesis of benzylglucosinolate.

This pathway can only be considered as tentative until studies on the formation of 1-nitro-2-phenylethane from the aldoxime and its subsequent conversion into

$$\longrightarrow C_6H_5\text{—}CH_2\text{—}CH{=}NOH \;(47) \longrightarrow C_6H_5\text{—}CH_2\text{—}CH_2\text{—}NO_2 \;(50) \rightleftharpoons$$

$$C_6H_5\text{—}CH_2\text{—}CH{=}\overset{+}{N}(OH)\text{—}O^- \;(51) \xrightarrow{RS^-} C_6H_5\text{—}CH_2\text{—}CH(SR)\text{—}N(OH)\text{—}O^- \;(52) \xrightarrow{-OH^-}$$

$$C_6H_5\text{—}CH_2\text{—}CH(SR)\text{—}N{=}O \;(53) \longrightarrow C_6H_5\text{—}CH_2\text{—}C(SR){=}NOH \;(54) \longrightarrow$$

the thiohydroximate are confirmed in cell-free systems and the efficiencies of other nitro compounds as precursors of glucosinolates are established.

It has generally been assumed that the *N*-sulphate ester of glucosinolates is derived by transfer of sulphate from 3′-phosphoadenosine 5′-sulphatophosphate to the desulphoglucosinolate. E. W. Underhill (unpublished work) has demonstrated the presence of a sulphotransferase in a number of glucosinolate-bearing plants which catalyses this reaction. The enzyme obtained from different sources utilized 3′-phosphoadenosine 5-sulphatophosphate as the sulphate donor but showed little or no specificity for the type of desulphoglucosinolate employed as substrate; each enzyme system was capable of catalysing the sulphation of desulphoglucosinolates not normally present in the plant from which the enzyme itself was isolated.

*Glucosinolate interconversions*

The co-existence in the same plant of glucosinolates whose aglycones differ only by one or more functional groups naturally leads to speculation that perhaps one glucosinolate is derived from another and consequently a number of glucosinolates could be derived from a single amino acid precursor.

Several glucosinolates are known whose aglycones contain phenolic and/or aliphatic hydroxyl groups and without exception their deoxy analogues are known. 2-Hydroxyisopropylglucosinolate was isolated by Kjaer & Christensen (1959) from *Sisymbrium austriacum* which also contains isopropylglucosinolate. They suggested that the former glucosinolate could be biogenetically derived from the latter by hydroxylation. Alternatively the introduction of the hydroxyl group could have occurred at the amino acid or subsequent stage of biosynthesis. 2-Hydroxy-3-butenylglucosinolate is efficiently produced from 5-methylthiovaleraldoxime (Lee & Serif, 1971), a result which suggests that hydroxylation occurs after the formation of the aldoxime.

Experimental data indicating that the aliphatic hydroxyl group is introduced at the glucosinolate stage of biosynthesis by hydroxylation of the deoxy glucosinolate comes from studies on the formation of glucosinolates in *Brassica napus* and in *Reseda luteola*. The glucosinolate content of *Brassica napus* cultivar Bronowski is low when compared with the cultivar Regina II. In a search to find the metabolic block that causes the low glucosinolate content, Josefsson (1971) fed a number of labelled precursors including desulpho-3-butenylglucosinolate. In Regina II the activity from labelled desulpho-3-butenylglucosinolate was incorporated into both 3-butenylglucosinolate (4.3 %) and 2-hydroxy-3-butenylglucosinolate (2.8 %) whereas in the cultivar Bronowski it was incorporated into 3-butenylglucosinolate (5.4 %) but only negligibly into 2-hydroxy-3-butenylglucosinolate. He concluded that 2-hydroxy-3-butenylglucosinolate is derived from its deoxy analogue in cultivar Regina II and that a metabolic block occurs at the hydroxylation step in cultivar Bronowski. The incorporation of $^{14}C$ into these rapeseed glucosinolates is low when compared with the efficiency of conversion of other desulphoglucosinolates into their corresponding glucosinolates which has exceeded 50 % (Underhill & Wetter, 1969; Chisholm & Matsuo, 1972; Mahadevan & Stowe, 1972). Although the data show that both sulphation and hydroxylation can take place after the terminal double bond has been formed,

Josefsson (1971) did not establish the natural occurrence of desulpho-3-butenylglucosinolate in these plants. Since the enzyme catalysing the sulphation of desulphoglucosinolates possesses little specificity for the aglycone moiety (E. W. Underhill, unpublished work) the formation of 3-butenylglucosinolate from its desulpho analogue may not be a transformation which occurs naturally in the plant.

Underhill & Kirkland (1972*a*) fed 2-amino-4-phenyl[2-$^{14}$C]butyric acid to *Reseda luteola* and from the specific radioactivities of the glucosinolates isolated, they concluded that 2-hydroxy-2-phenylethylglucosinolate was formed from 2-phenylethylglucosinolate. 2-[G-$^{14}$C]Phenylethylglucosinolate was then fed and its efficient incorporation into the hydroxylated glucosinolate was established. Although limited in scope these results strongly suggest that hydroxylation is the last step of the biosynthetic sequence.

It would appear that the introduction of phenolic hydroxyl groups occurs at a different stage in the biosynthesis. By using *Sinapis alba* it was demonstrated (Kindl, 1965; Kindl & Schiefer, 1969) that tyrosine and *p*-hydroxyphenylacetaldoxime were precursors of *p*-hydroxybenzylglucosinolate, whereas neither phenylacetaldoxime nor benzylglucosinolate was detectably incorporated into the *p*-hydroxy-substituted glucosinolate. From this, it was suggested that the phenolic hydroxyl group is introduced at the amino acid stage of biosynthesis.

Studies on the biosynthesis of allylglucosinolate have led to the suggestion that it may be derived from 3-methylthiopropylglucosinolate by elimination of the methylthio group or from a related glucosinolate containing the methylthio group at a higher level of oxidation (Chisholm & Wetter, 1966; Matsuo & Yamazaki, 1966). A similar suggestion has been made by Ettlinger & Kjaer (1968) who noted, for example, that 3-butenylglucosinolate, 2-hydroxy-3-butenylglucosinolate, 4-methylthiobutylglucosinolate and 4-methylsulphinylbutylglucosinolate co-exist in a number of *Brassica* species, and that similar glucosinolate combinations appear in *Allysum* and *Iberis* species. Distributional data such as these, combined with the results of biosynthetic studies on allylglucosinolate and its higher homologues strongly indicate that the terminal double bond in a straight-chain glucosinolate probably arises by elimination of methanethiol or an oxidized equivalent.

Experimental support for this proposal has been obtained by Chisholm & Matsuo (1972) who reported that 70% of both 3-methylthiopropylglucosinolate and 3-methylsulphinylpropylglucosinolate was converted into allylglucosinolate in *Armoracea lapathifolia*. They also found 3-methylthiopropylglucosinolate as a

$$CH_3{-}S{-}CH_2{-}CH_2{-}CH_2{-}C(=N{-}OSO_3^-){-}SGlc \quad (55)$$

$$\updownarrow$$

$$CH_3{-}S(=O){-}CH_2{-}CH_2{-}CH_2{-}C(=N{-}OSO_3^-){-}SGlc \quad (56)$$

$$(55),\ (56) \longrightarrow CH_2{=}CH{-}CH_2{-}C(=N{-}OSO_3^-){-}SGlc \quad (26)$$

naturally occurring glucosinolate in this plant, and that it was formed with high efficiency on feeding 3-methylsulphinylpropylglucosinolate whose natural occurrence in *Armoracea lapathifolia* was not, however, established. Since both glucosinolates were converted with equal efficiency into allylglucosinolate, it was not possible to distinguish the immediate precursor. Studies employing cell-free systems will be required before this question can be resolved, and here the prior elimination of the hydrolysing enzyme, thioglucosidase, from such cell-free systems will be required.

The reversible oxidation and reduction of 3-methylthiopropylglucosinolate and 3-methylsulphinylpropylglucosinolate has been demonstrated in *Cheiranthus kewensis* (Chisholm, 1972). When [2-$^{14}$C]homomethionine was fed the specific radioactivity was higher in 3-methylthiopropylglucosinolate than in 3-methylsulphinylpropylglucosinolate, results strongly suggesting that 3-methylthiopropylglucosinolate was formed first and then oxidized. Chisholm (1972) also fed 3-methylsulphinylpropyl[$^{14}$C=N]glucosinolate and found that it was reduced to 3-methylthiopropylglucosinolate with retention of the basic $CH_3$—S—C skeleton.

Studies have been made on the biosynthesis of the indole glucosinolates in *Isatis tinctoria* and from the specific radioactivities of the glucosinolates isolated after administration of [$^{35}$S]sulphate, Elliott & Stowe (1971*a*) concluded that both 1-methoxyindol-3-ylmethylglucosinolate and 1-sulphoindol-3-ylmethylglucosinolate were derived from indol-3-ylmethylglucosinolate. Mahadevan & Stowe (1972) have concluded from their feeding experiments with [$^{3}$H]indol-3-ylacetaldoxime and [$^{35}$S]cystine that 1-sulphoindol-3-ylmethylglucosinolate is derived from indol-3-ylmethylglucosinolate. In addition, their results indicated a rapid glucosinolate turnover with a half-life of about 2 days.

*New glucosinolates*

Since 1968 18 new glucosinolates have been identified (Table 1) and on the basis of current knowledge of glucosinolate biosynthesis reasonable speculation can be made regarding their formation.

In their studies on the biosynthesis of 2-hydroxy-2-methylpropylglucosinolate in *Conringia orientalis*, and on the premise that hydroxylated glucosinolates are derived from their deoxy analogues, Underhill & Kirkland (1972*b*) sought and found 2-methylpropylglucosinolate (57).

Kjaer & Schuster (1971) have identified *n*-butylglucosinolate (58), 3-hydroxybutylglucosinolate (59) and 4-hydroxybutylglucosinolate (60) in leaves of *Capparis flexuosa*. Biosynthetically *n*-butylglucosinolate is likely to be derived from 2-aminohexanoic acid and the hydroxylated analogues could then be formed by hydroxylation of the side chain at C-3 and C-4, respectively. Alternatively, they might be formed from 2-amino-5-hydroxyhexanoic acid and 2-amino-6-hydroxyhexanoic acid which in turn could be derived from threonine and homoserine respectively, by acetate extension of their carbon chains.

From *Erysium hieracifolium* Kjaer & Schuster (1970) have identified the isothiocyanates of 3-hydroxy-5-methylthiopentylglucosinolate (65), 3-hydroxy-5-methylsulphinylpentylglucosinolate (66) and 3-hydroxy-5-methylsulphonylpentylglucosinolate (67). This is the first reported instance of an aliphatic side

Table 1. *Glucosinolates isolated since* 1968

For glucosinolates isolated before 1968 see Ettlinger & Kjaer (1968).

| | Compound | Reference |
|---|---|---|
| (57) | $CH_3—CH(CH_3)—CH_2—C(=N—OSO_3^-)—SGlc$ | Underhill & Kirkland (1972*b*) |
| (58) | $CH_3—CH_2—CH_2—CH_2—C(=N—OSO_3^-)—SGlc$ | Kjaer & Schuster (1971) |
| (59) | $CH_3—CH(OH)—CH_2—CH_2—C(=N—OSO_3^-)—SGlc$ | Kjaer & Schuster (1971) |
| (60) | $HOCH_2—CH_2—CH_2—CH_2—C(=N—OSO_3^-)—SGlc$ | Kjaer & Schuster (1971) |
| (61) | $CH_3—S—[CH_2]_7—C(=N—OSO_3^-)—SGlc$ | Kjaer & Schuster (1972*a*) |
| (62) | $CH_3—S—[CH_2]_8—C(=N—OSO_3^-)—SGlc$ | Kjaer & Schuster (1972*a*) |
| (63) | $CH_3—S—[CH_2]_5—C(=O)—CH_2—CH_2—C(=N—OSO_3^-)—SGlc$ | Kjaer & Schuster (1972*a*) |
| (64) | $CH_3—S(=O)—[CH_2]_5—C(=O)—CH_2—CH_2—C(=N—OSO_3^-)—SGlc$ | Kjaer & Schuster (1972*a*) |
| (65) | $CH_3—S—CH_2—CH_2—CH(OH)—CH_2—CH_2—C(=N—OSO_3^-)—SGlc$ | Kjaer & Schuster (1970) |
| (66) | $CH_3—S(=O)—CH_2—CH_2—CH(OH)—CH_2—CH_2—C(=N—OSO_3^-)—SGlc$ | Kjaer & Schuster (1970) |
| (67) | $CH_3—S(=O)_2—CH_2—CH_2—CH(OH)—CH_2—CH_2—C(=N—OSO_3^-)—SGlc$ | Kjaer & Schuster (1970) |
| (68) | $CH_3—S(=O)—[CH_2]_7—C(=N—OSO_3^-)—SGlc$ | Gmelin *et al.* (1970) |
| (69) | $CH_3—S(=O)—[CH_2]_{11}—C(=N—OSO_3^-)—SGlc$ | Kjaer & Schuster (1972*b*) |
| (70) | 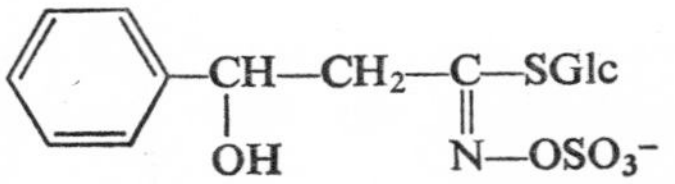 $C_6H_5—CH(OH)—CH_2—C(=N—OSO_3^-)—SGlc$ (*R*) | Gmelin *et al.* (1970) |

Table 1—continued

| | Compound | Reference |
|---|---|---|
| (71) | 3,4-(HO)$_2$C$_6$H$_3$—$CH_2$—C(=N—$OSO_3^-$)—SGlc | Danielak & Borkowski (1970) |
| (72) | $CH_3O$—C$_6$H$_4$—CH(OH)—$CH_2$—C(=N—$OSO_3^-$)—SGlc | Kjaer & Schuster (1972*a*) |
| (73) | 3,4,5-($CH_3O$)$_3$C$_6$H$_2$—$CH_2$—C(=N—$OSO_3^-$)—SGlc * | Kjaer *et al.* (1971) |
| (74) | N-$SO_3^-$-indol-3-yl—$CH_2$—C(=N—$OSO_3^-$)—SGlc * | Elliott & Stowe (1970) |

* Obtained as crystalline compounds.

chain bearing both a ω-methylthio and hydroxyl group. The biosynthesis of these three glucosinolates as well as glucosinolates (61, 62, 63, 64, 68 and 69) is likely to involve their derivation from methionine by chain-extension followed by oxidation and hydroxylation.

Danielak & Borkowski (1970) have reported the isolation of 3,4-dihydroxybenzylglucosinolate (71) from *Hesperis matronalis*; they also reported 3,4-dimethoxyphenylglucosinolate as a natural species which, if confirmed, would be the first glucosinolate possessing a C-N skeleton corresponding to phenylglycine. 2-Hydroxy-2-(*p*-methoxyphenyl)ethylglucosinolate (72) has been isolated from *Arabis hirsuta* (Kjaer & Schuster, 1972*a*) and 3,4,5-trimethoxybenzylglucosinolate (73) from *Lepidium hyssopifolium* (Kjaer *et al.*, 1971). From the limited biosynthetic data available on the derivation of phenolic and aliphatic hydroxyl groups it seems likely that the phenolic and/or methoxyl groups of these glucosinolates are introduced at the amino acid stage of biosynthesis (Kindl & Schiefer, 1969, 1971) whereas the aliphatic hydroxyl is derived by hydroxylation of the corresponding deoxyglucosinolate. Thus 3,4,5-trimethoxyphenylalanine may be the precursor of 3,4,5-trimethoxybenzylglucosinolate (73), and 2-(3,4-dimethoxyphenyl)ethylglucosinolate is likely to be the progenitor of 2-hydroxy-2-(3,4-dimethoxyphenyl)ethylglucosinolate.

This paper is issued as National Research Council of Canada publication no. 13473.

## References

Ahmad, A. & Spenser, I. D. (1961) *Can. J. Chem.* **39,** 1340–1359
Barothy, J. & Neukom, H. (1965) *Chem. Ind.* (*London*) 308–309
Benn, M. H. (1962) *Chem. Ind.* (*London*) 1907

Benn, M. H. & Meakin, D. (1965) *Can. J. Chem.* **43**, 1874–1877
Birch, A. J., McLoughlin, B. J., Smith, H. & Winter, J. (1960) *Chem. Ind.* (*London*) 840–841
Bowland, J. P., Clandinin, D. R. & Wetter, L. R. (1965) *Rapeseed Meal for Livestock & Poultry—A Review*: *Can. Dep. Agr. Publ.* 1257
Boyle, J. E. & Fowden, L. (1971) *Phytochemistry* **10**, 2671–2678
Burrows, B. F., Mills, S. D. & Turner, W. B. (1965) *Chem. Commun.* 75–76
Calderon, P., Pederson, C. S. & Mattick, L. R. (1966) *J. Agr. Food Chem.* **14**, 665–666
Challenger, F. (1959) *Aspects of the Organic Chemistry of Sulphur*, pp. 115–161, Butterworths Scientific Publications, London
Chisholm, M. D. (1972) *Phytochemistry* **11**, 197–202
Chisholm, M. D. & Matsuo, M. (1972) *Phytochemistry* **11**, 203–207
Chisholm, M. D. & Wetter, L. R. (1964) *Can. J. Biochem.* **42**, 1033–1040
Chisholm, M. D. & Wetter, L. R. (1966) *Can. J. Biochem.* **44**, 1625–1632
Chisholm, M. D. & Wetter, L. R. (1967) *Plant Physiol.* **42**, 1726–1730
Copenhaver, J. W. (1957) *U.S. Patent* 2 786 865; *Chem. Abstr.* **51**, 13920
Danielak, R. & Borkowski, B. (1970) *Diss. Pharm. Pharmacol.* **22**, 143–148; *Chem. Abstr.* (1971) **74**, 1006w
Elliott, M. C. & Stowe, B. B. (1970) *Phytochemistry* **9**, 1629–1632
Elliott, M. C. & Stowe, B. B. (1971*a*) *Plant Physiol.* **47**, 366–372
Elliott, M. C. & Stowe, B. B. (1971*b*) *Plant Physiol.* **48**, 498–503
Emery, T. F. (1966) *Biochemistry* **5**, 3694–3701
Ettlinger, M. G. & Dateo, G. P. Jr. (1961) *Studies of Mustard Oil Glucosides*, 12, Final Report Contract DA19-129-QM-1059, U.S. Army Natick Laboratories, Natick, Mass.
Ettlinger, M. G. & Kjaer, A. (1968) *Recent Advan. Phytochem.* **1**, 59–144
Ettlinger, M. G. & Lundeen, A. J. (1956) *J. Amer. Chem. Soc.* **78**, 4172–4173
Ettlinger, M. G., Dateo, G. P., Jr., Harrison, B. W., Mabry, T. J. & Thompson, C. P. (1961) *Proc. Nat. Acad. Sci. U.S.* **47**, 1875–1880
Fowden, L. & Mazelis, M. (1971) *Phytochemistry* **10**, 359–365
Gadamer, J. (1897) *Arch. Pharm.* (*Weinheim*) **235**, 44
Gaines, R. D. & Goering, K. J. (1960) *Biochem. Biophys. Res. Commun.* **2**, 207–212
Gaines, R. D. & Goering, K. J. (1962) *Arch. Biochem. Biophys.* **96**, 13–19
Gatenbeck, S. & Forsgren, B. (1964) *Acta Chem. Scand.* **18**, 1750–1754
Gildemeister, E. & Hofmann, F. (1927) *Die Äetherischen Öle*, 3rd edn., vol. 1, p. 145, Schimmel and Co., Leipzig
Giovanelli, J. & Mudd, S. H. (1967) *Biochem. Biophys. Res. Commun.* **27**, 150–156
Gmelin, R. (1969*a*) *Praep. Pharm.* **5**, 17–22
Gmelin, R. (1969*b*) *Praep. Pharm.* **5**, 33–41
Gmelin, R. & Virtanen, A. I. (1960) *Acta Chem. Scand.* **14**, 507–509
Gmelin, R., Kjaer, A. & Schuster, A. (1970) *Acta Chem. Scand.* **24**, 3031–3037
Ishimoto, M. & Yamashina, I. (1949) *Koso Kagaku Shimpojiumu* **2**, 36–42
Iversen, J. H. (1972) *J. Ultrastruct. Res.* **38**, 200
Josefsson, E. (1971) *Physiol. Plant.* **24**, 161–175
Kindl, H. (1964) *Monatsh. Chem.* **95**, 439–448
Kindl, H. (1965) *Monatsh. Chem.* **96**, 527–532
Kindl, H. (1968) *Hoppe-Seyler's Z. Physiol. Chem.* **349**, 519–520
Kindl, H. & Schiefer, S. (1969) *Monatsh. Chem.* **100**, 1773–1787
Kindl, H. & Schiefer, S. (1971) *Phytochemistry* **10**, 1795–1802
Kindl, H. & Underhill, E. W. (1968) *Phytochemistry* **7**, 745–756
Kingsbury, J. M. (1964) *Poisonous Plants of the United States and Canada*, Prentice–Hall Inc., Englewood Cliffs
Kjaer, A. (1954) *Acta Chem. Scand.* **8**, 1110
Kjaer, A. (1960) *Fortschr. Chem. Org. Naturst.* **18**, 122–176
Kjaer, A. (1963) *Pure Appl. Chem.* **7**, 229–245
Kjaer, A. & Boe Jensen, R. (1958) *Acta Chem. Scand.* **12**, 1746–1758
Kjaer, A. & Christensen, B. (1959) *Acta Chem. Scand.* **13**, 1575–1584
Kjaer, A. & Schuster, A. (1970) *Acta Chem. Scand.* **24**, 1631–1638
Kjaer, A. & Schuster, A. (1971) *Phytochemistry* **10**, 3155–3160
Kjaer, A. & Schuster, A. (1972*a*) *Acta Chem. Scand.* **26**, 8–14
Kjaer, A. & Schuster, A. (1972*b*) *Phytochemistry* **11**, 3045–3048
Kjaer, A., Schuster, A. & Park, R. J. (1971) *Phytochemistry* **10**, 455–457
Kosuge, T., Heskett, M. G. & Wilson, E. E. (1966) *J. Biol. Chem.* **241**, 3738–3744
Kutáček, M., Procházka, Z. & Vereš, K. (1962) *Nature* (*London*) **194**, 393–394
Kutáček, M., Spálený, J. & Opluštilová, K. (1966) *Experientia* **22**, 24–25

Lancini, G. C., Kluepfel, D., Lazzari, E. & Sartori, G. (1966) *Biochim. Biophys. Acta* **130**, 37–44
Larsen, L. M. (1968) *Dan. Tidsskr. Farm.* **42**, 272–283
Lee, C. J. & Serif, G. S. (1968) *Biochim. Biophys. Acta* **165**, 569–571
Lee, C. J. & Serif, G. S. (1970) *Biochemistry* **9**, 2068–2071
Lee, C. J. & Serif, G. S. (1971) *Biochim. Biophys. Acta* **230**, 462–467
MacGibbon, D. B. & Allison, R. M. (1971) *N.Z. J. Sci.* **14**, 134–140
Mahadevan, S. (1963) *Arch. Biochem. Biophys.* **100**, 557–558
Mahadevan, S. & Stowe, B. B. (1972) *Plant Physiol.* **50**, 43–50
Marsh, R. E. & Waser, J. (1970) *Acta Crystallogr. Sect. B* **26**, 1030–1037
Matsuo, M. (1963) *Nat. Inst. Radiol. Sci. Chiba, Annu. Rep.* 28
Matsuo, M. (1968*a*) *Chem. Pharm. Bull.* **16**, 1128–1129
Matsuo, M. (1968*b*) *Tetrahedron Lett.* 4101–4104
Matsuo, M. & Underhill, E. W. (1969) *Biochem. Biophys. Res. Commun.* **36**, 18–23
Matsuo, M. & Underhill, E. W. (1971) *Phytochemistry* **10**, 2279–2286
Matsuo, M. & Yamazaki, M. (1964) *Chem. Pharm. Bull.* **12**, 1388–1389
Matsuo, M. & Yamazaki, M. (1966) *Biochem. Biophys. Res. Commun.* **24**, 786–791
Matsuo, M., Kirkland, D. F. & Underhill, E. W. (1972) *Phytochemistry* **11**, 697–701
Mazelis, M. & Ingraham, L. L. (1962) *J. Biol. Chem.* **237**, 109–112
Meakin, D. (1965) Ph.D. Thesis, University of Calgary
Meakin, D. (1967) *Experientia* **23**, 174–175
Milborrow, B. V. (1963) *Biochem. J.* **87**, 255–258
Nagashima, Z. & Uchiyama, M. (1959) *Bull. Agr. Chem. Soc. Jap.* **23**, 555–556
Neuberg, C. & von Schoenebeck, O. (1933) *Biochem. Z.* **265**, 223–236
Oginsky, E. L., Stein, A. E. & Greer, M. A. (1965) *Proc. Soc. Exp. Biol. Med.* **119**, 360–364
Ohtsuru, M., Tsuruo, I. & Hata, T. (1969) *Agr. Biol. Chem.* **33**, 1309–1314
Reese, E. T., Clapp, R. C. & Mandels, M. (1958) *Arch. Biochem. Biophys.* **75**, 228–242
Sandberg, M. & Holley, O. M. (1932) *J. Biol. Chem.* **96**, 443–447
Sayre, F. W. & Greenberg, D. M. (1956) *J. Biol. Chem.* **220**, 787–799
Schraudolf, H. & Bergmann, F. (1965) *Planta* **67**, 75–95
Serif, G. S. & Schmotzer, L. A. (1968) *Phytochemistry* **7**, 1151–1157
Shaw, P. D. (1967) *Biochemistry* **6**, 2253–2260
Shaw, P. D. & McCloskey, J. A. (1967) *Biochemistry* **6**, 2247–2252
Shukla, P. S. & Mahadevan, S. (1968) *Arch. Biochem. Biophys.* **125**, 873–883
Spálený, J. & Kutáček, M. (1969) *Proc. Int. Congr. Nutr., 8th* 717–719; *Chem. Abstr.* **75**, 45735
Spenser, I. D. & Ahmad, A. (1961) *Proc. Chem. Soc. London* 375–376
Stadtman, E. R., Cohen, G. N., LeBras, G. & de Robichon-Szulmajster, H. (1961) *J. Biol. Chem.* **236**, 2033–2038
Stevens, R. L. & Emery, T. F. (1966) *Biochemistry* **5**, 74–81
Strassman, M. & Ceci, L. N. (1963) *J. Biol. Chem.* **238**, 2445–2452
Tapper, B. A. & Butler, G. W. (1967) *Arch. Biochem. Biophys.* **120**, 719–721
Tapper, B. A. & Butler, G. W. (1971) *Biochem. J.* **124**, 935–941
Tapper, B. A. & Butler, G. W. (1972) *Phytochemistry* **11**, 1041–1046
Tapper, B. A., Conn, E. E. & Butler, G. W. (1967) *Arch. Biochem. Biophys.* **119**, 593–595
Tsuruo, I. & Hata, T. (1967) *Agr. Biol. Chem.* **31**, 27–32
Tsuruo, I., Yoshida, M. & Hata, T. (1967) *Agr. Biol. Chem.* **31**, 18–26
Underhill, E. W. (1965*a*) *Can. J. Biochem.* **43**, 179–187
Underhill, E. W. (1965*b*) *Can. J. Biochem.* **43**, 189–198
Underhill, E. W. (1967) *Eur. J. Biochem.* **2**, 61–63
Underhill, E. W. (1968) *Can. J. Biochem.* **46**, 401–405
Underhill, E. W. & Chisholm, M. D. (1964) *Biochem. Biophys. Res. Commun.* **14**, 425–430
Underhill, E. W. & Kirkland, D. F. (1972*a*) *Phytochemistry* **11**, 1973–1979
Underhill, E. W. & Kirkland, D. F. (1972*b*) *Phytochemistry* **11**, 2085–2088
Underhill, E. W. & Wetter, L. R. (1966) *Colloq. Biosynthesis of Aromatic Compounds: Proc. FEBS Meet. 2nd* 129–137
Underhill, E. W. & Wetter, L. R. (1969) *Plant Physiol.* **44**, 584–590
Underhill, E. W., Chisholm, M. D. & Wetter, L. R. (1962) *Can. J. Biochem. Physiol.* **40**, 1505–1514
Van Etten, C. H. (1969) in *Toxic Constituents of Plant Foodstuffs* (Liener, I. E., ed.), pp. 103–142, Academic Press, London and New York
Van Etten, C. H., Daxenbichler, M. E. & Wolff, I. A. (1969) *J. Agr. Food Chem.* **17**, 483–491
Vaughan, J. G. & Hemingway, J. S. (1959) *Econ. Bot.* **13**, 196–204
Venulet, J. & Van Etten, R. L. (1970) in *The Chemistry of the Nitro and Nitroso Groups* (Feurer, H., ed.), pp. 201–287, Interscience Publishers, New York

Virtanen, A. I. (1962) *Arch. Biochem. Biophys. Suppl.* **1**, 200–208
Virtanen, A. I. (1965) *Phytochemistry* **4**, 207–228
von Euler, H. & Eriksson, S. E. (1926) *Fermentforschung* **8**, 518–523
Vose, J. R. (1972) *Phytochemistry* **11**, 1649–1653
Waser, J. & Watson, W. H. (1963) *Nature* (*London*) **198**, 1297–1298
Wetter, L. R. (1964) *Phytochemistry* **3**, 57–64
Wetter, L. R. & Chisholm, M. D. (1968) *Can. J. Biochem.* **46**, 931–935

# Author Index

Numbers in italics refer to pages in the references at the end of each chapter.

C

T

# Subject Index

D



E

K



L